# Electrical and Electronic Principles and Technology

*To Sue*

# Electrical and Electronic Principles and Technology

*Fourth edition*

John Bird, BSc (Hons), CEng, CSci, CMath, MIEE, FIIE, FIET, FIMA, FCollT

AMSTERDAM • BOSTON • HEIDELBERG • LONDON • NEW YORK • OXFORD
PARIS • SAN DIEGO • SAN FRANCISCO • SINGAPORE • SYDNEY • TOKYO

Newnes is an imprint of Elsevier

Newnes is an imprint of Elsevier
The Boulevard, Langford Lane, Kidlington, Oxford OX5 1GB, UK
30 Corporate Drive, Suite 400, Burlington, MA 01803, USA

First edition   2000 previously published as *Electrical Principles and Technology for Engineering*
Reprinted   2001
Second edition   2003
Reprinted   2004, 2005, 2006
Third edition   2007
Fourth edition   2010

**Notice**
No responsibility is assumed by the publisher for any injury and/or damage to persons or property as a matter of products liability, negligence or otherwise, or from any use or operation of any methods, products, instructions or ideas contained in the material herein. Because of rapid advances in the medical sciences, in particular, independent verification of diagnoses and drug dosages should be made.

**British Library Cataloguing-in-Publication Data**
A catalogue record for this book is available from the British Library.

**Library of Congress Cataloging-in-Publication Data**
A catalogue record for this book is available from the Library of Congress.

ISBN: 978-0-08-089056-2

For information on all Newnes publications
visit our Web site at *www.elsevierdirect.com*

*Typeset by*: diacriTech, India

Printed and bound in China
10 11 12 13 14   10 9 8 7 6 5 4 3 2 1

Working together to grow
libraries in developing countries

www.elsevier.com | www.bookaid.org | www.sabre.org

ELSEVIER   BOOK AID International   Sabre Foundation

# Contents

# Preface

'*Electrical and Electronic Principles and Technology 4th Edition*' introduces the principles which describe the operation of d.c. and a.c. circuits, covering both steady and transient states, and applies these principles to filter networks, operational amplifiers, three-phase supplies, transformers, d.c. machines and three-phase induction motors.

In this edition, new material has been added on resistor construction, the loading effect of instruments, potentiometers and rheostats, earth potential and short circuits, and electrical safety with insulation and fuses. In addition, a new chapter detailing some **10 practical laboratory experiments** has been included. (These may be downloaded and edited by tutors to suit local availability of equipment and components).

**This fourth edition of the textbook provides coverage of the following latest syllabuses:**

(i) **'Electrical and Electronic Principles'** (BTEC National Certificate and National Diploma, Unit 5) – see Chapters 1–10, 11(part), 13 (part), 14, 15 (part), 18(part), 21(part), 22(part).

(ii) **'Further Electrical Principles'** (BTEC National Certificate and National Diploma, Unit 67) – see Chapters 13, 15–18, 20, 22, 23.

(iii) Parts of the following BTEC National syllabuses: Electrical Applications, Three Phase Systems, Principles and Applications of Electronic Devices and Circuits, Aircraft Electrical Machines, and Telecommunications Principles.

(iv) Electrical part of 'Applied Electrical and Mechanical Science for Technicians' (BTEC First Certificate).

(v) Various parts of City & Guilds Technician Certificate/Diploma in Electrical and Electronic Principles/Telecommunication Systems, such as Electrical Engineering Principles, Power, and Science and Electronics.

(vi) 'Electrical and Electronic Principles' (EAL Advanced Diploma in Engineering and Technology).

(vii) Any introductory/Access/Foundation course involving Electrical and Electronic Engineering Principles.

The **text** is set out in four main sections:

**Section 1**, comprising Chapters 1 to 12, involves essential **Basic Electrical and Electronic Engineering Principles**, with chapters on electrical units and quantities, introduction to electric circuits, resistance variation, batteries and alternative sources of energy, series and parallel networks, capacitors and capacitance, magnetic circuits, electromagnetism, electromagnetic induction, electrical measuring instruments and measurements, semiconductors diodes and transistors.

**Section 2**, comprising Chapters 13 to 19, involves **Further Electrical and Electronic Principles,** with chapters on d.c. circuit theorems, alternating voltages and currents, single-phase series and parallel networks, filter networks, d.c. transients and operational amplifiers.

**Section 3**, comprising Chapters 20 to 23, involves **Electrical Power Technology**, with chapters on three-phase systems, transformers, d.c. machines and three-phase induction motors.

**Section 4**, comprising Chapter 24, detailing **10 practical laboratory experiments**.

Each topic considered in the text is presented in a way that assumes in the reader little previous knowledge of that topic. Theory is introduced in each chapter by a reasonably brief outline of essential information, definitions, formulae, procedures, etc. The theory is kept to a minimum, for problem solving is extensively used to establish and exemplify the theory. It is intended that readers will gain real understanding through seeing problems solved and then through solving similar problems themselves.

To aid tutors/lecturers/instructors, the following **free Internet downloads** are available with this edition (see page x for access details):

(i) a **sample of solutions** (some 410) of the 540 further problems contained in the book.

(ii) an **Instructors guide** detailing full worked solutions for the **Revision Tests**.

(iii) **10 practical laboratory experiments**, which may be edited.

(iv) **Suggested lesson plans for BTEC units 5 and 67**, together with **Practise Examination questions (with solution)** for revision purposes.

(v) a **PowerPoint presentation of all 538 illustrations** contained in the text.

'Electrical and Electronic Principles and Technology 4$^{th}$ Edition' contains **410 worked problems**, together with **341 multi-choice questions** (with answers at the back of the book). Also included are over **455 short answer questions**, the answers for which can be determined from the preceding material in that particular chapter, and some **540 further questions**, arranged in **146 Exercises**, all with answers, in brackets, immediately following each question; the Exercises appear at regular intervals - every 3 or 4 pages - throughout the text. **538 line diagrams** further enhance the understanding of the theory. All of the problems - multi-choice, short answer and further questions - mirror practical situations found in electrical and electronic engineering.

At regular intervals throughout the text are seven **Revision Tests** to check understanding. For example, Revision Test 1 covers material contained in Chapters 1 to 4, Revision Test 2 covers the material contained in Chapters 5 to 7, and so on. These Revision Tests do not have answers given since it is envisaged that lecturers/instructors could set the Tests for students to attempt as part of their course structure. Lecturers/instructors may obtain a free Internet download of full solutions of the Revision Tests in an **Instructor's Manual** – see next column.

A list of relevant **formulae** are included at the end of each of the three sections of the book.

**'Learning by Example'** is at the heart of 'Electrical and Electronic Principles and Technology 4$^{th}$ Edition'.

**JOHN BIRD**
**Royal Naval School of Marine Engineering,**
**HMS Sultan,**
**formerly University of Portsmouth**
**and Highbury College, Portsmouth**

### Free web downloads

A suite of five sets of support material is available to tutors/lecturers/instructors - only from Elsevier's textbook website.

To access material, please go to *http://www.booksite.elsevier.com/newnes/bird*, find the correct title, and click on to whichever of the following resource materials you need.

#### (i) Solutions manual

Within the text there are some 540 further problems arranged within 146 Exercises. A sample of about 410 worked solutions has been prepared for lecturers.

#### (ii) Instructor's manual

This manual provides full worked solutions and mark scheme for all 7 Revision Tests in this book.

#### (iii) Laboratory Experiments

In Chapter 24, 10 practical laboratory experiments are included. It maybe that tutors will want to edit these experiments to suit their own equipment/component availability. These have been made available on the website.

#### (iv) Lesson Plans and revision material

Typical 30-week lesson plans for 'Electrical and Electronic Principles', Unit 5, and 'Further Electrical Principles', Unit 67 are included, together with two practise examinations question papers (with solutions) for each of the modules.

#### (v) Illustrations

Lecturers can download electronic files for all 538 illustrations in this fourth edition.

# Basic Electrical and Electronic Engineering Principles

# Chapter 1

# Units associated with basic electrical quantities

At the end of this chapter you should be able to:

- state the basic SI units
- recognize derived SI units
- understand prefixes denoting multiplication and division
- state the units of charge, force, work and power and perform simple calculations involving these units
- state the units of electrical potential, e.m.f., resistance, conductance, power and energy and perform simple calculations involving these units

## 1.1 SI units

The system of units used in engineering and science is the Système Internationale d'Unités (International system of units), usually abbreviated to SI units, and is based on the metric system. This was introduced in 1960 and is now adopted by the majority of countries as the official system of measurement.

The basic units in the SI system are listed below with their symbols:

| Quantity | Unit |
|---|---|
| length | metre, m |
| mass | kilogram, kg |
| time | second, s |
| electric current | ampere, A |
| thermodynamic temperature | kelvin, K |
| luminous intensity | candela, cd |
| amount of substance | mole, mol |

**Derived SI units** use combinations of basic units and there are many of them. Two examples are:

Velocity – metres per second (m/s)

Acceleration – metres per second squared ($m/s^2$)

SI units may be made larger or smaller by using prefixes which denote multiplication or division by a particular amount. The six most common multiples, with their meaning, are listed below:

| Prefix | Name | Meaning |
|---|---|---|
| M | mega | multiply by $1\,000\,000$ (i.e. $\times 10^6$) |
| k | kilo | multiply by 1000 (i.e. $\times 10^3$) |
| m | milli | divide by 1000 (i.e. $\times 10^{-3}$) |
| μ | micro | divide by $1\,000\,000$ (i.e. $\times 10^{-6}$) |
| n | nano | divide by $1\,000\,000\,000$ (i.e. $\times 10^{-9}$) |
| p | pico | divide by $1\,000\,000\,000\,000$ (i.e. $\times 10^{-12}$) |

DOI: 10.1016/B978-0-08-089056-2.00001-2

## 1.2 Charge

The **unit of charge** is the coulomb (C) where one coulomb is one ampere second (1 coulomb = $6.24 \times 10^{18}$ electrons). The coulomb is defined as the quantity of electricity which flows past a given point in an electric circuit when a current of one ampere is maintained for one second. Thus,

$$\text{charge, in coulombs} \quad Q = It$$

where $I$ is the current in amperes and $t$ is the time in seconds.

> **Problem 1.** If a current of 5 A flows for 2 minutes, find the quantity of electricity transferred.

Quantity of electricity $Q = It$ coulombs

$$I = 5\,\text{A}, t = 2 \times 60 = 120\,\text{s}$$

Hence $Q = 5 \times 120 = \mathbf{600\,C}$

## 1.3 Force

The **unit of force** is the **newton (N)** where one newton is one kilogram metre per second squared. The newton is defined as the force which, when applied to a mass of one kilogram, gives it an acceleration of one metre per second squared. Thus,

$$\text{force, in newtons} \quad F = ma$$

where $m$ is the mass in kilograms and $a$ is the acceleration in metres per second squared. Gravitational force, or weight, is $mg$, where $g = 9.81\,\text{m/s}^2$.

> **Problem 2.** A mass of 5000 g is accelerated at $2\,\text{m/s}^2$ by a force. Determine the force needed.

Force = mass × acceleration

$$= 5\,\text{kg} \times 2\,\text{m/s}^2 = 10\,\text{kg\,m/s}^2 = \mathbf{10\,N}.$$

> **Problem 3.** Find the force acting vertically downwards on a mass of 200 g attached to a wire.

Mass = 200 g = 0.2 kg and acceleration due to gravity, $g = 9.81\,\text{m/s}^2$

$$\left.\begin{array}{r}\text{Force acting}\\ \text{downwards}\end{array}\right\} = \text{weight}$$

$$= \text{mass} \times \text{acceleration}$$

$$= 0.2\,\text{kg} \times 9.81\,\text{m/s}^2$$

$$= \mathbf{1.962\,N}$$

## 1.4 Work

The **unit of work or energy** is the **joule (J)** where one joule is one newton metre. The joule is defined as the work done or energy transferred when a force of one newton is exerted through a distance of one metre in the direction of the force. Thus

$$\text{work done on a body, in joules,} \quad W = Fs$$

where $F$ is the force in newtons and $s$ is the distance in metres moved by the body in the direction of the force. Energy is the capacity for doing work.

## 1.5 Power

The **unit of power** is the watt (W) where one watt is one joule per second. Power is defined as the rate of doing work or transferring energy. Thus,

$$\text{power, in watts,} \quad P = \frac{W}{t}$$

where $W$ is the work done or energy transferred, in joules, and $t$ is the time, in seconds. Thus,

$$\text{energy, in joules,} \quad W = Pt$$

> **Problem 4.** A portable machine requires a force of 200 N to move it. How much work is done if the machine is moved 20 m and what average power is utilized if the movement takes 25 s?

Work done = force × distance

$$= 200\,\text{N} \times 20\,\text{m}$$

$$= \mathbf{4000\,Nm\ or\ 4\,kJ}$$

$$\text{Power} = \frac{\text{work done}}{\text{time taken}}$$

$$= \frac{4000\,\text{J}}{25\,\text{s}} = 160\,\text{J/s} = \mathbf{160\,W}$$

> **Problem 5.** A mass of 1000 kg is raised through a height of 10 m in 20 s. What is (a) the work done and (b) the power developed?

(a) Work done = force × distance

and force = mass × acceleration

Hence,

$$\text{work done} = (1000\,\text{kg} \times 9.81\,\text{m/s}^2) \times (10\,\text{m})$$

$$= 98\,100\,\text{Nm}$$

$$= \mathbf{98.1\,kNm} \text{ or } \mathbf{98.1\,kJ}$$

(b)  Power $= \dfrac{\text{work done}}{\text{time taken}} = \dfrac{98100\,\text{J}}{20\,\text{s}}$

$$= 4905\,\text{J/s} = \mathbf{4905\,W} \text{ or } \mathbf{4.905\,kW}$$

**Now try the following exercise**

**Exercise 1    Further problems on charge, force, work and power**

(Take $g = 9.81\,\text{m/s}^2$ where appropriate)

1.  What quantity of electricity is carried by $6.24 \times 10^{21}$ electrons?    [1000 C]

2.  In what time would a current of 1 A transfer a charge of 30 C?    [30 s]

3.  A current of 3 A flows for 5 minutes. What charge is transferred?    [900 C]

4.  How long must a current of 0.1 A flow so as to transfer a charge of 30 C? [5 minutes]

5.  What force is required to give a mass of 20 kg an acceleration of 30 m/s²? [600 N]

6.  Find the accelerating force when a car having a mass of 1.7 Mg increases its speed with a constant acceleration of 3 m/s². [5.1 kN]

7.  A force of 40 N accelerates a mass at 5 m/s². Determine the mass.    [8 kg]

8.  Determine the force acting downwards on a mass of 1500 g suspended on a string. [14.72 N]

9.  A force of 4 N moves an object 200 cm in the direction of the force. What amount of work is done?    [8 J]

10. A force of 2.5 kN is required to lift a load. How much work is done if the load is lifted through 500 cm?    [12.5 kJ]

11. An electromagnet exerts a force of 12 N and moves a soft iron armature through a distance of 1.5 cm in 40 ms. Find the power consumed.    [4.5 W]

12. A mass of 500 kg is raised to a height of 6 m in 30 s. Find (a) the work done and (b) the power developed.
    [(a) 29.43 kNm (b) 981 W]

13. Rewrite the following as indicated:
    (a)  1000 pF = ...... nF
    (b)  0.02 μF = ...... pF
    (c)  5000 kHz = ...... MHz
    (d)  47 kΩ = ...... MΩ
    (e)  0.32 mA = ...... μA
        [(a) 1 nF (b) 20000 pF (c) 5 MHz
        (d) 0.047 MΩ (e) 320 μA]

## 1.6  Electrical potential and e.m.f.

The **unit of electric potential** is the volt (V), where one volt is one joule per coulomb. One volt is defined as the difference in potential between two points in a conductor which, when carrying a current of one ampere, dissipates a power of one watt, i.e.

$$\text{volts} = \frac{\text{watts}}{\text{amperes}} = \frac{\text{joules/second}}{\text{amperes}}$$

$$= \frac{\text{joules}}{\text{ampere seconds}} = \frac{\text{joules}}{\text{coulombs}}$$

A change in electric potential between two points in an electric circuit is called a **potential difference**. The **electromotive force (e.m.f.)** provided by a source of energy such as a battery or a generator is measured in volts.

## 1.7  Resistance and conductance

The **unit of electric resistance** is the **ohm**(Ω), where one ohm is one volt per ampere. It is defined as the resistance between two points in a conductor when a constant electric potential of one volt applied at the two points produces a current flow of one ampere in the conductor. Thus,

$$\textbf{resistance, in ohms } R = \frac{V}{I}$$

where $V$ is the potential difference across the two points, in volts, and $I$ is the current flowing between the two points, in amperes.

The reciprocal of resistance is called **conductance** and is measured in siemens (S). Thus

$$\text{conductance, in siemens } G = \frac{1}{R}$$

where $R$ is the resistance in ohms.

**Problem 6.** Find the conductance of a conductor of resistance: (a) $10\,\Omega$ (b) $5\,k\Omega$ (c) $100\,m\Omega$.

(a) Conductance $G = \dfrac{1}{R} = \dfrac{1}{10}$ siemen $= \mathbf{0.1\,S}$

(b) $G = \dfrac{1}{R} = \dfrac{1}{5 \times 10^3}\,S = 0.2 \times 10^{-3}\,S = \mathbf{0.2\,mS}$

(c) $G = \dfrac{1}{R} = \dfrac{1}{100 \times 10^{-3}}\,S = \dfrac{10^3}{100}\,S = \mathbf{10\,S}$

## 1.8 Electrical power and energy

When a direct current of $I$ amperes is flowing in an electric circuit and the voltage across the circuit is $V$ volts, then

$$\text{power, in watts } P = VI$$

$$\text{Electrical energy} = \text{Power} \times \text{time}$$

$$= VIt \text{ joules}$$

Although the unit of energy is the joule, when dealing with large amounts of energy, the unit used is the **kilowatt hour (kWh)** where

$$1\,kWh = 1000\,\text{watt hour}$$

$$= 1000 \times 3600\,\text{watt seconds or joules}$$

$$= 3\,600\,000\,J$$

**Problem 7.** A source e.m.f. of $5\,V$ supplies a current of $3\,A$ for $10$ minutes. How much energy is provided in this time?

Energy = power × time, and power = voltage × current. Hence

$$\textbf{Energy} = VIt = 5 \times 3 \times (10 \times 60)$$

$$= 9000\,\text{Ws or J} = \mathbf{9\,kJ}$$

**Problem 8.** An electric heater consumes $1.8\,MJ$ when connected to a $250\,V$ supply for $30$ minutes. Find the power rating of the heater and the current taken from the supply.

$$\text{Power} = \frac{\text{energy}}{\text{time}} = \frac{1.8 \times 10^6\,J}{30 \times 60\,s}$$

$$= 1000\,J/s = 1000\,W$$

i.e. **power rating of heater** $= \mathbf{1\,kW}$

$$\text{Power } P = VI, \text{ thus } I = \frac{P}{V} = \frac{1000}{250} = 4\,A$$

**Hence the current taken from the supply is $4\,A$.**

Now try the following exercise

**Exercise 2 Further problems on e.m.f., resistance, conductance, power and energy**

1. Find the conductance of a resistor of resistance (a) $10\,\Omega$ (b) $2\,k\Omega$ (c) $2\,m\Omega$ [(a) $0.1\,S$ (b) $0.5\,mS$ (c) $500\,S$]

2. A conductor has a conductance of $50\,\mu S$. What is its resistance? [$20\,k\Omega$]

3. An e.m.f. of $250\,V$ is connected across a resistance and the current flowing through the resistance is $4\,A$. What is the power developed? [$1\,kW$]

4. $450\,J$ of energy are converted into heat in $1$ minute. What power is dissipated? [$7.5\,W$]

5. A current of $10\,A$ flows through a conductor and $10\,W$ is dissipated. What p.d. exists across the ends of the conductor? [$1\,V$]

6. A battery of e.m.f. $12\,V$ supplies a current of $5\,A$ for $2$ minutes. How much energy is supplied in this time? [$7.2\,kJ$]

7. A d.c. electric motor consumes $36\,MJ$ when connected to a $250\,V$ supply for $1$ hour. Find the power rating of the motor and the current taken from the supply. [$10\,kW$, $40\,A$]

## 1.9 Summary of terms, units and their symbols

| Quantity | Quantity Symbol | Unit | Unit Symbol |
|---|---|---|---|
| Length | l | metre | m |
| Mass | m | kilogram | kg |
| Time | t | second | s |
| Velocity | v | metres per second | m/s or $\mathrm{m\,s^{-1}}$ |
| Acceleration | a | metres per second squared | m/s² or $\mathrm{m\,s^{-2}}$ |
| Force | F | newton | N |
| Electrical charge or quantity | Q | coulomb | C |
| Electric current | I | ampere | A |
| Resistance | R | ohm | Ω |
| Conductance | G | siemen | S |
| Electromotive force | E | volt | V |
| Potential difference | V | volt | V |
| Work | W | joule | J |
| Energy | E (or W) | joule | J |
| Power | P | watt | W |

**Now try the following exercises**

**Exercise 3    Short answer questions on units associated with basic electrical quantities**

1. What does 'SI units' mean?

2. Complete the following:
   Force = ...... × ......

3. What do you understand by the term 'potential difference'?

4. Define electric current in terms of charge and time

5. Name the units used to measure:
   (a) the quantity of electricity
   (b) resistance
   (c) conductance

6. Define the coulomb

7. Define electrical energy and state its unit

8. Define electrical power and state its unit

9. What is electromotive force?

10. Write down a formula for calculating the power in a d.c. circuit

11. Write down the symbols for the following quantities:
    (a) electric charge    (b) work
    (c) e.m.f.             (d) p.d.

12. State which units the following abbreviations refer to:
    (a) A    (b) C    (c) J    (d) N    (e) m

**Exercise 4    Multi-choice questions on units associated with basic electrical quantities (Answers on page 420)**

1. A resistance of 50 kΩ has a conductance of:
   (a) 20 S          (b) 0.02 S
   (c) 0.02 mS       (d) 20 kS

2. Which of the following statements is incorrect?
   (a) 1 N = 1 kg m/s²   (b) 1 V = 1 J/C
   (c) 30 mA = 0.03 A    (d) 1 J = 1 N/m

3. The power dissipated by a resistor of 10 Ω when a current of 2 A passes through it is:
   (a) 0.4 W  (b) 20 W  (c) 40 W  (d) 200 W

4. A mass of 1200 g is accelerated at 200 cm/s² by a force. The value of the force required is:
   (a) 2.4 N         (b) 2,400 N
   (c) 240 kN        (d) 0.24 N

5.   A charge of 240 C is transferred in 2 minutes. The current flowing is:
(a) 120 A   (b) 480 A   (c) 2 A   (d) 8 A

6.   A current of 2 A flows for 10 h through a 100 Ω resistor. The energy consumed by the resistor is:
(a) 0.5 kWh          (b) 4 kWh
(c) 2 kWh            (d) 0.02 kWh

7.   The unit of quantity of electricity is the:
(a) volt             (b) coulomb
(c) ohm              (d) joule

8.   Electromotive force is provided by:
(a) resistances
(b) a conducting path
(c) an electric current
(d) an electrical supply source

9.   The coulomb is a unit of:
(a) power
(b) voltage
(c) energy
(d) quantity of electricity

10.   In order that work may be done:
(a) a supply of energy is required
(b) the circuit must have a switch
(c) coal must be burnt
(d) two wires are necessary

11.   The ohm is the unit of:
(a) charge           (b) resistance
(c) power            (d) current

12.   The unit of current is the:
(a) volt             (b) coulomb
(c) joule            (d) ampere

# Chapter 2

# An introduction to electric circuits

At the end of this chapter you should be able to:

- appreciate that engineering systems may be represented by block diagrams
- recognize common electrical circuit diagram symbols
- understand that electric current is the rate of movement of charge and is measured in amperes
- appreciate that the unit of charge is the coulomb
- calculate charge or quantity of electricity $Q$ from $Q = It$
- understand that a potential difference between two points in a circuit is required for current to flow
- appreciate that the unit of p.d. is the volt
- understand that resistance opposes current flow and is measured in ohms
- appreciate what an ammeter, a voltmeter, an ohmmeter, a multimeter, an oscilloscope, a wattmeter, a bridge megger, a tachometer and stroboscope measure
- distinguish between linear and non-linear devices
- state Ohm's law as $V = IR$ or $I = V/R$ or $R = V/I$
- use Ohm's law in calculations, including multiples and sub-multiples of units
- describe a conductor and an insulator, giving examples of each
- appreciate that electrical power $P$ is given by $P = VI = I^2R = V^2/R$ watts
- calculate electrical power
- define electrical energy and state its unit
- calculate electrical energy
- state the three main effects of an electric current, giving practical examples of each
- explain the importance of fuses in electrical circuits
- appreciate the dangers of constant high current flow with insulation materials

## 2.1 Electrical/electronic system block diagrams

An electrical/electronic **system** is a group of components connected together to perform a desired function.

Figure 2.1 shows a simple public address system, where a microphone is used to collect acoustic energy in the form of sound pressure waves and converts this to electrical energy in the form of small voltages and currents; the signal from the microphone is then amplified by means of an electronic circuit containing

DOI: 10.1016/B978-0-08-089056-2.00002-4

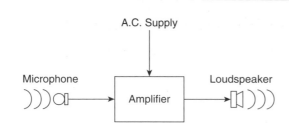

**Figure 2.1**

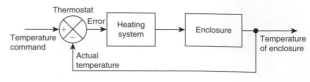

**Figure 2.3**

transistors/integrated circuits before it is applied to the loudspeaker.

A **sub-system** is a part of a system which performs an identified function within the whole system; the amplifier in Fig. 2.1 is an example of a sub-system.

A **component** or **element** is usually the simplest part of a system which has a specific and well-defined function – for example, the microphone in Fig. 2.1.

The illustration in Fig. 2.1 is called a block diagram and electrical/electronic systems, which can often be quite complicated, can be better understood when broken down in this way. It is not always necessary to know precisely what is inside each sub-system in order to know how the whole system functions.

As another example of an engineering system, Fig. 2.2 illustrates a temperature control system containing a heat source (such as a gas boiler), a fuel controller (such as an electrical solenoid valve), a thermostat and a source of electrical energy. The system of Fig. 2.2 can be shown in block diagram form as in Fig. 2.3; the thermostat compares the actual room temperature with the desired temperature and switches the heating on or off.

There are many types of engineering systems. A **communications system** is an example, where a local area network could comprise a file server, coaxial cable, network adapters, several computers and a laser printer; an **electromechanical system** is another example, where a car electrical system could comprise a battery, a starter motor, an ignition coil, a contact breaker and a distributor. All such systems as these may be represented by block diagrams.

## 2.2   Standard symbols for electrical components

Symbols are used for components in electrical circuit diagrams and some of the more common ones are shown in Fig. 2.4.

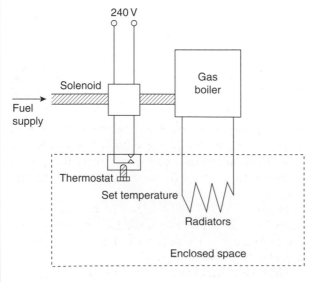

**Figure 2.2**

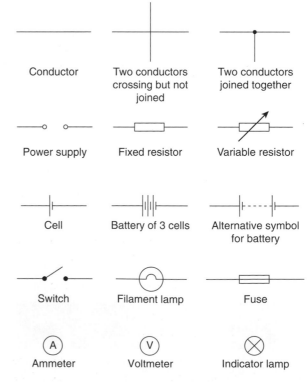

**Figure 2.4**

## 2.3   Electric current and quantity of electricity

All **atoms** consist of **protons, neutrons** and **electrons**. The protons, which have positive electrical charges, and the neutrons, which have no electrical charge, are contained within the **nucleus**. Removed from the nucleus are minute negatively charged particles called electrons. Atoms of different materials differ from one another by having different numbers of protons, neutrons and electrons. An equal number of protons and electrons exist within an atom and it is said to be electrically balanced, as the positive and negative charges cancel each other out. When there are more than two electrons in an atom the electrons are arranged into **shells** at various distances from the nucleus.

All atoms are bound together by powerful forces of attraction existing between the nucleus and its electrons. Electrons in the outer shell of an atom, however, are attracted to their nucleus less powerfully than are electrons whose shells are nearer the nucleus.

It is possible for an atom to lose an electron; the atom, which is now called an **ion**, is not now electrically balanced, but is positively charged and is thus able to attract an electron to itself from another atom. Electrons that move from one atom to another are called free electrons and such random motion can continue indefinitely. However, if an electric pressure or **voltage** is applied across any material there is a tendency for electrons to move in a particular direction. This movement of free electrons, known as **drift**, constitutes an electric current flow. **Thus current is the rate of movement of charge**. **Conductors** are materials that contain electrons that are loosely connected to the nucleus and can easily move through the material from one atom to another.
**Insulators** are materials whose electrons are held firmly to their nucleus.

The unit used to measure the **quantity of electrical charge Q** is called the **coulomb C** (where 1 coulomb $= 6.24 \times 10^{18}$ electrons)
If the drift of electrons in a conductor takes place at the rate of one coulomb per second the resulting current is said to be a current of one ampere.

Thus      1 ampere = 1 coulomb per second or
1 A = 1 C/s
Hence   1 coulomb = 1 ampere second or
1 C = 1 As

Generally, if $I$ is the current in amperes and $t$ the time in seconds during which the current flows, then $I \times t$ represents the quantity of electrical charge in coulombs,

i.e. quantity of electrical charge transferred,

$$Q = I \times t \text{ coulombs}$$

**Problem 1.**    What current must flow if 0.24 coulombs is to be transferred in 15 ms?

Since the quantity of electricity, $Q = It$, then

$$I = \frac{Q}{t} = \frac{0.24}{15 \times 10^{-3}} = \frac{0.24 \times 10^3}{15}$$

$$= \frac{240}{15} = \mathbf{16\,A}$$

**Problem 2.**    If a current of 10 A flows for four minutes, find the quantity of electricity transferred.

Quantity of electricity, $Q = It$ coulombs. $I = 10\,A$ and $t = 4 \times 60 = 240\,s$. Hence

$$Q = 10 \times 240 = \mathbf{2400\,C}$$

**Now try the following exercise**

**Exercise 5    Further problems on charge**

1.   In what time would a current of 10 A transfer a charge of 50 C?                                      [5 s]

2.   A current of 6 A flows for 10 minutes. What charge is transferred?                          [3600 C]

3.   How long must a current of 100 mA flow so as to transfer a charge of 80 C?
                                                      [13 min 20 s]

## 2.4   Potential difference and resistance

For a continuous current to flow between two points in a circuit a **potential difference (p.d.)** or **voltage**, $V$, is required between them; a complete conducting path is necessary to and from the source of electrical energy. The unit of p.d. is the **volt, V**.

Figure 2.5 shows a cell connected across a filament lamp. Current flow, by convention, is considered as flowing from the positive terminal of the cell, around the circuit to the negative terminal.

The flow of electric current is subject to friction. This friction, or opposition, is called **resistance $R$** and is the

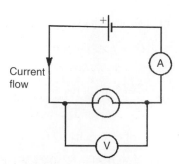

**Figure 2.5**

property of a conductor that limits current. The unit of resistance is the **ohm**; 1 ohm is defined as the resistance which will have a current of 1 ampere flowing through it when 1 volt is connected across it,

i.e.  $\text{resistance } R = \dfrac{\textbf{Potential difference}}{\textbf{current}}$

## 2.5  Basic electrical measuring instruments

An **ammeter** is an instrument used to measure current and must be connected **in series** with the circuit. Figure 2.5 shows an ammeter connected in series with the lamp to measure the current flowing through it. Since all the current in the circuit passes through the ammeter it must have a very **low resistance**.

A **voltmeter** is an instrument used to measure p.d. and must be connected **in parallel** with the part of the circuit whose p.d. is required. In Fig. 2.5, a voltmeter is connected in parallel with the lamp to measure the p.d. across it. To avoid a significant current flowing through it a voltmeter must have a very **high resistance**.

An **ohmmeter** is an instrument for measuring resistance.

A **multimeter**, or universal instrument, may be used to measure voltage, current and resistance. An 'Avometer' and 'Fluke' are typical examples.

The **oscilloscope** may be used to observe waveforms and to measure voltages and currents. The display of an oscilloscope involves a spot of light moving across a screen. The amount by which the spot is deflected from its initial position depends on the p.d. applied to the terminals of the oscilloscope and the range selected. The displacement is calibrated in 'volts per cm'. For example, if the spot is deflected 3 cm and the volts/cm switch is on 10 V/cm then the magnitude of the p.d. is 3 cm × 10 V/cm, i.e. 30 V.

A **wattmeter** is an instrument for the measurement of power in an electrical circuit.

A **BM80** or a **420 MIT megger** or a **bridge megger** may be used to measure both continuity and insulation resistance. **Continuity testing** is the measurement of the resistance of a cable to discover if the cable is continuous, i.e. that it has no breaks or high resistance joints. **Insulation resistance testing** is the measurement of resistance of the insulation between cables, individual cables to earth or metal plugs and sockets, and so on. An insulation resistance in excess of 1 MΩ is normally acceptable.

A **tachometer** is an instrument that indicates the speed, usually in revolutions per minute, at which an engine shaft is rotating.

A **stroboscope** is a device for viewing a rotating object at regularly recurring intervals, by means of either (a) a rotating or vibrating shutter, or (b) a suitably designed lamp which flashes periodically. If the period between successive views is exactly the same as the time of one revolution of the revolving object, and the duration of the view very short, the object will appear to be stationary. (See Chapter 10 for more detail about electrical measuring instruments and measurements.)

## 2.6  Linear and non-linear devices

Figure 2.6 shows a circuit in which current I can be varied by the variable resistor $R_2$. For various settings of $R_2$, the current flowing in resistor $R_1$, displayed on the ammeter, and the p.d. across $R_1$, displayed on the voltmeter, are noted and a graph is plotted of p.d. against current. The result is shown in Fig. 2.7(a) where the straight line graph passing through the origin indicates that current is directly proportional to the p.d. Since the gradient, i.e. (p.d.)/(current) is constant, resistance $R_1$ is constant. A resistor is thus an example of a **linear device**.

If the resistor $R_1$ in Fig. 2.6 is replaced by a component such as a lamp then the graph shown in Fig. 2.7(b) results when values of p.d. are noted for various current

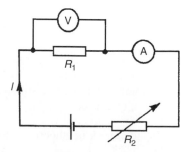

**Figure 2.6**

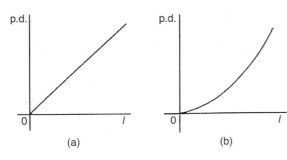

**Figure 2.7**

readings. Since the gradient is changing, the lamp is an example of a **non-linear device**.

## 2.7 Ohm's law

**Ohm's law** states that the current $I$ flowing in a circuit is directly proportional to the applied voltage $V$ and inversely proportional to the resistance $R$, provided the temperature remains constant. Thus,

$$I = \frac{V}{R} \text{ or } V = IR \text{ or } R = \frac{V}{I}$$

*For a practical laboratory experiment on Ohm's law, see Chapter 24, page 406.*

**Problem 3.** The current flowing through a resistor is 0.8 A when a p.d. of 20 V is applied. Determine the value of the resistance.

From Ohm's law,

$$\text{resistance } R = \frac{V}{I} = \frac{20}{0.8} = \frac{200}{8} = \mathbf{25\,\Omega}$$

## 2.8 Multiples and sub-multiples

Currents, voltages and resistances can often be very large or very small. Thus multiples and sub-multiples of units are often used, as stated in Chapter 1. The most common ones, with an example of each, are listed in Table 2.1.

**Problem 4.** Determine the p.d. which must be applied to a $2\,k\Omega$ resistor in order that a current of 10 mA may flow.

$$\text{Resistance } R = 2\,k\Omega = 2 \times 10^3 = 2000\,\Omega$$

$$\text{Current } I = 10\,mA = 10 \times 10^{-3}\,A$$

$$\text{or } \frac{10}{10^3}\,A \text{ or } \frac{10}{1000}\,A = 0.01\,A$$

From Ohm's law, potential difference,

$$V = IR = (0.01)(2000) = \mathbf{20\,V}$$

**Problem 5.** A coil has a current of 50 mA flowing through it when the applied voltage is 12 V. What is the resistance of the coil?

$$\text{Resistance, } R = \frac{V}{I} = \frac{12}{50 \times 10^{-3}}$$

$$= \frac{12 \times 10^3}{50} = \frac{12\,000}{50} = \mathbf{240\,\Omega}$$

**Problem 6.** A 100 V battery is connected across a resistor and causes a current of 5 mA to flow. Determine the resistance of the resistor. If the

**Table 2.1**

| Prefix | Name | Meaning | Example |
|--------|------|---------|---------|
| M | mega | multiply by 1 000 000 (i.e. $\times 10^6$) | $2\,M\Omega = 2\,000\,000$ ohms |
| k | kilo | multiply by 1000 (i.e. $\times 10^3$) | $10\,kV = 10\,000$ volts |
| m | milli | divide by 1000 (i.e. $\times 10^{-3}$) | $25\,mA = \dfrac{25}{1000}\,A$ $= 0.025$ amperes |
| $\mu$ | micro | divide by 1 000 000 (i.e. $\times 10^{-6}$) | $50\,\mu V = \dfrac{50}{1\,000\,000}\,V$ $= 0.000\,05$ volts |

voltage is now reduced to 25 V, what will be the new value of the current flowing?

$$\text{Resistance } R = \frac{V}{I} = \frac{100}{5 \times 10^{-3}} = \frac{100 \times 10^3}{5}$$

$$= 20 \times 10^3 = 20\,\text{k}\Omega$$

Current when voltage is reduced to 25 V,

$$I = \frac{V}{R} = \frac{25}{20 \times 10^3} = \frac{25}{20} \times 10^{-3} = 1.25\,\text{mA}$$

**Problem 7.** What is the resistance of a coil which draws a current of (a) 50 mA and (b) 200 μA from a 120 V supply?

(a) Resistance $R = \dfrac{V}{I} = \dfrac{120}{50 \times 10^{-3}}$

$$= \frac{120}{0.05} = \frac{12\,000}{5}$$

$$= 2400\,\Omega \text{ or } \mathbf{2.4\,k\Omega}$$

(b) Resistance $R = \dfrac{120}{200 \times 10^{-6}} = \dfrac{120}{0.0002}$

$$= \frac{1\,200\,000}{2} = 600\,000\,\Omega$$

$$\text{or } 600\,\text{k}\Omega \text{ or } \mathbf{0.6\,M\Omega}$$

**Problem 8.** The current/voltage relationship for two resistors A and B is as shown in Fig. 2.8. Determine the value of the resistance of each resistor.

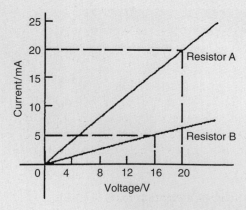

Figure 2.8

For resistor A,

$$R = \frac{V}{I} = \frac{20\,\text{V}}{20\,\text{mA}} = \frac{20}{0.02} = \frac{2000}{2}$$

$$= 1000\,\Omega \text{ or } \mathbf{1\,k\Omega}$$

For resistor B,

$$R = \frac{V}{I} = \frac{16\,\text{V}}{5\,\text{mA}} = \frac{16}{0.005} = \frac{16\,000}{5}$$

$$= 3200\,\Omega \text{ or } \mathbf{3.2\,k\Omega}$$

**Now try the following exercise**

**Exercise 6   Further problems on Ohm's law**

1.  The current flowing through a heating element is 5 A when a p.d. of 35 V is applied across it. Find the resistance of the element.
    [7 Ω]

2.  A 60 W electric light bulb is connected to a 240 V supply. Determine (a) the current flowing in the bulb and (b) the resistance of the bulb.    [(a) 0.25 A (b) 960 Ω]

3.  Graphs of current against voltage for two resistors P and Q are shown in Fig. 2.9. Determine the value of each resistor.
    [2 mΩ, 5 mΩ]

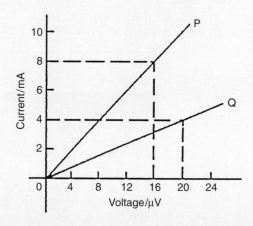

Figure 2.9

4.  Determine the p.d. which must be applied to a 5 kΩ resistor such that a current of 6 mA may flow.    [30 V]

5.  A 20 V source of e.m.f. is connected across a circuit having a resistance of 400 Ω. Calculate the current flowing.   [50 mA]

## 2.9  Conductors and insulators

A **conductor** is a material having a low resistance which allows electric current to flow in it. All metals are conductors and some examples include copper, aluminium, brass, platinum, silver, gold and carbon.

An **insulator** is a material having a high resistance which does not allow electric current to flow in it. Some examples of insulators include plastic, rubber, glass, porcelain, air, paper, cork, mica, ceramics and certain oils.

## 2.10  Electrical power and energy

### Electrical power

**Power** $P$ in an electrical circuit is given by the product of potential difference $V$ and current $I$, as stated in Chapter 1. The unit of power is the **watt, W**.

Hence  $$P = V \times I \text{ watts} \qquad (1)$$

From Ohm's law, $V = IR$. Substituting for $V$ in equation (1) gives:

$$P = (IR) \times I$$

i.e.  $$P = I^2R \text{ watts}$$

Also, from Ohm's law, $I = V/R$. Substituting for $I$ in equation (1) gives:

$$P = V \times \frac{V}{R}$$

i.e.  $$P = \frac{V^2}{R} \text{ watts}$$

There are thus three possible formulae which may be used for calculating power.

**Problem 9.** A 100 W electric light bulb is connected to a 250 V supply. Determine (a) the current flowing in the bulb, and (b) the resistance of the bulb.

Power $P = V \times I$, from which, current $I = \dfrac{P}{V}$

(a)  Current $I = \dfrac{100}{250} = \dfrac{10}{25} = \dfrac{2}{5} = \mathbf{0.4\,A}$

(b)  Resistance $R = \dfrac{V}{I} = \dfrac{250}{0.4} = \dfrac{2500}{4} = \mathbf{625\,\Omega}$

**Problem 10.** Calculate the power dissipated when a current of 4 mA flows through a resistance of 5 kΩ.

$$\textbf{Power } P = I^2R = (4 \times 10^{-3})^2(5 \times 10^3)$$
$$= 16 \times 10^{-6} \times 5 \times 10^3$$
$$= 80 \times 10^{-3}$$
$$= \mathbf{0.08\,W} \text{ or } \mathbf{80\,mW}$$

Alternatively, since $I = 4 \times 10^{-3}$ and $R = 5 \times 10^3$ then from Ohm's law, voltage

$$V = IR = 4 \times 10^{-3} \times 5 \times 10^3 = 20\,V$$

Hence,

$$\textbf{power } P = V \times I = 20 \times 4 \times 10^{-3}$$
$$= \mathbf{80\,mW}$$

**Problem 11.** An electric kettle has a resistance of 30 Ω. What current will flow when it is connected to a 240 V supply? Find also the power rating of the kettle.

$$\text{Current, } I = \frac{V}{R} = \frac{240}{30} = \mathbf{8\,A}$$

$$\text{Power, } P = VI = 240 \times 8 = 1920\,W$$
$$= \mathbf{1.92\,kW} = \text{power rating of kettle}$$

**Problem 12.** A current of 5 A flows in the winding of an electric motor, the resistance of the winding being 100 Ω. Determine (a) the p.d. across the winding, and (b) the power dissipated by the coil.

(a)  Potential difference across winding,

$$V = IR = 5 \times 100 = \mathbf{500\,V}$$

(b) Power dissipated by coil,

$$P = I^2 R = 5^2 \times 100$$

$$= \mathbf{2500\,W} \text{ or } \mathbf{2.5\,kW}$$

(Alternatively, $P = V \times I = 500 \times 5$

$$= \mathbf{2500\,W} \text{ or } \mathbf{2.5\,kW})$$

**Problem 13.** The hot resistance of a 240 V filament lamp is 960 Ω. Find the current taken by the lamp and its power rating.

From Ohm's law,

$$\text{current } I = \frac{V}{R} = \frac{240}{960}$$

$$= \frac{24}{96} = \frac{\mathbf{1}}{\mathbf{4}}\mathbf{A} \text{ or } \mathbf{0.25\,A}$$

Power rating $P = VI = (240)\left(\frac{1}{4}\right) = \mathbf{60\,W}$

## Electrical energy

**Electrical energy = power × time**

If the power is measured in watts and the time in seconds then the unit of energy is watt-seconds or **joules**. If the power is measured in kilowatts and the time in hours then the unit of energy is **kilowatt-hours**, often called the '**unit of electricity**'. The 'electricity meter' in the home records the number of kilowatt-hours used and is thus an energy meter.

**Problem 14.** A 12 V battery is connected across a load having a resistance of 40 Ω. Determine the current flowing in the load, the power consumed and the energy dissipated in 2 minutes.

$$\text{Current } I = \frac{V}{R} = \frac{12}{40} = \mathbf{0.3\,A}$$

Power consumed, $P = VI = (12)(0.3) = \mathbf{3.6\,W}$

Energy dissipated = power × time

$$= (3.6\,\text{W})(2 \times 60\,\text{s})$$

$$= \mathbf{432\,J} \text{ (since } 1\,\text{J} = 1\,\text{Ws)}$$

**Problem 15.** A source of e.m.f. of 15 V supplies a current of 2 A for 6 minutes. How much energy is provided in this time?

Energy = power × time, and power = voltage × current. Hence

$$\text{energy} = VIt = 15 \times 2 \times (6 \times 60)$$

$$= 10\,800\,\text{Ws or J} = \mathbf{10.8\,kJ}$$

**Problem 16.** Electrical equipment in an office takes a current of 13 A from a 240 V supply. Estimate the cost per week of electricity if the equipment is used for 30 hours each week and 1 kWh of energy costs 12.5p.

$$\text{Power} = VI \text{ watts} = 240 \times 13$$

$$= 3120\,\text{W} = 3.12\,\text{kW}$$

Energy used per week = power × time

$$= (3.12\,\text{kW}) \times (30\,\text{h})$$

$$= 93.6\,\text{kWh}$$

Cost at 12.5p per kWh = 93.6 × 12.5 = 1170p. Hence **weekly cost of electricity = £11.70**

**Problem 17.** An electric heater consumes 3.6 MJ when connected to a 250 V supply for 40 minutes. Find the power rating of the heater and the current taken from the supply.

$$\text{Power} = \frac{\text{energy}}{\text{time}} = \frac{3.6 \times 10^6}{40 \times 60}\frac{\text{J}}{\text{s}} \text{ (or W)} = 1500\,\text{W}$$

i.e. Power rating of heater = **1.5 kW**.

$$\text{Power } P = VI$$

$$\text{thus } I = \frac{P}{V} = \frac{1500}{250} = 6\,\text{A}$$

Hence the current taken from the supply is **6 A**.

**Problem 18.** Determine the power dissipated by the element of an electric fire of resistance 20 Ω when a current of 10 A flows through it. If the fire is on for 6 hours determine the energy used and the cost if 1 unit of electricity costs 13p.

$$\text{Power } P = I^2 R = 10^2 \times 20$$

$$= 100 \times 20 = \mathbf{2000\,W} \text{ or } \mathbf{2\,kW}.$$

(Alternatively, from Ohm's law,

$$V = IR = 10 \times 20 = 200\,\text{V},$$

hence power

$$P = V \times I = 200 \times 10 = 2000\,\text{W} = 2\,\text{kW}).$$

Energy used in 6 hours
$$= \text{power} \times \text{time} = 2\,\text{kW} \times 6\,\text{h} = \textbf{12\,kWh}.$$
1 unit of electricity $= 1\,\text{kWh}$; hence the number of units used is 12. Cost of energy $= 12 \times 13 = \textbf{£1.56p}$

---

**Problem 19.**  A business uses two 3 kW fires for an average of 20 hours each per week, and six 150 W lights for 30 hours each per week. If the cost of electricity is 14 p per unit, determine the weekly cost of electricity to the business.

---

Energy = power × time.
Energy used by one 3 kW fire in 20 hours
$$= 3\,\text{kW} \times 20\,\text{h} = 60\,\text{kWh}.$$
Hence weekly energy used by two 3 kW fires
$$= 2 \times 60 = 120\,\text{kWh}.$$
Energy used by one 150 W light for 30 hours
$$= 150\,\text{W} \times 30\,\text{h} = 4500\,\text{Wh} = 4.5\,\text{kWh}.$$
Hence weekly energy used by six 150 W lamps
$$= 6 \times 4.5 = 27\,\text{kWh}.$$
Total energy used per week $= 120 + 27 = 147\,\text{kWh}.$
1 unit of electricity $= 1\,\text{kWh}$ of energy. Thus weekly cost of energy at 14 p per kWh $= 14 \times 147 = 2058\,\text{p}$
$$= \textbf{£20.58}$$

---

**Now try the following exercise**

---

**Exercise 7    Further problems on power and energy**

1.  The hot resistance of a 250 V filament lamp is 625 Ω. Determine the current taken by the lamp and its power rating.    [0.4 A, 100 W]

2.  Determine the resistance of a coil connected to a 150 V supply when a current of (a) 75 mA (b) 300 μA flows through it.
[(a) 2 kΩ (b) 0.5 MΩ]

3.  Determine the resistance of an electric fire which takes a current of 12 A from a 240 V supply. Find also the power rating of the fire and the energy used in 20 h.
[20 Ω, 2.88 kW, 57.6 kWh]

4.  Determine the power dissipated when a current of 10 mA flows through an appliance having a resistance of 8 kΩ.    [0.8 W]

5.  85.5 J of energy are converted into heat in 9 s. What power is dissipated?    [9.5 W]

6.  A current of 4 A flows through a conductor and 10 W is dissipated. What p.d. exists across the ends of the conductor?    [2.5 V]

7.  Find the power dissipated when:
(a)  a current of 5 mA flows through a resistance of 20 kΩ
(b)  a voltage of 400 V is applied across a 120 kΩ resistor
(c)  a voltage applied to a resistor is 10 kV and the current flow is 4 m
[(a) 0.5 W (b) 1.33 W (c) 40 W]

8.  A battery of e.m.f. 15 V supplies a current of 2 A for 5 min. How much energy is supplied in this time?    [9 kJ]

9.  A d.c. electric motor consumes 72 MJ when connected to 400 V supply for 2 h 30 min. Find the power rating of the motor and the current taken from the supply.  [8 kW, 20 A]

10.  A p.d. of 500 V is applied across the winding of an electric motor and the resistance of the winding is 50 Ω. Determine the power dissipated by the coil.    [5 kW]

11.  In a household during a particular week three 2 kW fires are used on average 25 h each and eight 100 W light bulbs are used on average 35 h each. Determine the cost of electricity for the week if 1 unit of electricity costs 15 p.
[£26.70]

12.  Calculate the power dissipated by the element of an electric fire of resistance 30 Ω when a current of 10 A flows in it. If the fire is on for 30 hours in a week determine the energy used. Determine also the weekly cost of energy if electricity costs 13.5p per unit.
[3 kW, 90 kWh, £12.15]

---

## 2.11  Main effects of electric current

The three main effects of an electric current are:
(a)  magnetic effect
(b)  chemical effect
(c)  heating effect

Some practical applications of the effects of an electric current include:

**Magnetic effect:** bells, relays, motors, generators, transformers, telephones, car-ignition and lifting magnets (see Chapter 8)

**Chemical effect:** primary and secondary cells and electroplating (see Chapter 4)

**Heating effect:** cookers, water heaters, electric fires, irons, furnaces, kettles and soldering irons

## 2.12 Fuses

If there is a fault in a piece of equipment then excessive current may flow. This will cause overheating and possibly a fire; **fuses** protect against this happening. Current from the supply to the equipment flows through the fuse. The fuse is a piece of wire which can carry a stated current; if the current rises above this value it will melt. If the fuse melts (blows) then there is an open circuit and no current can then flow – thus protecting the equipment by isolating it from the power supply. The fuse must be able to carry slightly more than the normal operating current of the equipment to allow for tolerances and small current surges. With some equipment there is a very large surge of current for a short time at switch on. If a fuse is fitted to withstand this large current there would be no protection against faults which cause the current to rise slightly above the normal value. Therefore special anti-surge fuses are fitted. These can stand 10 times the rated current for 10 milliseconds. If the surge lasts longer than this the fuse will blow.

A circuit diagram symbol for a fuse is shown in Fig. 2.4 on page 10.

> **Problem 20.** If 5 A, 10 A and 13 A fuses are available, state which is most appropriate for the following appliances which are both connected to a 240 V supply: (a) Electric toaster having a power rating of 1 kW (b) Electric fire having a power rating of 3 kW.

Power $P = VI$, from which, current $I = \dfrac{P}{V}$

(a) For the toaster,

$$\text{current } I = \frac{P}{V} = \frac{1000}{240} = \frac{100}{24} = 4.17 \text{ A}$$

Hence a **5 A fuse** is most appropriate

(b) For the fire,

$$\text{current } I = \frac{P}{V} = \frac{3000}{240} = \frac{300}{24} = 12.5 \text{ A}$$

Hence a **13 A fuse** is most appropriate

---

**Now try the following exercises**

> **Exercise 8 Further problem on fuses**
>
> 1. A television set having a power rating of 120 W and electric lawnmower of power rating 1 kW are both connected to a 250 V supply. If 3 A, 5 A and 10 A fuses are available state which is the most appropriate for each appliance. [3 A, 5 A]

## 2.13 Insulation and the dangers of constant high current flow

The use of insulation materials on electrical equipment, whilst being necessary, also has the effect of preventing heat loss, i.e. the heat is not able to dissipate, thus creating the possible danger of fire. In addition, the insulating material has a maximum temperature rating – this is heat it can withstand without being damaged. The current rating for all equipment and electrical components is therefore limited to keep the heat generated within safe limits. In addition, the maximum voltage present needs to be considered when choosing insulation.

> **Exercise 9 Short answer questions on the introduction to electric circuits**
>
> 1. Draw the preferred symbols for the following components used when drawing electrical circuit diagrams:
>    (a) fixed resistor       (b) cell
>    (c) filament lamp        (d) fuse
>    (e) voltmeter
>
> 2. State the unit of
>    (a) current
>    (b) potential difference
>    (c) resistance
>
> 3. State an instrument used to measure
>    (a) current

(b) potential difference
(c) resistance

4. What is a multimeter?

5. State an instrument used to measure:
(a) engine rotational speed
(b) continuity and insulation testing
(c) electrical power

6. State Ohm's law

7. Give one example of
(a) a linear device
(b) a non-linear device

8. State the meaning of the following abbreviations of prefixes used with electrical units:
(a) k     (b) μ     (c) m     (d) M

9. What is a conductor? Give four examples

10. What is an insulator? Give four examples

11. Complete the following statement:
'An ammeter has a . . . resistance and must be connected . . . with the load'

12. Complete the following statement:
'A voltmeter has a . . . resistance and must be connected . . . with the load'

13. State the unit of electrical power. State three formulae used to calculate power

14. State two units used for electrical energy

15. State the three main effects of an electric current and give two examples of each

16. What is the function of a fuse in an electrical circuit?

---

**Exercise 10    Multi-choice problems on the introduction to electric circuits**
**(Answers on page 420)**

1. $60\,\mu s$ is equivalent to:
(a) 0.06           (b) 0.00006 s
(c) 1000 minutes   (d) 0.6 s

2. The current which flows when 0.1 coulomb is transferred in 10 ms is:
(a) 1 A             (b) 10 A
(c) 10 mA           (d) 100 mA

3. The p.d. applied to a $1\,k\Omega$ resistance in order that a current of $100\,\mu A$ may flow is:
(a) 1 V     (b) 100 V     (c) 0.1 V
(d) 10 V

4. Which of the following formulae for electrical power is incorrect?
(a) $VI$     (b) $\dfrac{V}{I}$     (c) $I^2R$     (d) $\dfrac{V^2}{R}$

5. The power dissipated by a resistor of $4\,\Omega$ when a current of 5 A passes through it is:
(a) 6.25 W              (b) 20 W
(c) 80 W                (d) 100 W

6. Which of the following statements is true?
(a) Electric current is measured in volts
(b) $200\,k\Omega$ resistance is equivalent to $2\,M\Omega$
(c) An ammeter has a low resistance and must be connected in parallel with a circuit
(d) An electrical insulator has a high resistance

7. A current of 3 A flows for 50 h through a $6\,\Omega$ resistor. The energy consumed by the resistor is:
(a) 0.9 kWh            (b) 2.7 kWh
(c) 9 kWh              (d) 27 kWh

8. What must be known in order to calculate the energy used by an electrical appliance?
(a) voltage and current
(b) current and time of operation
(c) power and time of operation
(d) current and resistance

9. Voltage drop is the:
(a) maximum potential
(b) difference in potential between two points
(c) voltage produced by a source
(d) voltage at the end of a circuit

10. A 240 V, 60 W lamp has a working resistance of:
(a) 1400 ohm           (b) 60 ohm
(c) 960 ohm            (d) 325 ohm

11.  The largest number of 100 W electric light bulbs which can be operated from a 240 V supply fitted with a 13 A fuse is:

(a) 2  (b) 7  (c) 31  (d) 18

12.  The energy used by a 1.5 kW heater in 5 minutes is:

(a) 5 J                     (b) 450 J
(c) 7500 J                  (d) 450 000 J

13.  When an atom loses an electron, the atom:

(a) becomes positively charged
(b) disintegrates
(c) experiences no effect at all
(d) becomes negatively charged

# Chapter 3

# Resistance variation

At the end of this chapter you should be able to:

- recognize three common methods of resistor construction
- appreciate that electrical resistance depends on four factors
- appreciate that resistance $R = \rho l/a$, where $\rho$ is the resistivity
- recognize typical values of resistivity and its unit
- perform calculations using $R = \rho l/a$
- define the temperature coefficient of resistance, $\alpha$
- recognize typical values for $\alpha$
- perform calculations using $R_\theta = R_0(1 + \alpha\theta)$
- determine the resistance and tolerance of a fixed resistor from its colour code
- determine the resistance and tolerance of a fixed resistor from its letter and digit code

## 3.1 Resistor construction

There is a wide range of resistor types. Three of the most common methods of construction are:

### (i) Wire wound resistors

A length of wire such as nichrome or manganin, whose resistive value per unit length is known, is cut to the desired value and wound around a ceramic former prior to being lacquered for protection. This type of resistor has a large physical size, which is a disadvantage; however, they can be made with a high degree of accuracy, and can have a **high power rating**.
Wire wound resistors are used in **power circuits** and **motor starters**.

### (ii) Metal oxide resistors

With a metal oxide resistor a thin coating of platinum is deposited on a glass plate; it is then fired and a thin track etched out. It is then totally enclosed in an outer tube.

Metal oxide resistors are used in **electronic equipment**.

### (iii) Carbon resistors

This type of resistor is made from a mixture of carbon black resin binder and a refractory powder that is pressed into shape and heated in a kiln to form a solid rod of standard length and width. The resistive value is predetermined by the ratio of the mixture. Metal end connections are crimped onto the rod to act as connecting points for electrical circuitry. This type of resistor is small and mass-produced cheaply; it has limited accuracy and a low power rating.
Carbon resistors are used in **electronic equipment**.

## 3.2 Resistance and resistivity

The resistance of an electrical conductor depends on four factors, these being: (a) the length of the conductor, (b) the cross-sectional area of the conductor, (c) the type of material and (d) the temperature of the material. Resistance, $R$, is directly proportional to length, $l$, of a

DOI: 10.1016/B978-0-08-089056-2.00003-6

conductor, i.e. $R \propto l$. Thus, for example, if the length of a piece of wire is doubled, then the resistance is doubled.

Resistance, $R$, is inversely proportional to cross-sectional area, $a$, of a conductor, i.e. $R \propto 1/a$. Thus, for example, if the cross-sectional area of a piece of wire is doubled then the resistance is halved.

Since $R \propto l$ and $R \propto 1/a$ then $R \propto l/a$. By inserting a constant of proportionality into this relationship the type of material used may be taken into account. The constant of proportionality is known as the **resistivity** of the material and is given the symbol $\rho$ (Greek rho). Thus,

$$\text{resistance} \quad R = \frac{\rho l}{a} \text{ ohms}$$

$\rho$ is measured in ohm metres ($\Omega$ m). The value of the resistivity is that resistance of a unit cube of the material measured between opposite faces of the cube.

Resistivity varies with temperature and some typical values of resistivities measured at about room temperature are given below:

Copper $1.7 \times 10^{-8} \Omega$ m (or $0.017 \mu\Omega$ m)

Aluminium $2.6 \times 10^{-8} \Omega$ m (or $0.026 \mu\Omega$ m)

Carbon (graphite) $10 \times 10^{-8} \Omega$ m ($0.10 \mu\Omega$ m)

Glass $1 \times 10^{10} \Omega$ m (or $10^4 \mu\Omega$ m)

Mica $1 \times 10^{13} \Omega$ m (or $10^7 \mu\Omega$ m)

Note that good conductors of electricity have a low value of resistivity and good insulators have a high value of resistivity.

**Problem 1.** The resistance of a 5 m length of wire is $600 \Omega$. Determine (a) the resistance of an 8 m length of the same wire, and (b) the length of the same wire when the resistance is $420 \Omega$.

(a) Resistance, $R$, is directly proportional to length, $l$, i.e. $R \propto l$. Hence, $600 \Omega \propto 5$ m or $600 = (k)(5)$, where $k$ is the coefficient of proportionality.

Hence, $k = \dfrac{600}{5} = 120$

When the length $l$ is 8 m, then resistance $R = kl = (120)(8) = \mathbf{960 \Omega}$

(b) When the resistance is $420 \Omega$, $420 = kl$, from which,

$$\text{length } l = \frac{420}{k} = \frac{420}{120} = \mathbf{3.5\,m}$$

**Problem 2.** A piece of wire of cross-sectional area $2 \text{ mm}^2$ has a resistance of $300 \Omega$. Find (a) the resistance of a wire of the same length and material if the cross-sectional area is $5 \text{ mm}^2$, (b) the cross-sectional area of a wire of the same length and material of resistance $750 \Omega$.

Resistance $R$ is inversely proportional to cross-sectional area, $a$, i.e. $R \propto l/a$

Hence $300 \Omega \propto \frac{1}{2} \text{ mm}^2$ or $300 = (k) (\frac{1}{2})$

from which, the coefficient of proportionality,

$$k = 300 \times 2 = 600$$

(a) When the cross-sectional area $a = 5 \text{ mm}^2$ then

$$R = (k)(\tfrac{1}{5}) = (600)(\tfrac{1}{5}) = \mathbf{120 \Omega}$$

(Note that resistance has decreased as the cross-sectional is increased.)

(b) When the resistance is $750 \Omega$ then

$$750 = (k) \left( \frac{1}{a} \right)$$

from which

$$\text{cross-sectional area, } a = \frac{k}{750} = \frac{600}{750}$$
$$= \mathbf{0.8 \text{ mm}^2}$$

**Problem 3.** A wire of length 8 m and cross-sectional area $3 \text{ mm}^2$ has a resistance of $0.16 \Omega$. If the wire is drawn out until its cross-sectional area is $1 \text{ mm}^2$, determine the resistance of the wire.

Resistance $R$ is directly proportional to length $l$, and inversely proportional to the cross-sectional area, $a$, i.e. $R \propto l/a$ or $R = k(l/a)$, where $k$ is the coefficient of proportionality.

Since $R = 0.16$, $l = 8$ and $a = 3$, then $0.16 = (k)(8/3)$, from which $k = 0.16 \times 3/8 = 0.06$

If the cross-sectional area is reduced to $1/3$ of its original area then the length must be tripled to $3 \times 8$, i.e. 24 m

$$\text{New resistance } R = k \left( \frac{l}{a} \right) = 0.06 \left( \frac{24}{1} \right) = \mathbf{1.44 \Omega}$$

**Problem 4.** Calculate the resistance of a 2 km length of aluminium overhead power cable if the cross-sectional area of the cable is $100\,\text{mm}^2$. Take the resistivity of aluminium to be $0.03 \times 10^{-6}\,\Omega\,\text{m}$.

Length $l = 2\,\text{km} = 2000\,\text{m}$,
area $a = 100\,\text{mm}^2 = 100 \times 10^{-6}\,\text{m}^2$
and resistivity $\rho = 0.03 \times 10^{-6}\,\Omega\,\text{m}$.

$$\text{Resistance } R = \frac{\rho l}{a}$$

$$= \frac{(0.03 \times 10^{-6}\,\Omega\,\text{m})(2000\,\text{m})}{(100 \times 10^{-6}\,\text{m}^2)}$$

$$= \frac{0.03 \times 2000}{100}\,\Omega = \mathbf{0.6\,\Omega}$$

**Problem 5.** Calculate the cross-sectional area, in $\text{mm}^2$, of a piece of copper wire, $40\,\text{m}$ in length and having a resistance of $0.25\,\Omega$. Take the resistivity of copper as $0.02 \times 10^{-6}\,\Omega\,\text{m}$.

Resistance $R = \rho l / a$ hence cross-sectional area

$$a = \frac{\rho l}{R} = \frac{(0.02 \times 10^{-6}\,\Omega\,\text{m})(40\,\text{m})}{0.25\,\Omega}$$

$$= 3.2 \times 10^{-6}\,\text{m}^2$$

$$= (3.2 \times 10^{-6}) \times 10^6\,\text{mm}^2 = \mathbf{3.2\,mm^2}$$

**Problem 6.** The resistance of $1.5\,\text{km}$ of wire of cross-sectional area $0.17\,\text{mm}^2$ is $150\,\Omega$. Determine the resistivity of the wire.

Resistance, $R = \rho l / a$ hence

$$\text{resistivity } \rho = \frac{Ra}{l}$$

$$= \frac{(150\,\Omega)(0.17 \times 10^{-6}\,\text{m}^2)}{(1500\,\text{m})}$$

$$= \mathbf{0.017 \times 10^{-6}\,\Omega\,m}$$

$$\text{or } \mathbf{0.017\,\mu\Omega\,m}$$

**Problem 7.** Determine the resistance of $1200\,\text{m}$ of copper cable having a diameter of $12\,\text{mm}$ if the resistivity of copper is $1.7 \times 10^{-8}\,\Omega\,\text{m}$.

Cross-sectional area of cable,

$$a = \pi r^2 = \pi \left(\frac{12}{2}\right)^2$$

$$= 36\pi\,\text{mm}^2 = 36\pi \times 10^{-6}\,\text{m}^2$$

$$\text{Resistance } R = \frac{\rho l}{a}$$

$$= \frac{(1.7 \times 10^{-8}\,\Omega\,\text{m})(1200\,\text{m})}{(36\pi \times 10^{-6}\,\text{m}^2)}$$

$$= \frac{1.7 \times 1200 \times 10^6}{10^8 \times 36\pi}\,\Omega$$

$$= \frac{1.7 \times 12}{36\pi}\,\Omega = \mathbf{0.180\,\Omega}$$

**Now try the following exercise**

**Exercise 11    Further problems on resistance and resistivity**

1. The resistance of a $2\,\text{m}$ length of cable is $2.5\,\Omega$. Determine (a) the resistance of a $7\,\text{m}$ length of the same cable and (b) the length of the same wire when the resistance is $6.25\,\Omega$.
   [(a) $8.75\,\Omega$ (b) $5\,\text{m}$]

2. Some wire of cross-sectional area $1\,\text{mm}^2$ has a resistance of $20\,\Omega$. Determine (a) the resistance of a wire of the same length and material if the cross-sectional area is $4\,\text{mm}^2$, and (b) the cross-sectional area of a wire of the same length and material if the resistance is $32\,\Omega$.
   [(a) $5\,\Omega$ (b) $0.625\,\text{mm}^2$]

3. Some wire of length $5\,\text{m}$ and cross-sectional area $2\,\text{mm}^2$ has a resistance of $0.08\,\Omega$. If the wire is drawn out until its cross-sectional area is $1\,\text{mm}^2$, determine the resistance of the wire.
   [$0.32\,\Omega$]

4. Find the resistance of $800\,\text{m}$ of copper cable of cross-sectional area $20\,\text{mm}^2$. Take the resistivity of copper as $0.02\,\mu\Omega\,\text{m}$.    [$0.8\,\Omega$]

5. Calculate the cross-sectional area, in $\text{mm}^2$, of a piece of aluminium wire $100\,\text{m}$ long and having a resistance of $2\,\Omega$. Take the resistivity of aluminium as $0.03 \times 10^{-6}\,\Omega\,\text{m}$.
   [$1.5\,\text{mm}^2$]

6. The resistance of $500\,\text{m}$ of wire of cross-sectional area $2.6\,\text{mm}^2$ is $5\,\Omega$. Determine the resistivity of the wire in $\mu\Omega\,\text{m}$.
   [$0.026\,\mu\Omega\,\text{m}$]

7.  Find the resistance of 1 km of copper cable having a diameter of 10 mm if the resistivity of copper is $0.017 \times 10^{-6}\,\Omega\,\mathrm{m}$.

[$0.216\,\Omega$]

## 3.3   Temperature coefficient of resistance

In general, as the temperature of a material increases, most conductors increase in resistance, insulators decrease in resistance, whilst the resistance of some special alloys remain almost constant.

The **temperature coefficient of resistance** of a material is the increase in the resistance of a $1\,\Omega$ resistor of that material when it is subjected to a rise of temperature of $1°C$. The symbol used for the temperature coefficient of resistance is $\alpha$ (Greek alpha). Thus, if some copper wire of resistance $1\,\Omega$ is heated through $1°C$ and its resistance is then measured as $1.0043\,\Omega$ then $\alpha = 0.0043\,\Omega/\Omega°C$ for copper. The units are usually expressed only as 'per $°C$', i.e. $\alpha = 0.0043/°C$ for copper. If the $1\,\Omega$ resistor of copper is heated through $100°C$ then the resistance at $100°C$ would be $1 + 100 \times 0.0043 = 1.43\,\Omega$. Some typical values of temperature coefficient of resistance measured at $0°C$ are given below:

| | |
|---|---|
| Copper | $0.0043/°C$ |
| Nickel | $0.0062/°C$ |
| Constantan | $0$ |
| Aluminium | $0.0038/°C$ |
| Carbon | $-0.00048/°C$ |
| Eureka | $0.00001/°C$ |

(Note that the negative sign for carbon indicates that its resistance falls with increase of temperature.)

If the resistance of a material at $0°C$ is known the resistance at any other temperature can be determined from:

$$R_\theta = R_0(1 + \alpha_0\theta)$$

where $R_0$ = resistance at $0°C$

$R_\theta$ = resistance at temperature $\theta°C$

$\alpha_0$ = temperature coefficient of resistance at $0°C$

**Problem 8.**   A coil of copper wire has a resistance of $100\,\Omega$ when its temperature is $0°C$. Determine its resistance at $70°C$ if the temperature coefficient of resistance of copper at $0°C$ is $0.0043/°C$.

Resistance $R_\theta = R_0(1 + \alpha_0\theta)$. Hence resistance at $100°C$,

$$R_{100} = 100[1 + (0.0043)(70)]$$
$$= 100[1 + 0.301]$$
$$= 100(1.301) = \mathbf{130.1\,\Omega}$$

**Problem 9.**   An aluminium cable has a resistance of $27\,\Omega$ at a temperature of $35°C$. Determine its resistance at $0°C$. Take the temperature coefficient of resistance at $0°C$ to be $0.0038/°C$.

Resistance at $\theta°C$, $R_\theta = R_0(1 + \alpha_0\theta)$. Hence resistance at $0°C$,

$$R_0 = \frac{R_\theta}{(1 + \alpha_0\theta)} = \frac{27}{[1 + (0.0038)(35)]}$$
$$= \frac{27}{1 + 0.133}$$
$$= \frac{27}{1.133} = \mathbf{23.83\,\Omega}$$

**Problem 10.**   A carbon resistor has a resistance of $1\,\mathrm{k}\Omega$ at $0°C$. Determine its resistance at $80°C$. Assume that the temperature coefficient of resistance for carbon at $0°C$ is $-0.0005/°C$.

Resistance at temperature $\theta°C$,

$$R_\theta = R_0(1 + \alpha_0\theta)$$

i.e.

$$R_\theta = 1000[1 + (-0.0005)(80)]$$
$$= 1000[1 - 0.040] = 1000(0.96) = \mathbf{960\,\Omega}$$

If the resistance of a material at room temperature (approximately $20°C$), $R_{20}$, and the temperature coefficient of resistance at $20°C$, $\alpha_{20}$, are known then the resistance $R_\theta$ at temperature $\theta°C$ is given by:

$$R_\theta = R_{20}[1 + \alpha_{20}(\theta - 20)]$$

**Problem 11.**   A coil of copper wire has a resistance of $10\,\Omega$ at $20°C$. If the temperature coefficient of resistance of copper at $20°C$ is $0.004/°C$ determine the resistance of the coil when the temperature rises to $100°C$.

Resistance at $\theta°C$,

$$R_\theta = R_{20}[1 + \alpha_{20}(\theta - 20)]$$

Hence resistance at 100°C,

$$R_{100} = 10[1 + (0.004)(100 - 20)]$$

$$= 10[1 + (0.004)(80)]$$

$$= 10[1 + 0.32]$$

$$= 10(1.32) = \textbf{13.2 }\Omega$$

**Problem 12.** The resistance of a coil of aluminium wire at 18°C is 200 $\Omega$. The temperature of the wire is increased and the resistance rises to 240 $\Omega$. If the temperature coefficient of resistance of aluminium is 0.0039/°C at 18°C determine the temperature to which the coil has risen.

Let the temperature rise to $\theta°C$. Resistance at $\theta°C$,

$$R_\theta = R_{18}[1 + \alpha_{18}(\theta - 18)]$$

i.e.

$$240 = 200[1 + (0.0039)(\theta - 18)]$$

$$240 = 200 + (200)(0.0039)(\theta - 18)$$

$$240 - 200 = 0.78(\theta - 18)$$

$$40 = 0.78(\theta - 18)$$

$$\frac{40}{0.78} = \theta - 18$$

$$51.28 = \theta - 18, \text{ from which,}$$

$$\theta = 51.28 + 18 = 69.28°C$$

**Hence the temperature of the coil increases to 69.28°C**

If the resistance at 0°C is not known, but is known at some other temperature $\theta_1$, then the resistance at any temperature can be found as follows:

$$R_1 = R_0(1 + \alpha_0\theta_1)$$
and
$$R_2 = R_0(1 + \alpha_0\theta_2)$$

Dividing one equation by the other gives:

$$\frac{R_1}{R_2} = \frac{1 + \alpha_0\theta_1}{1 + \alpha_0\theta_2}$$

where $R_2 =$ resistance at temperature $\theta_2$

**Problem 13.** Some copper wire has a resistance of 200 $\Omega$ at 20°C. A current is passed through the wire and the temperature rises to 90°C. Determine the resistance of the wire at 90°C, correct to the nearest ohm, assuming that the temperature coefficient of resistance is 0.004/°C at 0°C.

$$R_{20} = 200\,\Omega, \alpha_0 = 0.004/°C$$

and

$$\frac{R_{20}}{R_{90}} = \frac{[1 + \alpha_0(20)]}{[1 + \alpha_0(90)]}$$

Hence

$$R_{90} = \frac{R_{20}[1 + 90\alpha_0]}{[1 + 20\alpha_0]}$$

$$= \frac{200[1 + 90(0.004)]}{[1 + 20(0.004)]}$$

$$= \frac{200[1 + 0.36]}{[1 + 0.08]}$$

$$= \frac{200(1.36)}{(1.08)} = \textbf{251.85}\,\Omega$$

**i.e. the resistance of the wire at 90°C is 252 $\Omega$**, correct to the nearest ohm

**Now try the following exercise**

**Exercise 12    Further problems on the temperature coefficient of resistance**

1. A coil of aluminium wire has a resistance of 50 $\Omega$ when its temperature is 0°C. Determine its resistance at 100°C if the temperature coefficient of resistance of aluminium at 0°C is 0.0038/°C    [69 $\Omega$]

2. A copper cable has a resistance of 30 $\Omega$ at a temperature of 50°C. Determine its resistance at 0°C. Take the temperature coefficient of resistance of copper at 0°C as 0.0043/°C    [24.69 $\Omega$]

3. The temperature coefficient of resistance for carbon at 0°C is −0.00048/°C. What is the significance of the minus sign? A carbon resistor has a resistance of 500 $\Omega$ at 0°C. Determine its resistance at 50°C.    [488 $\Omega$]

4. A coil of copper wire has a resistance of $20\,\Omega$ at $18°C$. If the temperature coefficient of resistance of copper at $18°C$ is $0.004/°C$, determine the resistance of the coil when the temperature rises to $98°C$          $[26.4\,\Omega]$

5. The resistance of a coil of nickel wire at $20°C$ is $100\,\Omega$. The temperature of the wire is increased and the resistance rises to $130\,\Omega$. If the temperature coefficient of resistance of nickel is $0.006/°C$ at $20°C$, determine the temperature to which the coil has risen.

    $[70°C]$

6. Some aluminium wire has a resistance of $50\,\Omega$ at $20°C$. The wire is heated to a temperature of $100°C$. Determine the resistance of the wire at $100°C$, assuming that the temperature coefficient of resistance at $0°C$ is $0.004/°C$.

    $[64.8\,\Omega]$

7. A copper cable is $1.2\,km$ long and has a cross-sectional area of $5\,mm^2$. Find its resistance at $80°C$ if at $20°C$ the resistivity of copper is $0.02 \times 10^{-6}\,\Omega\,m$ and its temperature coefficient of resistance is $0.004/°C$.

    $[5.95\,\Omega]$

## 3.4 Resistor colour coding and ohmic values

### (a) Colour code for fixed resistors

The colour code for fixed resistors is given in Table 3.1

(i) For a **four-band fixed resistor** (i.e. resistance values with two significant figures): yellow-violet-orange-red indicates $47\,k\Omega$ with a tolerance of $\pm2\%$
    (Note that the first band is the one nearest the end of the resistor)

(ii) For a **five-band fixed resistor** (i.e. resistance values with three significant figures): red-yellow-white-orange-brown indicates $249\,k\Omega$ with a tolerance of $\pm1\%$
    (Note that the fifth band is 1.5 to 2 times wider than the other bands)

Problem 14. Determine the value and tolerance of a resistor having a colour coding of: orange-orange-silver-brown.

**Table 3.1**

| Colour | Significant Figures | Multiplier | Tolerance |
|---|---|---|---|
| Silver | – | $10^{-2}$ | $\pm10\%$ |
| Gold | – | $10^{-1}$ | $\pm5\%$ |
| Black | 0 | 1 | – |
| Brown | 1 | 10 | $\pm1\%$ |
| Red | 2 | $10^2$ | $\pm2\%$ |
| Orange | 3 | $10^3$ | – |
| Yellow | 4 | $10^4$ | – |
| Green | 5 | $10^5$ | $\pm0.5\%$ |
| Blue | 6 | $10^6$ | $\pm0.25\%$ |
| Violet | 7 | $10^7$ | $\pm0.1\%$ |
| Grey | 8 | $10^8$ | – |
| White | 9 | $10^9$ | – |
| None | – | – | $\pm20\%$ |

The first two bands, i.e. orange-orange, give 33 from Table 3.1.
The third band, silver, indicates a multiplier of $10^2$ from Table 3.1, which means that the value of the resistor is $33 \times 10^{-2} = 0.33\,\Omega$
The fourth band, i.e. brown, indicates a tolerance of $\pm1\%$ from Table 3.1. Hence a colour coding of orange-orange-silver-brown represents a resistor of value **$0.33\,\Omega$ with a tolerance of $\pm1\%$**

Problem 15. Determine the value and tolerance of a resistor having a colour coding of: brown-black-brown.

The first two bands, i.e. brown-black, give 10 from Table 3.1.
The third band, brown, indicates a multiplier of 10 from Table 3.1, which means that the value of the resistor is $10 \times 10 = 100\,\Omega$
There is no fourth band colour in this case; hence, from Table 3.1, the tolerance is $\pm20\%$. Hence a colour coding of brown-black-brown represents a resistor of value **$100\,\Omega$ with a tolerance of $\pm20\%$**

**Problem 16.** Between what two values should a resistor with colour coding brown-black-brown-silver lie?

From Table 3.1, brown-black-brown-silver indicates $10 \times 10$, i.e. $100\,\Omega$, with a tolerance of $\pm 10\%$
This means that the value could lie between

$$(100 - 10\% \text{ of } 100)\,\Omega$$

and $\qquad (100 + 10\% \text{ of } 100)\,\Omega$

i.e. brown-black-brown-silver indicates any value **between 90 Ω and 110 Ω**

**Problem 17.** Determine the colour coding for a $47\,\mathrm{k\Omega}$ having a tolerance of $\pm 5\%$.

From Table 3.1, $47\,\mathrm{k\Omega} = 47 \times 10^3$ has a colour coding of yellow-violet-orange. With a tolerance of $\pm 5\%$, the fourth band will be gold.
Hence $47\,\mathrm{k\Omega} \pm 5\%$ has a colour coding of:
**yellow-violet-orange-gold**

**Problem 18.** Determine the value and tolerance of a resistor having a colour coding of:
orange-green-red-yellow-brown.

Orange-green-red-yellow-brown is a five-band fixed resistor and from Table 3.1, indicates: $352 \times 10^4\,\Omega$ with a tolerance of $\pm 1\%$

$$352 \times 10^4\,\Omega = 3.52 \times 10^6\,\Omega, \text{ i.e. } 3.52\,\mathrm{M\Omega}$$

Hence orange-green-red-yellow-brown indicates **3.52 MΩ ± 1%**

## (b) Letter and digit code for resistors

Another way of indicating the value of resistors is the letter and digit code shown in Table 3.2.
**Tolerance** is indicated as follows: $F = \pm 1\%$, $G = \pm 2\%$, $J = \pm 5\%$, $K = \pm 10\%$ and $M = \pm 20\%$
Thus, for example,

$$R33M = 0.33\,\Omega \pm 20\%$$
$$4R7K = 4.7\,\Omega \pm 10\%$$
$$390RJ = 390\,\Omega \pm 5\%$$

**Problem 19.** Determine the value of a resistor marked as 6K8F.

**Table 3.2**

| Resistance Value | Marked as: |
|---|---|
| $0.47\,\Omega$ | R47 |
| $1\,\Omega$ | 1R0 |
| $4.7\,\Omega$ | 4R7 |
| $47\,\Omega$ | 47R |
| $100\,\Omega$ | 100R |
| $1\,\mathrm{k\Omega}$ | 1K0 |
| $10\,\mathrm{k\Omega}$ | 10K |
| $10\,\mathrm{M\Omega}$ | 10M |

From Table 3.2, 6K8F is equivalent to: **6.8 kΩ ± 1%**

**Problem 20.** Determine the value of a resistor marked as 4M7M.

From Table 3.2, 4M7M is equivalent to: **4.7 MΩ ± 20%**

**Problem 21.** Determine the letter and digit code for a resistor having a value of $68\,\mathrm{k\Omega} \pm 10\%$.

From Table 3.2, $68\,\mathrm{k\Omega} \pm 10\%$ has a letter and digit code of: **68 KK**

**Now try the following exercises**

**Exercise 13   Further problems on resistor colour coding and ohmic values**

1. Determine the value and tolerance of a resistor having a colour coding of: blue-grey-orange-red          [68 kΩ ± 2%]

2. Determine the value and tolerance of a resistor having a colour coding of: yellow-violet-gold          [4.7 Ω ± 20%]

3. Determine the value and tolerance of a resistor having a colour coding of: blue-white-black-black-gold          [690 Ω ± 5%]

4. Determine the colour coding for a $51\,k\Omega$ four-band resistor having a tolerance of $\pm 2\%$
[green-brown-orange-red]

5. Determine the colour coding for a $1\,M\Omega$ four-band resistor having a tolerance of $\pm 10\%$
[brown-black-green-silver]

6. Determine the range of values expected for a resistor with colour coding: red-black-green-silver        [$1.8\,M\Omega$ to $2.2\,M\Omega$]

7. Determine the range of values expected for a resistor with colour coding: yellow-black-orange-brown        [$39.6\,k\Omega$ to $40.4\,k\Omega$]

8. Determine the value of a resistor marked as (a) R22G (b) 4K7F
[(a) $0.22\,\Omega \pm 2\%$ (b) $4.7\,k\Omega \pm 1\%$]

9. Determine the letter and digit code for a resistor having a value of $100\,k\Omega \pm 5\%$
[100 KJ]

10. Determine the letter and digit code for a resistor having a value of $6.8\,M\Omega \pm 20\%$
[6 M8 M]

---

### Exercise 14    Short answer questions on resistance variation

1. Name three types of resistor construction and state one practical application of each

2. Name four factors which can effect the resistance of a conductor

3. If the length of a piece of wire of constant cross-sectional area is halved, the resistance of the wire is ......

4. If the cross-sectional area of a certain length of cable is trebled, the resistance of the cable is ......

5. What is resistivity? State its unit and the symbol used

6. Complete the following:

   Good conductors of electricity have a ...... value of resistivity and good insulators have a ...... value of resistivity

7. What is meant by the 'temperature coefficient of resistance? State its units and the symbols used

8. If the resistance of a metal at $0°C$ is $R_0$, $R_\theta$ is the resistance at $\theta°C$ and $\alpha_0$ is the temperature coefficient of resistance at $0°C$ then:
$R_\theta = \ldots\ldots$

9. Explain briefly the colour coding on resistors

10. Explain briefly the letter and digit code for resistors

---

### Exercise 15    Multi-choice questions on resistance variation
**(Answers on page 420)**

1. The unit of resistivity is:
   (a) ohms
   (b) ohm millimetre
   (c) ohm metre
   (d) ohm/metre

2. The length of a certain conductor of resistance $100\,\Omega$ is doubled and its cross-sectional area is halved. Its new resistance is:
   (a) $100\,\Omega$          (b) $200\,\Omega$
   (c) $50\,\Omega$          (d) $400\,\Omega$

3. The resistance of a $2\,km$ length of cable of cross-sectional area $2\,mm^2$ and resistivity of $2 \times 10^{-8}\,\Omega m$ is:
   (a) $0.02\,\Omega$          (b) $20\,\Omega$
   (c) $0.02\,m\Omega$          (d) $200\,\Omega$

4. A piece of graphite has a cross-sectional area of $10\,mm^2$. If its resistance is $0.1\,\Omega$ and its resistivity $10 \times 10^8\,\Omega m$, its length is:
   (a) $10\,km$          (b) $10\,cm$
   (c) $10\,mm$          (d) $10\,m$

5. The symbol for the unit of temperature coefficient of resistance is:
   (a) $\Omega/°C$          (b) $\Omega$
   (c) $°C$          (d) $\Omega/\Omega°C$

6. A coil of wire has a resistance of $10\,\Omega$ at $0°C$. If the temperature coefficient of resistance for the wire is $0.004/°C$, its resistance at $100°C$ is:
   (a) $0.4\,\Omega$          (b) $1.4\,\Omega$
   (c) $14\,\Omega$          (d) $10\,\Omega$

7. A nickel coil has a resistance of $13\,\Omega$ at $50°C$. If the temperature coefficient of

resistance at 0°C is 0.006/°C, the resistance at 0°C is:

(a)  16.9 Ω          (b)  10 Ω

(c)  43.3 Ω          (d)  0.1 Ω

8.  A colour coding of red-violet-black on a resistor indicates a value of:

(a)  27 Ω ± 20%      (b)  270 Ω

(c)  270 Ω ± 20%     (d)  27 Ω ± 10%

9.  A resistor marked as 4K7G indicates a value of:

(a)  47 Ω ± 20%      (b)  4.7 kΩ ± 20%

(c)  0.47 Ω ± 10%    (d)  4.7 kΩ ± 2%

# Chapter 4

# Batteries and alternative sources of energy

At the end of this chapter you should be able to:

- list practical applications of batteries
- understand electrolysis and its applications, including electroplating
- appreciate the purpose and construction of a simple cell
- explain polarisation and local action
- explain corrosion and its effects
- define the terms e.m.f., $E$, and internal resistance, $r$, of a cell
- perform calculations using $V = E - Ir$
- determine the total e.m.f. and total internal resistance for cells connected in series and in parallel
- distinguish between primary and secondary cells
- explain the construction and practical applications of the Leclanché, mercury, lead–acid and alkaline cells
- list the advantages and disadvantages of alkaline cells over lead–acid cells
- understand the term 'cell capacity' and state its unit
- understand the importance of safe battery disposal
- appreciate advantages of fuel cells and their likely future applications
- understand the implications of alternative energy sources and state five examples

## 4.1 Introduction to batteries

A battery is a device that **converts chemical energy to electricity**. If an appliance is placed between its terminals the current generated will power the device. Batteries are an indispensable item for many electronic devices and are essential for devices that require power when no mains power is available. For example, without the battery, there would be no mobile phones or laptop computers.

The battery is now over 200 years old and batteries are found almost everywhere in consumer and industrial products. Some **practical examples** where batteries are used include:

in laptops, in cameras, in mobile phones, in cars, in watches and clocks, for security equipment, in electronic meters, for smoke alarms, for meters used to read gas, water and electricity consumption at home, to power a camera for an endoscope looking internally at the body, and for transponders used for toll collection on highways throughout the world

Batteries tend to be split into two categories – **primary**, which are not designed to be electrically re-charged,

DOI: 10.1016/B978-0-08-089056-2.00004-8

i.e. are disposable (see Section 4.6), and **secondary batteries**, which are designed to be re-charged, such as those used in mobile phones (see Section 4.7).

In more recent years it has been necessary to design batteries with reduced size, but with increased lifespan and capacity.

If an application requires small size and high power then the 1.5 V battery is used. If longer lifetime is required then the 3 to 3.6 V battery is used. In the 1970s the 1.5 V **manganese battery** was gradually replaced by the **alkaline battery**. **Silver oxide batteries** were gradually introduced in the 1960s and are still the preferred technology for watch batteries today.

**Lithium-ion batteries** were introduced in the 1970s because of the need for longer lifetime applications. Indeed, some such batteries have been known to last well over 10 years before replacement, a characteristic that means that these batteries are still very much in demand today for digital cameras, and sometimes for watches and computer clocks. Lithium batteries are capable of delivering high currents but tend to be expensive.

More types of batteries and their uses are listed in Table 4.2 on page 37.

## 4.2  Some chemical effects of electricity

A material must contain **charged particles** to be able to conduct electric current. In **solids**, the current is carried by **electrons**. Copper, lead, aluminium, iron and carbon are some examples of solid conductors. In **liquids and gases**, the current is carried by the part of a molecule which has acquired an electric charge, called **ions**. These can possess a positive or negative charge, and examples include hydrogen ion $H^+$, copper ion $Cu^{++}$ and hydroxyl ion $OH^-$. Distilled water contains no ions and is a poor conductor of electricity, whereas salt water contains ions and is a fairly good conductor of electricity.

**Electrolysis** is the decomposition of a liquid compound by the passage of electric current through it. Practical applications of electrolysis include the electroplating of metals (see below), the refining of copper and the extraction of aluminium from its ore.

An **electrolyte** is a compound which will undergo electrolysis. Examples include salt water, copper sulphate and sulphuric acid.

The **electrodes** are the two conductors carrying current to the electrolyte. The positive-connected electrode is called the **anode** and the negative-connected electrode the **cathode**.

When two copper wires connected to a battery are placed in a beaker containing a salt water solution, current will flow through the solution. Air bubbles appear around the wires as the water is changed into hydrogen and oxygen by electrolysis.

**Electroplating** uses the principle of electrolysis to apply a thin coat of one metal to another metal. Some practical applications include the tin-plating of steel, silver-plating of nickel alloys and chromium-plating of steel. If two copper electrodes connected to a battery are placed in a beaker containing copper sulphate as the electrolyte it is found that the cathode (i.e. the electrode connected to the negative terminal of the battery) gains copper whilst the anode loses copper.

## 4.3  The simple cell

The purpose of an **electric cell** is to convert chemical energy into electrical energy.

A **simple cell** comprises two dissimilar conductors (electrodes) in an electrolyte. Such a cell is shown in Fig. 4.1, comprising copper and zinc electrodes. An electric current is found to flow between the electrodes. Other possible electrode pairs exist, including zinc–lead and zinc–iron. The electrode potential (i.e. the p.d. measured between the electrodes) varies for each pair of metals. By knowing the e.m.f. of each metal with respect to some standard electrode, the e.m.f. of any pair of metals may be determined. The standard used is the hydrogen electrode. The **electrochemical series** is a way of listing elements in order of electrical potential, and Table 4.1 shows a number of elements in such a series.

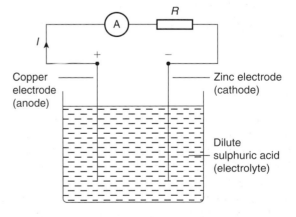

Figure 4.1

In a simple cell two faults exist – those due to **polarisation** and **local action**.

Table 4.1 Part of the electro-chemical series

| Potassium |
| --- |
| sodium |
| aluminium |
| zinc |
| iron |
| lead |
| hydrogen |
| copper |
| silver |
| carbon |

## Polarisation

If the simple cell shown in Fig. 4.1 is left connected for some time, the current $I$ decreases fairly rapidly. This is because of the formation of a film of hydrogen bubbles on the copper anode. This effect is known as the polarisation of the cell. The hydrogen prevents full contact between the copper electrode and the electrolyte and this increases the internal resistance of the cell. The effect can be overcome by using a chemical depolarising agent or depolariser, such as potassium dichromate which removes the hydrogen bubbles as they form. This allows the cell to deliver a steady current.

## Local action

When commercial zinc is placed in dilute sulphuric acid, hydrogen gas is liberated from it and the zinc dissolves. The reason for this is that impurities, such as traces of iron, are present in the zinc which set up small primary cells with the zinc. These small cells are short-circuited by the electrolyte, with the result that localised currents flow causing corrosion. This action is known as local action of the cell. This may be prevented by rubbing a small amount of mercury on the zinc surface, which forms a protective layer on the surface of the electrode.

When two metals are used in a simple cell the electrochemical series may be used to predict the behaviour of the cell:

(i) The metal that is higher in the series acts as the negative electrode, and vice versa. For example,

the zinc electrode in the cell shown in Fig. 4.1 is negative and the copper electrode is positive.

(ii) The greater the separation in the series between the two metals the greater is the e.m.f. produced by the cell.

The electrochemical series is representative of the order of reactivity of the metals and their compounds:

(i) The higher metals in the series react more readily with oxygen and vice-versa.

(ii) When two metal electrodes are used in a simple cell the one that is higher in the series tends to dissolve in the electrolyte.

**Corrosion** is the gradual destruction of a metal in a damp atmosphere by means of simple cell action. In addition to the presence of moisture and air required for rusting, an electrolyte, an anode and a cathode are required for corrosion. Thus, if metals widely spaced in the electrochemical series, are used in contact with each other in the presence of an electrolyte, corrosion will occur. For example, if a brass valve is fitted to a heating system made of steel, corrosion will occur.

The **effects of corrosion** include the weakening of structures, the reduction of the life of components and materials, the wastage of materials and the expense of replacement.

Corrosion may be **prevented** by coating with paint, grease, plastic coatings and enamels, or by plating with tin or chromium. Also, iron may be galvanised, i.e., plated with zinc, the layer of zinc helping to prevent the iron from corroding.

### 4.5   E.m.f. and internal resistance of a cell

The **electromotive force (e.m.f.), $E$**, of a cell is the p.d. between its terminals when it is not connected to a load (i.e. the cell is on 'no load').

The e.m.f. of a cell is measured by using a **high resistance voltmeter** connected in parallel with the cell. The voltmeter must have a high resistance otherwise it will pass current and the cell will not be on 'no-load'. For example, if the resistance of a cell is $1\,\Omega$ and that of a voltmeter $1\,M\Omega$ then the equivalent resistance of the circuit is $1\,M\Omega + 1\,\Omega$, i.e. approximately $1\,M\Omega$, hence no current flows and the cell is not loaded.

The voltage available at the terminals of a cell falls when a load is connected. This is caused by the **internal resistance** of the cell which is the opposition of the material of the cell to the flow of current. The internal resistance acts in series with other resistances in the circuit. Figure 4.2 shows a cell of e.m.f. $E$ volts and internal resistance, $r$, and $XY$ represents the terminals of the cell.

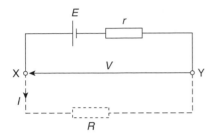

**Figure 4.2**

When a load (shown as resistance $R$) is not connected, no current flows and the terminal p.d., $V = E$. When $R$ is connected a current $I$ flows which causes a voltage drop in the cell, given by $Ir$. The p.d. available at the cell terminals is less than the e.m.f. of the cell and is given by:

$$V = E - Ir$$

Thus if a battery of e.m.f. 12 volts and internal resistance 0.01 Ω delivers a current of 100 A, the terminal p.d.,

$$V = 12 - (100)(0.01)$$
$$= 12 - 1 = 11\,\text{V}$$

When different values of potential difference $V$ across a cell or power supply are measured for different values of current $I$, a graph may be plotted as shown in Fig. 4.3. Since the e.m.f. $E$ of the cell or power supply is the p.d. across its terminals on no load (i.e. when $I = 0$), then $E$ is as shown by the broken line.

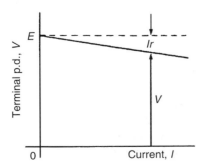

**Figure 4.3**

Since $V = E - Ir$ then the internal resistance may be calculated from

$$r = \frac{E - V}{I}$$

When a current is flowing in the direction shown in Fig. 4.2 the cell is said to be **discharging** ($E > V$). When a current flows in the opposite direction to that shown in Fig. 4.2 the cell is said to be **charging** ($V > E$). A **battery** is a combination of more than one cell. The cells in a battery may be connected in series or in parallel.

(i) **For cells connected in series:**
Total e.m.f. = sum of cell's e.m.f.s
Total internal resistance = sum of cell's internal resistances

(ii) **For cells connected in parallel:**
If each cell has the same e.m.f. and internal resistance:
Total e.m.f. = e.m.f. of one cell
Total internal resistance of $n$ cells
$$= \frac{1}{n} \times \text{internal resistance of one cell}$$

**Problem 1.** Eight cells, each with an internal resistance of 0.2 Ω and an e.m.f. of 2.2 V are connected (a) in series, (b) in parallel. Determine the e.m.f. and the internal resistance of the batteries so formed.

(a) When connected in series, total e.m.f.
= sum of cell's e.m.f.
$= 2.2 \times 8 = \textbf{17.6\,V}$
Total internal resistance
= sum of cell's internal resistance
$= 0.2 \times 8 = \textbf{1.6\,Ω}$

(b) When connected in parallel, total e.m.f
= e.m.f. of one cell
$= \textbf{2.2\,V}$
Total internal resistance of 8 cells
$= \frac{1}{8} \times \text{internal resistance of one cell}$
$= \frac{1}{8} \times 0.2 = \textbf{0.025\,Ω}$

**Problem 2.**   A cell has an internal resistance of $0.02\,\Omega$ and an e.m.f. of $2.0\,\mathrm{V}$. Calculate its terminal p.d. if it delivers (a) $5\,\mathrm{A}$ (b) $50\,\mathrm{A}$.

(a)   Terminal p.d. $V = E - Ir$ where $E$ = e.m.f. of cell, $I$ = current flowing and $r$ = internal resistance of cell

$E = 2.0\,\mathrm{V}$, $I = 5\,\mathrm{A}$ and $r = 0.02\,\Omega$

Hence **terminal p.d.**

$$V = 2.0 - (5)(0.02) = 2.0 - 0.1 = \mathbf{1.9\,V}$$

(b)   When the current is $50\,\mathrm{A}$, terminal p.d.,

$$V = E - Ir = 2.0 - 50(0.02)$$

i.e.      $$V = 2.0 - 1.0 = \mathbf{1.0\,V}$$

Thus the terminal p.d. decreases as the current drawn increases.

**Problem 3.**   The p.d. at the terminals of a battery is $25\,\mathrm{V}$ when no load is connected and $24\,\mathrm{V}$ when a load taking $10\,\mathrm{A}$ is connected. Determine the internal resistance of the battery.

When no load is connected the e.m.f. of the battery, $E$, is equal to the terminal p.d., $V$, i.e. $E = 25\,\mathrm{V}$
When current $I = 10\,\mathrm{A}$ and terminal p.d.

$$V = 24\,\mathrm{V}, \text{ then } V = E - Ir$$

i.e.      $$24 = 25 - (10)r$$

Hence, rearranging, gives

$$10r = 25 - 24 = 1$$

and the internal resistance,

$$\mathbf{r} = \frac{1}{10} = \mathbf{0.1\,\Omega}$$

**Problem 4.**   Ten $1.5\,\mathrm{V}$ cells, each having an internal resistance of $0.2\,\Omega$, are connected in series to a load of $58\,\Omega$. Determine (a) the current flowing in the circuit and (b) the p.d. at the battery terminals.

(a)   For ten cells, battery e.m.f., $E = 10 \times 1.5 = 15\,\mathrm{V}$, and the total internal resistance, $r = 10 \times 0.2 = 2\,\Omega$.

When connected to a $58\,\Omega$ load the circuit is as shown in Fig. 4.4

$$\text{Current } I = \frac{\text{e.m.f.}}{\text{total resistance}}$$

$$= \frac{15}{58 + 2}$$

$$= \frac{15}{60} = \mathbf{0.25\,A}$$

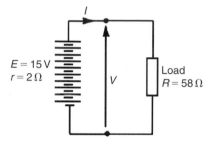

**Figure 4.4**

(b)   P.d. at battery terminals, $V = E - Ir$

i.e. $V = 15 - (0.25)(2) = \mathbf{14.5\,V}$

**Now try the following exercise**

**Exercise 16   Further problems on e.m.f. and internal resistance of a cell**

1.   Twelve cells, each with an internal resistance of $0.24\,\Omega$ and an e.m.f. of $1.5\,\mathrm{V}$ are connected (a) in series, (b) in parallel. Determine the e.m.f. and internal resistance of the batteries so formed.

[(a) $18\,\mathrm{V}$, $2.88\,\Omega$ (b) $1.5\,\mathrm{V}$, $0.02\,\Omega$]

2.   A cell has an internal resistance of $0.03\,\Omega$ and an e.m.f. of $2.2\,\mathrm{V}$. Calculate its terminal p.d. if it delivers

(a) $1\,\mathrm{A}$   (b) $20\,\mathrm{A}$   (c) $50\,\mathrm{A}$

[(a) $2.17\,\mathrm{V}$ (b) $1.6\,\mathrm{V}$ (c) $0.7\,\mathrm{V}$]

3.   The p.d. at the terminals of a battery is $16\,\mathrm{V}$ when no load is connected and $14\,\mathrm{V}$ when a load taking $8\,\mathrm{A}$ is connected. Determine the internal resistance of the battery.      [$0.25\,\Omega$]

4.   A battery of e.m.f. $20\,\mathrm{V}$ and internal resistance $0.2\,\Omega$ supplies a load taking $10\,\mathrm{A}$. Determine the p.d. at the battery terminals and the resistance of the load.      [$18\,\mathrm{V}$, $1.8\,\Omega$]

5. Ten 2.2 V cells, each having an internal resistance of 0.1 Ω are connected in series to a load of 21 Ω. Determine (a) the current flowing in the circuit, and (b) the p.d. at the battery terminals. [(a) 1 A (b) 21 V]

6. For the circuits shown in Fig. 4.5 the resistors represent the internal resistance of the batteries. Find, in each case:
   (i) the total e.m.f. across $PQ$
   (ii) the total equivalent internal resistances of the batteries.
   [(i) (a) 6 V (b) 2 V (ii) (a) 4 Ω (b) 0.25 Ω]

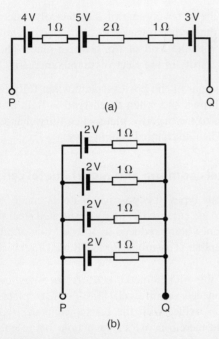

(a)

(b)

Figure 4.5

7. The voltage at the terminals of a battery is 52 V when no load is connected and 48.8 V when a load taking 80 A is connected. Find the internal resistance of the battery. What would be the terminal voltage when a load taking 20 A is connected? [0.04 Ω, 51.2 V]

## 4.6 Primary cells

**Primary cells** cannot be recharged, that is, the conversion of chemical energy to electrical energy is irreversible and the cell cannot be used once the chemicals

are exhausted. Examples of primary cells include the Leclanché cell and the mercury cell.

## Lechlanché cell

A typical dry Lechlanché cell is shown in Fig. 4.6. Such a cell has an e.m.f. of about 1.5 V when new, but this falls rapidly if in continuous use due to polarisation. The hydrogen film on the carbon electrode forms faster than can be dissipated by the depolariser. The Lechlanché cell is suitable only for intermittent use, applications including torches, transistor radios, bells, indicator circuits, gas lighters, controlling switch-gear, and so on. The cell is the most commonly used of primary cells, is cheap, requires little maintenance and has a shelf life of about 2 years.

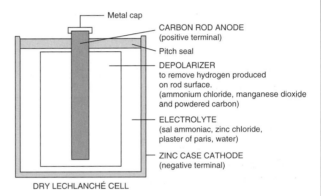

DRY LECHLANCHÉ CELL

Figure 4.6

## Mercury cell

A typical mercury cell is shown in Fig. 4.7. Such a cell has an e.m.f. of about 1.3 V which remains constant for a relatively long time. Its main advantages over the Lechlanché cell is its smaller size and its long shelf life. Typical practical applications include hearing aids, medical electronics, cameras and for guided missiles.

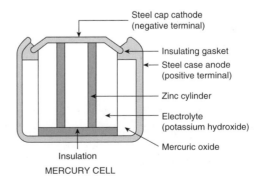

MERCURY CELL

Figure 4.7

## 4.7 Secondary cells

**Secondary cells** can be recharged after use, that is, the conversion of chemical energy to electrical energy is reversible and the cell may be used many times. Examples of secondary cells include the lead–acid cell and the nickel cadmium and nickel–metal cells. Practical applications of such cells include car batteries, telephone circuits and for traction purposes – such as milk delivery vans and fork lift trucks.

### Lead–acid cell

A typical lead–acid cell is constructed of:

(i) A container made of glass, ebonite or plastic.

(ii) **Lead plates**

  (a) the negative plate (cathode) consists of spongy lead

  (b) the positive plate (anode) is formed by pressing lead peroxide into the lead grid.

The plates are interleaved as shown in the plan view of Fig. 4.8 to increase their effective cross-sectional area and to minimise internal resistance.

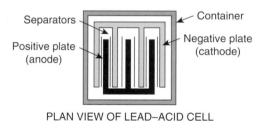

PLAN VIEW OF LEAD–ACID CELL

Figure 4.8

(iii) **Separators** made of glass, celluloid or wood.

(iv) An **electrolyte** which is a mixture of sulphuric acid and distilled water.

The relative density (or specific gravity) of a lead–acid cell, which may be measured using a hydrometer, varies between about 1.26 when the cell is fully charged to about 1.19 when discharged. The terminal p.d. of a lead–acid cell is about 2 V.

When a cell supplies current to a load it is said to be **discharging**. During discharge:

(i) the lead peroxide (positive plate) and the spongy lead (negative plate) are converted into lead sulphate, and

(ii) the oxygen in the lead peroxide combines with hydrogen in the electrolyte to form water. The electrolyte is therefore weakened and the relative density falls.

The terminal p.d. of a lead–acid cell when fully discharged is about 1.8 V. A cell is **charged** by connecting a d.c. supply to its terminals, the positive terminal of the cell being connected to the positive terminal of the supply. The charging current flows in the reverse direction to the discharge current and the chemical action is reversed. During charging:

(i) the lead sulphate on the positive and negative plates is converted back to lead peroxide and lead respectively, and

(ii) the water content of the electrolyte decreases as the oxygen released from the electrolyte combines with the lead of the positive plate. The relative density of the electrolyte thus increases.

The colour of the positive plate when fully charged is dark brown and when discharged is light brown. The colour of the negative plate when fully charged is grey and when discharged is light grey.

### Nickel cadmium and nickel–metal cells

In both types the positive plate is made of nickel hydroxide enclosed in finely perforated steel tubes, the resistance being reduced by the addition of pure nickel or graphite. The tubes are assembled into nickel–steel plates.

In the nickel–metal cell, (sometimes called the **Edison cell** or **nife cell**), the negative plate is made of iron oxide, with the resistance being reduced by a little mercuric oxide, the whole being enclosed in perforated steel tubes and assembled in steel plates. In the nickel cadmium cell the negative plate is made of cadmium. The electrolyte in each type of cell is a solution of potassium hydroxide which does not undergo any chemical change and thus the quantity can be reduced to a minimum. The plates are separated by insulating rods and assembled in steel containers which are then enclosed in a non-metallic crate to insulate the cells from one another. The average discharge p.d. of an alkaline cell is about 1.2 V.

**Advantages** of a nickel cadmium cell or a nickel–metal cell over a lead–acid cell include:

(i) More robust construction

(ii) Capable of withstanding heavy charging and discharging currents without damage

**Table 4.2**

| Type of battery | Common uses | Hazardous component | Disposal recycling options |
|---|---|---|---|
| *Wet cell* (i.e. a primary cell that has a liquid electrolyte) | | | |
| Lead acid batteries | Electrical energy supply for vehicles including cars, trucks, boats, tractors and motorcycles. Small sealed lead acid batteries are used for emergency lighting and uninterruptible power supplies | Sulphuric acid and lead | Recycle – most petrol stations and garages accept old car batteries, and council waste facilities have collection points for lead acid batteries |
| *Dry cell: Non-chargeable – single use* (for example, AA, AAA, C, D, lantern and miniature watch sizes) | | | |
| Zinc carbon | Torches, clocks, shavers, radios, toys and smoke alarms | Zinc | Not classed as hazardous waste – can be disposed with household waste |
| Zinc chloride | Torches, clocks, shavers, radios, toys and smoke alarms | Zinc | Not classed as hazardous waste – can be disposed with household waste |
| Alkaline manganese | Personal stereos and radio/cassette players | Manganese | Not classed as hazardous waste – can be disposed with household waste |
| *Primary button cells* (i.e. a small flat battery shaped like a 'button' used in small electronic devices) | | | |
| Mercuric oxide | Hearing aids, pacemakers and cameras | Mercury | Recycle at council waste facility, if available |
| Zinc air | Hearing aids, pagers and cameras | Zinc | Recycle at council waste facility, if available |
| Silver oxide | Calculators, watches and cameras | Silver | Recycle at council waste facility, if available |
| Lithium | Computers, watches and cameras | Lithium (explosive and flammable) | Recycle at council waste facility, if available |
| *Dry cell rechargeable – secondary batteries* | | | |
| Nickel cadmium (NiCd) | Mobile phones, cordless power tools, laptop computers, shavers, motorised toys, personal stereos | Cadmium | Recycle at council waste facility, if available |
| Nickel–metal hydride (NiMH) | Alternative to NiCd batteries, but longer life | Nickel | Recycle at council waste facility, if available |
| Lithium-ion (Li-ion) | Alternative to NiCd and NiMH batteries, but greater energy storage capacity | Lithium | Recycle at council waste facility, if available |

(iii)  Has a longer life

(iv)  For a given capacity is lighter in weight

(v)  Can be left indefinitely in any state of charge or discharge without damage

(vi)  Is not self-discharging

**Disadvantages** of nickel cadmium and nickel–metal cells over a lead–acid cell include:

(i)  Is relatively more expensive

(ii)  Requires more cells for a given e.m.f.

(iii)  Has a higher internal resistance

(iv)  Must be kept sealed

(v)  Has a lower efficiency

Nickel cells may be used in extremes of temperature, in conditions where vibration is experienced or where duties require long idle periods or heavy discharge currents. Practical examples include traction and marine work, lighting in railway carriages, military portable radios and for starting diesel and petrol engines. See also Table 4.2, page 37.

## 4.8  Cell capacity

The **capacity** of a cell is measured in ampere-hours (Ah). A fully charged 50 Ah battery rated for 10 h discharge can be discharged at a steady current of 5 A for 10 h, but if the load current is increased to 10 A then the battery is discharged in 3–4 h, since the higher the discharge current, the lower is the effective capacity of the battery. Typical discharge characteristics for a lead–acid cell are shown in Fig. 4.9

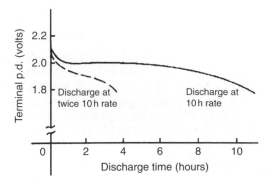

Figure 4.9

## 4.9  Safe disposal of batteries

Battery disposal has become a topical subject in the UK because of greater awareness of the dangers and implications of depositing up to 300 million batteries per annum – a waste stream of over 20 000 tonnes – into landfill sites.

Certain batteries contain substances which can be a hazard to humans, wildlife and the environment, as well a posing a fire risk. Other batteries can be recycled for their metal content.

Waste batteries are a concentrated source of toxic heavy metals such as mercury, lead and cadmium. If batteries containing heavy metals are disposed of incorrectly, the metals can leach out and pollute the soil and groundwater, endangering humans and wildlife. Long-term exposure to cadmium, a known human carcinogen (i.e. a substance producing cancerous growth), can cause liver and lung disease. Mercury can cause damage to the human brain, spinal system, kidneys and liver. Sulphuric acid in lead acid batteries can cause severe skin burns or irritation upon contact. It is increasingly important to correctly dispose of all types of batteries.

Table 4.2 lists types of batteries, their common uses, their hazardous components and disposal recycling options.

Battery disposal has become more regulated since the Landfill Regulations 2002 and Hazardous Waste Regulations 2005. From the Waste Electrical and Electronic Equipment (WEEE) Regulations 2006, commencing July 2007 all producers (manufacturers and importers) of electrical and electronic equipment will be responsible for the cost of collection, treatment and recycling of obligated WEEE generated in the UK.

## 4.10  Fuel cells

A **fuel cell** is an electrochemical energy conversion device, similar to a battery, but differing from the latter in that it is designed for continuous replenishment of the reactants consumed, i.e. it produces electricity from an external source of fuel and oxygen, as opposed to the limited energy storage capacity of a battery. Also, the electrodes within a battery react and change as a battery is charged or discharged, whereas a fuel cells' electrodes are catalytic (i.e. not permanently changed) and relatively stable.

Typical reactants used in a fuel cell are hydrogen on the anode side and oxygen on the cathode side (i.e. a **hydrogen cell**). Usually, reactants flow in and reaction products flow out. Virtually continuous long-term operation is feasible as long as these flows are maintained.

Fuel cells are very attractive in modern applications for their high efficiency and ideally emission-free use, in contrast to currently more modern fuels such as methane or natural gas that generate carbon dioxide. The only by-product of a fuel cell operating on pure hydrogen is water vapour.

Currently, fuel cells are a very expensive alternative to internal combustion engines. However, continued research and development is likely to make fuel cell vehicles available at market prices within a few years.

Fuel cells are very useful as power sources in remote locations, such as spacecraft, remote weather stations, and in certain military applications. A fuel cell running on hydrogen can be compact, lightweight and has no moving parts.

## 4.11   Alternative and renewable energy sources

**Alternative energy** refers to energy sources which could replace coal, traditional gas and oil, all of which increase the atmospheric carbon when burned as fuel. **Renewable energy** implies that it is derived from a source which is automatically replenished or one that is effectively infinite so that it is not depleted as it is used. Coal, gas and oil are not renewable because, although the fields may last for generations, their time span is finite and will eventually run out.

There are many means of harnessing energy which have less damaging impacts on our environment and include the following:

1.   **Solar energy** is one of the most resourceful sources of energy for the future. The reason for this is that the total energy received each year from the sun is around 35 000 times the total energy used by man. However, about one third of this energy is either absorbed by the outer atmosphere or reflected back into space. Solar energy could be used to run cars, power plants and space ships. **Solar panels** on roofs capture heat in water storage systems. **Photovoltaic cells**, when suitably positioned, convert sunlight to electricity.

2.   **Wind power** is another alternative energy source that can be used without producing by-products that are harmful to nature. The fins of a windmill rotate in a vertical plane which is kept vertical to the wind by means of a tail fin and as wind flow crosses the blades of the windmill it is forced to rotate and can be used to generate electricity (see Chapter 9). Like solar power, harnessing the wind is highly dependent upon weather and location. The average wind velocity of Earth is around 9 m/s, and the power that could be produced when a windmill is facing a wind of 10 m.p.h. (i.e. around 4.5 m/s) is around 50 watts.

3.   **Hydroelectricity** is achieved by the damming of rivers and utilising the potential energy in the water. As the water stored behind a dam is released at high pressure, its kinetic energy is transferred onto turbine blades and used to generate electricity. The system has enormous initial costs but has relatively low maintenance costs and provides power quite cheaply.

4.   **Tidal power** utilises the natural motion of the tides to fill reservoirs which are then slowly discharged through electricity-producing turbines.

5.   **Geothermal energy** is obtained from the internal heat of the planet and can be used to generate steam to run a steam turbine which, in turn, generates electricity. The radius of the Earth is about 4000 miles with an internal core temperature of around 4000°C at the centre. Drilling 3 miles from the surface of the Earth, a temperature of 100°C is encountered; this is sufficient to boil water to run a steam-powered electric power plant. Although drilling 3 miles down is possible, it is not easy. Fortunately, however, volcanic features called **geothermal hotspots** are found all around the world. These are areas which transmit excess internal heat from the interior of the Earth to the outer crust which can be used to generate electricity.

**Now try the following exercises**

> **Exercise 17    Short answer questions on the chemical effects of electricity**
>
> 1.   Define a battery
>
> 2.   State five practical applications of batteries
>
> 3.   State advantages of lithium-ion batteries over alkaline batteries
>
> 4.   What is electrolysis?
>
> 5.   What is an electrolyte?

6. Conduction in electrolytes is due to ......

7. A positive-connected electrode is called the ...... and the negative-connected electrode the ......

8. State two practical applications of electrolysis

9. The purpose of an electric cell is to convert ...... to ......

10. Make a labelled sketch of a simple cell

11. What is the electrochemical series?

12. With reference to a simple cell, explain briefly what is meant by
(a) polarisation (b) local action

13. What is corrosion? Name two effects of corrosion and state how they may be prevented

14. What is meant by the e.m.f. of a cell? How may the e.m.f. of a cell be measured?

15. Define internal resistance

16. If a cell has an e.m.f. of $E$ volts, an internal resistance of $r$ ohms and supplies a current $I$ amperes to a load, the terminal p.d. $V$ volts is given by: $V = ......$

17. Name the two main types of cells

18. Explain briefly the difference between primary and secondary cells

19. Name two types of primary cells

20. Name two types of secondary cells

21. State three typical applications of primary cells

22. State three typical applications of secondary cells

23. In what unit is the capacity of a cell measured?

24. Why is safe disposal of batteries important?

25. Name any six types of battery and state three common applications for each

26. What is a 'fuel cell'? How does it differ from a battery?

27. State the advantages of fuel cells

28. State three practical applications of fuel cells

29. What is meant by (a) alternative energy (b) renewable energy

30. State five alternative energy sources and briefly describe each

---

**Exercise 18    Multi-choice questions on the chemical effects of electricity (Answers on page 420)**

1. A battery consists of:
   (a) a cell             (b) a circuit
   (c) a generator        (d) a number of cells

2. The terminal p.d. of a cell of e.m.f. 2 V and internal resistance $0.1\,\Omega$ when supplying a current of 5 A will be:
   (a) 1.5 V              (b) 2 V
   (c) 1.9 V              (d) 2.5 V

3. Five cells, each with an e.m.f. of 2 V and internal resistance $0.5\,\Omega$ are connected in series. The resulting battery will have:
   (a) an e.m.f. of 2 V and an internal resistance of $0.5\,\Omega$
   (b) an e.m.f. of 10 V and an internal resistance of $2.5\,\Omega$
   (c) an e.m.f. of 2 V and an internal resistance of $0.1\,\Omega$
   (d) an e.m.f. of 10 V and an internal resistance of $0.1\,\Omega$

4. If the five cells of question 3 are connected in parallel the resulting battery will have:
   (a) an e.m.f. of 2 V and an internal resistance of $0.5\,\Omega$
   (b) an e.m.f. of 10 V and an internal resistance of $2.5\,\Omega$
   (c) an e.m.f. of 2 V and an internal resistance of $0.1\,\Omega$
   (d) an e.m.f. of 10 V and an internal resistance of $0.1\,\Omega$

5. Which of the following statements is false?
   (a) A Leclanché cell is suitable for use in torches
   (b) A nickel–cadmium cell is an example of a primary cell

(c)  When a cell is being charged its terminal p.d. exceeds the cell e.m.f.
(d)  A secondary cell may be recharged after use

6.  Which of the following statements is false? When two metal electrodes are used in a simple cell, the one that is higher in the electrochemical series:
(a)  tends to dissolve in the electrolyte
(b)  is always the negative electrode
(c)  reacts most readily with oxygen
(d)  acts as an anode

7.  Five 2 V cells, each having an internal resistance of $0.2\,\Omega$ are connected in series to a load of resistance $14\,\Omega$. The current flowing in the circuit is:
(a) 10 A   (b) 1.4 A
(c) 1.5 A   (d) $\frac{2}{3}$ A

8.  For the circuit of question 7, the p.d. at the battery terminals is:
(a) 10 V   (b) $9\frac{1}{3}$ V
(c) 0 V   (d) $10\frac{2}{3}$ V

9.  Which of the following statements is true?
(a)  The capacity of a cell is measured in volts
(b)  A primary cell converts electrical energy into chemical energy

(c)  Galvanising iron helps to prevent corrosion
(d)  A positive electrode is termed the cathode

10.  The greater the internal resistance of a cell:
(a)  the greater the terminal p.d.
(b)  the less the e.m.f.
(c)  the greater the e.m.f.
(d)  the less the terminal p.d.

11.  The negative pole of a dry cell is made of:
(a)  carbon
(b)  copper
(c)  zinc
(d)  mercury

12.  The energy of a secondary cell is usually renewed:
(a)  by passing a current through it
(b)  it cannot be renewed at all
(c)  by renewing its chemicals
(d)  by heating it

13.  Which of the following statements is true?
(a)  A zinc carbon battery is rechargeable and is not classified as hazardous
(b)  A nickel cadmium battery is not rechargeable and is classified as hazardous
(c)  A lithium battery is used in watches and is not rechargeable
(d)  An alkaline manganese battery is used in torches and is classified as hazardous

## Revision Test 1

This revision test covers the material contained in Chapters 1 to 4. *The marks for each question are shown in brackets at the end of each question.*

1. An electromagnet exerts a force of 15 N and moves a soft iron armature through a distance of 12 mm in 50 ms. Determine the power consumed. (5)

2. A d.c. motor consumes 47.25 MJ when connected to a 250 V supply for 1 hour 45 minutes. Determine the power rating of the motor and the current taken from the supply. (5)

3. A 100 W electric light bulb is connected to a 200 V supply. Calculate (a) the current flowing in the bulb, and (b) the resistance of the bulb. (4)

4. Determine the charge transferred when a current of 5 mA flows for 10 minutes. (2)

5. A current of 12 A flows in the element of an electric fire of resistance 25 $\Omega$. Determine the power dissipated by the element. If the fire is on for 5 hours every day, calculate for a one week period (a) the energy used, and (b) cost of using the fire if electricity cost 13.5p per unit. (6)

6. Calculate the resistance of 1200 m of copper cable of cross-sectional area 15 mm$^2$. Take the resistivity of copper as 0.02 $\mu\,\Omega$m (5)

7. At a temperature of 40°C, an aluminium cable has a resistance of 25 $\Omega$. If the temperature coefficient of resistance at 0°C is 0.0038/°C, calculate its resistance at 0°C (5)

8. (a) Determine the values of the resistors with the following colour coding:
   (i) red-red-orange-silver
   (ii) orange-orange-black-blue-green
   (b) What is the value of a resistor marked as 47 KK? (6)

9. Four cells, each with an internal resistance of 0.40 $\Omega$ and an e.m.f. of 2.5 V are connected in series to a load of 38.4 $\Omega$. (a) Determine the current flowing in the circuit and the p.d. at the battery terminals. (b) If the cells are connected in parallel instead of in series, determine the current flowing and the p.d. at the battery terminals. (10)

10. (a) State six typical applications of primary cells
    (b) State six typical applications of secondary cells
    (c) State the advantages of a fuel cell over a conventional battery and state three practical applications. (12)

11. Name five alternative, renewable energy sources, and give a brief description of each. (15)

# Chapter 5

# Series and parallel networks

At the end of this chapter you should be able to:

- calculate unknown voltages, current and resistances in a series circuit
- understand voltage division in a series circuit
- calculate unknown voltages, currents and resistances in a parallel network
- calculate unknown voltages, currents and resistances in series-parallel networks
- understand current division in a two-branch parallel network
- appreciate the loading effect of a voltmeter
- understand the difference between potentiometers and rheostats
- perform calculations to determine load currents and voltages in potentiometers and rheostats
- understand and perform calculations on relative and absolute voltages
- state three causes of short circuits in electrical circuits
- describe the advantages and disadvantages of series and parallel connection of lamps

## 5.1 Series circuits

Figure 5.1 shows three resistors $R_1$, $R_2$ and $R_3$ connected end to end, i.e. in series, with a battery source of $V$ volts. Since the circuit is closed a current $I$ will flow and the p.d. across each resistor may be determined from the voltmeter readings $V_1$, $V_2$ and $V_3$.

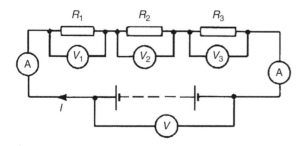

Figure 5.1

**In a series circuit**

(a) the current $I$ is the same in all parts of the circuit and hence the same reading is found on each of the ammeters shown, and

(b) the sum of the voltages $V_1$, $V_2$ and $V_3$ is equal to the total applied voltage, $V$,

i.e. $$V = V_1 + V_2 + V_3$$

From Ohm's law: $V_1 = IR_1$, $V_2 = IR_2$, $V_3 = IR_3$ and $V = IR$ where $R$ is the total circuit resistance. Since $V = V_1 + V_2 + V_3$ then $IR = IR_1 + IR_2 + IR_3$. Dividing throughout by $I$ gives

$$R = R_1 + R_2 + R_3$$

Thus for a series circuit, the total resistance is obtained by adding together the values of the separate resistances.

DOI: 10.1016/B978-0-08-089056-2.00005-X

**Problem 1.** For the circuit shown in Fig. 5.2, determine (a) the battery voltage $V$, (b) the total resistance of the circuit, and (c) the values of resistors $R_1$, $R_2$ and $R_3$, given that the p.d.'s across $R_1$, $R_2$ and $R_3$ are 5 V, 2 V and 6 V respectively.

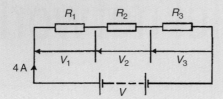

Figure 5.2

(a) Battery voltage $V = V_1 + V_2 + V_3$
$$= 5 + 2 + 6 = \mathbf{13\,V}$$

(b) Total circuit resistance $R = \dfrac{V}{I} = \dfrac{13}{4} = \mathbf{3.25\,\Omega}$

(c) Resistance $R_1 = \dfrac{V_1}{I} = \dfrac{5}{4} = \mathbf{1.25\,\Omega}$

Resistance $R_2 = \dfrac{V_2}{I} = \dfrac{2}{4} = \mathbf{0.5\,\Omega}$

Resistance $R_3 = \dfrac{V_3}{I} = \dfrac{6}{4} = \mathbf{1.5\,\Omega}$

(Check: $R_1 + R_2 + R_3 = 1.25 + 0.5 + 1.5$
$$= 3.25\,\Omega = R)$$

**Problem 2.** For the circuit shown in Fig. 5.3, determine the p.d. across resistor $R_3$. If the total resistance of the circuit is 100 $\Omega$, determine the current flowing through resistor $R_1$. Find also the value of resistor $R_2$.

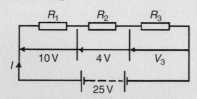

Figure 5.3

P.d. across $R_3$, $V_3 = 25 - 10 - 4 = \mathbf{11\,V}$

Current $I = \dfrac{V}{R} = \dfrac{25}{100} = \mathbf{0.25\,A}$,

which is the current flowing in each resistor

Resistance $R_2 = \dfrac{V_2}{I} = \dfrac{4}{0.25} = \mathbf{16\,\Omega}$

**Problem 3.** A 12 V battery is connected in a circuit having three series-connected resistors having resistances of 4 $\Omega$, 9 $\Omega$ and 11 $\Omega$. Determine the current flowing through, and the p.d. across the 9 $\Omega$ resistor. Find also the power dissipated in the 11 $\Omega$ resistor.

The circuit diagram is shown in Fig. 5.4

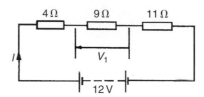

Figure 5.4

Total resistance $R = 4 + 9 + 11 = 24\,\Omega$

Current $I = \dfrac{V}{R} = \dfrac{12}{24} = \mathbf{0.5\,A}$,

which is the current in the 9 $\Omega$ resistor.
P.d. across the 9 $\Omega$ resistor,

$$V_1 = I \times 9 = 0.5 \times 9 = \mathbf{4.5\,V}$$

Power dissipated in the 11 $\Omega$ resistor,

$$P = I^2 R = (0.5)^2 (11)$$
$$= (0.25)(11) = \mathbf{2.75\,W}$$

## 5.2 Potential divider

The voltage distribution for the circuit shown in Fig. 5.5(a) is given by:

$$V_1 = \left(\frac{R_1}{R_1 + R_2}\right) V \text{ and } V_2 = \left(\frac{R_2}{R_1 + R_2}\right) V$$

The circuit shown in Fig. 5.5(b) is often referred to as a **potential divider** circuit. Such a circuit can consist of a number of similar elements in series connected across a voltage source, voltages being taken from connections between the elements. Frequently the divider consists of two resistors as shown in Fig. 5.5(b), where

$$V_{\text{OUT}} = \left(\frac{R_2}{R_1 + R_2}\right) V_{\text{IN}}$$

A potential divider is the simplest way of producing a source of lower e.m.f. from a source of higher e.m.f., and is the basic operating mechanism of the **potentiometer**,

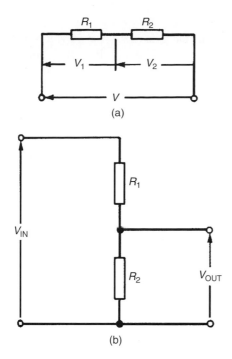

Figure 5.5

a measuring device for accurately measuring potential differences (see page 135).

**Problem 4.**    Determine the value of voltage $V$ shown in Fig. 5.6

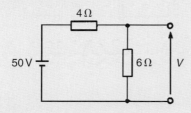

**Figure 5.6**

Figure 5.6 may be redrawn as shown in Fig. 5.7, and

$$\text{voltage } V = \left(\frac{6}{6+4}\right)(50) = \mathbf{30\,V}$$

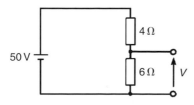

**Figure 5.7**

**Problem 5.**    Two resistors are connected in series across a 24 V supply and a current of 3 A flows in the circuit. If one of the resistors has a resistance of $2\,\Omega$ determine (a) the value of the other resistor, and (b) the p.d. across the $2\,\Omega$ resistor. If the circuit is connected for 50 hours, how much energy is used?

The circuit diagram is shown in Fig. 5.8

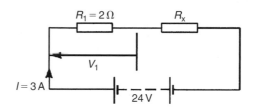

**Figure 5.8**

(a)    Total circuit resistance

$$R = \frac{V}{I} = \frac{24}{3} = 8\,\Omega$$

Value of unknown resistance,

$$R_x = 8 - 2 = \mathbf{6\,\Omega}$$

(b)    P.d. across $2\,\Omega$ resistor,

$$V_1 = IR_1 = 3 \times 2 = \mathbf{6\,V}$$

Alternatively, from above,

$$V_1 = \left(\frac{R_1}{R_1 + R_x}\right) V$$

$$= \left(\frac{2}{2+6}\right)(24) = 6\,V$$

Energy used = power × time

$$= (V \times I) \times t$$

$$= (24 \times 3\,\text{W})(50\,\text{h})$$

$$= 3600\,\text{Wh} = \mathbf{3.6\,kWh}$$

**Now try the following exercise**

**Exercise 19    Further problems on series circuits**

1.    The p.d.'s measured across three resistors connected in series are 5 V, 7 V and 10 V, and the

supply current is 2 A. Determine (a) the supply voltage, (b) the total circuit resistance, and (c) the values of the three resistors.

[(a) 22 V (b) 11 $\Omega$ (c) 2.5 $\Omega$, 3.5 $\Omega$, 5 $\Omega$]

2. For the circuit shown in Fig. 5.9, determine the value of $V_1$. If the total circuit resistance is 36 $\Omega$ determine the supply current and the value of resistors $R_1$, $R_2$ and $R_3$

[10 V, 0.5 A, 20 $\Omega$, 10 $\Omega$, 6 $\Omega$]

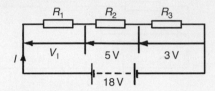

Figure 5.9

3. When the switch in the circuit in Fig. 5.10 is closed the reading on voltmeter 1 is 30 V and that on voltmeter 2 is 10 V. Determine the reading on the ammeter and the value of resistor $R_x$          [4 A, 2.5 $\Omega$]

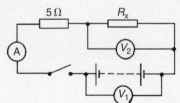

Figure 5.10

4. Calculate the value of voltage $V$ in Fig. 5.11          [45 V]

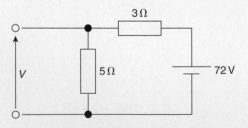

Figure 5.11

5. Two resistors are connected in series across an 18 V supply and a current of 5 A flows. If one of the resistors has a value of 2.4 $\Omega$ determine (a) the value of the other resistor and (b) the p.d. across the 2.4 $\Omega$ resistor.

[(a) 1.2 $\Omega$ (b) 12 V]

6. An arc lamp takes 9.6 A at 55 V. It is operated from a 120 V supply. Find the value of the stabilising resistor to be connected in series.

[6.77 $\Omega$]

7. An oven takes 15 A at 240 V. It is required to reduce the current to 12 A. Find (a) the resistor which must be connected in series, and (b) the voltage across the resistor.

[(a) 4 $\Omega$ (b) 48 V]

## 5.3 Parallel networks

Figure 5.12 shows three resistors, $R_1$, $R_2$ and $R_3$ connected across each other, i.e. in parallel, across a battery source of $V$ volts.

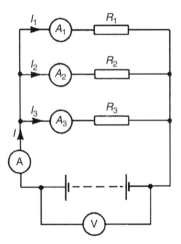

Figure 5.12

### In a parallel circuit:

(a) the sum of the currents $I_1$, $I_2$ and $I_3$ is equal to the total circuit current, $I$,

i.e.          $I = I_1 + I_2 + I_3$          and

(b) the source p.d., $V$ volts, is the same across each of the resistors.

From Ohm's law:

$$I_1 = \frac{V}{R_1}, \quad I_2 = \frac{V}{R_2}, \quad I_3 = \frac{V}{R_3} \quad \text{and} \quad I = \frac{V}{R}$$

where $R$ is the total circuit resistance. Since

$$I = I_1 + I_2 + I_3 \text{ then } \frac{V}{R} = \frac{V}{R_1} + \frac{V}{R_2} + \frac{V}{R_3}$$

Dividing throughout by $V$ gives:

$$\frac{1}{R} = \frac{1}{R_1} + \frac{1}{R_2} + \frac{1}{R_3}$$

This equation must be used when finding the total resistance $R$ of a parallel circuit. For the special case of **two resistors in parallel**

$$\frac{1}{R} = \frac{1}{R_1} + \frac{1}{R_2} = \frac{R_2 + R_1}{R_1 R_2}$$

Hence
$$R = \frac{R_1 R_2}{R_1 + R_2} \quad \left(\text{i.e. } \frac{\text{product}}{\text{sum}}\right)$$

**Problem 6.** For the circuit shown in Fig. 5.13, determine (a) the reading on the ammeter, and (b) the value of resistor $R_2$.

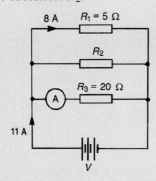

Figure 5.13

P.d. across $R_1$ is the same as the supply voltage $V$. Hence supply voltage, $V = 8 \times 5 = 40\,\text{V}$

(a) Reading on ammeter,

$$I = \frac{V}{R_3} = \frac{40}{20} = 2\,\text{A}$$

(b) Current flowing through $R_2 = 11 - 8 - 2 = 1\,\text{A}$. Hence

$$R_2 = \frac{V}{I_2} = \frac{40}{1} = 40\,\Omega$$

**Problem 7.** Two resistors, of resistance $3\,\Omega$ and $6\,\Omega$, are connected in parallel across a battery having a voltage of 12 V. Determine (a) the total circuit resistance and (b) the current flowing in the $3\,\Omega$ resistor.

The circuit diagram is shown in Fig. 5.14

(a) The total circuit resistance $R$ is given by

$$\frac{1}{R} = \frac{1}{R_1} + \frac{1}{R_2} = \frac{1}{3} + \frac{1}{6} = \frac{2+1}{6} = \frac{3}{6}$$

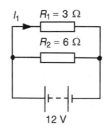

Figure 5.14

Since $\dfrac{1}{R} = \dfrac{3}{6}$ then $R = 2\,\Omega$

(Alternatively,

$$R = \frac{R_1 R_2}{R_1 + R_2} = \frac{3 \times 6}{3+6} = \frac{18}{9} = 2\,\Omega)$$

(b) Current in the $3\,\Omega$ resistance,

$$I_1 = \frac{V}{R_1} = \frac{12}{3} = 4\,\text{A}$$

**Problem 8.** For the circuit shown in Fig. 5.15, find (a) the value of the supply voltage $V$ and (b) the value of current $I$.

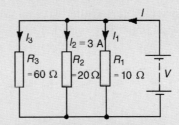

Figure 5.15

(a) P.d. across $20\,\Omega$ resistor $= I_2 R_2 = 3 \times 20 = 60\,\text{V}$, hence supply voltage **$V = 60\,\text{V}$** since the circuit is connected in parallel

(b) Current $I_1 = \dfrac{V}{R_1} = \dfrac{60}{10} = 6\,\text{A}, \quad I_2 = 3\,\text{A}$

and $\quad I_3 = \dfrac{V}{R_3} = \dfrac{60}{60} = 1\,\text{A}$

Current $I = I_1 + I_2 + I_3$ hence
$$I = 6 + 3 + 1 = 10\,\text{A}.$$
Alternatively,

$$\frac{1}{R} = \frac{1}{60} + \frac{1}{20} + \frac{1}{10} = \frac{1+3+6}{60} = \frac{10}{60}$$

Hence total resistance

$$R = \frac{60}{10} = 6\,\Omega, \text{ and current}$$

$$I = \frac{V}{R} = \frac{60}{6} = 10\,\text{A}$$

**Problem 9.** Given four $1\,\Omega$ resistors, state how they must be connected to give an overall resistance of (a) $\frac{1}{4}\,\Omega$ (b) $1\,\Omega$ (c) $1\frac{1}{3}\,\Omega$ (d) $2\frac{1}{2}\,\Omega$, all four resistors being connected in each case.

(a) **All four in parallel** (see Fig. 5.16), since

$$\frac{1}{R} = \frac{1}{1} + \frac{1}{1} + \frac{1}{1} + \frac{1}{1} = \frac{4}{1} \text{ i.e. } R = \frac{1}{4}\,\Omega$$

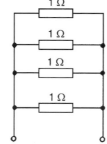

Figure 5.16

(b) **Two in series, in parallel with another two in series** (see Fig. 5.17), since $1\,\Omega$ and $1\,\Omega$ in series gives $2\,\Omega$, and $2\,\Omega$ in parallel with $2\,\Omega$ gives

$$\frac{2 \times 2}{2 + 2} = \frac{4}{4} = 1\,\Omega$$

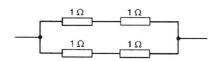

Figure 5.17

(c) **Three in parallel, in series with one** (see Fig. 5.18), since for the three in parallel,

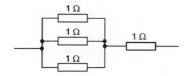

Figure 5.18

$$\frac{1}{R} = \frac{1}{1} + \frac{1}{1} + \frac{1}{1} = \frac{3}{1}$$

i.e. $R = \frac{1}{3}\,\Omega$ and $\frac{1}{3}\,\Omega$ in series with $1\,\Omega$ gives $1\frac{1}{3}\,\Omega$

(d) **Two in parallel, in series with two in series** (see Fig. 5.19), since for the two in parallel

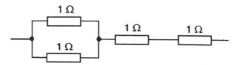

Figure 5.19

$$R = \frac{1 \times 1}{1 + 1} = \frac{1}{2}\,\Omega$$

and $\frac{1}{2}\,\Omega$, $1\,\Omega$ and $1\,\Omega$ in series gives $2\frac{1}{2}\,\Omega$

**Problem 10.** Find the equivalent resistance for the circuit shown in Fig. 5.20

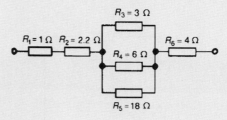

Figure 5.20

$R_3$, $R_4$ and $R_5$ are connected in parallel and their equivalent resistance $R$ is given by

$$\frac{1}{R} = \frac{1}{3} + \frac{1}{6} + \frac{1}{18} = \frac{6 + 3 + 1}{18} = \frac{10}{18}$$

hence $R = (18/10) = 1.8\,\Omega$. The circuit is now equivalent to four resistors in series and the **equivalent circuit resistance** $= 1 + 2.2 + 1.8 + 4 = \textbf{9}\,\Omega$

**Problem 11.** Resistances of $10\,\Omega$, $20\,\Omega$ and $30\,\Omega$ are connected (a) in series and (b) in parallel to a 240 V supply. Calculate the supply current in each case.

(a) The series circuit is shown in Fig. 5.21

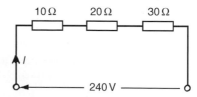

Figure 5.21

The equivalent resistance
$R_T = 10\,\Omega + 20\,\Omega + 30\,\Omega = 60\,\Omega$

$$\text{Supply current } I = \frac{V}{R_T} = \frac{240}{60} = \textbf{4\,A}$$

Section 1

(b) The parallel circuit is shown in Fig. 5.22.
The equivalent resistance $R_T$ of $10\,\Omega$, $20\,\Omega$
and $30\,\Omega$ resistance's connected in parallel is

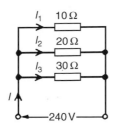

**Figure 5.22**

given by:

$$\frac{1}{R_T} = \frac{1}{10} + \frac{1}{20} + \frac{1}{30} = \frac{6+3+2}{60} = \frac{11}{60}$$

hence $R_T = \dfrac{60}{11}\,\Omega$

Supply current

$$I = \frac{V}{R_T} = \frac{240}{\frac{60}{11}} = \frac{240 \times 11}{60} = \mathbf{44\,A}$$

(Check:

$$I_1 = \frac{V}{R_1} = \frac{240}{10} = 24\,A,$$

$$I_2 = \frac{V}{R_2} = \frac{240}{20} = 12\,A$$

and $I_3 = \dfrac{V}{R_3} = \dfrac{240}{30} = 8\,A$

For a parallel circuit $I = I_1 + I_2 + I_3$
$= 24 + 12 + 8 = \mathbf{44\,A}$, as above)

## 5.4   Current division

For the circuit shown in Fig. 5.23, the total circuit
resistance, $R_T$ is given by

$$R_T = \frac{R_1 R_2}{R_1 + R_2}$$

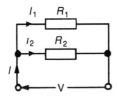

**Figure 5.23**

and

$$V = IR_T = I\left(\frac{R_1 R_2}{R_1 + R_2}\right)$$

Current

$$I_1 = \frac{V}{R_1} = \frac{I}{R_1}\left(\frac{R_1 R_2}{R_1 + R_2}\right)$$

$$= \left(\frac{\mathbf{R_2}}{\mathbf{R_1 + R_2}}\right)\mathbf{(I)}$$

Similarly,

current

$$I_2 = \frac{V}{R_2} = \frac{I}{R_2}\left(\frac{R_1 R_2}{R_1 + R_2}\right)$$

$$= \left(\frac{\mathbf{R_1}}{\mathbf{R_1 + R_2}}\right)\mathbf{(I)}$$

Summarising, with reference to Fig. 5.23

$$I_1 = \left(\frac{\mathbf{R_2}}{\mathbf{R_1 + R_2}}\right)\mathbf{(I)}$$

and

$$I_2 = \left(\frac{\mathbf{R_1}}{\mathbf{R_1 + R_2}}\right)\mathbf{(I)}$$

**Problem 12.**   For the series-parallel arrangement
shown in Fig. 5.24, find (a) the supply current,
(b) the current flowing through each resistor and
(c) the p.d. across each resistor.

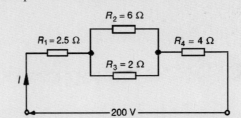

**Figure 5.24**

(a) The equivalent resistance $R_x$ of $R_2$ and $R_3$ in
parallel is:

$$R_x = \frac{6 \times 2}{6 + 2} = 1.5\,\Omega$$

The equivalent resistance $R_T$ of $R_1$, $R_x$ and $R_4$ in
series is:

$$R_T = 2.5 + 1.5 + 4 = 8\,\Omega$$

Supply current

$$I = \frac{V}{R_T} = \frac{200}{8} = \mathbf{25\,A}$$

(b) The current flowing through $R_1$ and $R_4$ is 25 A. The current flowing through $R_2$

$$= \left( \frac{R_3}{R_2 + R_3} \right) I = \left( \frac{2}{6+2} \right) 25$$

$$= \mathbf{6.25\,A}$$

The current flowing through $R_3$

$$= \left( \frac{R_2}{R_2 + R_3} \right) I$$

$$= \left( \frac{6}{6+2} \right) 25 = \mathbf{18.75\,A}$$

(Note that the currents flowing through $R_2$ and $R_3$ must add up to the total current flowing into the parallel arrangement, i.e. 25 A)

(c) The equivalent circuit of Fig. 5.24 is shown in Fig. 5.25

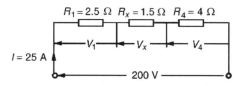

**Figure 5.25**

p.d. across $R_1$, i.e.

$$V_1 = IR_1 = (25)(2.5) = \mathbf{62.5\,V}$$

p.d. across $R_x$, i.e.

$$V_x = IR_x = (25)(1.5) = \mathbf{37.5\,V}$$

p.d. across $R_4$, i.e.

$$V_4 = IR_4 = (25)(4) = \mathbf{100\,V}$$

Hence the p.d. across $R_2$

$$= \text{p.d. across } R_3 = \mathbf{37.5\,V}$$

**Problem 13.** For the circuit shown in Fig. 5.26 calculate (a) the value of resistor $R_x$ such that the

total power dissipated in the circuit is 2.5 kW, (b) the current flowing in each of the four resistors.

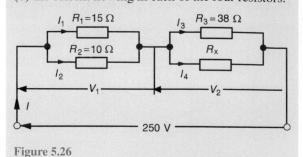

**Figure 5.26**

(a) Power dissipated $P = VI$ watts, hence

$$2500 = (250)(I)$$

i.e.

$$I = \frac{2500}{250} = 10\,A$$

From Ohm's law,

$$R_T = \frac{V}{I} = \frac{250}{10} = 25\,\Omega,$$

where $R_T$ is the equivalent circuit resistance. The equivalent resistance of $R_1$ and $R_2$ in parallel is

$$\frac{15 \times 10}{15 + 10} = \frac{150}{25} = 6\,\Omega$$

The equivalent resistance of resistors $R_3$ and $R_x$ in parallel is equal to $25\,\Omega - 6\,\Omega$, i.e. $19\,\Omega$. There are three methods whereby $R_x$ can be determined.

**Method 1**

The voltage $V_1 = IR$, where $R$ is $6\,\Omega$, from above, i.e. $V_1 = (10)(6) = 60\,V$. Hence

$$V_2 = 250\,V - 60\,V = 190\,V$$

$$= \text{p.d. across } R_3$$

$$= \text{p.d. across } R_x$$

$$I_3 = \frac{V_2}{R_3} = \frac{190}{38} = 5\,A.$$

Thus $I_4 = 5\,A$ also, since $I = 10\,A$. Thus

$$\mathbf{R_x} = \frac{V_2}{I_4} = \frac{190}{5} = \mathbf{38\,\Omega}$$

**Method 2**

Since the equivalent resistance of $R_3$ and $R_x$ in parallel is $19\,\Omega$,

$$19 = \frac{38 R_x}{38 + R_x} \quad \left( \text{i.e. } \frac{\text{product}}{\text{sum}} \right)$$

Hence

$$19(38 + R_x) = 38R_x$$

$$722 + 19R_x = 38R_x$$

$$722 = 38R_x - 19R_x = 19R_x$$

$$= 19R_x$$

Thus
$$\mathbf{R_x} = \frac{722}{19} = \mathbf{38\,\Omega}$$

## Method 3

When two resistors having the same value are connected in parallel the equivalent resistance is always half the value of one of the resistors. Thus, in this case, since $R_T = 19\,\Omega$ and $R_3 = 38\,\Omega$, then $R_x = 38\,\Omega$ could have been deduced on sight.

(b)  Current $I_1 = \left(\dfrac{R_2}{R_1 + R_2}\right) I$

$$= \left(\frac{10}{15 + 10}\right)(10)$$

$$= \left(\frac{2}{5}\right)(10) = \mathbf{4\,A}$$

Current $I_2 = \left(\dfrac{R_1}{R_1 + R_2}\right) I = \left(\dfrac{15}{15 + 10}\right)(10)$

$$= \left(\frac{3}{5}\right)(10) = \mathbf{6\,A}$$

From part (a), method 1, $\mathbf{I_3 = I_4 = 5\,A}$

**Problem 14.**  For the arrangement shown in Fig. 5.27, find the current $I_x$.

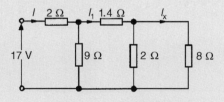

Figure 5.27

Commencing at the right-hand side of the arrangement shown in Fig. 5.27, the circuit is gradually reduced in stages as shown in Fig. 5.28(a)–(d).
From Fig. 5.28(d),

$$I = \frac{17}{4.25} = 4\,A$$

From Fig. 5.28(b),

$$I_1 = \left(\frac{9}{9+3}\right)(I) = \left(\frac{9}{12}\right)(4) = 3\,A$$

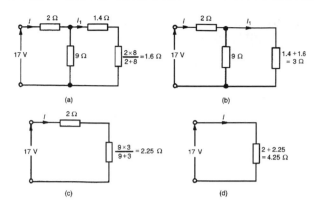

Figure 5.28

From Fig. 5.27

$$I_x = \left(\frac{2}{2+8}\right)(I_1) = \left(\frac{2}{10}\right)(3) = \mathbf{0.6\,A}$$

*For a practical laboratory experiment on series-parallel dc circuits, see Chapter 24, page 407.*

**Now try the following exercise**

**Exercise 20   Further problems on parallel networks**

1.  Resistances of $4\,\Omega$ and $12\,\Omega$ are connected in parallel across a 9 V battery. Determine (a) the equivalent circuit resistance, (b) the supply current, and (c) the current in each resistor.
    [(a) $3\,\Omega$ (b) 3 A (c) 2.25 A, 0.75 A]

2.  For the circuit shown in Fig. 5.29 determine (a) the reading on the ammeter, and (b) the value of resistor $R$.      [2.5 A, 2.5 $\Omega$]

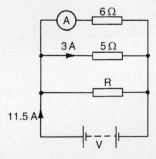

Figure 5.29

3.  Find the equivalent resistance when the following resistances are connected (a) in series (b) in parallel (i) $3\,\Omega$ and $2\,\Omega$ (ii) $20\,k\Omega$ and

40 kΩ (iii) 4 Ω, 8 Ω and 16 Ω (iv) 800 Ω, 4 kΩ and 1500 Ω.

$$[(a) \quad (i) \ 5\,\Omega \qquad (ii) \ 60\,\mathrm{k}\Omega$$
$$(iii) \ 28\,\Omega \quad (iv) \ 6.3\,\Omega$$
$$(b) \quad (i) \ 1.2\,\Omega \quad (ii) \ 13.33\,\mathrm{k}\Omega$$
$$(iii) \ 2.29\,\Omega \ (iv) \ 461.54\,\Omega]$$

4. Find the total resistance between terminals A and B of the circuit shown in Fig. 5.30(a). [8 Ω]

5. Find the equivalent resistance between terminals C and D of the circuit shown in Fig. 5.30(b). [27.5 Ω]

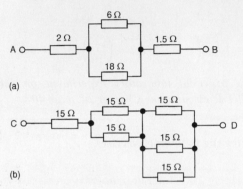

(a)

(b)

Figure 5.30

6. Resistors of 20 Ω, 20 Ω and 30 Ω are connected in parallel. What resistance must be added in series with the combination to obtain a total resistance of 10 Ω. If the complete circuit expends a power of 0.36 kW, find the total current flowing. [2.5 Ω, 6 A]

7. (a) Calculate the current flowing in the 30 Ω resistor shown in Fig. 5.31. (b) What additional value of resistance would have to be placed in parallel with the 20 Ω and 30 Ω resistors to change the supply current to 8 A, the supply voltage remaining constant.
[(a) 1.6 A (b) 6 Ω]

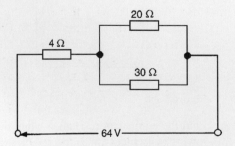

Figure 5.31

8. For the circuit shown in Fig. 5.32, find (a) $V_1$, (b) $V_2$, without calculating the current flowing. [(a) 30 V (b) 42 V]

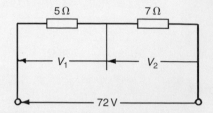

Figure 5.32

9. Determine the currents and voltages indicated in the circuit shown in Fig. 5.33.

$$[I_1 = 5\,\mathrm{A}, \ I_2 = 2.5\,\mathrm{A}, \ I_3 = 1\tfrac{2}{3}\,\mathrm{A}, \ I_4 = \tfrac{5}{6}\,\mathrm{A}$$
$$I_5 = 3\,\mathrm{A}, \ I_6 = 2\,\mathrm{A}, \ V_1 = 20\,\mathrm{V}, \ V_2 = 5\,\mathrm{V},$$
$$V_3 = 6\,\mathrm{V}]$$

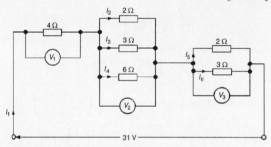

Figure 5.33

10. Find the current $I$ in Fig. 5.34. [1.8 A]

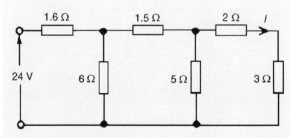

Figure 5.34

11. A resistor of 2.4 Ω is connected in series with another of 3.2 Ω. What resistance must be placed across the one of 2.4 Ω so that the total resistance of the circuit shall be 5 Ω?
[7.2 Ω]

12. A resistor of 8 Ω is connected in parallel with one of 12 Ω and the combination is connected in series with one of 4 Ω. A p.d. of 10 V is applied to the circuit. The 8 Ω resistor is now

placed across the $4\,\Omega$ resistor. Find the p.d. required to send the same current through the $8\,\Omega$ resistor.                    [30 V]

## 5.5 Loading effect

Loading effect is the terminology used when a measuring instrument such as an oscilloscope or voltmeter is connected across a component and the current drawn by the instrument upsets the circuit under test. The best way of demonstrating loading effect is by a numerical example.

In the simple circuit of Fig. 5.35, the voltage across each of the resistors can be calculated using voltage division, or by inspection. In this case, the voltage shown as $V$ should be 20 V.

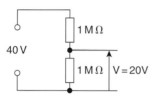

Figure 5.35

Using a voltmeter having a resistance of, say, $600\,\text{k}\Omega$, places $600\,\text{k}\Omega$ in parallel with the $1\,\text{M}\Omega$ resistor, as shown in Fig. 5.36.

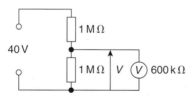

Figure 5.36

Resistance of parallel section

$$= \frac{1 \times 10^6 \times 600 \times 10^3}{(1 \times 10^6 + 600 \times 10^3)}$$

$$= 375\,\text{k}\Omega \text{ (using product/sum)}$$

The voltage $V$ now equals

$$= \frac{375 \times 10^3}{(1 \times 10^6 + 375 \times 10^3)} \times 40$$

$$= \mathbf{10.91\,V} \text{ (by voltage division)}$$

The voltmeter has loaded the circuit by drawing current for its operation, and by so doing, reduces the voltage

across the $1\,\text{M}\Omega$ resistor from the correct value of 20 V to 10.91 V.

Using a Fluke (or multimeter) which has a set internal resistance of, say, $10\,\text{M}\Omega$, as shown in Fig. 5.37, produces a much better result and the loading effect is minimal, as shown below.

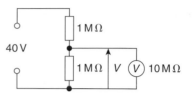

Figure 5.37

Resistance of parallel section

$$= \frac{1 \times 10^6 \times 10 \times 10^6}{(1 \times 10^6 + 10 \times 10^6)} = 0.91\,\text{M}\Omega$$

The voltage $V$ now equals

$$= \frac{0.91 \times 10^6}{(0.91 \times 10^6 + 1 \times 10^6)} \times 40 = \mathbf{19.06\,V}$$

When taking measurements, it is vital that the loading effect is understood and kept in mind at all times. An incorrect voltage reading may be due to this loading effect rather than the equipment under investigation being defective. Ideally, **the resistance of a voltmeter should be infinite**.

## 5.6 Potentiometers and rheostats

It is frequently desirable to be able to **vary the value of a resistor** in a circuit. A simple example of this is the volume control of a radio or television set.
Voltages and currents may be varied in electrical circuits by using **potentiometers** and **rheostats**.

### Potentiometers

When a variable resistor uses **three terminals**, it is known as a **potentiometer**. The potentiometer provides an adjustable voltage divider circuit, which is useful as a means of obtaining **various voltages** from a fixed potential difference. Consider the potentiometer circuit shown in Fig. 5.38 incorporating a lamp and supply voltage $V$.

In the circuit of Fig. 5.38, the input voltage is applied across points A and B at the ends of the potentiometer, while the output is tapped off between the sliding contact S and the fixed end B. It will be seen that with the slider at the far left-hand end of the resistor, the full voltage

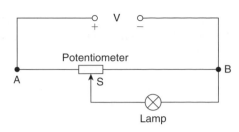

**Figure 5.38**

will appear across the lamp, and as the slider is moved towards point B the lamp brightness will reduce. When S is at the far right of the potentiometer, the lamp is short-circuited, no current will flow through it, and the lamp will be fully off.

> **Problem 15.**   Calculate the volt drop across the $60\,\Omega$ load in the circuit shown in Fig. 5.39, when the slider S is at the halfway point of the $200\,\Omega$ potentiometer.
>
>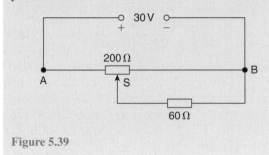
>
> **Figure 5.39**

With the slider halfway, the equivalent circuit is shown in Fig. 5.40.

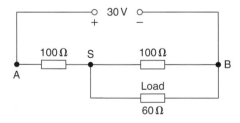

**Figure 5.40**

For the parallel resistors, total resistance,

$$R_P = \frac{100 \times 60}{100 + 60} = \frac{100 \times 60}{160} = 37.5\,\Omega$$

$$\text{(or use} \quad \frac{1}{R_P} = \frac{1}{100} + \frac{1}{60} \quad \text{to determine } R_P)$$

The equivalent circuit is now as shown in Fig. 5.41. The volt drop across the $37.5\,\Omega$ resistor in Fig. 5.41 is the same as the volt drop across both of the parallel resistors in Fig. 5.40.

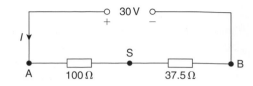

**Figure 5.41**

There are two methods for determining the volt drop $V_{SB}$:

## Method 1

Total circuit resistance,

$$R_T = 100 + 37.5 = 137.5\,\Omega$$

Hence, supply current,   $I = \dfrac{30}{137.5} = 0.2182A$

Thus, volt drop,   $V_{SB} = I \times 37.5 = 0.2182 \times 37.5$

$$= \mathbf{8.18\,V}$$

## Method 2

By the principle of voltage division,

$$V_{SB} = \left( \frac{37.5}{100 + 37.5} \right)(30) = 8.18\,V$$

Hence, **the volt drop across the $60\,\Omega$ load of Fig. 5.39 is 8.18 V.**

## Rheostats

A variable resistor where only **two terminals** are used, one fixed and one sliding, is known as a **rheostat**. The rheostat circuit, shown in Fig. 5.42, similar in construction to the potentiometer, is used to **control current flow**. The rheostat also acts as a dropping resistor, reducing the voltage across the load, but is more effective at controlling current.

For this reason **the resistance of the rheostat should be greater than that of the load**, otherwise it will have little or no effect. Typical uses are in a train set or Scalextric. Another practical example is in varying the brilliance of the panel lighting controls in a car.

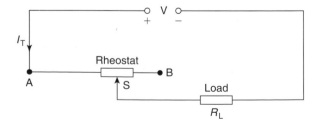

**Figure 5.42**

The rheostat resistance is connected in series with the load circuit, $R_L$, with the slider arm tapping off an amount of resistance (i.e. that between A and S) to provide the current flow required. With the slider at the far left-hand end, the load receives maximum current; with the slider at the far right-hand end, minimum current flows. The current flowing can be calculated by finding the total resistance of the circuit (i.e. $R_T = R_{AS} + R_L$), then by applying Ohm's law, $I_T = \dfrac{V}{R_{AS} + R_L}$

Calculations involved with the rheostat circuit are simpler than those for the potentiometer circuit.

**Problem 16.** In the circuit of Fig. 5.43, calculate the current flowing in the 100 Ω load, when the sliding point S is 2/3 of the way from A to B.

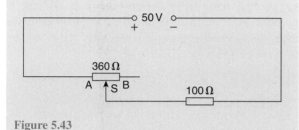

Figure 5.43

Resistance,                $R_{AS} = \dfrac{2}{3} \times 360 = 240\,\Omega$

Total circuit resistance,   $R_T = R_{AS} + R_L = 240 + 100$

$$= 340\,\Omega$$

**Current flowing in load, $I = \dfrac{V}{R_T} = \dfrac{50}{340}$**

$$= \mathbf{0.147\,A \ or \ 147\,mA}$$

## Summary

A **potentiometer** (a) has three terminals, and (b) is used for voltage control.
A **rheostat** (a) has two terminals, and (b) is used for current control.

A rheostat is not suitable if the load resistance is higher than the rheostat resistance; rheostat resistance must be higher than the load resistance to be able to influence current flow.

## Now try the following exercise

**Exercise 21    Further problems on potentiometers and rheostats**

1.  For the circuit shown in Fig. 5.44, AS is 3/5 of AB. Determine the voltage across the 120 Ω load. Is this a potentiometer or a rheostat circuit?

[44.44 V, potentiometer]

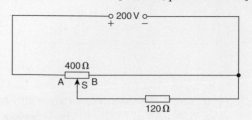

Figure 5.44

2.  For the circuit shown in Fig. 5.45, calculate the current flowing in the 25 Ω load and the voltage drop across the load when (a) AS is half of AB, and (b) point S coincides with point B. Is this a potentiometer or a rheostat?

[(a) 0.545 A, 13.64 V
(b) 0.286 A, 7.14 V rheostat]

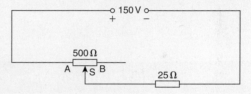

Figure 5.45

3.  For the circuit shown in Fig. 5.46, calculate the voltage across the 600 Ω load when point S splits AB in the ratio 1:3.          [136.4 V]

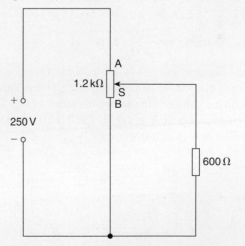

Figure 5.46

4.  For the circuit shown in Fig. 5.47, the slider S is set at halfway. Calculate the voltage drop across the 120 Ω load.          [9.68 V]

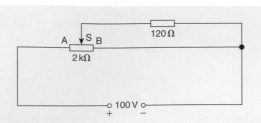

**Figure 5.47**

5.  For the potentiometer circuit shown in Fig. 5.48, AS is 60% of AB. Calculate the voltage across the 70 Ω load.   [63.40 V]

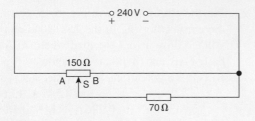

**Figure 5.48**

## 5.7   Relative and absolute voltages

In an electrical circuit, the voltage at any point can be quoted as being 'with reference to' (w.r.t.) any other point in the circuit. Consider the circuit shown in Fig. 5.49. The total resistance,

$$R_T = 30 + 50 + 5 + 15 = 100 \, \Omega \text{ and}$$

current, $I = \dfrac{200}{100} = 2 \, \text{A}$

If a voltage at point A is quoted with reference to point B then the voltage is written as $V_{AB}$. This is known as a '**relative voltage**'. In the circuit shown in Fig. 5.49, the voltage at A w.r.t. B is $I \times 50$, i.e. $2 \times 50 = 100 \, \text{V}$ and is written as $V_{AB} = 100 \, \text{V}$.

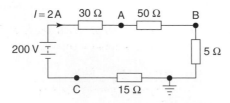

**Figure 5.49**

It must also be indicated whether the voltage at A w.r.t. B is closer to the positive terminal or the negative terminal of the supply source. Point A is nearer to the

positive terminal than B so is written as $V_{AB} = 100 \, \text{V}$ or $V_{AB} = +100 \, \text{V}$ or $V_{AB} = 100 \, \text{V} + \text{ve}$.

If no positive or negative is included, then the voltage is always taken to be positive.

If the voltage at B w.r.t. A is required, then $V_{BA}$ is negative and written as $V_{BA} = -100 \, \text{V}$ or $V_{BA} = 100 \, \text{V} - \text{ve}$. If the reference point is changed to the **earth point** then any voltage taken w.r.t. the earth is known as an '**absolute potential**'. If the absolute voltage of A in Fig. 5.49 is required, then this will be the sum of the voltages across the 50 Ω and 5 Ω resistors, i.e. $100 + 10 = 110 \, \text{V}$ and is written as $V_A = 110 \, \text{V}$ or $V_A = +110 \, \text{V}$ or $V_A = 110 \, \text{V} + \text{ve}$, positive since moving from the earth point to point A is moving towards the positive terminal of the source. If the voltage is negative w.r.t. earth then this must be indicated; for example, $V_C = 30 \, \text{V}$ negative w.r.t. earth, and is written as $V_C = -30 \, \text{V}$ or $V_C = 30 \, \text{V} - \text{ve}$.

---

**Problem 17.**   For the circuit shown in Fig. 5.50, calculate (a) the voltage drop across the 4 kΩ resistor, (b) the current through the 5 kΩ resistor, (c) the power developed in the 1.5 kΩ resistor, (d) the voltage at point X w.r.t. earth, and (e) the absolute voltage at point X.

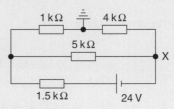

**Figure 5.50**

---

(a)   Total circuit resistance, $R_T = [(1 + 4) \text{k}\Omega \text{ in parallel}$ with 5 kΩ] in series with 1.5 kΩ

i.e.   $R_T = \dfrac{5 \times 5}{5 + 5} + 1.5 = 4 \, \text{k}\Omega$

Total circuit current, $I_T = \dfrac{V}{R_T} = \dfrac{24}{4 \times 10^3} = 6 \, \text{mA}$

By current division, current in top branch

$$= \left(\dfrac{5}{5 + 1 + 4}\right) \times 6 = 3 \, \text{mA}$$

Hence, **volt drop across 4 kΩ resistor**
$$= 3 \times 10^{-3} \times 4 \times 10^3 = \mathbf{12 \, V}$$

(b)   **Current through the 5 kΩ resistor**
$$= \left(\dfrac{1 + 4}{5 + 1 + 4}\right) \times 6 = \mathbf{3 \, mA}$$

(c)    **Power in the 1.5 kΩ resistor**
$$= I_T^2 R = (6 \times 10^{-3})^2 (1.5 \times 10^3) = \mathbf{54\,mW}$$

(d)    The voltage at the earth point is 0 volts. The volt drop across the 4 kΩ is 12 V, from part (a). Since moving from the earth point to point X is moving towards the negative terminal of the voltage source, the voltage at point X w.r.t. earth is **−12 V**

(e)    The 'absolute voltage at point X' means the 'voltage at point X w.r.t. earth', hence **the absolute voltage at point X is −12 V**. Questions (d) and (e) mean the same thing.

**Now try the following exercise**

**Exercise 22    Further problems on relative and absolute voltages**

1.    For the circuit of Fig. 5.51, calculate (a) the absolute voltage at points A, B and C, (b) the voltage at A relative to B and C, and (c) the voltage at D relative to B and A.

[(a) +40 V, +29.6 V, +24 V (b) +10.4 V, +16 V (c) −5.6 V, −16 V]

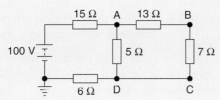

Figure 5.51

2.    For the circuit shown in Fig. 5.52, calculate (a) the voltage drop across the 7 Ω resistor, (b) the current through the 30 Ω resistor, (c) the power developed in the 8 Ω resistor, (d) the voltage at point X w.r.t. earth, and (e) the absolute voltage at point X.

[(a) 1.68 V (b) 0.16 A (c) 460.8 mW (d) +2.88 V (e) +2.88 V]

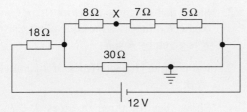

Figure 5.52

3.    In the bridge circuit of Fig. 5.53 calculate (a) the absolute voltages at points A and B, and (b) the voltage at A relative to B.

[(a) 10 V, 10 V (b) 0 V]

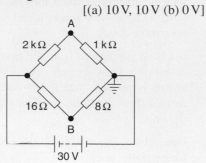

Figure 5.53

## 5.8    Earth potential and short circuits

The earth, and hence the sea, is at a potential of zero volts. Items connected to the earth (or sea), i.e. circuit wiring and electrical components, are said to be earthed or at earth potential. This means that there is no difference of potential between the item and earth. A ships' hull, being immersed in the sea, is at earth potential and therefore at zero volts. Earth faults, or short circuits, are caused by low resistance between the current-carrying conductor and earth. This occurs when the insulation resistance of the circuit wiring decreases, and is normally caused by:

1.    Dampness.

2.    Insulation becoming hard or brittle with age or heat.

3.    Accidental damage.

## 5.9    Wiring lamps in series and in parallel

**Series connection**

Figure 5.54 shows three lamps, each rated at 240 V, connected in series across a 240 V supply.

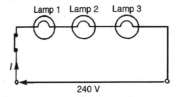

Figure 5.54

(i)  Each lamp has only $(240/3)$ V, i.e. 80 V across it and thus each lamp glows dimly.

(ii)  If another lamp of similar rating is added in series with the other three lamps then each lamp now has $(240/4)$ V, i.e. 60 V across it and each now glows even more dimly.

(iii)  If a lamp is removed from the circuit or if a lamp develops a fault (i.e. an open circuit) or if the switch is opened, then the circuit is broken, no current flows, and the remaining lamps will not light up.

(iv)  Less cable is required for a series connection than for a parallel one.

The series connection of lamps is usually limited to decorative lighting such as for Christmas tree lights.

## Parallel connection

Figure 5.55 shows three similar lamps, each rated at 240 V, connected in parallel across a 240 V supply.

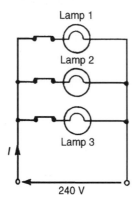

Lamp 1

Lamp 2

Lamp 3

$I$

240 V

**Figure 5.55**

(i)  Each lamp has 240 V across it and thus each will glow brilliantly at their rated voltage.

(ii)  If any lamp is removed from the circuit or develops a fault (open circuit) or a switch is opened, the remaining lamps are unaffected.

(iii)  The addition of further similar lamps in parallel does not affect the brightness of the other lamps.

(iv)  More cable is required for parallel connection than for a series one.

The parallel connection of lamps is the most widely used in electrical installations.

**Problem 18.** If three identical lamps are connected in parallel and the combined resistance is 150 Ω, find the resistance of one lamp.

Let the resistance of one lamp be $R$, then

$$\frac{1}{150} = \frac{1}{R} + \frac{1}{R} + \frac{1}{R} = \frac{3}{R}$$

from which, $R = 3 \times 150 = \mathbf{450\,\Omega}$

**Problem 19.** Three identical lamps A, B and C are connected in series across a 150 V supply. State (a) the voltage across each lamp, and (b) the effect of lamp C failing.

(a)  Since each lamp is identical and they are connected in series there is $150/3$ V, i.e. **50 V** across each.

(b)  If lamp C fails, i.e. open circuits, no current will flow and **lamps A and B will not operate**.

**Now try the following exercises**

**Exercise 23    Further problems on wiring lamps in series and in parallel**

1.  If four identical lamps are connected in parallel and the combined resistance is 100 Ω, find the resistance of one lamp.            [400 Ω]

2.  Three identical filament lamps are connected (a) in series, (b) in parallel across a 210 V supply. State for each connection the p.d. across each lamp.            [(a) 70 V (b) 210 V]

**Exercise 24    Short answer questions on series and parallel networks**

1.  Name three characteristics of a series circuit

2.  Show that for three resistors $R_1$, $R_2$ and $R_3$ connected in series the equivalent resistance $R$ is given by $R = R_1 + R_2 + R_3$

3.  Name three characteristics of a parallel network

4.  Show that for three resistors $R_1$, $R_2$ and $R_3$ connected in parallel the equivalent

resistance $R$ is given by

$$\frac{1}{R} = \frac{1}{R_1} + \frac{1}{R_2} + \frac{1}{R_3}$$

5. Explain the potential divider circuit

6. Describe, using a circuit diagram, the method of operation of a potentiometer

7. State the main use of a potentiometer

8. Describe, using a circuit diagram, the method of operation of a rheostat

9. State the main use of a rheostat

10. Explain the difference between relative and absolute voltages

11. State three causes of short circuits in electrical circuits

12. Compare the merits of wiring lamps in (a) series (b) parallel

---

**Exercise 25   Multi-choice questions on series and parallel networks**
**(Answers on page 420)**

1. If two $4\,\Omega$ resistors are connected in series the effective resistance of the circuit is:
   (a) $8\,\Omega$   (b) $4\,\Omega$   (c) $2\,\Omega$   (d) $1\,\Omega$

2. If two $4\,\Omega$ resistors are connected in parallel the effective resistance of the circuit is:
   (a) $8\,\Omega$   (b) $4\,\Omega$   (c) $2\,\Omega$   (d) $1\,\Omega$

3. With the switch in Fig. 5.56 closed, the ammeter reading will indicate:
   (a) $1\,A$   (b) $75\,A$   (c) $\frac{1}{3}\,A$   (d) $3\,A$

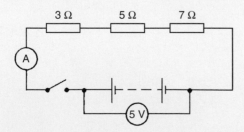

**Figure 5.56**

4. The effect of connecting an additional parallel load to an electrical supply source is to increase the
   (a) resistance of the load
   (b) voltage of the source
   (c) current taken from the source
   (d) p.d. across the load

5. The equivalent resistance when a resistor of $\frac{1}{3}\,\Omega$ is connected in parallel with a $\frac{1}{4}\,\Omega$ resistance is:
   (a) $\frac{1}{7}\,\Omega$   (b) $7\,\Omega$   (c) $\frac{1}{12}\,\Omega$   (d) $\frac{3}{4}\,\Omega$

6. With the switch in Fig. 5.57 closed the ammeter reading will indicate:
   (a) $108\,A$   (b) $\frac{1}{3}\,A$   (c) $3\,A$   (d) $4\frac{3}{5}\,A$

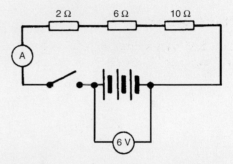

**Figure 5.57**

7. A $6\,\Omega$ resistor is connected in parallel with the three resistors of Fig. 5.57. With the switch closed the ammeter reading will indicate:
   (a) $\frac{3}{4}\,A$   (b) $4\,A$   (c) $\frac{1}{4}\,A$   (d) $1\frac{1}{3}\,A$

8. A $10\,\Omega$ resistor is connected in parallel with a $15\,\Omega$ resistor and the combination in series with a $12\,\Omega$ resistor. The equivalent resistance of the circuit is:
   (a) $37\,\Omega$   (b) $18\,\Omega$   (c) $27\,\Omega$   (d) $4\,\Omega$

9. When three $3\,\Omega$ resistors are connected in parallel, the total resistance is:
   (a) $3\,\Omega$   (b) $9\,\Omega$
   (c) $1\,\Omega$   (d) $0.333\,\Omega$

10. The total resistance of two resistors $R_1$ and $R_2$ when connected in parallel is given by:
    (a) $R_1 + R_2$   (b) $\dfrac{1}{R_1} + \dfrac{1}{R_2}$
    (c) $\dfrac{R_1 + R_2}{R_1 R_2}$   (d) $\dfrac{R_1 R_2}{R_1 + R_2}$

**Section 1**

11. If in the circuit shown in Fig. 5.58, the reading on the voltmeter is 5 V and the reading on the ammeter is 25 mA, the resistance of resistor $R$ is:
    (a) $0.005\,\Omega$       (b) $5\,\Omega$
    (c) $125\,\Omega$       (d) $200\,\Omega$

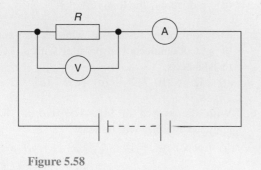

**Figure 5.58**

12. A variable resistor has a range of 0 to $5\,k\Omega$. If the slider is set at halfway, the value of current flowing through a $750\,\Omega$ load, when connected to a 100 V supply and used as a potentiometer, is:
    (a) $25\,mA$       (b) $40\,mA$
    (c) $17.39\,mA$       (d) $20\,mA$

# Chapter 6

# Capacitors and capacitance

At the end of this chapter you should be able to:

- appreciate some applications of capacitors

- describe an electrostatic field

- appreciate Coulomb's law

- define electric field strength $E$ and state its unit

- define capacitance and state its unit

- describe a capacitor and draw the circuit diagram symbol

- perform simple calculations involving $C = Q/V$ and $Q = It$

- define electric flux density $D$ and state its unit

- define permittivity, distinguishing between $\varepsilon_0$, $\varepsilon_r$ and $\varepsilon$

- perform simple calculations involving

$$D = \frac{Q}{A}, \quad E = \frac{V}{d} \quad \text{and} \quad \frac{D}{E} = \varepsilon_0 \varepsilon_r$$

- understand that for a parallel plate capacitor,

$$C = \frac{\varepsilon_0 \varepsilon_r A(n-1)}{d}$$

- perform calculations involving capacitors connected in parallel and in series

- define dielectric strength and state its unit

- state that the energy stored in a capacitor is given by $W = \frac{1}{2}CV^2$ joules

- describe practical types of capacitor

- understand the precautions needed when discharging capacitors

## 6.1 Introduction to capacitors

A capacitor is an electrical device that is used to store electrical energy. Next to the resistor, the capacitor is the most commonly encountered component in electrical circuits. Capacitors are used extensively in electrical and electronic circuits. For example, capacitors are used to smooth rectified a.c. outputs, they are used in telecommunication equipment – such as radio receivers – for tuning to the required frequency, they are used in time delay circuits, in electrical filters, in oscillator circuits, and in magnetic resonance imaging (MRI) in medical body scanners, to name but a few practical applications.

DOI: 10.1016/B978-0-08-089056-2.00006-1

## 6.2 Electrostatic field

Figure 6.1 represents two parallel metal plates, $A$ and $B$, charged to different potentials. If an electron that has a negative charge is placed between the plates, a force will act on the electron tending to push it away from the negative plate $B$ towards the positive plate, $A$. Similarly, a positive charge would be acted on by a force tending to move it toward the negative plate. Any region such as that shown between the plates in Fig. 6.1, in which an electric charge experiences a force, is called an **electrostatic field**. The direction of the field is defined as that of the force acting on a positive charge placed in the field. In Fig. 6.1, the direction of the force is from the positive plate to the negative plate. Such a field may be represented in magnitude and direction by **lines of electric force** drawn between the charged surfaces. The closeness of the lines is an indication of the field strength. Whenever a p.d. is established between two points, an electric field will always exist.

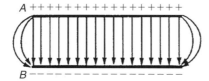

Figure 6.1

Figure 6.2(a) shows a typical field pattern for an isolated point charge, and Fig. 6.2(b) shows the field pattern for adjacent charges of opposite polarity. Electric lines of force (often called electric flux lines) are continuous and start and finish on point charges; also, the lines cannot cross each other. When a charged body is placed close to an uncharged body, an induced charge of opposite sign appears on the surface of the uncharged body. This is because lines of force from the charged body terminate on its surface.

The concept of field lines or lines of force is used to illustrate the properties of an electric field. However, it should be remembered that they are only aids to the imagination.

The **force of attraction or repulsion** between two electrically charged bodies is proportional to the magnitude of their charges and inversely proportional to the square of the distance separating them, i.e.

$$\text{force} \propto \frac{q_1 q_2}{d^2}$$

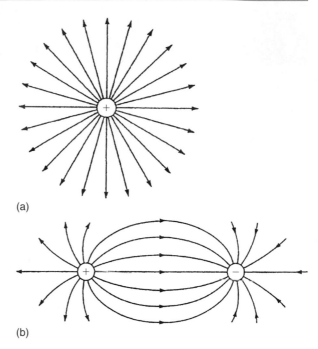

(a)

(b)

Figure 6.2

or

$$\text{force} = k\frac{q_1 q_2}{d^2}$$

where constant $k \approx 9 \times 10^9$. This is known as **Coulomb's law**.

Hence the force between two charged spheres in air with their centres 16 mm apart and each carrying a charge of $+1.6\,\mu\text{C}$ is given by:

$$\text{force} = k\frac{q_1 q_2}{d^2} \approx (9 \times 10^9)\frac{(1.6 \times 10^{-6})^2}{(16 \times 10^{-3})^2}$$

$$= \textbf{90 newtons}$$

## 6.3 Electric field strength

Figure 6.3 shows two parallel conducting plates separated from each other by air. They are connected to opposite terminals of a battery of voltage $V$ volts. There is therefore an electric field in the space between the plates. If the plates are close together, the electric lines of force will be straight and parallel and equally spaced, except near the edge where fringing will occur (see Fig. 6.1). Over the area in which there is negligible fringing,

**Electric field strength,** $E = \dfrac{V}{d}$ **volts/metre**

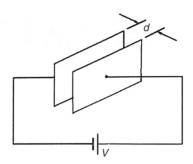

**Figure 6.3**

where $d$ is the distance between the plates. Electric field strength is also called **potential gradient**.

## 6.4  Capacitance

Static electric fields arise from electric charges, electric field lines beginning and ending on electric charges. Thus the presence of the field indicates the presence of equal positive and negative electric charges on the two plates of Fig. 6.3. Let the charge be $+Q$ coulombs on one plate and $-Q$ coulombs on the other. The property of this pair of plates which determines how much charge corresponds to a given p.d. between the plates is called their capacitance:

$$\text{capacitance } C = \frac{Q}{V}$$

The **unit of capacitance** is the **farad** $F$ (or more usually $\mu F = 10^{-6}\,F$ or $pF = 10^{-12}\,F$), which is defined as the capacitance when a p.d. of one volt appears across the plates when charged with one coulomb.

## 6.5  Capacitors

Every system of electrical conductors possesses capacitance. For example, there is capacitance between the conductors of overhead transmission lines and also between the wires of a telephone cable. In these examples the capacitance is undesirable but has to be accepted, minimised or compensated for. There are other situations where capacitance is a desirable property.

Devices specially constructed to possess capacitance are called **capacitors** (or condensers, as they used to be called). In its simplest form a capacitor consists of two plates which are separated by an insulating material known as a **dielectric**. A capacitor has the ability to store a quantity of static electricity.

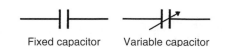

Fixed capacitor     Variable capacitor

**Figure 6.4**

The symbols for a fixed capacitor and a variable capacitor used in electrical circuit diagrams are shown in Fig. 6.4

The **charge Q** stored in a capacitor is given by:

$$Q = I \times t \text{ coulombs}$$

where $I$ is the current in amperes and $t$ the time in seconds.

---

**Problem 1.**  (a) Determine the p.d. across a $4\,\mu F$ capacitor when charged with $5\,mC$. (b) Find the charge on a $50\,pF$ capacitor when the voltage applied to it is $2\,kV$.

(a)  $C = 4\,\mu F = 4 \times 10^{-6}\,F$ and
$Q = 5\,mC = 5 \times 10^{-3}\,C$.

Since $C = \dfrac{Q}{V}$  then  $V = \dfrac{Q}{C} = \dfrac{5 \times 10^{-3}}{4 \times 10^{-6}}$

$$= \frac{5 \times 10^{6}}{4 \times 10^{3}} = \frac{5000}{4}$$

**Hence p.d.  $V = 1250\,V$ or $1.25\,kV$**

(b)  $C = 50\,pF = 50 \times 10^{-12}\,F$ and
$V = 2\,kV = 2000\,V$

$$Q = CV = 50 \times 10^{-12} \times 2000$$

$$= \frac{5 \times 2}{10^{8}} = 0.1 \times 10^{-6}$$

**Hence, charge  $Q = 0.1\,\mu C$**

---

**Problem 2.**  A direct current of $4\,A$ flows into a previously uncharged $20\,\mu F$ capacitor for $3\,ms$. Determine the p.d. between the plates.

$I = 4\,A$, $C = 20\,\mu F = 20 \times 10^{-6}\,F$ and
$t = 3\,ms = 3 \times 10^{-3}\,s$.
$Q = It = 4 \times 3 \times 10^{-3}\,C$.

$$V = \frac{Q}{C} = \frac{4 \times 3 \times 10^{-3}}{20 \times 10^{-6}}$$

$$= \frac{12 \times 10^{6}}{20 \times 10^{3}} = 0.6 \times 10^{3} = 600\,V$$

**Hence, the p.d. between the plates is $600\,V$**

**Problem 3.** A $5\,\mu F$ capacitor is charged so that the p.d. between its plates is 800 V. Calculate how long the capacitor can provide an average discharge current of 2 mA.

$C = 5\,\mu F = 5 \times 10^{-6}\,F$, $V = 800\,V$ and
$I = 2\,mA = 2 \times 10^{-3}\,A$.

$Q = CV = 5 \times 10^{-6} \times 800 = 4 \times 10^{-3}\,C$
Also, $Q = It$. Thus,

$$t = \frac{Q}{I} = \frac{4 \times 10^{-3}}{2 \times 10^{-3}} = 2\,s$$

Hence, the capacitor can provide an average discharge current of 2 mA for 2 s.

Now try the following exercise

**Exercise 26   Further problems on capacitors and capacitance**

1. Find the charge on a $10\,\mu F$ capacitor when the applied voltage is 250 V.        [2.5 mC]

2. Determine the voltage across a 1000 pF capacitor to charge it with $2\,\mu C$.        [2 kV]

3. The charge on the plates of a capacitor is 6 mC when the potential between them is 2.4 kV. Determine the capacitance of the capacitor.        [$2.5\,\mu F$]

4. For how long must a charging current of 2 A be fed to a $5\,\mu F$ capacitor to raise the p.d. between its plates by 500 V.        [1.25 ms]

5. A direct current of 10 A flows into a previously uncharged $5\,\mu F$ capacitor for 1 ms. Determine the p.d. between the plates.        [2 kV]

6. A $16\,\mu F$ capacitor is charged at a constant current of $4\,\mu A$ for 2 min. Calculate the final p.d. across the capacitor and the corresponding charge in coulombs.        [30 V, $480\,\mu C$]

7. A steady current of 10 A flows into a previously uncharged capacitor for 1.5 ms when the p.d. between the plates is 2 kV. Find the capacitance of the capacitor.        [$7.5\,\mu F$]

## 6.6   Electric flux density

Unit flux is defined as emanating from a positive charge of 1 coulomb. Thus electric flux $\psi$ is measured in coulombs, and for a charge of $Q$ coulombs, the flux $\psi = Q$ coulombs.

Electric flux density $D$ is the amount of flux passing through a defined area $A$ that is perpendicular to the direction of the flux:

$$\text{electric flux density, } D = \frac{Q}{A} \text{ coulombs/metre}^2$$

Electric flux density is also called **charge density**, $\sigma$.

## 6.7   Permittivity

At any point in an electric field, the electric field strength $E$ maintains the electric flux and produces a particular value of electric flux density $D$ at that point. For a field established in **vacuum** (or for practical purposes in air), the ratio $D/E$ is a constant $\varepsilon_0$, i.e.

$$\frac{D}{E} = \varepsilon_0$$

where $\varepsilon_0$ is called the **permittivity of free space** or the free space constant. The value of $\varepsilon_0$ is $8.85 \times 10^{-12}\,F/m$.

When an insulating medium, such as mica, paper, plastic or ceramic, is introduced into the region of an electric field the ratio of $D/E$ is modified:

$$\frac{D}{E} = \varepsilon_0 \varepsilon_r$$

where $\varepsilon_r$, the **relative permittivity** of the insulating material, indicates its insulating power compared with that of vacuum:

relative permittivity,

$$\varepsilon_r = \frac{\text{flux density in material}}{\text{flux density in vacuum}}$$

$\varepsilon_r$ has no unit. Typical values of $\varepsilon_r$ include air, 1.00; polythene, 2.3; mica, 3–7; glass, 5–10; water, 80; ceramics, 6–1000.

The product $\varepsilon_0 \varepsilon_r$ is called the **absolute permittivity**, $\varepsilon$, i.e.

$$\varepsilon = \varepsilon_0 \varepsilon_r$$

The insulating medium separating charged surfaces is called a **dielectric**. Compared with conductors, dielectric materials have very high resistivities. They are therefore used to separate conductors at different potentials, such as capacitor plates or electric power lines.

> **Problem 4.** Two parallel rectangular plates measuring 20 cm by 40 cm carry an electric charge of $0.2\,\mu C$. Calculate the electric flux density. If the plates are spaced 5 mm apart and the voltage between them is 0.25 kV determine the electric field strength.

Area $= 20\,cm \times 40\,cm = 800\,cm^2 = 800 \times 10^{-4}\,m^2$ and charge $Q = 0.2\,\mu C = 0.2 \times 10^{-6}\,C$,
**Electric flux density**

$$D = \frac{Q}{A} = \frac{0.2 \times 10^{-6}}{800 \times 10^{-4}} = \frac{0.2 \times 10^4}{800 \times 10^6}$$

$$= \frac{2000}{800} \times 10^{-6} = \textbf{2.5}\,\boldsymbol{\mu}\textbf{C/m}^2$$

Voltage $V = 0.25\,kV = 250\,V$ and plate spacing, $d = 5\,mm = 5 \times 10^{-3}\,m$.
**Electric field strength**

$$E = \frac{V}{d} = \frac{250}{5 \times 10^{-3}} = \textbf{50}\,\textbf{kV/m}$$

> **Problem 5.** The flux density between two plates separated by mica of relative permittivity 5 is $2\,\mu C/m^2$. Find the voltage gradient between the plates.

Flux density $D = 2\,\mu C/m^2 = 2 \times 10^{-6}\,C/m^2$,
$\varepsilon_0 = 8.85 \times 10^{-12}\,F/m$ and $\varepsilon_r = 5$.
$D/E = \varepsilon_0 \varepsilon_r$, hence **voltage gradient**,

$$E = \frac{D}{\varepsilon_0 \varepsilon_r} = \frac{2 \times 10^{-6}}{8.85 \times 10^{-12} \times 5}\,V/m$$

$$= \textbf{45.2}\,\textbf{kV/m}$$

> **Problem 6.** Two parallel plates having a p.d. of 200 V between them are spaced 0.8 mm apart. What is the electric field strength? Find also the electric flux density when the dielectric between the plates is (a) air, and (b) polythene of relative permittivity 2.3.

**Electric field strength**

$$E = \frac{V}{d} = \frac{200}{0.8 \times 10^{-3}} = \textbf{250}\,\textbf{kV/m}$$

(a)  For air: $\varepsilon_r = 1$ and $\dfrac{D}{E} = \varepsilon_0 \varepsilon_r$

Hence **electric flux density**

$$D = E\varepsilon_0 \varepsilon_r$$

$$= (250 \times 10^3 \times 8.85 \times 10^{-12} \times 1)\,C/m^2$$

$$= \textbf{2.213}\,\boldsymbol{\mu}\textbf{C/m}^2$$

(b)  For polythene, $\varepsilon_r = 2.3$

**Electric flux density**

$$D = E\varepsilon_0 \varepsilon_r$$

$$= (250 \times 10^3 \times 8.85 \times 10^{-12} \times 2.3)\,C/m^2$$

$$= \textbf{5.089}\,\boldsymbol{\mu}\textbf{C/m}^2$$

---

**Now try the following exercise**

> **Exercise 27    Further problems on electric field strength, electric flux density and permittivity**
>
> (Where appropriate take $\varepsilon_0$ as $8.85 \times 10^{-12}\,F/m$)
>
> 1.  A capacitor uses a dielectric 0.04 mm thick and operates at 30 V. What is the electric field strength across the dielectric at this voltage?
>     [750 kV/m]
>
> 2.  A two-plate capacitor has a charge of 25 C. If the effective area of each plate is $5\,cm^2$ find the electric flux density of the electric field.
>     [$50\,kC/m^2$]
>
> 3.  A charge of $1.5\,\mu C$ is carried on two parallel rectangular plates each measuring 60 mm by 80 mm. Calculate the electric flux density. If the plates are spaced 10 mm apart and the voltage between them is 0.5 kV determine the electric field strength.
>     [$312.5\,\mu C/m^2$, 50 kV/m]
>
> 4.  Two parallel plates are separated by a dielectric and charged with $10\,\mu C$. Given that the

area of each plate is $50\,\text{cm}^2$, calculate the electric flux density in the dielectric separating the plates.

[$2\,\text{mC/m}^2$]

5. The electric flux density between two plates separated by polystyrene of relative permittivity 2.5 is $5\,\mu\text{C/m}^2$. Find the voltage gradient between the plates. [$226\,\text{kV/m}$]

6. Two parallel plates having a p.d. of $250\,\text{V}$ between them are spaced $1\,\text{mm}$ apart. Determine the electric field strength. Find also the electric flux density when the dielectric between the plates is (a) air and (b) mica of relative permittivity 5.

[$250\,\text{kV/m}$ (a) $2.213\,\mu\text{C/m}^2$
(b) $11.063\,\mu\text{C/m}^2$]

## 6.8  The parallel plate capacitor

For a parallel plate capacitor, as shown in Fig. 6.5(a), experiments show that capacitance $C$ is proportional to the area $A$ of a plate, inversely proportional to the plate spacing $d$ (i.e. the dielectric thickness) and depends on the nature of the dielectric:

$$\textbf{Capacitance, } C = \frac{\varepsilon_0 \varepsilon_r A}{d} \textbf{ farads}$$

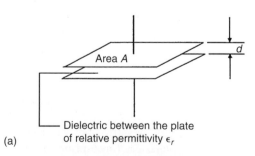

Area $A$

$d$

Dielectric between the plate
of relative permittivity $\epsilon_r$

(a)

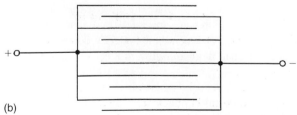

(b)

**Figure 6.5**

where $\varepsilon_0 = 8.85 \times 10^{-12}\,\text{F/m}$ (constant)

$\varepsilon_r$ = relative permittivity

$A$ = area of one of the plates, in $m^2$, and

$d$ = thickness of dielectric in $m$

Another method used to increase the capacitance is to interleave several plates as shown in Fig. 6.5(b). Ten plates are shown, forming nine capacitors with a capacitance nine times that of one pair of plates. If such an arrangement has $n$ plates then capacitance $C \propto (n-1)$. Thus capacitance

$$C = \frac{\varepsilon_0 \varepsilon_r A(n-1)}{d} \text{ farads}$$

**Problem 7.** (a) A ceramic capacitor has an effective plate area of $4\,\text{cm}^2$ separated by $0.1\,\text{mm}$ of ceramic of relative permittivity 100. Calculate the capacitance of the capacitor in picofarads. (b) If the capacitor in part (a) is given a charge of $1.2\,\mu\text{C}$ what will be the p.d. between the plates?

(a) Area $A = 4\,\text{cm}^2 = 4 \times 10^{-4}\,\text{m}^2$,
$d = 0.1\,\text{mm} = 0.1 \times 10^{-3}\,\text{m}$,
$\varepsilon_0 = 8.85 \times 10^{-12}\,\text{F/m}$ and $\varepsilon_r = 100$
**Capacitance,**

$$C = \frac{\varepsilon_0 \varepsilon_r A}{d} \text{ farads}$$

$$= \frac{8.85 \times 10^{-12} \times 100 \times 4 \times 10^{-4}}{0.1 \times 10^{-3}}\,\text{F}$$

$$= \frac{8.85 \times 4}{10^{10}}\,\text{F}$$

$$= \frac{8.85 \times 4 \times 10^{12}}{10^{10}}\,\text{pF} = \textbf{3540\,pF}$$

(b) $Q = CV$ thus

$$V = \frac{Q}{C} = \frac{1.2 \times 10^{-6}}{3540 \times 10^{-12}}\,\text{V} = \textbf{339\,V}$$

**Problem 8.** A waxed paper capacitor has two parallel plates, each of effective area $800\,\text{cm}^2$. If the capacitance of the capacitor is $4425\,\text{pF}$ determine the effective thickness of the paper if its relative permittivity is 2.5

$A = 800\,\text{cm}^2 = 800 \times 10^{-4}\,\text{m}^2 = 0.08\,\text{m}^2$,
$C = 4425\,\text{pF} = 4425 \times 10^{-12}\,\text{F}, \varepsilon_0 = 8.85 \times 10^{-12}\,\text{F/m}$

and $\varepsilon_r = 2.5$. Since

$$C = \frac{\varepsilon_0 \varepsilon_A A}{d} \quad \text{then} \quad d = \frac{\varepsilon_0 \varepsilon_r A}{C}$$

i.e.

$$d = \frac{8.85 \times 10^{-12} \times 2.5 \times 0.08}{4425 \times 10^{-12}}$$

$$= 0.0004 \, \text{m}$$

**Hence, the thickness of the paper is 0.4 mm**.

**Problem 9.** A parallel plate capacitor has nineteen interleaved plates each 75 mm by 75 mm separated by mica sheets 0.2 mm thick. Assuming the relative permittivity of the mica is 5, calculate the capacitance of the capacitor

$n = 19$ thus $n - 1 = 18$, $A = 75 \times 75 = 5625 \, \text{mm}^2 = 5625 \times 10^{-6} \, \text{m}^2$, $\varepsilon_r = 5$, $\varepsilon_0 = 8.85 \times 10^{-12} \, \text{F/m}$ and $d = 0.2 \, \text{mm} = 0.2 \times 10^{-3} \, \text{m}$. Capacitance,

$$C = \frac{\varepsilon_0 \varepsilon_r A (n-1)}{d}$$

$$= \frac{8.85 \times 10^{-12} \times 5 \times 5625 \times 10^{-6} \times 18}{0.2 \times 10^{-3}} \, \text{F}$$

$$= \mathbf{0.0224 \, \mu F \ or \ 22.4 \, nF}$$

**Now try the following exercise**

**Exercise 28   Further problems on parallel plate capacitors**

(Where appropriate take $\varepsilon_0$ as $8.85 \times 10^{-12} \, \text{F/m}$)

1.  A capacitor consists of two parallel plates each of area 0.01 m², spaced 0.1 mm in air. Calculate the capacitance in picofarads.  [885 pF]

2.  A waxed paper capacitor has two parallel plates, each of effective area 0.2 m². If the capacitance is 4000 pF determine the effective thickness of the paper if its relative permittivity is 2  [0.885 mm]

3.  Calculate the capacitance of a parallel plate capacitor having 5 plates, each 30 mm by 20 mm and separated by a dielectric 0.75 mm thick having a relative permittivity of 2.3.  [65.14 pF]

4.  How many plates has a parallel plate capacitor having a capacitance of 5 nF, if each plate is 40 mm by 40 mm and each dielectric is 0.102 mm thick with a relative permittivity of 6.  [7]

5.  A parallel plate capacitor is made from 25 plates, each 70 mm by 120 mm interleaved with mica of relative permittivity 5. If the capacitance of the capacitor is 3000 pF determine the thickness of the mica sheet.  [2.97 mm]

6.  A capacitor is constructed with parallel plates and has a value of 50 pF. What would be the capacitance of the capacitor if the plate area is doubled and the plate spacing is halved?  [200 pF]

7.  The capacitance of a parallel plate capacitor is 1000 pF. It has 19 plates, each 50 mm by 30 mm separated by a dielectric of thickness 0.40 mm. Determine the relative permittivity of the dielectric.  [1.67]

8.  The charge on the square plates of a multiplate capacitor is 80 μC when the potential between them is 5 kV. If the capacitor has twenty-five plates separated by a dielectric of thickness 0.102 mm and relative permittivity 4.8, determine the width of a plate.  [40 mm]

9.  A capacitor is to be constructed so that its capacitance is 4250 pF and to operate at a p.d. of 100 V across its terminals. The dielectric is to be polythene ($\varepsilon_r = 2.3$) which, after allowing a safety factor, has a dielectric strength of 20 MV/m. Find (a) the thickness of the polythene needed, and (b) the area of a plate.  [(a) 0.005 mm (b) 10.44 cm²]

## 6.9  Capacitors connected in parallel and series

### (a) Capacitors connected in parallel

Figure 6.6 shows three capacitors, $C_1$, $C_2$ and $C_3$, connected in parallel with a supply voltage $V$ applied across the arrangement.

When the charging current $I$ reaches point $A$ it divides, some flowing into $C_1$, some flowing into $C_2$ and some into $C_3$. Hence the total charge $Q_T (= I \times t)$ is divided between the three capacitors. The capacitors

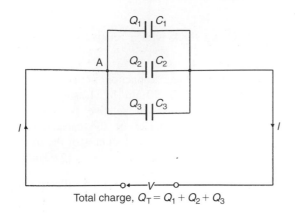

Total charge, $Q_T = Q_1 + Q_2 + Q_3$

Figure 6.6

each store a charge and these are shown as $Q_1$, $Q_2$ and $Q_3$ respectively. Hence

$$Q_T = Q_1 + Q_2 + Q_3$$

But $Q_T = CV$, $Q_1 = C_1 V$, $Q_2 = C_2 V$ and $Q_3 = C_3 V$. Therefore $CV = C_1 V + C_2 V + C_3 V$ where $C$ is the total equivalent circuit capacitance, i.e.

$$C = C_1 + C_2 + C_3$$

It follows that for $n$ parallel-connected capacitors,

$$C = C_1 + C_2 + C_3 \cdots\cdots + C_n$$

i.e. the equivalent capacitance of a group of parallel-connected capacitors is the sum of the capacitances of the individual capacitors. (Note that this formula is similar to that used for **resistors** connected in **series**).

### (b) Capacitors connected in series

Figure 6.7 shows three capacitors, $C_1$, $C_2$ and $C_3$, connected in series across a supply voltage $V$. Let the p.d. across the individual capacitors be $V_1$, $V_2$ and $V_3$ respectively as shown.

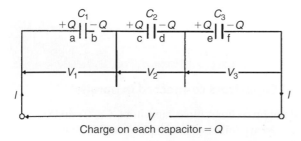

Charge on each capacitor $= Q$

Figure 6.7

Let the charge on plate '$a$' of capacitor $C_1$ be $+Q$ coulombs. This induces an equal but opposite charge

of $-Q$ coulombs on plate '$b$'. The conductor between plates '$b$' and '$c$' is electrically isolated from the rest of the circuit so that an equal but opposite charge of $+Q$ coulombs must appear on plate '$c$', which, in turn, induces an equal and opposite charge of $-Q$ coulombs on plate '$d$', and so on.

Hence when capacitors are connected in series the charge on each is the same. In a series circuit:

$$V = V_1 + V_2 + V_3$$

Since $V = \dfrac{Q}{C}$ then $\dfrac{Q}{C} = \dfrac{Q}{C_1} + \dfrac{Q}{C_2} + \dfrac{Q}{C_3}$

where $C$ is the total equivalent circuit capacitance, i.e.

$$\frac{1}{C} = \frac{1}{C_1} + \frac{1}{C_2} + \frac{1}{C_3}$$

It follows that for $n$ series-connected capacitors:

$$\frac{1}{C} = \frac{1}{C_1} + \frac{1}{C_2} + \frac{1}{C_3} + \cdots + \frac{1}{C_n}$$

i.e. for series-connected capacitors, the reciprocal of the equivalent capacitance is equal to the sum of the reciprocals of the individual capacitances. (Note that this formula is similar to that used for **resistors** connected in **parallel**.)

For the special case of **two capacitors in series**:

$$\frac{1}{C} = \frac{1}{C_1} + \frac{1}{C_2} = \frac{C_2 + C_1}{C_1 C_2}$$

Hence

$$C = \frac{C_1 C_2}{C_1 + C_2} \quad \left(\text{i.e.} \ \frac{\text{product}}{\text{sum}}\right)$$

**Problem 10.** Calculate the equivalent capacitance of two capacitors of $6\,\mu F$ and $4\,\mu F$ connected (a) in parallel and (b) in series.

(a) In parallel, equivalent capacitance,
$$C = C_1 + C_2 = 6\,\mu F + 4\,\mu F = \mathbf{10\,\mu F}$$

(b) In series, equivalent capacitance $C$ is given by:

$$C = \frac{C_1 C_2}{C_1 + C_2}$$

This formula is used for the special case of **two** capacitors in series. Thus

$$C = \frac{6 \times 4}{6 + 4} = \frac{24}{10} = \mathbf{2.4\,\mu F}$$

**Problem 11.** What capacitance must be connected in series with a $30\,\mu\text{F}$ capacitor for the equivalent capacitance to be $12\,\mu\text{F}$?

Let $C = 12\,\mu\text{F}$ (the equivalent capacitance), $C_1 = 30\,\mu\text{F}$ and $C_2$ be the unknown capacitance. For two capacitors in series

$$\frac{1}{C} = \frac{1}{C_1} + \frac{1}{C_2}$$

Hence

$$\frac{1}{C_2} = \frac{1}{C} - \frac{1}{C_1} = \frac{C_1 - C}{CC_1}$$

and

$$C_2 = \frac{CC_1}{C_1 - C} = \frac{12 \times 30}{30 - 12} = \frac{360}{18} = \mathbf{20\,\mu F}$$

**Problem 12.** Capacitances of $1\,\mu\text{F}$, $3\,\mu\text{F}$, $5\,\mu\text{F}$ and $6\,\mu\text{F}$ are connected in parallel to a direct voltage supply of $100\,\text{V}$. Determine (a) the equivalent circuit capacitance, (b) the total charge, and (c) the charge on each capacitor.

(a)  The equivalent capacitance $C$ for four capacitors in parallel is given by:

$$C = C_1 + C_2 + C_3 + C_4$$

i.e. $C = 1 + 3 + 5 + 6 = \mathbf{15\,\mu F}$

(b)  Total charge $Q_T = CV$ where $C$ is the equivalent circuit capacitance i.e.

$$Q_T = 15 \times 10^{-6} \times 100 = 1.5 \times 10^{-3}C$$
$$= \mathbf{1.5\,mC}$$

(c)  The charge on the $1\,\mu\text{F}$ capacitor
$$Q_1 = C_1 V = 1 \times 10^{-6} \times 100 = \mathbf{0.1\,mC}$$
The charge on the $3\,\mu\text{F}$ capacitor
$$Q_2 = C_2 V = 3 \times 10^{-6} \times 100 = \mathbf{0.3\,mC}$$
The charge on the $5\,\mu\text{F}$ capacitor
$$Q_3 = C_3 V = 5 \times 10^{-6} \times 100 = \mathbf{0.5\,mC}$$
The charge on the $6\,\mu\text{F}$ capacitor
$$Q_4 = C_4 V = 6 \times 10^{-6} \times 100 = \mathbf{0.6\,mC}$$
[Check: In a parallel circuit

$$Q_T = Q_1 + Q_2 + Q_3 + Q_4$$
$$Q_1 + Q_2 + Q_3 + Q_4 = 0.1 + 0.3 + 0.5 + 0.6$$
$$= 1.5\,\text{mC} = Q_T]$$

**Problem 13.** Capacitances of $3\,\mu\text{F}$, $6\,\mu\text{F}$ and $12\,\mu\text{F}$ are connected in series across a $350\,\text{V}$ supply. Calculate (a) the equivalent circuit capacitance, (b) the charge on each capacitor, and (c) the p.d. across each capacitor.

The circuit diagram is shown in Fig. 6.8.

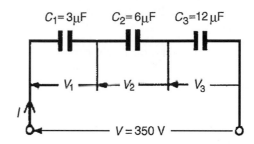

**Figure 6.8**

(a)  The equivalent circuit capacitance $C$ for three capacitors in series is given by:

$$\frac{1}{C} = \frac{1}{C_1} + \frac{1}{C_2} + \frac{1}{C_3}$$

i.e. $\dfrac{1}{C} = \dfrac{1}{3} + \dfrac{1}{6} + \dfrac{1}{12} = \dfrac{4 + 2 + 1}{12} = \dfrac{7}{12}$

**Hence the equivalent circuit capacitance**

$$C = \frac{12}{7} = 1\frac{5}{7}\,\mu\text{F or } \mathbf{1.714\,\mu F}$$

(b)  Total charge $Q_T = CV$, hence

$$Q_T = \frac{12}{7} \times 10^{-6} \times 350$$
$$= 600\,\mu\text{C or } 0.6\,\text{mC}$$

**Since the capacitors are connected in series $0.6\,\text{mC}$ is the charge on each of them.**

(c)  The voltage across the $3\,\mu\text{F}$ capacitor,

$$V_1 = \frac{Q}{C_1}$$
$$= \frac{0.6 \times 10^{-3}}{3 \times 10^{-6}} = \mathbf{200\,V}$$

The voltage across the $6\,\mu\text{F}$ capacitor,

$$V_2 = \frac{Q}{C_2}$$
$$= \frac{0.6 \times 10^{-3}}{6 \times 10^{-6}} = \mathbf{100\,V}$$

The voltage across the $12\,\mu\mathrm{F}$ capacitor,

$$V_3 = \frac{Q}{C_3}$$

$$= \frac{0.6 \times 10^{-3}}{12 \times 10^{-6}} = \mathbf{50\,V}$$

[Check: In a series circuit $V = V_1 + V_2 + V_3$. $V_1 + V_2 + V_3 = 200 + 100 + 50 = 350\,V =$ supply voltage]

In practice, capacitors are rarely connected in series unless they are of the same capacitance. The reason for this can be seen from the above problem where the lowest valued capacitor (i.e. $3\,\mu\mathrm{F}$) has the highest p.d. across it (i.e. $200\,V$) which means that if all the capacitors have an identical construction they must all be rated at the highest voltage.

---

**Problem 14.** For the arrangement shown in Fig. 6.9 find (a) the equivalent capacitance of the circuit, (b) the voltage across $QR$, and (c) the charge on each capacitor.

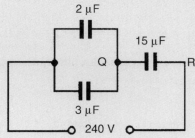

Figure 6.9

---

(a) $2\,\mu\mathrm{F}$ in parallel with $3\,\mu\mathrm{F}$ gives an equivalent capacitance of $2\,\mu\mathrm{F} + 3\,\mu\mathrm{F} = 5\,\mu\mathrm{F}$. The circuit is now as shown in Fig. 6.10.

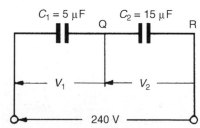

Figure 6.10

The **equivalent capacitance** of $5\,\mu\mathrm{F}$ in series with $15\,\mu\mathrm{F}$ is given by

$$\frac{5 \times 15}{5 + 15}\,\mu\mathrm{F} = \frac{75}{20}\,\mu\mathrm{F} = \mathbf{3.75\,\mu F}$$

---

(b) The charge on each of the capacitors shown in Fig. 6.10 will be the same since they are connected in series. Let this charge be $Q$ coulombs.

Then $\qquad\qquad Q = C_1 V_1 = C_2 V_2$

i.e. $\qquad\qquad 5V_1 = 15V_2$

$$V_1 = 3V_2 \qquad\qquad (1)$$

Also $\qquad V_1 + V_2 = 240\,V$

Hence $3V_2 + V_2 = 240\,V$ from equation (1)

Thus $\qquad\qquad V_2 = 60\,V$ and $V_1 = 180\,V$

**Hence the voltage across $QR$ is $60\,V$**

(c) The charge on the $15\,\mu\mathrm{F}$ capacitor is

$$C_2 V_2 = 15 \times 10^{-6} \times 60 = \mathbf{0.9\,mC}$$

The charge on the $2\,\mu\mathrm{F}$ capacitor is

$$2 \times 10^{-6} \times 180 = \mathbf{0.36\,mC}$$

The charge on the $3\,\mu\mathrm{F}$ capacitor is

$$3 \times 10^{-6} \times 180 = \mathbf{0.54\,mC}$$

---

**Now try the following exercise**

**Exercise 29   Further problems on capacitors in parallel and series**

1. Capacitors of $2\,\mu\mathrm{F}$ and $6\,\mu\mathrm{F}$ are connected (a) in parallel and (b) in series. Determine the equivalent capacitance in each case.
   [(a) $8\,\mu\mathrm{F}$ (b) $1.5\,\mu\mathrm{F}$]

2. Find the capacitance to be connected in series with a $10\,\mu\mathrm{F}$ capacitor for the equivalent capacitance to be $6\,\mu\mathrm{F}$. [$15\,\mu\mathrm{F}$]

3. What value of capacitance would be obtained if capacitors of $0.15\,\mu\mathrm{F}$ and $0.10\,\mu\mathrm{F}$ are connected (a) in series and (b) in parallel?
   [(a) $0.06\,\mu\mathrm{F}$ (b) $0.25\,\mu\mathrm{F}$]

4. Two $6\,\mu\mathrm{F}$ capacitors are connected in series with one having a capacitance of $12\,\mu\mathrm{F}$. Find the total equivalent circuit capacitance. What

capacitance must be added in series to obtain a capacitance of $1.2\,\mu F$?

[$2.4\,\mu F$, $2.4\,\mu F$]

5.  Determine the equivalent capacitance when the following capacitors are connected (a) in parallel and (b) in series:

(i)   $2\,\mu F$, $4\,\mu F$ and $8\,\mu F$

(ii)   $0.02\,\mu F$, $0.05\,\mu F$ and $0.10\,\mu F$

(iii)   $50\,pF$ and $450\,pF$

(iv)   $0.01\,\mu F$ and $200\,pF$

[(a)   (i)   $14\,\mu F$   (ii)   $0.17\,\mu F$

(iii)   $500\,pF$   (iv)   $0.0102\,\mu F$

(b)   (i)   $1.143\,\mu F$   (ii)   $0.0125\,\mu F$

(iii)   $45\,pF$   (iv)   $196.1\,pF$]

6.  For the arrangement shown in Fig. 6.11 find (a) the equivalent circuit capacitance and (b) the voltage across a $4.5\,\mu F$ capacitor.

[(a) $1.2\,\mu F$ (b) $100\,V$]

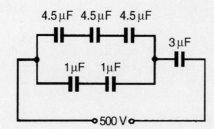

**Figure 6.11**

7.  Three $12\,\mu F$ capacitors are connected in series across a $750\,V$ supply. Calculate (a) the equivalent capacitance, (b) the charge on each capacitor, and (c) the p.d. across each capacitor.   [(a) $4\,\mu F$ (b) $3\,mC$ (c) $250\,V$]

8.  If two capacitors having capacitances of $3\,\mu F$ and $5\,\mu F$ respectively are connected in series across a $240\,V$ supply, determine (a) the p.d. across each capacitor and (b) the charge on each capacitor.

[(a) $150\,V$, $90\,V$ (b) $0.45\,mC$ on each]

9.  In Fig. 6.12 capacitors $P$, $Q$ and $R$ are identical and the total equivalent capacitance of the circuit is $3\,\mu F$. Determine the values of $P$, $Q$ and $R$   [$4.2\,\mu F$ each]

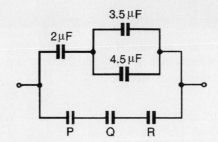

**Figure 6.12**

10.  Capacitances of $4\,\mu F$, $8\,\mu F$ and $16\,\mu F$ are connected in parallel across a $200\,V$ supply. Determine (a) the equivalent capacitance, (b) the total charge, and (c) the charge on each capacitor.   [(a) $28\,\mu F$ (b) $5.6\,mC$ (c) $0.8\,mC$, $1.6\,mC$, $3.2\,mC$]

11.  A circuit consists of two capacitors $P$ and $Q$ in parallel, connected in series with another capacitor $R$. The capacitances of $P$, $Q$ and $R$ are $4\,\mu F$, $12\,\mu F$ and $8\,\mu F$ respectively. When the circuit is connected across a $300\,V$ d.c. supply find (a) the total capacitance of the circuit, (b) the p.d. across each capacitor, and (c) the charge on each capacitor.

[(a) $5.33\,\mu F$ (b) $100\,V$ across $P$, $100\,V$ across $Q$, $200\,V$ across $R$ (c) $0.4\,mC$ on $P$, $1.2\,mC$ on $Q$, $1.6\,mC$ on $R$]

12.  For the circuit shown in Fig. 6.13, determine (a) the total circuit capacitance, (b) the total energy in the circuit, and (c) the charges in the capacitors shown as $C_1$ and $C_2$

[(a) $0.857\,\mu F$ (b) $1.071\,mJ$ (c) $42.85\,\mu C$ on each]

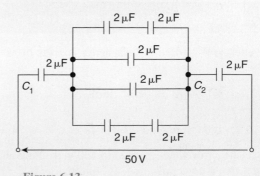

**Figure 6.13**

## 6.10 Dielectric strength

The maximum amount of field strength that a dielectric can withstand is called the dielectric strength of the material. Dielectric strength,

$$E_m = \frac{V_m}{d}$$

> **Problem 15.** A capacitor is to be constructed so that its capacitance is $0.2\,\mu F$ and to take a p.d. of $1.25\,kV$ across its terminals. The dielectric is to be mica which, after allowing a safety factor of 2, has a dielectric strength of $50\,MV/m$. Find (a) the thickness of the mica needed, and (b) the area of a plate assuming a two-plate construction. (Assume $\varepsilon_r$ for mica to be 6.)

(a) Dielectric strength,

$$E = \frac{V}{d}$$

i.e. $\quad d = \dfrac{V}{E} = \dfrac{1.25 \times 10^3}{50 \times 10^6}$ m

$$= 0.025\,mm$$

(b) Capacitance,

$$C = \frac{\varepsilon_0 \varepsilon_r A}{d}$$

hence

$$\text{area } A = \frac{Cd}{\varepsilon_0 \varepsilon_r} = \frac{0.2 \times 10^{-6} \times 0.025 \times 10^{-3}}{8.85 \times 10^{-12} \times 6} \text{ m}^2$$

$$= 0.09416\,m^2 = \mathbf{941.6\,cm^2}$$

## 6.11 Energy stored in capacitors

The energy, $W$, stored by a capacitor is given by

$$W = \frac{1}{2}CV^2 \text{ joules}$$

> **Problem 16.** (a) Determine the energy stored in a $3\,\mu F$ capacitor when charged to $400\,V$. (b) Find also the average power developed if this energy is dissipated in a time of $10\,\mu s$.

(a) **Energy stored**

$$W = \frac{1}{2}CV^2 \text{ joules} = \frac{1}{2} \times 3 \times 10^{-6} \times 400^2$$

$$= \frac{3}{2} \times 16 \times 10^{-2} = \mathbf{0.24\,J}$$

(b) **Power** $= \dfrac{\text{energy}}{\text{time}} = \dfrac{0.24}{10 \times 10^{-6}} \text{ W} = \mathbf{24\,kW}$

> **Problem 17.** A $12\,\mu F$ capacitor is required to store $4\,J$ of energy. Find the p.d. to which the capacitor must be charged.

Energy stored

$$W = \frac{1}{2}CV^2$$

hence $\quad V^2 = \dfrac{2W}{C}$

and $\quad$ p.d. $V = \sqrt{\dfrac{2W}{C}} = \sqrt{\dfrac{2 \times 4}{12 \times 10^{-6}}}$

$$= \sqrt{\dfrac{2 \times 10^6}{3}} = \mathbf{816.5\,V}$$

> **Problem 18.** A capacitor is charged with $10\,mC$. If the energy stored is $1.2\,J$ find (a) the voltage and (b) the capacitance.

Energy stored $W = \frac{1}{2}CV^2$ and $C = Q/V$. Hence

$$W = \frac{1}{2}\left(\frac{Q}{V}\right)V^2$$

$$= \frac{1}{2}QV$$

from which $\quad V = \dfrac{2W}{Q}$

$$Q = 10\,mC = 10 \times 10^{-3}\,C$$

and $\quad W = 1.2\,J$

(a) Voltage

$$V = \frac{2W}{Q} = \frac{2 \times 1.2}{10 \times 10^{-3}} = \mathbf{0.24\,kV} \text{ or } \mathbf{240\,V}$$

(b) Capacitance

$$C = \frac{Q}{V} = \frac{10 \times 10^{-3}}{240}\,F = \frac{10 \times 10^6}{240 \times 10^3}\,\mu F$$

$$= \mathbf{41.67\,\mu F}$$

**Now try the following exercise**

## 6.12    Practical types of capacitor

Practical types of capacitor are characterized by the material used for their dielectric. The main types include: variable air, mica, paper, ceramic, plastic, titanium oxide and electrolytic.

1.  **Variable air capacitors**. These usually consist of two sets of metal plates (such as aluminium), one fixed, the other variable. The set of moving plates rotate on a spindle as shown by the end view of Fig. 6.14.

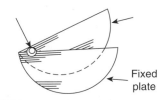

**Figure 6.14**

As the moving plates are rotated through half a revolution, the meshing, and therefore the capacitance, varies from a minimum to a maximum value. Variable air capacitors are used in radio and electronic circuits where very low losses are required, or where a variable capacitance is needed. The maximum value of such capacitors is between 500 pF and 1000 pF.

2.  **Mica capacitors**. A typical older type construction is shown in Fig. 6.15.

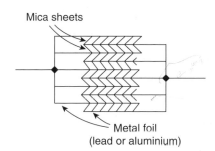

**Figure 6.15**

Usually the whole capacitor is impregnated with wax and placed in a bakelite case. Mica is easily obtained in thin sheets and is a good insulator.
However, mica is expensive and is not used in capacitors above about 0.2 µF. A modified form of mica capacitor is the silvered mica type. The mica is coated on both sides with a thin layer of silver which forms the plates. Capacitance is stable and less likely to change with age. Such capacitors have a constant capacitance with change of temperature, a high working voltage rating and a long service life and are used in high frequency circuits with fixed values of capacitance up to about 1000 pF.

3.  **Paper capacitors**. A typical paper capacitor is shown in Fig. 6.16 where the length of the roll corresponds to the capacitance required.

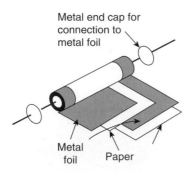

Figure 6.16

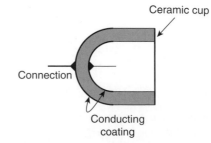

Figure 6.18

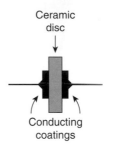

Figure 6.19

The whole is usually impregnated with oil or wax to exclude moisture, and then placed in a plastic or aluminium container for protection. Paper capacitors are made in various working voltages up to about 150 kV and are used where loss is not very important. The maximum value of this type of capacitor is between 500 pF and 10 μF. Disadvantages of paper capacitors include variation in capacitance with temperature change and a shorter service life than most other types of capacitor.

4. **Ceramic capacitors**. These are made in various forms, each type of construction depending on the value of capacitance required. For high values, a tube of ceramic material is used as shown in the cross-section of Fig. 6.17. For smaller values the cup construction is used as shown in Fig. 6.18, and for still smaller values the disc construction shown in Fig. 6.19 is used. Certain ceramic materials have a very high permittivity and this enables capacitors of high capacitance to be made which are of small physical size with a high working voltage rating. Ceramic capacitors are available in the range 1 pF to 0.1 μF and may be used in high frequency electronic circuits subject to a wide range of temperatures.

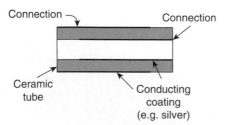

Figure 6.17

5. **Plastic capacitors**. Some plastic materials such as polystyrene and Teflon can be used as dielectrics. Construction is similar to the paper capacitor but using a plastic film instead of paper. Plastic capacitors operate well under conditions of high temperature, provide a precise value of capacitance, a very long service life and high reliability.

6. **Titanium oxide capacitors** have a very high capacitance with a small physical size when used at a low temperature.

7. **Electrolytic capacitors**. Construction is similar to the paper capacitor with aluminium foil used for the plates and with a thick absorbent material, such as paper, impregnated with an electrolyte (ammonium borate), separating the plates. The finished capacitor is usually assembled in an aluminium container and hermetically sealed. Its operation depends on the formation of a thin aluminium oxide layer on the positive plate by electrolytic action when a suitable direct potential is maintained between the plates. This oxide layer is very thin and forms the dielectric. (The absorbent paper between the plates is a conductor and does not act as a dielectric.) Such capacitors **must always be used on d.c.** and must be connected with the correct polarity; if this is not done the capacitor will be destroyed since the oxide layer will be destroyed. Electrolytic capacitors are manufactured with working voltage from 6 V to 600 V, although accuracy is generally not very high. These capacitors possess a much larger capacitance than other types of capacitors of similar dimensions due to the oxide film being only a few microns thick. The fact that they can be used only on d.c. supplies limit their usefulness.

## 6.13 Discharging capacitors

When a capacitor has been disconnected from the supply it may still be charged and it may retain this charge for some considerable time. Thus precautions must be taken to ensure that the capacitor is automatically discharged after the supply is switched off. This is done by connecting a high value resistor across the capacitor terminals.

**Now try the following exercises**

**Exercise 31   Short answer questions on capacitors and capacitance**

1. What is a capacitor?

2. State five practical applications of capacitors

3. Explain the term 'electrostatics'

4. Complete the statements:
   Like charges ......; unlike charges ......

5. How can an 'electric field' be established between two parallel metal plates?

6. What is capacitance?

7. State the unit of capacitance

8. Complete the statement:
   $$\text{Capacitance} = \frac{\cdots\cdots}{\cdots\cdots}$$

9. Complete the statements:
   (a)  $1\,\mu F = \ldots F$    (b)  $1\,pF = \ldots F$

10. Complete the statement:
    $$\text{Electric field strength } E = \frac{\cdots\cdots}{\cdots\cdots}$$

11. Complete the statement:
    $$\text{Electric flux density } D = \frac{\cdots\cdots}{\cdots\cdots}$$

12. Draw the electrical circuit diagram symbol for a capacitor

13. Name two practical examples where capacitance is present, although undesirable

14. The insulating material separating the plates of a capacitor is called the ......

15. 10 volts applied to a capacitor results in a charge of 5 coulombs. What is the capacitance of the capacitor?

16. Three $3\,\mu F$ capacitors are connected in parallel. The equivalent capacitance is. ...

17. Three $3\,\mu F$ capacitors are connected in series. The equivalent capacitance is. ...

18. State a disadvantage of series-connected capacitors

19. Name three factors upon which capacitance depends

20. What does 'relative permittivity' mean?

21. Define 'permittivity of free space'

22. What is meant by the 'dielectric strength' of a material?

23. State the formula used to determine the energy stored by a capacitor

24. Name five types of capacitor commonly used

25. Sketch a typical rolled paper capacitor

26. Explain briefly the construction of a variable air capacitor

27. State three advantages and one disadvantage of mica capacitors

28. Name two disadvantages of paper capacitors

29. Between what values of capacitance are ceramic capacitors normally available

30. What main advantages do plastic capacitors possess?

31. Explain briefly the construction of an electrolytic capacitor

32. What is the main disadvantage of electrolytic capacitors?

33. Name an important advantage of electrolytic capacitors

34. What safety precautions should be taken when a capacitor is disconnected from a supply?

**Exercise 32** **Multi-choice questions on capacitors and capacitance**

**(Answers on page 420)**

1. Electrostatics is a branch of electricity concerned with
   (a) energy flowing across a gap between conductors
   (b) charges at rest
   (c) charges in motion
   (d) energy in the form of charges

2. The capacitance of a capacitor is the ratio
   (a) charge to p.d. between plates
   (b) p.d. between plates to plate spacing
   (c) p.d. between plates to thickness of dielectric
   (d) p.d. between plates to charge

3. The p.d. across a $10\,\mu F$ capacitor to charge it with $10\,mC$ is
   (a) $10\,V$          (b) $1\,kV$
   (c) $1\,V$           (d) $10\,V$

4. The charge on a $10\,pF$ capacitor when the voltage applied to it is $10\,kV$ is
   (a) $100\,\mu C$     (b) $0.1\,C$
   (c) $0.1\,\mu C$     (d) $0.01\,\mu C$

5. Four $2\,\mu F$ capacitors are connected in parallel. The equivalent capacitance is
   (a) $8\,\mu F$       (b) $0.5\,\mu F$
   (c) $2\,\mu F$       (d) $6\,\mu F$

6. Four $2\,\mu F$ capacitors are connected in series. The equivalent capacitance is
   (a) $8\,\mu F$       (b) $0.5\,\mu F$
   (c) $2\,\mu F$       (d) $6\,\mu F$

7. State which of the following is false.
   The capacitance of a capacitor
   (a) is proportional to the cross-sectional area of the plates
   (b) is proportional to the distance between the plates
   (c) depends on the number of plates
   (d) is proportional to the relative permittivity of the dielectric

8. Which of the following statement is false?
   (a) An air capacitor is normally a variable type
   (b) A paper capacitor generally has a shorter service life than most other types of capacitor
   (c) An electrolytic capacitor must be used only on a.c. supplies
   (d) Plastic capacitors generally operate satisfactorily under conditions of high temperature

9. The energy stored in a $10\,\mu F$ capacitor when charged to $500\,V$ is
   (a) $1.25\,mJ$       (b) $0.025\,\mu J$
   (c) $1.25\,J$        (d) $1.25\,C$

10. The capacitance of a variable air capacitor is at maximum when
    (a) the movable plates half overlap the fixed plates
    (b) the movable plates are most widely separated from the fixed plates
    (c) both sets of plates are exactly meshed
    (d) the movable plates are closer to one side of the fixed plate than to the other

11. When a voltage of $1\,kV$ is applied to a capacitor, the charge on the capacitor is $500\,nC$. The capacitance of the capacitor is:
    (a) $2 \times 10^9\,F$   (b) $0.5\,pF$
    (c) $0.5\,mF$            (d) $0.5\,nF$

# Chapter 7

# Magnetic circuits

At the end of this chapter you should be able to:

- appreciate some applications of magnets
- describe the magnetic field around a permanent magnet
- state the laws of magnetic attraction and repulsion for two magnets in close proximity
- define magnetic flux, $\Phi$, and magnetic flux density, $B$, and state their units
- perform simple calculations involving $B = \Phi/A$
- define magnetomotive force, $F_m$, and magnetic field strength, $H$, and state their units
- perform simple calculations involving $F_m = NI$ and $H = NI/l$
- define permeability, distinguishing between $\mu_0$, $\mu_r$ and $\mu$
- understand the B–H curves for different magnetic materials
- appreciate typical values of $\mu_r$
- perform calculations involving $B = \mu_0\mu_r H$
- define reluctance, $S$, and state its units
- perform calculations involving

$$S = \frac{\text{m.m.f.}}{\Phi} = \frac{l}{\mu_0\mu_r A}$$

- perform calculations on composite series magnetic circuits
- compare electrical and magnetic quantities
- appreciate how a hysteresis loop is obtained and that hysteresis loss is proportional to its area

## 7.1 Introduction to magnetism and magnetic circuits

The study of magnetism began in the thirteenth century with many eminent scientists and physicists such as William Gilbert, Hans Christian Oersted, Michael Faraday, James Maxwell, André Ampère and Wilhelm Weber all having some input on the subject since. The association between electricity and magnetism is a fairly recent finding in comparison with the very first understanding of basic magnetism.

Today, magnets have **many varied practical applications**. For example, they are used in motors and generators, telephones, relays, loudspeakers, computer hard drives and floppy disks, anti-lock brakes, cameras, fishing reels, electronic ignition systems, keyboards, t.v. and radio components and in transmission equipment.

DOI: 10.1016/B978-0-08-089056-2.00007-3

The full theory of magnetism is one of the most complex of subjects; this chapter provides an introduction to the topic.

## 7.2 Magnetic fields

A **permanent magnet** is a piece of ferromagnetic material (such as iron, nickel or cobalt) which has properties of attracting other pieces of these materials. A permanent magnet will position itself in a north and south direction when freely suspended. The north-seeking end of the magnet is called the **north pole, N**, and the south-seeking end the **south pole, S**.

The area around a magnet is called the **magnetic field** and it is in this area that the effects of the **magnetic force** produced by the magnet can be detected. A magnetic field cannot be seen, felt, smelt or heard and therefore is difficult to represent. Michael Faraday suggested that the magnetic field could be represented pictorially, by imagining the field to consist of **lines of magnetic flux**, which enables investigation of the distribution and density of the field to be carried out.

The distribution of a magnetic field can be investigated by using some iron filings. A bar magnet is placed on a flat surface covered by, say, cardboard, upon which is sprinkled some iron filings. If the cardboard is gently tapped the filings will assume a pattern similar to that shown in Fig. 7.1. If a number of magnets of different strength are used, it is found that the stronger the field the closer are the lines of magnetic flux and vice versa. Thus a magnetic field has the property of exerting a force, demonstrated in this case by causing the iron filings to move into the pattern shown. The strength of the magnetic field decreases as we move away from the magnet. It should be realised, of course, that the magnetic field is three dimensional in its effect, and not acting in one plane as appears to be the case in this experiment.

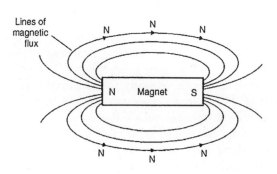

**Figure 7.1**

If a compass is placed in the magnetic field in various positions, the direction of the lines of flux may be determined by noting the direction of the compass pointer. The direction of a magnetic field at any point is taken as that in which the north-seeking pole of a compass needle points when suspended in the field. The direction of a line of flux is from the north pole to the south pole on the outside of the magnet and is then assumed to continue through the magnet back to the point at which it emerged at the north pole. Thus such lines of flux always form complete closed loops or paths, they never intersect and always have a definite direction.

The laws of magnetic attraction and repulsion can be demonstrated by using two bar magnets. In Fig. 7.2(a), **with unlike poles adjacent, attraction takes place**. Lines of flux are imagined to contract and the magnets try to pull together. The magnetic field is strongest in between the two magnets, shown by the lines of flux being close together. In Fig. 7.2(b), **with similar poles adjacent (i.e. two north poles), repulsion occurs**, i.e. the two north poles try to push each other apart, since magnetic flux lines running side by side in the same direction repel.

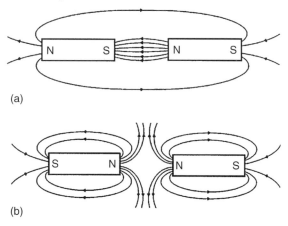

(a)

(b)

**Figure 7.2**

## 7.3 Magnetic flux and flux density

**Magnetic flux** is the amount of magnetic field (or the number of lines of force) produced by a magnetic source. The symbol for magnetic flux is $\Phi$ (Greek letter 'phi'). The unit of magnetic flux is the **weber, Wb**.

**Magnetic flux density** is the amount of flux passing through a defined area that is perpendicular to the direction of the flux:

$$\text{Magnetic flux density} = \frac{\text{magnetic flux}}{\text{area}}$$

The symbol for magnetic flux density is $B$. The unit of magnetic flux density is the tesla, $T$, where $1\,T = 1\,\text{Wb/m}^2$. Hence

$$B = \frac{\Phi}{A} \text{ tesla}$$

where $A(\text{m}^2)$ is the area

> **Problem 1.** A magnetic pole face has a rectangular section having dimensions 200 mm by 100 mm. If the total flux emerging from the pole is $150\,\mu\text{Wb}$, calculate the flux density.

Flux $\Phi = 150\,\mu\text{Wb} - 150 \times 10^{-6}\,\text{Wb}$
Cross-sectional area $A = 200 \times 100 = 20\,000\,\text{mm}^2$
$\qquad\qquad\qquad = 20\,000 \times 10^{-6}\,\text{m}^2$.

$$\text{Flux density, } B = \frac{\Phi}{A} = \frac{150 \times 10^{-6}}{20\,000 \times 10^{-6}}$$

$$= \textbf{0.0075\,T or 7.5\,mT}$$

> **Problem 2.** The maximum working flux density of a lifting electromagnet is 1.8 T and the effective area of a pole face is circular in cross-section. If the total magnetic flux produced is 353 mWb, determine the radius of the pole face.

Flux density $B = 1.8\,\text{T}$ and
flux $\Phi = 353\,\text{mWb} = 353 \times 10^{-3}\,\text{Wb}$.
Since $B = \Phi/A$, cross-sectional area $A = \Phi/B$

i.e. $\qquad A = \dfrac{353 \times 10^{-3}}{1.8}\,\text{m}^2 = 0.1961\,\text{m}^2$

The pole face is circular, hence area$= \pi r^2$, where $r$ is the radius. Hence $\pi r^2 = 0.1961$ from which, $r^2 = 0.1961/\pi$ and radius $r = \sqrt{(0.1961/\pi)} = 0.250\,\text{m}$
i.e. **the radius of the pole face is 250 mm**.

## 7.4 Magnetomotive force and magnetic field strength

**Magnetomotive force (m.m.f.)** is the cause of the existence of a magnetic flux in a magnetic circuit,

$$\textbf{m.m.f. } F_\text{m} = NI \textbf{ amperes}$$

where $N$ is the number of conductors (or turns) and $I$ is the current in amperes. The unit of m.m.f is sometimes expressed as 'ampere-turns'. However since 'turns' have no dimensions, the S.I. unit of m.m.f. is the ampere.

**Magnetic field strength** (or **magnetising force**),

$$H = \frac{NI}{l} \textbf{ ampere per metre}$$

where $l$ is the mean length of the flux path in metres. Thus

$$\textbf{m.m.f.} = NI = Hl \textbf{ amperes}$$

> **Problem 3.** A magnetising force of 8000 A/m is applied to a circular magnetic circuit of mean diameter 30 cm by passing a current through a coil wound on the circuit. If the coil is uniformly wound around the circuit and has 750 turns, find the current in the coil.

$H = 8000\,\text{A/m}, l = \pi d = \pi \times 30 \times 10^{-2}\,\text{m}$ and $N = 750$ turns. Since $H = NI/l$, then

$$I = \frac{Hl}{N} = \frac{8000 \times \pi \times 30 \times 10^{-2}}{750}$$

Thus, **current $I = 10.05\,\text{A}$**

---

**Now try the following exercise**

> **Exercise 33    Further problems on magnetic circuits**
>
> 1. What is the flux density in a magnetic field of cross-sectional area $20\,\text{cm}^2$ having a flux of 3 mWb? [1.5 T]
>
> 2. Determine the total flux emerging from a magnetic pole face having dimensions 5 cm by 6 cm, if the flux density is 0.9 T [2.7 mWb]
>
> 3. The maximum working flux density of a lifting electromagnet is 1.9 T and the effective area of a pole face is circular in cross-section. If the total magnetic flux produced is 611 mWb determine the radius of the pole face. [32 cm]
>
> 4. A current of 5 A is passed through a 1000-turn coil wound on a circular magnetic circuit of radius 120 mm. Calculate (a) the magnetomotive force, and (b) the magnetic field strength. [(a) 5000 A (b) 6631 A/m]
>
> 5. An electromagnet of square cross-section produces a flux density of 0.45 T. If the magnetic

flux is $720\,\mu\text{Wb}$ find the dimensions of the electromagnet cross-section.

[4 cm by 4 cm]

6. Find the magnetic field strength applied to a magnetic circuit of mean length 50 cm when a coil of 400 turns is applied to it carrying a current of 1.2 A                  [960 A/m]

7. A solenoid 20 cm long is wound with 500 turns of wire. Find the current required to establish a magnetising force of 2500 A/m inside the solenoid.                  [1 A]

8. A magnetic field strength of 5000 A/m is applied to a circular magnetic circuit of mean diameter 250 mm. If the coil has 500 turns find the current in the coil.                  [7.85 A]

## 7.5  Permeability and B–H curves

For air, or any non-magnetic medium, the ratio of magnetic flux density to magnetising force is a constant, i.e. $B/H =$ a constant. This constant is $\mu_0$, the **permeability of free space** (or the magnetic space constant) and is equal to $4\pi \times 10^{-7}$ H/m, i.e. **for air, or any non-magnetic medium**, the ratio

$$\frac{B}{H} = \mu_0$$

(Although all non-magnetic materials, including air, exhibit slight magnetic properties, these can effectively be neglected.)
**For all media other than free space,**

$$\frac{B}{H} = \mu_0 \mu_r$$

where $\mu_r$ is the relative permeability, and is defined as

$$\mu_r = \frac{\text{flux density in material}}{\text{flux density in a vacuum}}$$

$\mu_r$ varies with the type of magnetic material and, since it is a ratio of flux densities, it has no unit. From its definition, $\mu_r$ for a vacuum is 1.
$\mu_0 \mu_r = \mu$, called the **absolute permeability**
By plotting measured values of flux density $B$ against magnetic field strength $H$, a **magnetisation curve** (or **B–H curve**) is produced. For non-magnetic materials this is a straight line. Typical curves for four magnetic materials are shown in Fig. 7.3

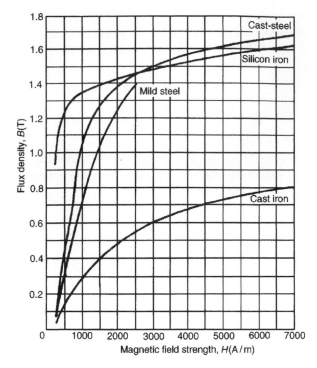

**Figure 7.3**

The **relative permeability** of a ferromagnetic material is proportional to the slope of the B–H curve and thus varies with the magnetic field strength. The approximate range of values of relative permeability $\mu_r$ for some common magnetic materials are:

Cast iron      $\mu_r = 100–250$
Mild steel     $\mu_r = 200–800$
Silicon iron   $\mu_r = 1000–5000$
Cast steel     $\mu_r = 300–900$
Mumetal        $\mu_r = 200–5000$
Stalloy        $\mu_r = 500–6000$

**Problem 4.** A flux density of 1.2 T is produced in a piece of cast steel by a magnetising force of 1250 A/m. Find the relative permeability of the steel under these conditions.

For a magnetic material: $B = \mu_0 \mu_r H$

i.e.      $\mu_r = \dfrac{B}{\mu_0 H} = \dfrac{1.2}{(4\pi \times 10^{-7})(1250)} = \mathbf{764}$

**Problem 5.** Determine the magnetic field strength and the m.m.f. required to produce a flux density of 0.25 T in an air gap of length 12 mm.

For air: $B = \mu_0 H$ (since $\mu_r = 1$)

**Magnetic field strength,**

$$H = \frac{B}{\mu_0} = \frac{0.25}{4\pi \times 10^{-7}} = 198\,940\,\text{A/m}$$

**m.m.f.** $= Hl = 198\,940 \times 12 \times 10^{-3} = 2387\,\text{A}$

**Problem 6.** A coil of 300 turns is wound uniformly on a ring of non-magnetic material. The ring has a mean circumference of 40 cm and a uniform cross-sectional area of 4 cm². If the current in the coil is 5 A, calculate (a) the magnetic field strength, (b) the flux density and (c) the total magnetic flux in the ring.

(a) Magnetic field strength

$$H = \frac{NI}{l} = \frac{300 \times 5}{40 \times 10^{-2}}$$
$$= 3750\,\text{A/m}$$

(b) For a non-magnetic material $\mu_r = 1$, thus flux density $B = \mu_0 H$

i.e $B = 4\pi \times 10^{-7} \times 3750$
$$= 4.712\,\text{mT}$$

(c) Flux $\Phi = BA = (4.712 \times 10^{-3})(4 \times 10^{-4})$
$$= 1.885\,\mu\text{Wb}$$

**Problem 7.** An iron ring of mean diameter 10 cm is uniformly wound with 2000 turns of wire. When a current of 0.25 A is passed through the coil a flux density of 0.4 T is set up in the iron. Find (a) the magnetising force and (b) the relative permeability of the iron under these conditions.

$l = \pi d = \pi \times 10\,\text{cm} = \pi \times 10 \times 10^{-2}\,\text{m}$,
$N = 2000$ turns, $I = 0.25\,\text{A}$ and $B = 0.4\,\text{T}$

(a) $H = \dfrac{NI}{l} = \dfrac{2000 \times 0.25}{\pi \times 10 \times 10^{-2}}$
$$= 1592\,\text{A/m}$$

(b) $B = \mu_0 \mu_r H$, hence
$$\mu_r = \frac{B}{\mu_0 H} = \frac{0.4}{(4\pi \times 10^{-7})(1592)} = 200$$

**Problem 8.** A uniform ring of cast iron has a cross-sectional area of 10 cm² and a mean circumference of 20 cm. Determine the m.m.f. necessary to produce a flux of 0.3 mWb in the ring. The magnetisation curve for cast iron is shown on page 80.

$A = 10\,\text{cm}^2 = 10 \times 10^{-4}\,\text{m}^2$, $l = 20\,\text{cm} = 0.2\,\text{m}$ and $\Phi = 0.3 \times 10^{-3}\,\text{Wb}$.

Flux density $B = \dfrac{\Phi}{A} = \dfrac{0.3 \times 10^{-3}}{10 \times 10^{-4}} = 0.3\,\text{T}$

From the magnetisation curve for cast iron on page 80, when $B = 0.3\,\text{T}$, $H = 1000\,\text{A/m}$, hence
**m.m.f.** $= Hl = 1000 \times 0.2 = 200\,\text{A}$
A tabular method could have been used in this problem. Such a solution is shown below in Table 7.1.

**Problem 9.** From the magnetisation curve for cast iron, shown on page 80, derive the curve of $\mu_r$ against $H$.

$B = \mu_0 \mu_r H$, hence
$$\mu_r = \frac{B}{\mu_0 H} = \frac{1}{\mu_0} \times \frac{B}{H}$$
$$= \frac{10^7}{4\pi} \times \frac{B}{H}$$

A number of co-ordinates are selected from the B–H curve and $\mu_r$ is calculated for each as shown in Table 7.2. $\mu_r$ is plotted against $H$ as shown in Fig. 7.4. The curve demonstrates the change that occurs in the relative permeability as the magnetising force increases.

**Table 7.1**

| Part of circuit | Material | $\Phi$(Wb) | $A(m^2)$ | $B = \dfrac{\Phi}{A}$(T) | $H$ from graph | $l(m)$ | m.m.f.$=$ $Hl(A)$ |
|---|---|---|---|---|---|---|---|
| Ring | Cast iron | $0.3 \times 10^{-3}$ | $10 \times 10^{-4}$ | 0.3 | 1000 | 0.2 | 200 |

**Table 7.2**

| $B\,(T)$ | 0.04 | 0.13 | 0.17 | 0.30 | 0.41 | 0.49 | 0.60 | 0.68 | 0.73 | 0.76 | 0.79 |
|---|---|---|---|---|---|---|---|---|---|---|---|
| $H\,(A/m)$ | 200 | 400 | 500 | 1000 | 1500 | 2000 | 3000 | 4000 | 5000 | 6000 | 7000 |
| $\mu_r = \dfrac{10^7}{4\pi} \times \dfrac{B}{H}$ | 159 | 259 | 271 | 239 | 218 | 195 | 159 | 135 | 116 | 101 | 90 |

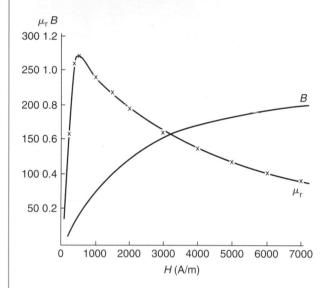

**Figure 7.4**

**Now try the following exercise**

**Exercise 34    Further problems on magnetic circuits**

(Where appropriate, assume $\mu_0 = 4\pi \times 10^{-7}\,H/m$)

1. Find the magnetic field strength and the magnetomotive force needed to produce a flux density of 0.33 T in an air gap of length 15 mm.
   [(a) 262 600 A/m (b) 3939 A]

2. An air gap between two pole pieces is 20 mm in length and the area of the flux path across the gap is 5 cm$^2$. If the flux required in the air gap is 0.75 mWb find the m.m.f. necessary.
   [23 870 A]

3. (a) Determine the flux density produced in an air-cored solenoid due to a uniform magnetic field strength of 8000 A/m. (b) Iron having a relative permeability of 150 at 8000 A/m is inserted into the solenoid of part (a). Find the flux density now in the solenoid.
   [(a) 10.05 mT (b) 1.508 T]

4. Find the relative permeability of a material if the absolute permeability is $4.084 \times 10^{-4}\,H/m$.
   [325]

5. Find the relative permeability of a piece of silicon iron if a flux density of 1.3 T is produced by a magnetic field strength of 700 A/m.
   [1478]

6. A steel ring of mean diameter 120 mm is uniformly wound with 1500 turns of wire. When a current of 0.30 A is passed through the coil a flux density of 1.5 T is set up in the steel. Find the relative permeability of the steel under these conditions.
   [1000]

7. A uniform ring of cast steel has a cross-sectional area of 5 cm$^2$ and a mean circumference of 15 cm. Find the current required in a coil of 1200 turns wound on the ring to produce a flux of 0.8 mWb. (Use the magnetisation curve for cast steel shown on page 80.)
   [0.60 A]

8. (a) A uniform mild steel ring has a diameter of 50 mm and a cross-sectional area of 1 cm$^2$. Determine the m.m.f. necessary to produce a flux of 50 μWb in the ring. (Use the B–H curve for mild steel shown on page 80.) (b) If a coil of 440 turns is wound uniformly around the ring in Part (a) what current would be required to produce the flux?
   [(a) 110 A (b) 0.25 A]

9. From the magnetisation curve for mild steel shown on page 80, derive the curve of relative permeability against magnetic field strength. From your graph determine (a) the value of $\mu_r$ when the magnetic field strength is 1200 A/m, and (b) the value of the magnetic field strength when $\mu_r$ is 500.
   [(a) 590–600 (b) 2000]

## 7.6 Reluctance

Reluctance $S$ (or $R_M$) is the 'magnetic resistance' of a magnetic circuit to the presence of magnetic flux. **Reluctance**,

$$S = \frac{F_M}{\Phi} = \frac{NI}{\Phi} = \frac{Hl}{BA} = \frac{l}{(B/H)A} = \frac{l}{\mu_0 \mu_r A}$$

The unit of reluctance is $1/H$ (or $H^{-1}$) or A/Wb.

**Ferromagnetic materials** have a low reluctance and can be used as **magnetic screens** to prevent magnetic fields affecting materials within the screen.

---

**Problem 10.** Determine the reluctance of a piece of mumetal of length 150 mm and cross-sectional area 1800 mm$^2$ when the relative permeability is 4000. Find also the absolute permeability of the mumetal.

**Reluctance**,

$$S = \frac{l}{\mu_0 \mu_r A}$$

$$= \frac{150 \times 10^{-3}}{(4\pi \times 10^{-7})(4000)(1800 \times 10^{-6})}$$

$$= \mathbf{16\,580/H}$$

**Absolute permeability**,

$$\mu = \mu_0 \mu_r = (4\pi \times 10^{-7})(4000)$$

$$= \mathbf{5.027 \times 10^{-3}\,H/m}$$

---

**Problem 11.** A mild steel ring has a radius of 50 mm and a cross-sectional area of 400 mm$^2$. A current of 0.5 A flows in a coil wound uniformly around the ring and the flux produced is 0.1 mWb. If the relative permeability at this value of current is 200 find (a) the reluctance of the mild steel and (b) the number of turns on the coil.

$l = 2\pi r = 2 \times \pi \times 50 \times 10^{-3}$ m,     $A = 400 \times 10^{-6}$ m$^2$,
$I = 0.5$ A, $\Phi = 0.1 \times 10^{-3}$ Wb and $\mu_r = 200$

(a) **Reluctance**,

$$S = \frac{l}{\mu_0 \mu_r A}$$

$$= \frac{2 \times \pi \times 50 \times 10^{-3}}{(4\pi \times 10^{-7})(200)(400 \times 10^{-6})}$$

$$= \mathbf{3.125 \times 10^6/H}$$

(b) $S = \dfrac{\text{m.m.f.}}{\Phi}$ from which

   m.m.f. $= S\Phi$   i.e.   $NI = S\Phi$

   Hence, number of terms

$$N = \frac{S\Phi}{I} = \frac{3.125 \times 10^6 \times 0.1 \times 10^{-3}}{0.5}$$

$$= \mathbf{625\,turns}$$

---

**Now try the following exercise**

---

**Exercise 35  Further problems on magnetic circuits**

(Where appropriate, assume $\mu_0 = 4\pi \times 10^{-7}$ H/m)

1. Part of a magnetic circuit is made from steel of length 120 mm, cross-sectional area 15 cm$^2$ and relative permeability 800. Calculate (a) the reluctance and (b) the absolute permeability of the steel.

   [(a) 79 580 /H (b) 1 mH/m]

2. A mild steel closed magnetic circuit has a mean length of 75 mm and a cross-sectional area of 320.2 mm$^2$. A current of 0.40 A flows in a coil wound uniformly around the circuit and the flux produced is 200 μWb. If the relative permeability of the steel at this value of current is 400 find (a) the reluctance of the material and (b) the number of turns of the coil.

   [(a) 466 000 /H (b) 233]

---

## 7.7 Composite series magnetic circuits

For a series magnetic circuit having $n$ parts, the **total reluctance $S$** is given by: $S = S_1 + S_2 + \cdots + S_n$ (This is similar to resistors connected in series in an electrical circuit).

---

**Problem 12.** A closed magnetic circuit of cast steel contains a 6 cm long path of cross-sectional area 1 cm$^2$ and a 2 cm path of cross-sectional area 0.5 cm$^2$. A coil of 200 turns is wound around the 6 cm length of the circuit and a current of 0.4 A flows. Determine the flux density in the 2 cm path, if the relative permeability of the cast steel is 750.

## For the 6 cm long path:

Reluctance $S_1 = \dfrac{l_1}{\mu_0 \mu_r A_1}$

$= \dfrac{6 \times 10^{-2}}{(4\pi \times 10^{-7})(750)(1 \times 10^{-4})}$

$= 6.366 \times 10^5/\text{H}$

## For the 2 cm long path:

Reluctance $S_2 = \dfrac{l_2}{\mu_0 \mu_r A_2}$

$= \dfrac{2 \times 10^{-2}}{(4\pi \times 10^{-7})(750)(0.5 \times 10^{-4})}$

$= 4.244 \times 10^5/\text{H}$

Total circuit reluctance $S = S_1 + S_2$

$= (6.366 + 4.244) \times 10^5$

$= 10.61 \times 10^5/\text{H}$

$S = \dfrac{\text{m.m.f.}}{\Phi}$ i.e. $\Phi = \dfrac{\text{m.m.f.}}{S} = \dfrac{NI}{S}$

$= \dfrac{200 \times 0.4}{10.61 \times 10^5} = 7.54 \times 10^{-5}\,\text{Wb}$

Flux density in the 2 cm path,

$$B = \dfrac{\Phi}{A} = \dfrac{7.54 \times 10^{-5}}{0.5 \times 10^{-4}} = \mathbf{1.51\,T}$$

---

**Problem 13.** A silicon iron ring of cross-sectional area $5\,\text{cm}^2$ has a radial air gap of 2 mm cut into it. If the mean length of the silicon iron path is 40 cm calculate the magnetomotive force to produce a flux of 0.7 mWb. The magnetisation curve for silicon is shown on page 80.

---

There are two parts to the circuit – the silicon iron and the air gap. The total m.m.f. will be the sum of the m.m.f.'s of each part.

## For the silicon iron:

$$B = \dfrac{\Phi}{A} = \dfrac{0.7 \times 10^{-3}}{5 \times 10^{-4}} = 1.4\,\text{T}$$

From the B–H curve for silicon iron on page 80, when $B = 1.4\,\text{T}$, $H = 1650\,\text{A/m}$. Hence the m.m.f. for the iron path $= Hl = 1650 \times 0.4 = 660\,\text{A}$

## For the air gap:

The flux density will be the same in the air gap as in the iron, i.e. 1.4 T (This assumes no leakage or fringing occurring). For air,

$$H = \dfrac{B}{\mu_0} = \dfrac{1.4}{4\pi \times 10^{-7}} = 1\,114\,000\,\text{A/m}$$

Hence the m.m.f. for the air gap
$= Hl = 1\,114\,000 \times 2 \times 10^{-3} = 2228\,\text{A}$.
**Total m.m.f. to produce a flux of 0.6 mWb**
$= 660 + 2228 = \mathbf{2888\,A}$.

A tabular method could have been used as shown in Table 7.3 at top of next page.

---

**Problem 14.** Figure 7.5 shows a ring formed with two different materials – cast steel and mild steel.

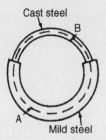

**Figure 7.5**

The dimensions are:

|  | mean length | cross-sectional area |
|---|---|---|
| Mild steel | 400 mm | $500\,\text{mm}^2$ |
| Cast steel | 300 mm | $312.5\,\text{mm}^2$ |

Find the total m.m.f. required to cause a flux of $500\,\mu\text{Wb}$ in the magnetic circuit. Determine also the total circuit reluctance.

---

A tabular solution is shown in Table 7.4 on page 85.

$\left.\begin{array}{c}\textbf{Total circuit}\\\textbf{reluctance}\end{array}\right\} S = \dfrac{\text{m.m.f.}}{\Phi}$

$= \dfrac{2000}{500 \times 10^{-6}} = \mathbf{4 \times 10^6/\text{H}}$

---

**Problem 15.** A section through a magnetic circuit of uniform cross-sectional area $2\,\text{cm}^2$ is

Table 7.3

| Part of circuit | Material | $\Phi$(Wb) | $A(m^2)$ | $B(T)$ | $H(A/m)$ | $l(m)$ | m.m.f. $=Hl(A)$ |
|---|---|---|---|---|---|---|---|
| Ring | Silicon iron | $0.7 \times 10^{-3}$ | $5 \times 10^{-4}$ | 1.4 | 1650 (from graph) | 0.4 | 660 |
| Air gap | Air | $0.7 \times 10^{-3}$ | $5 \times 10^{-4}$ | 1.4 | $\dfrac{1.4}{4\pi \times 10^{-7}}$ $= 1\,114\,000$ | $2 \times 10^{-3}$ | 2228 |
| | | | | | | Total: | **2888 A** |

Table 7.4

| Part of circuit | Material | $\Phi$(Wb) | $A(m^2)$ | $B(T)$ $(=\Phi/A)$ | $H(A/m)$ (from graphs page 80) | $l(m)$ | m.m.f. $=Hl(A)$ |
|---|---|---|---|---|---|---|---|
| A | Mild steel | $500 \times 10^{-6}$ | $500 \times 10^{-6}$ | 1.0 | 1400 | $400 \times 10^{-3}$ | 560 |
| B | Cast steel | $500 \times 10^{-6}$ | $312.5 \times 10^{-6}$ | 1.6 | 4800 | $300 \times 10^{-3}$ | 1440 |
| | | | | | | Total: | **2000 A** |

shown in Fig. 7.6. The cast steel core has a mean length of 25 cm. The air gap is 1 mm wide and the coil has 5000 turns. The B–H curve for cast steel is shown on page 80. Determine the current in the coil to produce a flux density of 0.80 T in the air gap, assuming that all the flux passes through both parts of the magnetic circuit.

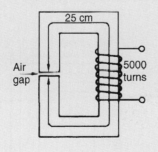

**Figure 7.6**

For the cast steel core, when $B=0.80\,\text{T}$, $H=750\,\text{A/m}$ (from page 80).

Reluctance of core $S_1 = \dfrac{l_1}{\mu_0 \mu_r A_1}$ and

since $B = \mu_0 \mu_r H$, then $\mu_r = \dfrac{B}{\mu_0 H}$

$$S_1 = \frac{l_1}{\mu_0 \left(\dfrac{B}{\mu_0 H}\right) A_1} = \frac{l_1 H}{B A_1}$$

$$= \frac{(25 \times 10^{-2})(750)}{(0.8)(2 \times 10^{-4})} = 1\,172\,000/\text{H}$$

**For the air gap:**

Reluctance, $S_2 = \dfrac{l_2}{\mu_0 \mu_r A_2}$

$$= \frac{l_2}{\mu_0 A_2} \quad (\text{since } \mu_r = 1 \text{ for air})$$

$$= \frac{1 \times 10^{-3}}{(4\pi \times 10^{-7})(2 \times 10^{-4})}$$

$$= 3\,979\,000/\text{H}$$

Total circuit reluctance

$$S = S_1 + S_2 = 1\,172\,000 + 3\,979\,000$$

$$= 5\,151\,000/\text{H}$$

Flux $\Phi = BA = 0.80 \times 2 \times 10^{-4} = 1.6 \times 10^{-4}\,\text{Wb}$

$$S = \frac{\text{m.m.f.}}{\Phi}$$

thus

$$\text{m.m.f.} = S\Phi \text{ hence } NI = S\Phi$$

and

$$\text{current } I = \frac{S\Phi}{N} = \frac{(5\,151\,000)(1.6 \times 10^{-4})}{5000}$$

$$= \mathbf{0.165\,A}$$

**Now try the following exercise**

**Exercise 36    Further problems on composite series magnetic circuits**

1. A magnetic circuit of cross-sectional area $0.4\,\text{cm}^2$ consists of one part $3\,\text{cm}$ long, of material having relative permeability 1200, and a second part $2\,\text{cm}$ long of material having relative permeability 750. With a 100 turn coil carrying $2\,\text{A}$, find the value of flux existing in the circuit.    [0.195 mWb]

2. (a) A cast steel ring has a cross-sectional area of $600\,\text{mm}^2$ and a radius of $25\,\text{mm}$. Determine the m.m.f. necessary to establish a flux of $0.8\,\text{mWb}$ in the ring. Use the B–H curve for cast steel shown on page 80. (b) If a radial air gap $1.5\,\text{mm}$ wide is cut in the ring of part (a) find the m.m.f. now necessary to maintain the same flux in the ring.

    [(a) 270 A (b)1860 A]

3. A closed magnetic circuit made of silicon iron consists of a $40\,\text{mm}$ long path of cross-sectional area $90\,\text{mm}^2$ and a $15\,\text{mm}$ long path of cross-sectional area $70\,\text{mm}^2$. A coil of 50 turns is wound around the $40\,\text{mm}$ length of the circuit and a current of $0.39\,\text{A}$ flows. Find the flux density in the $15\,\text{mm}$ length path if the relative permeability of the silicon iron at this value of magnetising force is 3000.

    [1.59 T]

4. For the magnetic circuit shown in Fig. 7.7 find the current $I$ in the coil needed to produce a flux of $0.45\,\text{mWb}$ in the air gap. The silicon iron magnetic circuit has a uniform cross-sectional area of $3\,\text{cm}^2$ and its magnetisation curve is as shown on page 80.    [0.83 A]

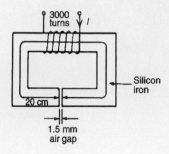

**Figure 7.7**

5. A ring forming a magnetic circuit is made from two materials; one part is mild steel of mean length $25\,\text{cm}$ and cross-sectional area $4\,\text{cm}^2$, and the remainder is cast iron of mean length $20\,\text{cm}$ and cross-sectional area $7.5\,\text{cm}^2$. Use a tabular approach to determine the total m.m.f. required to cause a flux of $0.30\,\text{mWb}$ in the magnetic circuit. Find also the total reluctance of the circuit. Use the magnetisation curves shown on page 80.

    [550 A, $1.83 \times 10^6$/H]

6. Figure 7.8 shows the magnetic circuit of a relay. When each of the air gaps are $1.5\,\text{mm}$ wide find the m.m.f. required to produce a flux density of $0.75\,\text{T}$ in the air gaps. Use the B–H curves shown on page 80.    [2970 A]

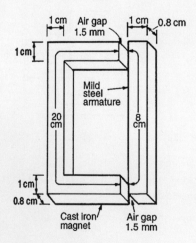

**Figure 7.8**

## 7.8    Comparison between electrical and magnetic quantities

| Electrical circuit | Magnetic circuit |
|---|---|
| e.m.f. $E$ (V) | m.m.f. $F_m$ (A) |
| current $I$ (A) | flux $\Phi$ (Wb) |
| resistance $R$ ($\Omega$) | reluctance $S$ ($H^{-1}$) |
| $I = \dfrac{E}{R}$ | $\Phi = \dfrac{\text{m.m.f.}}{S}$ |
| $R = \dfrac{\rho l}{A}$ | $S = \dfrac{l}{\mu_0 \mu_r A}$ |

## 7.9    Hysteresis and hysteresis loss

### Hysteresis loop

Let a ferromagnetic material which is completely demagnetised, i.e. one in which $B = H = 0$ be subjected to increasing values of magnetic field strength $H$ and the corresponding flux density $B$ measured. The resulting relationship between $B$ and $H$ is shown by the curve Oab in Fig. 7.9. At a particular value of $H$, shown as $Oy$, it becomes difficult to increase the flux density any further. The material is said to be saturated. Thus **by** is the **saturation flux density**.

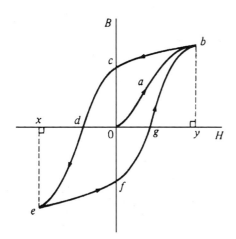

**Figure 7.9**

If the value of $H$ is now reduced it is found that the flux density follows curve **bc**. When $H$ is reduced to zero, flux remains in the iron. This **remanent flux density** or **remanence** is shown as **Oc** in Fig. 7.9. When $H$ is increased in the opposite direction, the flux density decreases until, at a value shown as **Od**, the flux density has been reduced to zero. The magnetic field strength **Od** required to remove the residual magnetism, i.e. reduce $B$ to zero, is called the **coercive force**.

Further increase of $H$ in the reverse direction causes the flux density to increase in the reverse direction until saturation is reached, as shown by curve **dc**. If $H$ is varied backwards from **Ox** to **Oy**, the flux density follows the curve **efgb**, similar to curve **bcde**.

It is seen from Fig. 7.9 that the flux density changes lag behind the changes in the magnetic field strength. This effect is called **hysteresis**. The closed figure **bcdefgb** is called the **hysteresis loop** (or the $B/H$ loop).

### Hysteresis loss

A disturbance in the alignment of the domains (i.e. groups of atoms) of a ferromagnetic material causes energy to be expended in taking it through a cycle of magnetisation. This energy appears as heat in the specimen and is called the **hysteresis loss**.

**The energy loss associated with hysteresis is proportional to the area of the hysteresis loop.**

The area of a hysteresis loop varies with the type of material. The area, and thus the energy loss, is much greater for hard materials than for soft materials.

Figure 7.10 shows typical hysteresis loops for:

(a)  **hard material**, which has a high remanence Oc and a large coercivity **Od**

(b)  **soft steel**, which has a large remanence and small coercivity

(c)  **ferrite**, this being a ceramic-like magnetic substance made from oxides of iron, nickel, cobalt, magnesium, aluminium and manganese; the hysteresis of ferrite is very small.

For a.c.-excited devices the hysteresis loop is repeated every cycle of alternating current. Thus a hysteresis loop with a large area (as with hard steel) is often unsuitable since the energy loss would be considerable. Silicon steel has a narrow hysteresis loop, and thus small hysteresis loss, and is suitable for transformer cores and rotating machine armatures.

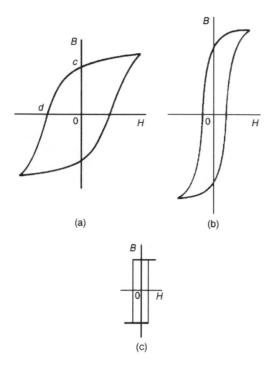

**Figure 7.10**

## Now try the following exercises

### Exercise 37 Short answer questions on magnetic circuits

1.  State six practical applications of magnets

2.  What is a permanent magnet?

3.  Sketch the pattern of the magnetic field associated with a bar magnet. Mark the direction of the field.

4.  Define magnetic flux

5.  The symbol for magnetic flux is ... and the unit of flux is the ...

6.  Define magnetic flux density

7.  The symbol for magnetic flux density is ... and the unit of flux density is ...

8.  The symbol for m.m.f. is ... and the unit of m.m.f. is the ...

9.  Another name for the magnetising force is ......; its symbol is ... and its unit is ...

10. Complete the statement:

$$\frac{\text{flux density}}{\text{magnetic field strength}} = \ldots$$

11. What is absolute permeability?

12. The value of the permeability of free space is ...

13. What is a magnetisation curve?

14. The symbol for reluctance is ... and the unit of reluctance is ...

15. Make a comparison between magnetic and electrical quantities

16. What is hysteresis?

17. Draw a typical hysteresis loop and on it identify:
    (a) saturation flux density
    (b) remanence
    (c) coercive force

18. State the units of (a) remanence (b) coercive force

19. How is magnetic screening achieved?

20. Complete the statement: magnetic materials have a ... reluctance; non-magnetic materials have a ... reluctance

21. What loss is associated with hysteresis?

### Exercise 38 Multi-choice questions on magnetic circuits
### (Answers on page 420)

1.  The unit of magnetic flux density is the:

    (a) weber          (b) weber per metre
    (c) ampere per metre    (d) tesla

2.  The total flux in the core of an electrical machine is 20 mWb and its flux density is 1 T. The cross-sectional area of the core is:

    (a) $0.05\,\text{m}^2$        (b) $0.02\,\text{m}^2$
    (c) $20\,\text{m}^2$          (d) $50\,\text{m}^2$

3.  If the total flux in a magnetic circuit is 2 mWb and the cross-sectional area of the circuit is $10 \, cm^2$, the flux density is:
    (a)  0.2 T          (b)  2 T
    (c)  20 T           (d)  20 mT

    Questions 4 to 8 refer to the following data: A coil of 100 turns is wound uniformly on a wooden ring. The ring has a mean circumference of 1 m and a uniform cross-sectional area of $10 \, cm^2$. The current in the coil is 1 A.

4.  The magnetomotive force is:
    (a)  1 A            (b)  10 A
    (c)  100 A          (d)  1000 A

5.  The magnetic field strength is:
    (a)  1 A/m          (b)  10 A/m
    (c)  100 A/m        (d)  1000 A/m

6.  The magnetic flux density is:
    (a)  800 T          (b)  $8.85 \times 10^{-10} \, T$
    (c)  $4\pi \times 10^{-7} \, T$   (d)  $40\pi \, \mu T$

7.  The magnetic flux is:
    (a)  $0.04\pi \, \mu Wb$    (b)  0.01 Wb
    (c)  $8.85 \, \mu Wb$       (d)  $4\pi \, \mu Wb$

8.  The reluctance is:
    (a)  $\dfrac{10^8}{4\pi} \, H^{-1}$         (b)  $1000 \, H^{-1}$
    (c)  $\dfrac{2.5}{\pi} \times 10^9 \, H^{-1}$   (d)  $\dfrac{10^8}{8.85} \, H^{-1}$

9.  Which of the following statements is false?
    (a)  For non-magnetic materials reluctance is high
    (b)  Energy loss due to hysteresis is greater for harder magnetic materials than for softer magnetic materials
    (c)  The remanence of a ferrous material is measured in ampere/metre
    (d)  Absolute permeability is measured in henrys per metre

10. The current flowing in a 500 turn coil wound on an iron ring is 4 A. The reluctance of the circuit is $2 \times 10^6 \, H$. The flux produced is:
    (a)  1 Wb           (b)  1000 Wb
    (c)  1 mWb          (d)  $62.5 \, \mu Wb$

11. A comparison can be made between magnetic and electrical quantities. From the following list, match the magnetic quantities with their equivalent electrical quantities.
    (a)  current        (b)  reluctance
    (c)  e.m.f.         (d)  flux
    (e)  m.m.f.         (f)  resistance

12. The effect of an air gap in a magnetic circuit is to:
    (a)  increase the reluctance
    (b)  reduce the flux density
    (c)  divide the flux
    (d)  reduce the magnetomotive force

13. Two bar magnets are placed parallel to each other and about 2 cm apart, such that the south pole of one magnet is adjacent to the north pole of the other. With this arrangement, the magnets will:
    (a)  attract each other
    (b)  have no effect on each other
    (c)  repel each other
    (d)  lose their magnetism

## Revision Test 2

This revision test covers the material contained in Chapters 5 to 7. *The marks for each question are shown in brackets at the end of each question.*

1. Resistances of $5\Omega$, $7\Omega$, and $8\Omega$ are connected in series. If a $10V$ supply voltage is connected across the arrangement determine the current flowing through and the p.d. across the $7\Omega$ resistor. Calculate also the power dissipated in the $8\Omega$ resistor. (6)

2. For the series-parallel network shown in Fig. RT2.1, find (a) the supply current, (b) the current flowing through each resistor, (c) the p.d. across each resistor, (d) the total power dissipated in the circuit, (e) the cost of energy if the circuit is connected for 80 hours. Assume electrical energy costs 14p per unit. (15)

3. The charge on the plates of a capacitor is $8mC$ when the potential between them is $4kV$. Determine the capacitance of the capacitor. (2)

4. Two parallel rectangular plates measuring $80mm$ by $120mm$ are separated by $4mm$ of mica and carry an electric charge of $0.48\mu C$. The voltage between the plates is $500V$. Calculate (a) the electric flux density (b) the electric field strength, and (c) the capacitance of the capacitor, in picofarads, if the relative permittivity of mica is 5. (7)

5. A $4\mu F$ capacitor is connected in parallel with a $6\mu F$ capacitor. This arrangement is then connected in series with a $10\mu F$ capacitor. A supply p.d. of $250V$ is connected across the circuit. Find (a) the equivalent capacitance of the circuit, (b) the voltage across the $10\mu F$ capacitor, and (c) the charge on each capacitor. (7)

6. A coil of 600 turns is wound uniformly on a ring of non-magnetic material. The ring has a uniform cross-sectional area of $200mm^2$ and a mean circumference of $500mm$. If the current in the coil is $4A$, determine (a) the magnetic field strength,

(b) the flux density, and (c) the total magnetic flux in the ring. (5)

7. A mild steel ring of cross-sectional area $4cm^2$ has a radial air gap of $3mm$ cut into it. If the mean length of the mild steel path is $300mm$, calculate the magnetomotive force to produce a flux of $0.48mWb$. (Use the B–H curve on page 80) (8)

8. In the circuit shown in Fig. RT2.2, the slider $S$ is at the half-way point.

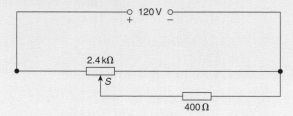

Figure RT2.2

   (a) Calculate the p.d. across and the current flowing in the $400\Omega$ load resistor.
   (b) Is the circuit a potentiometer or a rheostat? (5)

9. For the circuit shown in Fig. RT2.3, calculate the current flowing in the $50\Omega$ load and the voltage drop across the load when

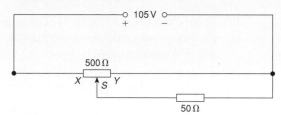

Figure RT2.3

   (a) $XS$ is 3/5 of $XY$
   (b) point $S$ coincides with point $Y$ (5)

$R_3 = 2\Omega$

$R_1 = 2.4\Omega$    $R_2 = 5\Omega$    $R_5 = 11\Omega$

$R_4 = 8\Omega$

$I$

$100V$

Figure RT2.1

# Chapter 8

# Electromagnetism

At the end of this chapter you should be able to:

- understand that magnetic fields are produced by electric currents
- apply the screw rule to determine direction of magnetic field
- recognize that the magnetic field around a solenoid is similar to a magnet
- apply the screw rule or grip rule to a solenoid to determine magnetic field direction
- recognize and describe practical applications of an electromagnet, i.e. electric bell, relay, lifting magnet, telephone receiver
- appreciate factors upon which the force $F$ on a current-carrying conductor depends
- perform calculations using $F = BIl$ and $F = BIl\sin\theta$
- recognize that a loudspeaker is a practical application of force $F$
- use Fleming's left-hand rule to pre-determine direction of force in a current-carrying conductor
- describe the principle of operation of a simple d.c. motor
- describe the principle of operation and construction of a moving coil instrument
- appreciate that force $F$ on a charge in a magnetic field is given by $F = QvB$
- perform calculations using $F = QvB$

## 8.1 Magnetic field due to an electric current

Magnetic fields can be set up not only by permanent magnets, as shown in Chapter 7, but also by electric currents.

Let a piece of wire be arranged to pass vertically through a horizontal sheet of cardboard on which is placed some iron filings, as shown in Fig. 8.1(a). If a current is now passed through the wire, then the iron filings will form a definite circular field pattern with the wire at the centre, when the cardboard is gently tapped. By placing a compass in different positions the lines of flux are seen to have a definite direction as shown in Fig. 8.1(b).

If the current direction is reversed, the direction of the lines of flux is also reversed. The effect on both the iron filings and the compass needle disappears when

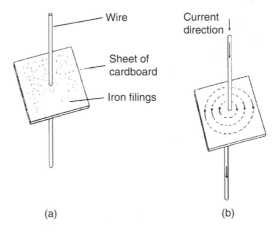

Figure 8.1

the current is switched off. The magnetic field is thus produced by the electric current. The magnetic flux

DOI: 10.1016/B978-0-08-089056-2.00008-5

produced has the same properties as the flux produced by a permanent magnet. If the current is increased the strength of the field increases and, as for the permanent magnet, the field strength decreases as we move away from the current-carrying conductor.

In Fig. 8.1, the effect of only a small part of the magnetic field is shown. If the whole length of the conductor is similarly investigated it is found that the magnetic field round a straight conductor is in the form of concentric cylinders as shown in Fig. 8.2, the field direction depending on the direction of the current flow.

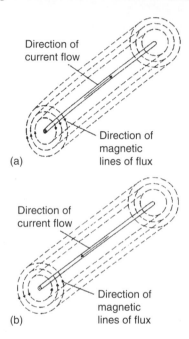

Figure 8.2

When dealing with magnetic fields formed by electric current it is usual to portray the effect as shown in Fig. 8.3. The convention adopted is:

(a) Current flowing away from viewer    (b) Current flowing towards viewer

Figure 8.3

(i)   Current flowing away from the viewer, i.e. into the paper, is indicated by $\oplus$. This may be thought of as the feathered end of the shaft of an arrow. See Fig. 8.3(a).

(ii)  Current flowing towards the viewer, i.e. out of the paper, is indicated by $\odot$. This may be thought of as the point of an arrow. See Fig. 8.3(b).

The direction of the magnetic lines of flux is best remembered by the **screw rule** which states that:

*If a normal right-hand thread screw is screwed along the conductor in the direction of the current, the direction of rotation of the screw is in the direction of the magnetic field.*

For example, with current flowing away from the viewer (Fig. 8.3(a)) a right-hand thread screw driven into the paper has to be rotated clockwise. Hence the direction of the magnetic field is clockwise.

A magnetic field set up by a long coil, or **solenoid**, is shown in Fig. 8.4(a) and is seen to be similar to that of a bar magnet. If the solenoid is wound on an iron bar, as shown in Fig. 8.4(b), an even stronger magnetic field is produced, the iron becoming magnetised and behaving like a permanent magnet. The direction of the magnetic field produced by the current I in the solenoid may be found by either of two methods, i.e. the screw rule or the grip rule.

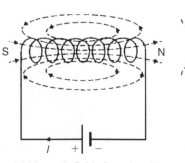  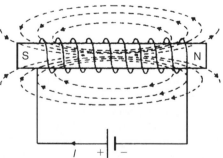

(a) Magnetic field of a solenoid    (b) Magnetic field of an iron cored solenoid

Figure 8.4

(a) **The screw rule** states that if a normal right-hand thread screw is placed along the axis of the solenoid and is screwed in the direction of the current it moves in the direction of the magnetic field **inside** the solenoid. The direction of the magnetic field **inside** the solenoid is from south to north. Thus in Figs 8.4(a) and (b) the north pole is to the right.

(b) **The grip rule** states that if the coil is gripped with the **right** hand, with the fingers pointing in the direction of the current, then the thumb, outstretched parallel to the axis of the solenoid, points in the direction of the magnetic field **inside** the solenoid.

**Problem 1.** Figure 8.5 shows a coil of wire wound on an iron core connected to a battery. Sketch the magnetic field pattern associated with the current-carrying coil and determine the polarity of the field.

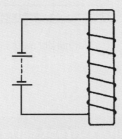

Figure 8.5

The magnetic field associated with the solenoid in Fig. 8.5 is similar to the field associated with a bar magnet and is as shown in Fig. 8.6. The polarity of the field is determined either by the screw rule or by the grip rule. Thus the north pole is at the bottom and the south pole at the top.

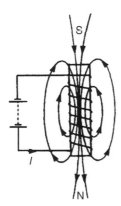

Figure 8.6

## 8.2 Electromagnets

The solenoid is very important in electromagnetic theory since the magnetic field inside the solenoid is practically uniform for a particular current, and is also versatile, inasmuch that a variation of the current can alter the strength of the magnetic field. An electromagnet, based on the solenoid, provides the basis of many items of electrical equipment, examples of which include electric bells, relays, lifting magnets and telephone receivers.

### (i) Electric bell

There are various types of electric bell, including the single-stroke bell, the trembler bell, the buzzer and a continuously ringing bell, but all depend on the attraction exerted by an electromagnet on a soft iron armature. A typical single stroke bell circuit is shown in Fig. 8.7. When the push button is operated a current passes through the coil. Since the iron-cored coil is energised the soft iron armature is attracted to the electromagnet. The armature also carries a striker which hits the gong. When the circuit is broken the coil becomes demagnetised and the spring steel strip pulls the armature back to its original position. The striker will only operate when the push button is operated.

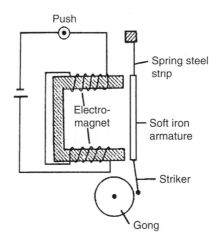

Figure 8.7

### (ii) Relay

A relay is similar to an electric bell except that contacts are opened or closed by operation instead of a gong being struck. A typical simple relay is shown in Fig. 8.8, which consists of a coil wound on a soft iron core. When the coil is energised the hinged soft iron armature is

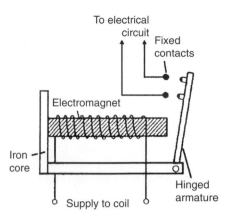

**Figure 8.8**

attracted to the electromagnet and pushes against two fixed contacts so that they are connected together, thus closing some other electrical circuit.

### (iii) Lifting magnet

Lifting magnets, incorporating large electromagnets, are used in iron and steel works for lifting scrap metal. A typical robust lifting magnet, capable of exerting large attractive forces, is shown in the elevation and plan view of Fig. 8.9 where a coil, C, is wound round a central core, P, of the iron casting. Over the face of the electromagnet is placed a protective non-magnetic sheet of material, R. The load, Q, which must be of magnetic material is lifted when the coils are energised, the magnetic flux paths, M, being shown by the broken lines.

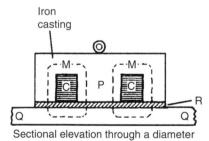

Sectional elevation through a diameter

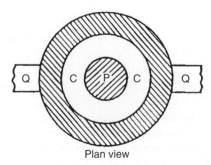

Plan view

**Figure 8.9**

### (iv) Telephone receiver

Whereas a transmitter or microphone changes sound waves into corresponding electrical signals, a telephone receiver converts the electrical waves back into sound waves. A typical telephone receiver is shown in Fig. 8.10 and consists of a permanent magnet with coils wound on its poles. A thin, flexible diaphragm of magnetic material is held in position near to the magnetic poles but not touching them. Variation in current from the transmitter varies the magnetic field and the diaphragm consequently vibrates. The vibration produces sound variations corresponding to those transmitted.

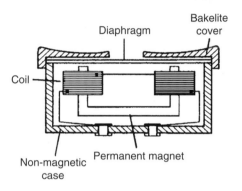

**Figure 8.10**

## 8.3 Force on a current-carrying conductor

If a current-carrying conductor is placed in a magnetic field produced by permanent magnets, then the fields due to the current-carrying conductor and the permanent magnets interact and cause a force to be exerted on the conductor. The force on the current-carrying conductor in a magnetic field depends upon:

(a) the flux density of the field, $B$ teslas,

(b) the strength of the current, $I$ amperes,

(c) the length of the conductor perpendicular to the magnetic field, $l$ metres, and

(d) the directions of the field and the current.

When the magnetic field, the current and the conductor are mutually at right angles then:

$$\text{Force } F = BIl \text{ newtons}$$

When the conductor and the field are at an angle $\theta°$ to each other then:

$$\text{Force } F = BIl \sin \theta \text{ newtons}$$

Since when the magnetic field, current and conductor are mutually at right angles, $F = BIl$, the magnetic flux density $B$ may be defined by $B = (F)/(Il)$, i.e. the flux density is 1 T if the force exerted on 1 m of a conductor when the conductor carries a current of 1 A is 1 N.

## Loudspeaker

A simple application of the above force is the moving coil loudspeaker. The loudspeaker is used to convert electrical signals into sound waves.

Figure 8.11 shows a typical loudspeaker having a magnetic circuit comprising a permanent magnet and soft iron pole pieces so that a strong magnetic field is available in the short cylindrical air gap. A moving coil, called the voice or speech coil, is suspended from the end of a paper or plastic cone so that it lies in the gap. When an electric current flows through the coil it produces a force which tends to move the cone backwards and forwards according to the direction of the current. The cone acts as a piston, transferring this force to the air, and producing the required sound waves.

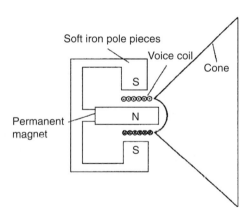

Figure 8.11

Problem 2.    A conductor carries a current of 20 A and is at right-angles to a magnetic field having a flux density of 0.9 T. If the length of the conductor in the field is 30 cm, calculate the force acting on the conductor. Determine also the value of the force if the conductor is inclined at an angle of 30° to the direction of the field.

$B = 0.9$ T, $I = 20$ A and $l = 30$ cm $= 0.30$ m
Force $F = BIl = (0.9)(20)(0.30)$ newtons when the conductor is at right-angles to the field, as shown in Fig. 8.12(a), i.e. $F = 5.4$ N.

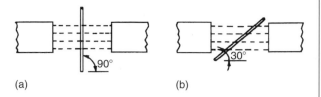

Figure 8.12

When the conductor is inclined at 30° to the field, as shown in Fig. 8.12(b), then

$$\text{Force } F = BIl \sin\theta$$

$$= (0.9)(20)(0.30)\sin 30°$$

i.e. **F = 2.7 N**

If the current-carrying conductor shown in Fig. 8.3(a) is placed in the magnetic field shown in Fig. 8.13(a), then the two fields interact and cause a force to be exerted on the conductor as shown in Fig. 8.13(b). The field is strengthened above the conductor and weakened below, thus tending to move the conductor downwards. This is the basic principle of operation of the electric motor (see Section 8.4) and the moving-coil instrument (see Section 8.5).

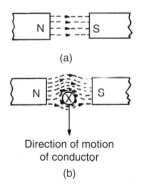

Direction of motion
of conductor

(b)

Figure 8.13

The direction of the force exerted on a conductor can be pre-determined by using **Fleming's left-hand rule** (often called the motor rule) which states:

*Let the thumb, first finger and second finger of the left hand be extended such that they are all at right-angles to each other (as shown in Fig. 8.14). If the first finger points in the direction of the magnetic field, the second finger points in the direction of the current, then the thumb will point in the direction of the motion of the conductor.*

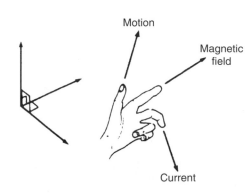

**Figure 8.14**

Summarising:

First finger – Field

SeCond finger – Current

ThuMb – Motion

**Problem 3.** Determine the current required in a 400 mm length of conductor of an electric motor, when the conductor is situated at right-angles to a magnetic field of flux density 1.2 T, if a force of 1.92 N is to be exerted on the conductor. If the conductor is vertical, the current flowing downwards and the direction of the magnetic field is from left to right, what is the direction of the force?

Force $= 1.92$ N, $l = 400$ mm $= 0.40$ m and $B = 1.2$ T. Since $F = BIl$, then $I = F/Bl$ hence

$$\textbf{current } I = \frac{1.92}{(1.2)(0.4)} = \textbf{4 A}$$

If the current flows downwards, the direction of its magnetic field due to the current alone will be clockwise when viewed from above. The lines of flux will reinforce (i.e. strengthen) the main magnetic field at the back of the conductor and will be in opposition in the front (i.e. weaken the field). **Hence the force on the conductor will be from back to front (i.e. toward the viewer)**. This direction may also have been deduced using Fleming's left-hand rule.

**Problem 4.** A conductor 350 mm long carries a current of 10 A and is at right-angles to a magnetic field lying between two circular pole faces each of radius 60 mm. If the total flux between the pole faces is 0.5 mWb, calculate the magnitude of the force exerted on the conductor.

$l = 350$ mm $= 0.35$ m, $I = 10$ A, area of pole face $A = \pi r^2 = \pi (0.06)^2$ m$^2$ and $\Phi = 0.5$ mWb $= 0.5 \times 10^{-3}$ Wb

Force $F = BIl$, and $B = \dfrac{\Phi}{A}$ hence

$$\text{force } F = \frac{\Phi}{A} Il$$

$$= \frac{(0.5 \times 10^{-3})}{\pi (0.06)^2}(10)(0.35) \text{ newtons}$$

i.e. **force** $= \textbf{0.155 N}$

**Problem 5.** With reference to Fig. 8.15 determine (a) the direction of the force on the conductor in Fig. 8.15(a), (b) the direction of the force on the conductor in Fig. 8.15(b), (c) the direction of the current in Fig. 8.15(c), (d) the polarity of the magnetic system in Fig. 8.15(d).

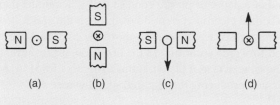

**Figure 8.15**

(a) The direction of the main magnetic field is from north to south, i.e. left to right. The current is flowing towards the viewer, and using the screw rule, the direction of the field is anticlockwise. Hence either by Fleming's left-hand rule, or by sketching the interacting magnetic field as shown in Fig. 8.16(a), the direction of the force on the conductor is seen to be upward.

(b) Using a similar method to part (a) it is seen that the force on the conductor is to the right – see Fig. 8.16(b).

(c) Using Fleming's left-hand rule, or by sketching as in Fig. 8.16(c), it is seen that the current is toward the viewer, i.e. out of the paper.

(d) Similar to part (c), the polarity of the magnetic system is as shown in Fig. 8.16(d).

**Problem 6.** A coil is wound on a rectangular former of width 24 mm and length 30 mm. The former is pivoted about an axis passing through the middle of the two shorter sides and is placed in a uniform magnetic field of flux density 0.8 T, the axis being perpendicular to the field. If the coil carries a current of 50 mA, determine the force on each coil side (a) for a single-turn coil, (b) for a coil wound with 300 turns.

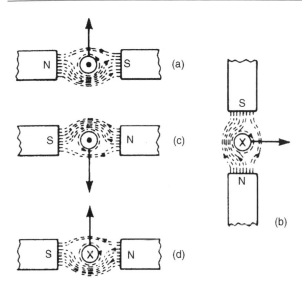

**Figure 8.16**

(a)   Flux density $B = 0.8$ T, length of conductor lying at right-angles to field $l = 30$ mm $= 30 \times 10^{-3}$ m and current $I = 50$ mA $= 50 \times 10^{-3}$ A. For a single-turn coil, force on each coil side

$$F = BIl = 0.8 \times 50 \times 10^{-3} \times 30 \times 10^{-3}$$
$$= \mathbf{1.2 \times 10^{-3}\,N} \text{ or } \mathbf{0.0012\,N}$$

(b)   When there are 300 turns on the coil there are effectively 300 parallel conductors each carrying a current of 50 mA. Thus the total force produced by the current is 300 times that for a single-turn coil. Hence force on coil side,
$F = 300\ BIl = 300 \times 0.0012 = \mathbf{0.36\,N}$

**Now try the following exercise**

**Exercise 39    Further problems on the force on a current-carrying conductor**

1.   A conductor carries a current of 70 A at right-angles to a magnetic field having a flux density of 1.5 T. If the length of the conductor in the field is 200 mm calculate the force acting on the conductor. What is the force when the conductor and field are at an angle of 45°?
[21.0 N, 14.8 N]

2.   Calculate the current required in a 240 mm length of conductor of a d.c. motor when the conductor is situated at right-angles to the magnetic field of flux density 1.25 T, if a force of 1.20 N is to be exerted on the conductor.
[4.0 A]

3.   A conductor 30 cm long is situated at right-angles to a magnetic field. Calculate the flux density of the magnetic field if a current of 15 A in the conductor produces a force on it of 3.6 N.
[0.80 T]

4.   A conductor 300 mm long carries a current of 13 A and is at right-angles to a magnetic field between two circular pole faces, each of diameter 80 mm. If the total flux between the pole faces is 0.75 mWb calculate the force exerted on the conductor.
[0.582 N]

5.   (a) A 400 mm length of conductor carrying a current of 25 A is situated at right-angles to a magnetic field between two poles of an electric motor. The poles have a circular cross-section. If the force exerted on the conductor is 80 N and the total flux between the pole faces is 1.27 mWb, determine the diameter of a pole face.

(b) If the conductor in part (a) is vertical, the current flowing downwards and the direction of the magnetic field is from left to right, what is the direction of the 80 N force?
[(a) 14.2 mm (b) towards the viewer]

6.   A coil is wound uniformly on a former having a width of 18 mm and a length of 25 mm. The former is pivoted about an axis passing through the middle of the two shorter sides and is placed in a uniform magnetic field of flux density 0.75 T, the axis being perpendicular to the field. If the coil carries a current of 120 mA, determine the force exerted on each coil side, (a) for a single-turn coil, (b) for a coil wound with 400 turns.
[(a) $2.25 \times 10^{-3}$ N (b) 0.9 N]

## 8.4  Principle of operation of a simple d.c. motor

A rectangular coil which is free to rotate about a fixed axis is shown placed inside a magnetic field produced by permanent magnets in Fig. 8.17. A direct current is fed

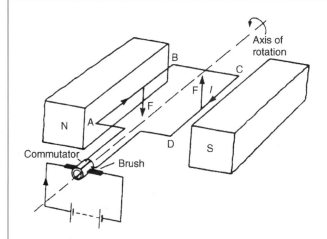

Figure 8.17

into the coil via carbon brushes bearing on a commutator, which consists of a metal ring split into two halves separated by insulation. When current flows in the coil a magnetic field is set up around the coil which interacts with the magnetic field produced by the magnets. This causes a force $F$ to be exerted on the current-carrying conductor which, by Fleming's left-hand rule, is downwards between points A and B and upward between C and D for the current direction shown. This causes a torque and the coil rotates anticlockwise. When the coil has turned through 90° from the position shown in Fig. 8.17 the brushes connected to the positive and negative terminals of the supply make contact with different halves of the commutator ring, thus reversing the direction of the current flow in the conductor. If the current is not reversed and the coil rotates past this position the forces acting on it change direction and it rotates in the opposite direction thus never making more than half a revolution. The current direction is reversed every time the coil swings through the vertical position and thus

the coil rotates anticlockwise for as long as the current flows. This is the principle of operation of a d.c. motor which is thus a device that takes in electrical energy and converts it into mechanical energy.

## 8.5 Principle of operation of a moving-coil instrument

A moving-coil instrument operates on the motor principle. When a conductor carrying current is placed in a magnetic field, a force $F$ is exerted on the conductor, given by $F = BIl$. If the flux density B is made constant (by using permanent magnets) and the conductor is a fixed length (say, a coil) then the force will depend only on the current flowing in the conductor.

In a moving-coil instrument a coil is placed centrally in the gap between shaped pole pieces as shown by the front elevation in Fig. 8.18(a). (The air gap is kept as small as possible, although for clarity it is shown exaggerated in Fig. 8.18). The coil is supported by steel pivots, resting in jewel bearings, on a cylindrical iron core. Current is led into and out of the coil by two phosphor bronze spiral hairsprings which are wound in opposite directions to minimise the effect of temperature change and to limit the coil swing (i.e. to **control** the movement) and return the movement to zero position when no current flows. Current flowing in the coil produces forces as shown in Fig. 8.18(b), the directions being obtained by Fleming's left-hand rule. The two forces, $F_A$ and $F_B$, produce a torque which will move the coil in a clockwise direction, i.e. move the pointer from left to right. Since force is proportional to current the scale is linear.

When the aluminium frame, on which the coil is wound, is rotated between the poles of the magnet,

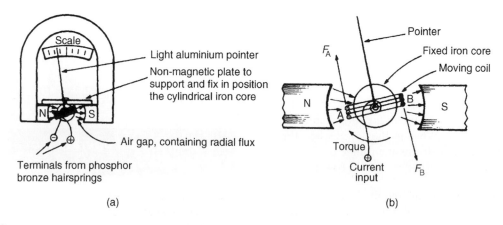

(a)

(b)

Figure 8.18

small currents (called eddy currents) are induced into the frame, and this provides automatically the necessary **damping** of the system due to the reluctance of the former to move within the magnetic field. The moving-coil instrument will measure only direct current or voltage and the terminals are marked positive and negative to ensure that the current passes through the coil in the correct direction to deflect the pointer 'up the scale'.

The range of this sensitive instrument is extended by using shunts and multipliers (see Chapter 10).

## 8.6   Force on a charge

When a charge of $Q$ coulombs is moving at a velocity of $v$ m/s in a magnetic field of flux density $B$ teslas, the charge moving perpendicular to the field, then the magnitude of the force $F$ exerted on the charge is given by:

$$F = QvB \text{ newtons}$$

**Problem 7.**   An electron in a television tube has a charge of $1.6 \times 10^{-19}$ coulombs and travels at $3 \times 10^7$ m/s perpendicular to a field of flux density $18.5 \, \mu$T. Determine the force exerted on the electron in the field.

From above, force $F = QvB$ newtons, where $Q$ = charge in coulombs = $1.6 \times 10^{-19}$ C, $v$ = velocity of charge $-3 \times 10^7$ m/s, and $B$ = flux density = $18.5 \times 10^{-6}$ T.
Hence force on electron,

$$F = 1.6 \times 10^{-19} \times 3 \times 10^7 \times 18.5 \times 10^{-6}$$
$$= 1.6 \times 3 \times 18.5 \times 10^{-18}$$
$$= 88.8 \times 10^{-18} = \mathbf{8.88 \times 10^{-17} \, N}$$

**Now try the following exercises**

### Exercise 40   Further problems on the force on a charge

1.   Calculate the force exerted on a charge of $2 \times 10^{-18}$ C travelling at $2 \times 10^6$ m/s perpendicular to a field of density $2 \times 10^{-7}$ T
   $[8 \times 10^{-19} \text{N}]$

2.   Determine the speed of a $10^{-19}$ C charge travelling perpendicular to a field of flux density $10^{-7}$ T, if the force on the charge is $10^{-20}$ N
   $[10^6 \text{ m/s}]$

### Exercise 41   Short answer questions on electromagnetism

1.   The direction of the magnetic field around a current-carrying conductor may be remembered using the ...... rule

2.   Sketch the magnetic field pattern associated with a solenoid connected to a battery and wound on an iron bar. Show the direction of the field

3.   Name three applications of electromagnetism

4.   State what happens when a current-carrying conductor is placed in a magnetic field between two magnets

5.   The force on a current-carrying conductor in a magnetic field depends on four factors. Name them

6.   The direction of the force on a conductor in a magnetic field may be predetermined using Fleming's ...... rule

7.   State three applications of the force on a current-carrying conductor

8.   Figure 8.19 shows a simplified diagram of a section through the coil of a moving-coil instrument. For the direction of current flow shown in the coil determine the direction that the pointer will move

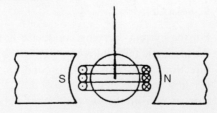

Figure 8.19

9.   Explain, with the aid of a sketch, the action of a simplified d.c. motor

10. Sketch and label the movement of a moving-coil instrument. Briefly explain the principle of operation of such an instrument

**Exercise 42    Multi-choice questions on electromagnetism**
**(Answers on page 420)**

1. A conductor carries a current of 10 A at right-angles to a magnetic field having a flux density of 500 mT. If the length of the conductor in the field is 20 cm, the force on the conductor is:
   (a) 100 kN          (b) 1 kN
   (c) 100 N           (d) 1 N

2. If a conductor is horizontal, the current flowing from left to right and the direction of the surrounding magnetic field is from above to below, the force exerted on the conductor is:
   (a) from left to right
   (b) from below to above
   (c) away from the viewer
   (d) towards the viewer

3. For the current-carrying conductor lying in the magnetic field shown in Fig. 8.20(a), the direction of the force on the conductor is:
   (a) to the left          (b) upwards
   (c) to the right         (d) downwards

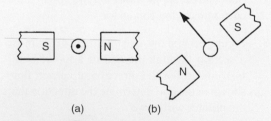

(a)                (b)

Figure 8.20

4. For the current-carrying conductor lying in the magnetic field shown in Fig. 8.20(b), the direction of the current in the conductor is:
   (a) towards the viewer
   (b) away from the viewer

5. Figure 8.21 shows a rectangular coil of wire placed in a magnetic field and free to rotate about axis AB. If the current flows into the coil at C, the coil will:

   (a) commence to rotate anti-clockwise
   (b) commence to rotate clockwise
   (c) remain in the vertical position
   (d) experience a force towards the north pole

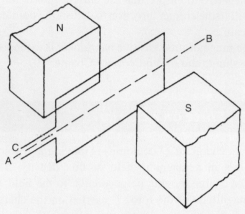

Figure 8.21

6. The force on an electron travelling at $10^7$ m/s in a magnetic field of density $10\,\mu$T is $1.6 \times 10^{-17}$ N. The electron has a charge of:
   (a) $1.6 \times 10^{-28}$ C     (b) $1.6 \times 10^{-15}$ C
   (c) $1.6 \times 10^{-19}$ C     (d) $1.6 \times 10^{-25}$ C

7. An electric bell depends for its action on:
   (a) a permanent magnet
   (b) reversal of current
   (c) a hammer and a gong
   (d) an electromagnet

8. A relay can be used to:
   (a) decrease the current in a circuit
   (b) control a circuit more readily
   (c) increase the current in a circuit
   (d) control a circuit from a distance

9. There is a force of attraction between two current-carrying conductors when the current in them is:
   (a) in opposite directions
   (b) in the same direction
   (c) of different magnitude
   (d) of the same magnitude

10. The magnetic field due to a current-carrying conductor takes the form of:
    (a) rectangles
    (b) concentric circles
    (c) wavy lines
    (d) straight lines radiating outwards

# Chapter 9

# Electromagnetic induction

At the end of this chapter you should be able to:

- understand how an e.m.f. may be induced in a conductor
- state Faraday's laws of electromagnetic induction
- state Lenz's law
- use Fleming's right-hand rule for relative directions
- appreciate that the induced e.m.f., $E = Blv$ or $E = Blv\sin\theta$
- calculate induced e.m.f. given $B, l, v$ and $\theta$ and determine relative directions
- understand and perform calculations on rotation of a loop in a magnetic field
- define inductance $L$ and state its unit
- define mutual inductance
- appreciate that e.m.f.

$$E = -N\frac{d\Phi}{dt} = -L\frac{dI}{dt}$$

- calculate induced e.m.f. given $N, t, L$, change of flux or change of current
- appreciate factors which affect the inductance of an inductor
- draw the circuit diagram symbols for inductors
- calculate the energy stored in an inductor using $W = \frac{1}{2}LI^2$ joules
- calculate inductance $L$ of a coil, given $L = \frac{N\Phi}{I}$ and $L = \frac{N^2}{S}$
- calculate mutual inductance using $E_2 = -M\frac{dI_1}{dt}$ and $M = \frac{N_1 N_2}{S}$

## 9.1 Introduction to electromagnetic induction

When a conductor is moved across a magnetic field so as to cut through the lines of force (or flux), an electromotive force (e.m.f.) is produced in the conductor. If the conductor forms part of a closed circuit then the e.m.f. produced causes an electric current to flow round the circuit. Hence an e.m.f. (and thus current) is 'induced' in

the conductor as a result of its movement across the magnetic field. This effect is known as '**electromagnetic induction**'.

Figure 9.1(a) shows a coil of wire connected to a centre-zero galvanometer, which is a sensitive ammeter with the zero-current position in the centre of the scale.

(a)  When the magnet is moved at constant speed towards the coil (Fig. 9.1(a)), a deflection is noted

DOI: 10.1016/B978-0-08-089056-2.00009-7

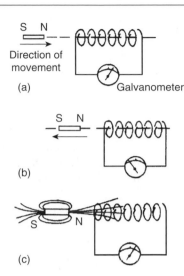

**Figure 9.1**

on the galvanometer showing that a current has been produced in the coil.

(b) When the magnet is moved at the same speed as in (a) but away from the coil the same deflection is noted but is in the opposite direction (see Fig. 9.1(b)).

(c) When the magnet is held stationary, even within the coil, no deflection is recorded.

(d) When the coil is moved at the same speed as in (a) and the magnet held stationary the same galvanometer deflection is noted.

(e) When the relative speed is, say, doubled, the galvanometer deflection is doubled.

(f) When a stronger magnet is used, a greater galvanometer deflection is noted.

(g) When the number of turns of wire of the coil is increased, a greater galvanometer deflection is noted.

Figure 9.1(c) shows the magnetic field associated with the magnet. As the magnet is moved towards the coil, the magnetic flux of the magnet moves across, or cuts, the coil. **It is the relative movement of the magnetic flux and the coil that causes an e.m.f. and thus current, to be induced in the coil**. This effect is known as electromagnetic induction. The laws of electromagnetic induction stated in Section 9.2 evolved from experiments such as those described above.

## 9.2 Laws of electromagnetic induction

Faraday's laws of electromagnetic induction state:

(i) *An induced e.m.f. is set up whenever the magnetic field linking that circuit changes.*

(ii) *The magnitude of the induced e.m.f. in any circuit is proportional to the rate of change of the magnetic flux linking the circuit.*

Lenz's law states:

*The direction of an induced e.m.f. is always such that it tends to set up a current opposing the motion or the change of flux responsible for inducing that e.m.f.*

An alternative method to Lenz's law of determining relative directions is given by **Fleming's <u>R</u>ight-hand rule** (often called the gene<u>R</u>ator rule) which states:

*Let the thumb, first finger and second finger of the right hand be extended such that they are all at right angles to each other (as shown in Fig. 9.2). If the first finger points in the direction of the magnetic field and the thumb points in the direction of motion of the conductor relative to the magnetic field, then the second finger will point in the direction of the induced e.m.f.*

Summarising:

<u>F</u>irst finger – <u>F</u>ield

Thu<u>M</u>b – <u>M</u>otion

S<u>E</u>cond finger – <u>E</u>.m.f.

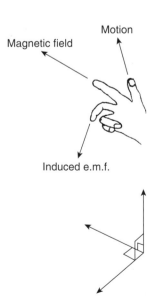

**Figure 9.2**

In a generator, conductors forming an electric circuit are made to move through a magnetic field. By Faraday's law an e.m.f. is induced in the conductors and thus a source of e.m.f. is created. A generator converts mechanical energy into electrical energy. (The action of a simple a.c. generator is described in Chapter 14).

The induced e.m.f. $E$ set up between the ends of the conductor shown in Fig. 9.3 is given by:

$$E = Blv \text{ volts}$$

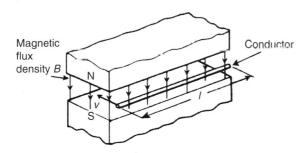

Magnetic flux density $B$

Conductor

**Figure 9.3**

where $B$, the flux density, is measured in teslas, $l$, the length of conductor in the magnetic field, is measured in metres, and $v$, the conductor velocity, is measured in metres per second.

If the conductor moves at an angle $\theta°$ to the magnetic field (instead of at 90° as assumed above) then

$$E = Blv \sin \theta \text{ volts}$$

**Problem 1.** A conductor 300 mm long moves at a uniform speed of 4 m/s at right-angles to a uniform magnetic field of flux density 1.25 T. Determine the current flowing in the conductor when (a) its ends are open-circuited, (b) its ends are connected to a load of 20 Ω resistance.

When a conductor moves in a magnetic field it will have an e.m.f. induced in it but this e.m.f. can only produce a current if there is a closed circuit. Induced e.m.f.

$$E = Blv = (1.25)\left(\frac{300}{1000}\right)(4) = 1.5 \text{ V}$$

(a) If the ends of the conductor are open circuited **no current will flow** even though 1.5 V has been induced.

(b) From Ohm's law,

$$I = \frac{E}{R} = \frac{1.5}{20} = 0.075 \text{ A or } 75 \text{ mA}$$

**Problem 2.** At what velocity must a conductor 75 mm long cut a magnetic field of flux density 0.6 T if an e.m.f. of 9 V is to be induced in it? Assume the conductor, the field and the direction of motion are mutually perpendicular.

Induced e.m.f. $E = Blv$, hence velocity $v = E/Bl$
Thus

$$v = \frac{9}{(0.6)(75 \times 10^{-3})}$$
$$= \frac{9 \times 10^3}{0.6 \times 75}$$
$$= 200 \text{ m/s}$$

**Problem 3.** A conductor moves with a velocity of 15 m/s at an angle of (a) 90° (b) 60° and (c) 30° to a magnetic field produced between two square-faced poles of side length 2 cm. If the flux leaving a pole face is 5 μWb, find the magnitude of the induced e.m.f. in each case.

$v = 15$ m/s, length of conductor in magnetic field, $l = 2 \text{ cm} = 0.02 \text{ m}$, $A = 2 \times 2 \text{ cm}^2 = 4 \times 10^{-4} \text{ m}^2$ and $\Phi = 5 \times 10^{-6} \text{ Wb}$

(a) $E_{90} = Blv \sin 90°$
$$= \left(\frac{\Phi}{A}\right) lv \sin 90°$$
$$= \left(\frac{5 \times 10^{-6}}{4 \times 10^{-4}}\right)(0.02)(15)(1)$$
$$= 3.75 \text{ mV}$$

(b) $E_{60} = Blv \sin 60° = E_{90} \sin 60°$
$$= 3.75 \sin 60° = 3.25 \text{ mV}$$

(c) $E_{30} = Blv \sin 30° = E_{90} \sin 30°$
$$= 3.75 \sin 30° = 1.875 \text{ mV}$$

**Problem 4.** The wing span of a metal aeroplane is 36 m. If the aeroplane is flying at 400 km/h, determine the e.m.f. induced between its wing tips. Assume the vertical component of the earth's magnetic field is 40 μT.

Induced e.m.f. across wing tips, $E = Blv$
$B = 40\,\mu\text{T} = 40 \times 10^{-6}$ T, $l = 36$ m and

$$v = 400\,\frac{\text{km}}{\text{h}} \times 1000\,\frac{\text{m}}{\text{km}} \times \frac{1\,\text{h}}{60 \times 60\,\text{s}}$$

$$= \frac{(400)(1000)}{3600}$$

$$= \frac{4000}{36}\,\text{m/s}$$

Hence

$$E = Blv = (40 \times 10^{-6})(36)\left(\frac{4000}{36}\right)$$

$$= \mathbf{0.16\,V}$$

**Problem 5.** The diagrams shown in Fig. 9.4 represents the generation of e.m.f.'s. Determine (i) the direction in which the conductor has to be moved in Fig. 9.4(a), (ii) the direction of the induced e.m.f. in Fig. 9.4(b), (iii) the polarity of the magnetic system in Fig. 9.4(c).

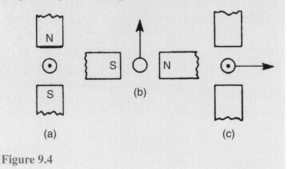

(a)          (b)          (c)

Figure 9.4

The direction of the e.m.f., and thus the current due to the e.m.f. may be obtained by either Lenz's law or Fleming's <u>R</u>ight-hand rule (i.e. Gene<u>R</u>ator rule).

(i)  Using Lenz's law: The field due to the magnet and the field due to the current-carrying conductor are shown in Fig. 9.5(a) and are seen to reinforce to the left of the conductor. Hence the force on the conductor is to the right. However Lenz's law states that the direction of the induced e.m.f. is always such as to oppose the effect producing it. **Thus the conductor will have to be moved to the left.**

(ii)  Using Fleming's right-hand rule:

<u>F</u>irst finger – <u>F</u>ield,

i.e. N → S, or right to left;

Thu<u>M</u>b – <u>M</u>otion, i.e. upwards;

S<u>E</u>cond finger – <u>E</u>.m.f.

i.e. **towards the viewer or out of the paper**, as shown in Fig. 9.5(b)

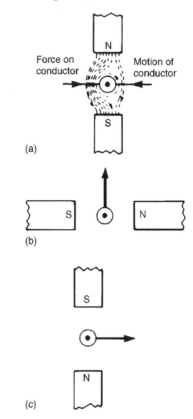

(a)

(b)

(c)

Figure 9.5

(iii)  The polarity of the magnetic system of Fig. 9.4(c) is shown in Fig. 9.5(c) and is obtained using Fleming's right-hand rule.

**Now try the following exercise**

**Exercise 43    Further problems on induced e.m.f.**

1.  A conductor of length 15 cm is moved at 750 mm/s at right-angles to a uniform flux density of 1.2 T. Determine the e.m.f. induced in the conductor.               [0.135 V]

2.  Find the speed that a conductor of length 120 mm must be moved at right-angles to a magnetic field of flux density 0.6 T to induce in it an e.m.f. of 1.8 V.          [25 m/s]

3. A 25 cm long conductor moves at a uniform speed of 8 m/s through a uniform magnetic field of flux density 1.2 T. Determine the current flowing in the conductor when (a) its ends are open-circuited, (b) its ends are connected to a load of 15 ohms resistance.

[(a) 0 (b) 0.16 A]

4. A straight conductor 500 mm long is moved with constant velocity at right-angles both to its length and to a uniform magnetic field. Given that the e.m.f. induced in the conductor is 2.5 V and the velocity is 5 m/s, calculate the flux density of the magnetic field. If the conductor forms part of a closed circuit of total resistance 5 ohms, calculate the force on the conductor.          [1 T, 0.25 N]

5. A car is travelling at 80 km/h. Assuming the back axle of the car is 1.76 m in length and the vertical component of the earth's magnetic field is 40 μT, find the e.m.f. generated in the axle due to motion.          [1.56 mV]

6. A conductor moves with a velocity of 20 m/s at an angle of (a) 90° (b) 45° (c) 30°, to a magnetic field produced between two square-faced poles of side length 2.5 cm. If the flux on the pole face is 60 mWb, find the magnitude of the induced e.m.f. in each case.

[(a) 48 V (b) 33.9 V (c) 24 V]

7. A conductor 400 mm long is moved at 70° to a 0.85 T magnetic field. If it has a velocity of 115 km/h, calculate (a) the induced voltage, and (b) force acting on the conductor if connected to an 8 Ω resistor.

[(a) 10.21 V (b) 0.408 N]

## 9.3   Rotation of a loop in a magnetic field

Figure 9.6 shows a view of a looped conductor whose sides are moving across a magnetic field.

The left-hand side is moving in an upward direction (check using Fleming's right-hand rule), with length $l$ cutting the lines of flux which are travelling from left to right. By definition, the induced e.m.f. will be equal to $Blv \sin \theta$ and flowing into the page.

The right-hand side is moving in a downward direction (again, check using Fleming's right-hand rule), with

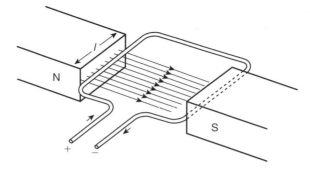

**Figure 9.6**

length $l$ cutting the same lines of flux as above. The induced e.m.f. will also be equal to $Blv \sin \theta$ but flowing out of the page.

Therefore the total e.m.f. for the loop conductor = $2Blv \sin \theta$

Now consider a coil made up of a number of turns $N$

The total e.m.f. $E$ for the loop conductor is now given by:

$$E = 2NBlv \sin \theta$$

**Problem 6.**   A rectangular coil of sides 12 cm and 8 cm is rotated in a magnetic field of flux density 1.4 T, the longer side of the coil actually cutting this flux. The coil is made up of 80 turns and rotates at 1200 rev/min. (a) Calculate the maximum generated e.m.f. (b) If the coil generates 90 V, at what speed will the coil rotate?

(a) Generated e.m.f. $E = 2NBLv \sin \theta$

where number of turns, $N - 80$, flux density, $B = 1.4$ T,

length of conductor in magnetic field, $l = 12$ cm $= 0.12$ m,

velocity, $v = \omega r = \left(\dfrac{1200}{60} \times 2\pi \text{ rad/s}\right)\left(\dfrac{0.08}{2} \text{ m}\right)$

$= 1.6\pi$ m/s,

and for maximum e.m.f. induced, $\theta = 90°$, from which, $\sin \theta = 1$

Hence, **maximum e.m.f. induced,**

$E = 2NBlv \sin \theta$

$= 2 \times 80 \times 1.4 \times 0.12 \times 1.6\pi \times 1$

$= \textbf{135.1 volts}$

(b) Since          $E = 2NBlv \sin \theta$

then          $90 = 2 \times 80 \times 1.4 \times 0.12 \times v \times 1$

from which, $v = \dfrac{90}{2 \times 80 \times 1.4 \times 0.12} = 3.348$ m/s

$v = \omega r$ hence, angular velocity, $\omega = \dfrac{v}{r} = \dfrac{3.348}{\dfrac{0.08}{2}}$

$$= 83.7 \, \text{rad/s}$$

**Speed of coil in rev/min** $= \dfrac{83.7 \times 60}{2\pi}$

$$= 799 \, \text{rev/min}$$

An **alternative method** of determining (b) is by **direct proportion**.

Since $E = 2NBlv\sin\theta$, then with $N$, $B$, $l$ and $\theta$ being constant, $E \propto v$

If from (a), 135.1 V is produced by a speed of 1200 rev/min,

then 1 V would be produced by a speed of $\dfrac{1200}{135.1} = 8.88 \, \text{rev/min}$

Hence, 90 V would be produced by a speed of $90 \times 8.88 = 799 \, \text{rev/min}$

**Now try the following exercise**

---

**Exercise 44    Further problems on induced e.m.f. in a coil**

1.  A rectangular coil of sides 8 cm by 6 cm is rotating in a magnetic field such that the longer sides cut the magnetic field. Calculate the maximum generated e.m.f. if there are 60 turns on the coil, the flux density is 1.6 T and the coil rotates at 1500 rev/min.    [72.38 V]

2.  A generating coil on a former 100 mm long has 120 turns and rotates in a 1.4 T magnetic field. Calculate the maximum e.m.f. generated if the coil, having a diameter of 60 mm, rotates at 450 rev/min.    [47.50 V]

3.  If the coils in problems 1 and 2 generates 60 V, calculate (a) the new speed for each coil, and (b) the flux density required if the speed is unchanged.
    [(a) 1243 rev/min, 568 rev/min
    (b) 1.33 T, 1.77 T]

---

## 9.4   Inductance

**Inductance** is the name given to the property of a circuit whereby there is an e.m.f. induced into the circuit by the change of flux linkages produced by a current change.

When the e.m.f. is induced in the same circuit as that in which the current is changing, the property is called **self inductance, L**. When the e.m.f. is induced in a circuit by a change of flux due to current changing in an adjacent circuit, the property is called **mutual inductance, M**. The unit of inductance is the **henry, H**.

*A circuit has an inductance of one henry when an e.m.f. of one volt is induced in it by a current changing at the rate of one ampere per second*
Induced e.m.f. in a coil of $N$ turns,

$$E = -N\frac{d\Phi}{dt} \, \text{volts}$$

where $d\Phi$ is the change in flux in Webers, and $dt$ is the time taken for the flux to change in seconds (i.e. $\frac{d\Phi}{dt}$ is the rate of change of flux).

Induced e.m.f. in a coil of inductance $L$ henrys,

$$E = -L\frac{dI}{dt} \, \text{volts}$$

where $dI$ is the change in current in amperes and $dt$ is the time taken for the current to change in seconds (i.e. $\frac{dI}{dt}$ is the rate of change of current). The minus sign in each of the above two equations remind us of its direction (given by Lenz's law).

---

**Problem 7.**   Determine the e.m.f. induced in a coil of 200 turns when there is a change of flux of 25 mWb linking with it in 50 ms.

Induced e.m.f. $E = -N\dfrac{d\Phi}{dt} = -(200)\left(\dfrac{25 \times 10^{-3}}{50 \times 10^{-3}}\right)$

$$= -100 \, \text{volts}$$

---

**Problem 8.**   A flux of 400 μWb passing through a 150-turn coil is reversed in 40 ms. Find the average e.m.f. induced.

Since the flux reverses, the flux changes from $+400\,\mu\text{Wb}$ to $-400\,\mu\text{Wb}$, a total change of flux of $800\,\mu\text{Wb}$.

Induced e.m.f. $E = -N\dfrac{d\Phi}{dt} = -(150)\left(\dfrac{800 \times 10^{-6}}{40 \times 10^{-3}}\right)$

$$= -\frac{150 \times 800 \times 10^{3}}{40 \times 10^{6}}$$

Hence, **the average e.m.f. induced, E = −3 volts**

**Problem 9.** Calculate the e.m.f. induced in a coil of inductance 12 H by a current changing at the rate of 4 A/s.

$$\text{Induced e.m.f. } E = -L\frac{dI}{dt} = -(12)(4)$$

$$= -48 \text{ volts}$$

**Problem 10.** An e.m.f. of 1.5 kV is induced in a coil when a current of 4 A collapses uniformly to zero in 8 ms. Determine the inductance of the coil.

Change in current, $dI = (4-0) = 4$ A, $dt = 8$ ms $= 8 \times 10^{-3}$ s,

$$\frac{dI}{dt} = \frac{4}{8 \times 10^{-3}} = \frac{4000}{8}$$

$$= 500 \text{ A/s}$$

and        $E = 1.5 \text{ kV} = 1500$ V

Since        $|E| = L\frac{dI}{dt}$

inductance, $L = \frac{|E|}{(dI/dt)} = \frac{1500}{500} = \mathbf{3\,H}$

(Note that $|E|$ means the 'magnitude of $E$' which disregards the minus sign)

**Problem 11.** An average e.m.f. of 40 V is induced in a coil of inductance 150 mH when a current of 6 A is reversed. Calculate the time taken for the current to reverse.

$|E| = 40$ V, $L = 150$ mH $= 0.15$ H and change in current, $dI = 6 - (-6) = 12$ A (since the current is reversed).

Since $|E| = \dfrac{dI}{dt}$

$$\text{time } dt = \frac{L\,dI}{|E|} = \frac{(0.15)(12)}{40}$$

$$= \mathbf{0.045\,s} \text{ or } \mathbf{45\,ms}$$

**Now try the following exercise**

**Exercise 45    Further problems on inductance**

1. Find the e.m.f. induced in a coil of 200 turns when there is a change of flux of 30 mWb linking with it in 40 ms.        [−150 V]

2. An e.m.f. of 25 V is induced in a coil of 300 turns when the flux linking with it changes by 12 mWb. Find the time, in milliseconds, in which the flux makes the change.        [144 ms]

3. An ignition coil having 10000 turns has an e.m.f. of 8 kV induced in it. What rate of change of flux is required for this to happen?        [0.8 Wb/s]

4. A flux of 0.35 mWb passing through a 125-turn coil is reversed in 25 ms. Find the magnitude of the average e.m.f. induced.        [3.5 V]

5. Calculate the e.m.f. induced in a coil of inductance 6 H by a current changing at a rate of 15 A/s.        [−90 V]

## 9.5  Inductors

A component called an inductor is used when the property of inductance is required in a circuit. The basic form of an inductor is simply a coil of wire. Factors which affect the inductance of an inductor include:

(i)   the number of turns of wire – the more turns the higher the inductance

(ii)   the cross-sectional area of the coil of wire – the greater the cross-sectional area the higher the inductance

(iii)   the presence of a magnetic core – when the coil is wound on an iron core the same current sets up a more concentrated magnetic field and the inductance is increased

(iv)   the way the turns are arranged – a short thick coil of wire has a higher inductance than a long thin one.

Two examples of practical inductors are shown in Fig. 9.7, and the standard electrical circuit diagram symbols for air-cored and iron-cored inductors are shown in Fig. 9.8.

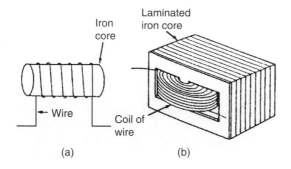

Figure 9.7

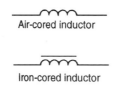

Figure 9.8

An iron-cored inductor is often called a **choke** since, when used in a.c. circuits, it has a choking effect, limiting the current flowing through it.

Inductance is often undesirable in a circuit. To reduce inductance to a minimum the wire may be bent back on itself, as shown in Fig. 9.9, so that the magnetising effect of one conductor is neutralised by that of the adjacent conductor. The wire may be coiled around an insulator, as shown, without increasing the inductance. Standard resistors may be non-inductively wound in this manner.

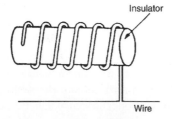

Figure 9.9

## 9.6 Energy stored

An inductor possesses an ability to store energy. The energy stored, $W$, in the magnetic field of an inductor is given by:

$$W = \frac{1}{2}LI^2 \text{ joules}$$

**Problem 12.** An 8 H inductor has a current of 3 A flowing through it. How much energy is stored in the magnetic field of the inductor?

Energy stored,

$$W = \frac{1}{2}LI^2 = \frac{1}{2}(8)(3)^2 = \textbf{36 joules}$$

**Now try the following exercise**

**Exercise 46  Further problems on energy stored**

1. An inductor of 20 H has a current of 2.5 A flowing in it. Find the energy stored in the magnetic field of the inductor.  [62.5 J]

2. Calculate the value of the energy stored when a current of 30 mA is flowing in a coil of inductance 400 mH.  [0.18 mJ]

3. The energy stored in the magnetic field of an inductor is 80 J when the current flowing in the inductor is 2 A. Calculate the inductance of the coil.  [40 H]

## 9.7 Inductance of a coil

If a current changing from 0 to $I$ amperes, produces a flux change from 0 to $\Phi$ webers, then $dI = I$ and $d\Phi = \Phi$. Then, from Section 9.3,

$$\text{induced e.m.f. } E = \frac{N\Phi}{t} = \frac{LI}{t}$$

from which, **inductance of coil**,

$$L = \frac{N\Phi}{I} \text{ henrys}$$

Since $E = -L\dfrac{dI}{dt} = -N\dfrac{d\Phi}{dt}$ then $L = N\dfrac{d\Phi}{dt}\left(\dfrac{dt}{dI}\right)$

i.e. $L = N\dfrac{d\Phi}{dI}$

From Chapter 7, m.m.f. $= \Phi S$ from which, $\Phi = \dfrac{\text{m.m.f}}{S}$

Substituting into $L = N\dfrac{d\Phi}{dI}$ gives

$$L = N\frac{d}{dI}\left(\frac{\text{m.m.f.}}{S}\right)$$

i.e. $\qquad L = \dfrac{N}{S}\dfrac{d(NI)}{dI}$ since m.m.f. $= NI$

i.e. $\qquad L = \dfrac{N^2}{S}\dfrac{dI}{dI}$ and since $\dfrac{dI}{dI} = 1$,

$$L = \frac{N^2}{S} \text{ henrys}$$

---

**Problem 13.** Calculate the coil inductance when a current of 4 A in a coil of 800 turns produces a flux of 5 mWb linking with the coil.

---

For a coil, inductance

$$L = \frac{N\Phi}{I} = \frac{(800)(5 \times 10^{-3})}{4} = \mathbf{1\,H}$$

---

**Problem 14.** A flux of 25 mWb links with a 1500 turn coil when a current of 3 A passes through the coil. Calculate (a) the inductance of the coil, (b) the energy stored in the magnetic field, and (c) the average e.m.f. induced if the current falls to zero in 150 ms.

---

(a) **Inductance,**

$$L = \frac{N\Phi}{I} = \frac{(1500)(25 \times 10^{-3})}{3} = \mathbf{12.5\,H}$$

(b) **Energy stored,**

$$W = \frac{1}{2}LI^2 = \frac{1}{2}(12.5)(3)^2 = \mathbf{56.25\,J}$$

(c) **Induced e.m.f.,**

$$E = -L\frac{dI}{dt} = -(12.5)\left(\frac{3-0}{150 \times 10^{-3}}\right)$$

$$= \mathbf{-250\,V}$$

(Alternatively,

$$E = -N\frac{d\Phi}{dt}$$

$$= -(1500)\left(\frac{25 \times 10^{-3}}{150 \times 10^{-3}}\right)$$

$$= \mathbf{-250\,V}$$

since if the current falls to zero so does the flux)

---

**Problem 15.** When a current of 1.5 A flows in a coil the flux linking with the coil is 90 μWb. If the coil inductance is 0.60 H, calculate the number of turns of the coil.

---

For a coil, $L = \dfrac{N\Phi}{I}$

Thus $\qquad N = \dfrac{LI}{\Phi} = \dfrac{(0.6)(1.5)}{90 \times 10^{-6}} = \mathbf{10\,000\,turns}$

---

**Problem 16.** A 750 turn coil of inductance 3 H carries a current of 2 A. Calculate the flux linking the coil and the e.m.f. induced in the coil when the current collapses to zero in 20 ms.

---

Coil inductance, $L = \dfrac{N\Phi}{I}$ from which,

flux $\qquad \Phi = \dfrac{LI}{N} = \dfrac{(3)(2)}{750} = 8 \times 10^{-3} = \mathbf{8\,mWb}$

Induced e.m.f.

$$E = -L\frac{dI}{dt} = -(3)\left(\frac{2-0}{20 \times 10^{-3}}\right)$$

$$= \mathbf{-300\,V}$$

(Alternatively,

$$E = -N\frac{d\Phi}{dt} = -(750)\left(\frac{8 \times 10^{-3}}{20 \times 10^{-3}}\right)$$

$$= \mathbf{-300\,V})$$

---

**Problem 17.** A silicon iron ring is wound with 800 turns, the ring having a mean diameter of 120 mm and a cross-sectional area of 400 mm². If when carrying a current of 0.5 A the relative permeability is found to be 3000, calculate (a) the self-inductance of the coil, (b) the induced e.m.f. if the current is reduced to zero in 80 ms.

---

The ring is shown sketched in Fig. 9.10.

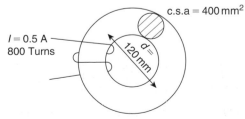

**Figure 9.10**

(a)   Inductance, $L = \dfrac{N^2}{S}$ and from Chapter 7,

reluctance,   $S = \dfrac{l}{\mu_0 \mu_r A}$

i.e. $S = \dfrac{\pi \times 120 \times 10^{-3}}{4\pi \times 10^{-7} \times 3000 \times 400 \times 10^{-6}}$

$= 250 \times 10^3 \text{ A/Wb}$

Hence, **self-inductance**, $L = \dfrac{N^2}{S} = \dfrac{800^2}{250 \times 10^3}$

$= \mathbf{2.56\,H}$

(b)   **Induced e.m.f.,** $E = -L\dfrac{dI}{dt} = -(2.56)\dfrac{(0.5 - 0)}{80 \times 10^{-3}}$

$= \mathbf{-16\,V}$

---

**Now try the following exercise**

**Exercise 47    Further problems on the inductance of a coil**

1.   A flux of 30 mWb links with a 1200 turn coil when a current of 5 A is passing through the coil. Calculate (a) the inductance of the coil, (b) the energy stored in the magnetic field, and (c) the average e.m.f. induced if the current is reduced to zero in 0.20 s.
     [(a) 7.2 H (b) 90 J (c) 180 V]

2.   An e.m.f. of 2 kV is induced in a coil when a current of 5 A collapses uniformly to zero in 10 ms. Determine the inductance of the coil.
     [4 H]

3.   An average e.m.f. of 60 V is induced in a coil of inductance 160 mH when a current of 7.5 A is reversed. Calculate the time taken for the current to reverse.
     [40 ms]

4.   A coil of 2500 turns has a flux of 10 mWb linking with it when carrying a current of 2 A. Calculate the coil inductance and the e.m.f. induced in the coil when the current collapses to zero in 20 ms.
     [12.5 H, 1.25 kV]

5.   Calculate the coil inductance when a current of 5 A in a coil of 1000 turns produces a flux of 8 mWb linking with the coil.
     [1.6 H]

6.   A coil is wound with 600 turns and has a self inductance of 2.5 H. What current must flow to set up a flux of 20 mWb?
     [4.8 A]

7.   When a current of 2 A flows in a coil, the flux linking with the coil is 80 μWb. If the coil inductance is 0.5 H, calculate the number of turns of the coil.
     [12 500]

8.   A coil of 1200 turns has a flux of 15 mWb linking with it when carrying a current of 4 A. Calculate the coil inductance and the e.m.f. induced in the coil when the current collapses to zero in 25 ms.
     [4.5 H, 720 V]

9.   A coil has 300 turns and an inductance of 4.5 mH. How many turns would be needed to produce a 0.72 mH coil assuming the same core is used?
     [48 turns]

10.   A steady current of 5 A when flowing in a coil of 1000 turns produces a magnetic flux of 500 μWb. Calculate the inductance of the coil. The current of 5 A is then reversed in 12.5 ms. Calculate the e.m.f. induced in the coil.
     [0.1 H, 80 V]

11.   An iron ring has a cross-sectional area of 500 mm² and a mean length of 300 mm. It is wound with 100 turns and its relative permeability is 1600. Calculate (a) the current required to set up a flux of 500 μWb in the coil, (b) the inductance of the system, and (c) the induced e.m.f. if the field collapses in 1 ms.
     [(a) 1.492 A (b) 33.51 mH (c) −50 V]

## 9.8   Mutual inductance

Mutually induced e.m.f. in the second coil,

$$E_2 = -M\dfrac{dI_1}{dt} \text{ volts}$$

where $M$ is the **mutual inductance** between two coils, in henrys, and $(dI_1/dt)$ is the rate of change of current in the first coil.

The phenomenon of mutual inductance is used in **transformers** (see Chapter 21, page 333)

**Another expression for M**

Let an iron ring have two coils, A and B, wound on it. If the fluxes $\Phi_1$ and $\Phi_2$ are produced from currents $I_1$ and $I_2$ in coils A and B respectively, then the reluctance could be expressed as:

$$S = \dfrac{I_1 N_1}{\Phi_1} = \dfrac{I_2 N_2}{\Phi_2}$$

If the flux in coils A and B are the same and produced from the current $I_1$ in coil A only, assuming 100% coupling, then the mutual inductance can be expressed as:

$$M = \frac{N_2 \Phi_1}{I_1}$$

Multiplying by $\left(\dfrac{N_1}{N_1}\right)$ gives:

$$M = \frac{N_2 \Phi_1 N_1}{I_1 N_1}$$

However,        $S = \dfrac{I_1 N_1}{\Phi_1}$

Thus, **mutual inductance,** $M = \dfrac{N_1 N_2}{S}$

---

**Problem 18.**   Calculate the mutual inductance between two coils when a current changing at 200 A/s in one coil induces an e.m.f. of 1.5 V in the other.

Induced e.m.f. $|E_2| = M \, dI_1/dt$, i.e. $1.5 = M(200)$. Thus **mutual inductance,**

$$M = \frac{1.5}{200} = \textbf{0.0075 H or 7.5 mH}$$

---

**Problem 19.**   The mutual inductance between two coils is 18 mH. Calculate the steady rate of change of current in one coil to induce an e.m.f. of 0.72 V in the other.

Induced e.m.f. $|E_2| = M \dfrac{dI_1}{dt}$

Hence rate of change of current,

$$\frac{dI_1}{dt} = \frac{|E_2|}{M} = \frac{0.72}{0.018} = \textbf{40 A/s}$$

---

**Problem 20.**   Two coils have a mutual inductance of 0.2 H. If the current in one coil is changed from 10 A to 4 A in 10 ms, calculate (a) the average induced e.m.f. in the second coil, (b) the change of flux linked with the second coil if it is wound with 500 turns.

(a)  Induced e.m.f.

$$|E_2| = -M \frac{dI_1}{dt}$$
$$= -(0.2)\left(\frac{10-4}{10 \times 10^{-3}}\right) = \textbf{-120 V}$$

(b)  Induced e.m.f.

$$|E_2| = N \frac{d\Phi}{dt}, \text{ hence } d\Phi = \frac{|E_2| dt}{N}$$

Thus the change of flux,

$$d\Phi = \frac{(120)(10 \times 10^{-3})}{500} = \textbf{2.4 mWb}$$

---

**Problem 21.**   In the device shown in Fig. 9.11, when the current in the primary coil of 1000 turns increases linearly from 1 A to 6 A in 200 ms, an e.m.f. of 15 V is induced into the secondary coil of 480 turns, which is left open circuited. Determine (a) the mutual inductance of the two coils, (b) the reluctance of the former, and (c) the self-inductance of the primary coil.

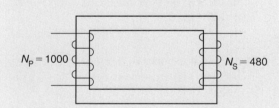

$N_P = 1000$        $N_S = 480$

**Figure 9.11**

(a)  $E_S = M \dfrac{dI_p}{dt}$ from which,

mutual inductance, $M = \dfrac{E_S}{\dfrac{dI_P}{dt}} = \dfrac{15}{\left(\dfrac{6-1}{200 \times 10^{-3}}\right)}$

$$= \frac{15}{25} = \textbf{0.60 H}$$

(b)  $M = \dfrac{N_P N_S}{S}$ from which,

reluctance, $S = \dfrac{N_P N_S}{M} = \dfrac{(1000)(480)}{0.60}$

$$= \textbf{800 000 A/Wb or 800 kA/Wb}$$

(c)  Primary self-inductance, $L_P = \dfrac{N_P^2}{S} = \dfrac{(1000)^2}{800\,000}$

$$= \textbf{1.25 H}$$

Section 1

### Now try the following exercises

#### Exercise 48　Further problems on mutual inductance

1. The mutual inductance between two coils is 150 mH. Find the magnitude of the e.m.f. induced in one coil when the current in the other is increasing at a rate of 30 A/s.

   [4.5 V]

2. Determine the mutual inductance between two coils when a current changing at 50 A/s in one coil induces an e.m.f. of 80 mV in the other.

   [1.6 mH]

3. Two coils have a mutual inductance of 0.75 H. Calculate the magnitude of the e.m.f. induced in one coil when a current of 2.5 A in the other coil is reversed in 15 ms.　[250 V]

4. The mutual inductance between two coils is 240 mH. If the current in one coil changes from 15 A to 6 A in 12 ms, calculate (a) the average e.m.f. induced in the other coil, (b) the change of flux linked with the other coil if it is wound with 400 turns.

   [(a) −180 V (b) 5.4 mWb]

5. A mutual inductance of 0.06 H exists between two coils. If a current of 6 A in one coil is reversed in 0.8 s calculate (a) the average e.m.f. induced in the other coil, (b) the number of turns on the other coil if the flux change linking with the other coil is 5 mWb.

   [(a) −0.9 V (b) 144]

6. When the current in the primary coil of 400 turns of a magnetic circuit increases linearly from 10 mA to 35 mA in 100 ms, an e.m.f. of 75 mV is induced into the secondary coil of 240 turns, which is left open circuited. Determine (a) the mutual inductance of the two coils, (b) the reluctance of the former, and (c) the self-inductance of the secondary coil.

   [(a) 0.30 H (b) 320 kA/Wb (c) 0.18 H]

#### Exercise 49　Short answer questions on electromagnetic induction

1. What is electromagnetic induction?

2. State Faraday's laws of electromagnetic induction

3. State Lenz's law

4. Explain briefly the principle of the generator

5. The direction of an induced e.m.f. in a generator may be determined using Fleming's ...... rule

6. The e.m.f. $E$ induced in a moving conductor may be calculated using the formula $E = Blv$. Name the quantities represented and their units

7. The total e.m.f., $E$, for a loop conductor with $N$ turns is given by: $E = \ldots\ldots\ldots\ldots$

8. What is self-inductance? State its symbol

9. State and define the unit of inductance

10. When a circuit has an inductance $L$ and the current changes at a rate of $(di/dt)$ then the induced e.m.f. $E$ is given by $E = \ldots\ldots$ volts

11. If a current of $I$ amperes flowing in a coil of $N$ turns produces a flux of $\Phi$ webers, the coil inductance $L$ is given by $L = \ldots\ldots$ henrys

12. The energy $W$ stored by an inductor is given by $W = \ldots\ldots$ joules

13. If the number of turns of a coil is $N$ and its reluctance is $S$, then the inductance, $L$, is given by: $L = \ldots\ldots\ldots$

14. What is mutual inductance? State its symbol

15. The mutual inductance between two coils is $M$. The e.m.f. $E_2$ induced in one coil by the current changing at $(dI_1/dt)$ in the other is given by $E_2 = \ldots\ldots$ volts

16. Two coils wound on an iron ring of reluctance $S$ have $N_A$ and $N_B$ turns respectively. The mutual inductance, $M$, is given by: $M = \ldots\ldots\ldots$

#### Exercise 50　Multi-choice questions on electromagnetic induction (Answers on page 420)

1. A current changing at a rate of 5 A/s in a coil of inductance 5 H induces an e.m.f. of:

(a) 25 V in the same direction as the applied voltage

(b) 1 V in the same direction as the applied voltage

(c) 25 V in the opposite direction to the applied voltage

(d) 1 V in the opposite direction to the applied voltage

2. A bar magnet is moved at a steady speed of 1.0 m/s towards a coil of wire which is connected to a centre-zero galvanometer. The magnet is now withdrawn along the same path at 0.5 m/s. The deflection of the galvanometer is in the:

(a) same direction as previously, with the magnitude of the deflection doubled

(b) opposite direction as previously, with the magnitude of the deflection halved

(c) same direction as previously, with the magnitude of the deflection halved

(d) opposite direction as previously, with the magnitude of the deflection doubled

3. When a magnetic flux of 10 Wb links with a circuit of 20 turns in 2 s, the induced e.m.f. is:
(a) 1 V                (b) 4 V
(c) 100 V              (d) 400 V

4. A current of 10 A in a coil of 1000 turns produces a flux of 10 mWb linking with the coil. The coil inductance is:
(a) $10^6$ H           (b) 1 H
(c) 1 μH               (d) 1 mH

5. An e.m.f. of 1 V is induced in a conductor moving at 10 cm/s in a magnetic field of 0.5 T. The effective length of the conductor in the magnetic field is:
(a) 20 cm              (b) 5 m
(c) 20 m               (d) 50 m

6. Which of the following is false?

(a) Fleming's left-hand rule or Lenz's law may be used to determine the direction of an induced e.m.f.

(b) An induced e.m.f. is set up whenever the magnetic field linking that circuit changes

(c) The direction of an induced e.m.f. is always such as to oppose the effect producing it

(d) The induced e.m.f. in any circuit is proportional to the rate of change of the magnetic flux linking the circuit

7. The effect of inductance occurs in an electrical circuit when:
(a) the resistance is changing
(b) the flux is changing
(c) the current is changing

8. Which of the following statements is false? The inductance of an inductor increases:
(a) with a short, thick coil
(b) when wound on an iron core
(c) as the number of turns increases
(d) as the cross-sectional area of the coil decreases

9. The mutual inductance between two coils, when a current changing at 20 A/s in one coil induces an e.m.f. of 10 mV in the other, is:
(a) 0.5 H              (b) 200 mH
(c) 0.5 mH            (d) 2 H

10. A strong permanent magnet is plunged into a coil and left in the coil. What is the effect produced on the coil after a short time?

(a) There is no effect

(b) The insulation of the coil burns out

(c) A high voltage is induced

(d) The coil winding becomes hot

11. Self-inductance occurs when:

(a) the current is changing

(b) the circuit is changing

(c) the flux is changing

(d) the resistance is changing

12. Faraday's laws of electromagnetic induction are related to:

(a) the e.m.f. of a chemical cell

(b) the e.m.f. of a generator

(c) the current flowing in a conductor

(d) the strength of a magnetic field

# Chapter 10

# Electrical measuring instruments and measurements

At the end of this chapter you should be able to:

- recognize the importance of testing and measurements in electric circuits
- appreciate the essential devices comprising an analogue instrument
- explain the operation of an attraction and a repulsion type of moving-iron instrument
- explain the operation of a moving-coil rectifier instrument
- compare moving-coil, moving-iron and moving-coil rectifier instruments
- calculate values of shunts for ammeters and multipliers for voltmeters
- understand the advantages of electronic instruments
- understand the operation of an ohmmeter/megger
- appreciate the operation of multimeters/Avometers /Flukes
- understand the operation of a wattmeter
- appreciate instrument 'loading' effect
- understand the operation of an oscilloscope for d.c. and a.c. measurements
- calculate periodic time, frequency, peak-to-peak values from waveforms on an oscilloscope
- appreciate virtual test and measuring instruments
- recognize harmonics present in complex waveforms
- determine ratios of powers, currents and voltages in decibels
- understand null methods of measurement for a Wheatstone bridge and d.c. potentiometer
- understand the operation of a.c. bridges
- understand the operation of a Q-meter
- appreciate the most likely source of errors in measurements
- appreciate calibration accuracy of instruments

DOI: 10.1016/B978-0-08-089056-2.00010-3

## 10.1    Introduction

Tests and measurements are important in designing, evaluating, maintaining and servicing electrical circuits and equipment. In order to detect electrical quantities such as current, voltage, resistance or power, it is necessary to transform an electrical quantity or condition into a visible indication. This is done with the aid of instruments (or meters) that indicate the magnitude of quantities either by the position of a pointer moving over a graduated scale (called an analogue instrument) or in the form of a decimal number (called a digital instrument).

The digital instrument has, in the main, become the instrument of choice in recent years; in particular, computer-based instruments are rapidly replacing items of conventional test equipment, with the virtual storage test instrument, the **digital storage oscilloscope**, being the most common. This is explained later in this chapter, but before that some analogue instruments, which are still used in some installations, are explored.

## 10.2    Analogue instruments

All analogue electrical indicating instruments require three essential devices:

(a)  **A deflecting or operating device.** A mechanical force is produced by the current or voltage which causes the pointer to deflect from its zero position.

(b)  **A controlling device.** The controlling force acts in opposition to the deflecting force and ensures that the deflection shown on the meter is always the same for a given measured quantity. It also prevents the pointer always going to the maximum deflection. There are two main types of controlling device – spring control and gravity control.

(c)  **A damping device.** The damping force ensures that the pointer comes to rest in its final position quickly and without undue oscillation. There are three main types of damping used – eddy-current damping, air-friction damping and fluid-friction damping.

There are basically **two types of scale** – linear and non-linear. A **linear scale** is shown in Fig. 10.1(a), where the divisions or graduations are evenly spaced. The voltmeter shown has a range 0–100 V, i.e. a full-scale deflection (f.s.d.) of 100 V. A **non-linear scale** is shown in Fig. 10.1(b) where the scale is cramped at the

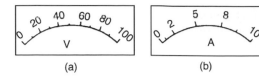

Figure 10.1

beginning and the graduations are uneven throughout the range. The ammeter shown has a f.s.d. of 10 A.

## 10.3    Moving-iron instrument

(a)  An **attraction type** of moving-iron instrument is shown diagrammatically in Fig. 10.2(a). When current flows in the solenoid, a pivoted soft-iron disc is attracted towards the solenoid and the movement causes a pointer to move across a scale.

(b)  In the **repulsion type** moving-iron instrument shown diagrammatically in Fig. 10.2(b), two pieces of iron are placed inside the solenoid, one being fixed, and the other attached to the spindle carrying the pointer. When current passes through the solenoid, the two pieces of iron are magnetised in the same direction and therefore repel each other. The pointer thus moves across the scale. The force moving the pointer is, in each type, proportional to $I^2$ and because of this the

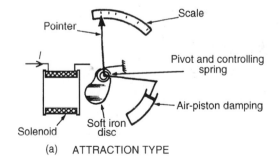

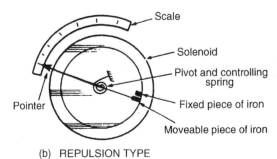

Figure 10.2

direction of current does not matter. The moving-iron instrument can be used on d.c. or a.c.; the scale, however, is non-linear.

## 10.4 The moving-coil rectifier instrument

A moving-coil instrument, which measures only d.c., may be used in conjunction with a bridge rectifier circuit as shown in Fig. 10.3 to provide an indication of alternating currents and voltages (see Chapter 14). The average value of the full wave rectified current is 0.637 $I_\mathrm{m}$. However, a meter being used to measure a.c. is usually calibrated in r.m.s. values. For sinusoidal quantities the indication is $(0.707 I_\mathrm{m})/(0.637 I_\mathrm{m})$ i.e. 1.11 times the mean value. Rectifier instruments have scales calibrated in r.m.s. quantities and it is assumed by the manufacturer that the a.c. is sinusoidal.

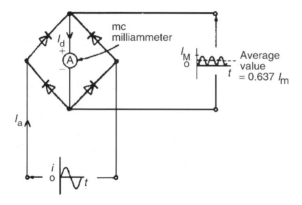

Figure 10.3

## 10.5 Comparison of moving-coil, moving-iron and moving-coil rectifier instruments

See Table at top of next page. (For the principle of operation of a moving-coil instrument, see Chapter 8, page 98.)

## 10.6 Shunts and multipliers

An **ammeter**, which measures current, has a low resistance (ideally zero) and must be connected in series with the circuit.

A **voltmeter**, which measures p.d., has a high resistance (ideally infinite) and must be connected in parallel with the part of the circuit whose p.d. is required.

There is no difference between the basic instrument used to measure current and voltage since both use a milliammeter as their basic part. This is a sensitive instrument which gives f.s.d. for currents of only a few milliamperes. When an ammeter is required to measure currents of larger magnitude, a proportion of the current is diverted through a low-value resistance connected in parallel with the meter. Such a diverting resistor is called a **shunt**.

From Fig. 10.4(a), $V_\mathrm{PQ} = V_\mathrm{RS}$.

Hence $I_\mathrm{a} r_\mathrm{a} = I_\mathrm{s} R_\mathrm{s}$. Thus the value of the shunt,

$$R_\mathrm{s} = \frac{I_\mathrm{a} r_\mathrm{a}}{I_\mathrm{s}} \text{ ohms}$$

The milliammeter is converted into a voltmeter by connecting a high value resistance (called a **multiplier**) in series with it as shown in Fig. 10.4(b). From Fig. 10.4(b),

$$V = V_\mathrm{a} + V_\mathrm{M} = I r_\mathrm{a} + I R_\mathrm{M}$$

Thus the value of the multiplier,

$$R_\mathrm{M} = \frac{V - I r_\mathrm{a}}{I} \text{ ohms}$$

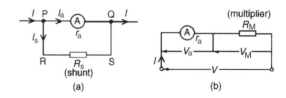

Figure 10.4

Problem 1. A moving-coil instrument gives a f.s.d. when the current is 40 mA and its resistance is 25 Ω. Calculate the value of the shunt to be connected in parallel with the meter to enable it to be used as an ammeter for measuring currents up to 50 A.

The circuit diagram is shown in Fig. 10.5, where $r_\mathrm{a}$ = resistance of instrument = 25 Ω, $R_\mathrm{s}$ = resistance

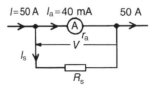

Figure 10.5

| Type of instrument | Moving-coil | Moving-iron | Moving-coil rectifier |
|---|---|---|---|
| Suitable for measuring | Direct current and voltage | Direct and alternating currents and voltage (reading in r.m.s. value) | Alternating current and voltage (reads average value but scale is adjusted to give r.m.s. value for sinusoidal waveforms) |
| Scale | Linear | Non-linear | Linear |
| Method of control | Hairsprings | Hairsprings | Hairsprings |
| Method of damping | Eddy current | Air | Eddy current |
| Frequency limits | — | 20–200 Hz | 20–100 kHz |
| Advantages | 1. Linear scale<br>2. High sensitivity<br>3. Well shielded from stray magnetic fields<br>4. Low power consumption | 1. Robust construction<br>2. Relatively cheap<br>3. Measures dc and ac<br>4. In frequency range 20–100 Hz reads r.m.s. correctly regardless of supply wave-form | 1. Linear scale<br>2. High sensitivity<br>3. Well shielded from stray magnetic fields<br>4. Lower power consumption<br>5. Good frequency range |
| Disadvantages | 1. Only suitable for dc<br>2. More expensive than moving iron type<br>3. Easily damaged | 1. Non-linear scale<br>2. Affected by stray magnetic fields<br>3. Hysteresis errors in dc circuits<br>4. Liable to temperature errors<br>5. Due to the inductance of the solenoid, readings can be affected by variation of frequency | 1. More expensive than moving iron type<br>2. Errors caused when supply is non-sinusoidal |

of shunt, $I_a$ = maximum permissible current flowing in instrument = 40 mA = 0.04 A, $I_s$ = current flowing in shunt and $I$ = total circuit current required to give f.s.d. = 50 A.

$$\text{Since } I = I_a + I_s \text{ then } I_s = I - I_a$$

i.e. $$I_s = 50 - 0.04 = 49.96\,A$$

$$V = I_a r_a = I_s R_s, \text{ hence}$$

$$R_s = \frac{I_a r_a}{I_s} = \frac{(0.04)(25)}{49.96} = 0.02002\,\Omega$$

$$= 20.02\,m\Omega$$

Thus for the moving-coil instrument to be used as an ammeter with a range 0–50 A, a resistance of value 20.02 mΩ needs to be connected in parallel with the instrument.

**Problem 2.** A moving-coil instrument having a resistance of 10 Ω, gives a f.s.d. when the current is 8 mA. Calculate the value of the multiplier to be connected in series with the instrument so that it can be used as a voltmeter for measuring p.d.s. up to 100 V.

The circuit diagram is shown in Fig. 10.6, where $r_a$ = resistance of instrument = 10 Ω, $R_M$ = resistance

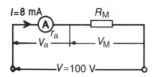

**Figure 10.6**

of multiplier $I$ = total permissible instrument current = 8 mA = 0.008 A, $V$ = total p.d. required to give f.s.d. = 100 V

$$V = V_a + V_M = Ir_a + IR_M$$

i.e. $100 = (0.008)(10) + (0.008)R_M$
or $100 - 0.08 = 0.008\,R_M$, thus

$$R_M = \frac{99.92}{0.008} = 12490\,\Omega = \textbf{12.49 k}\boldsymbol{\Omega}$$

Hence for the moving-coil instrument to be used as a voltmeter with a range 0–100 V, a resistance of value 12.49 kΩ needs to be connected in series with the instrument.

**Now try the following exercise**

**Exercise 51 Further problems on shunts and multipliers**

1. A moving-coil instrument gives f.s.d. for a current of 10 mA. Neglecting the resistance of the instrument, calculate the approximate value of series resistance needed to enable the instrument to measure up to (a) 20 V (b) 100 V (c) 250 V. [(a) 2 kΩ (b) 10 kΩ (c) 25 kΩ]

2. A meter of resistance 50 Ω has a f.s.d. of 4 mA. Determine the value of shunt resistance required in order that f.s.d. should be (a) 15 mA (b) 20 A (c) 100 A.
[(a) 18.18 Ω (b) 10.00 mΩ (c) 2.00 mΩ]

3. A moving-coil instrument having a resistance of 20 Ω, gives a f.s.d. when the current is 5 mA. Calculate the value of the multiplier to be connected in series with the instrument so that it can be used as a voltmeter for measuring p.d.'s up to 200 V. [39.98 kΩ]

4. A moving-coil instrument has a f.s.d. of 20 mA and a resistance of 25 Ω. Calculate the values of resistance required to enable the instrument to be used (a) as a 0–10 A ammeter, and (b) as a 0–100 V voltmeter. State the mode of resistance connection in each case.
[(a) 50.10 mΩ in parallel
(b) 4.975 kΩ in series]

5. A meter has a resistance of 40 Ω and registers a maximum deflection when a current of 15 mA flows. Calculate the value of resistance that converts the movement into (a) an ammeter with a maximum deflection of 50 A, and (b) a voltmeter with a range 0–250 V.
[(a) 12.00 mΩ in parallel
(b) 16.63 kΩ in series]

## 10.7 Electronic instruments

Electronic measuring instruments have advantages over instruments such as the moving-iron or moving-coil meters, in that they have a much higher input resistance (some as high as 1000 MΩ) and can handle a much wider range of frequency (from d.c. up to MHz).

The digital voltmeter (DVM) is one which provides a digital display of the voltage being measured. Advantages of a DVM over analogue instruments include higher accuracy and resolution, no observational or parallex errors (see Section 10.22) and a very high input resistance, constant on all ranges.

A digital multimeter is a DVM with additional circuitry which makes it capable of measuring a.c. voltage, d.c. and a.c. current and resistance.

Instruments for a.c. measurements are generally calibrated with a sinusoidal alternating waveform to indicate r.m.s. values when a sinusoidal signal is applied to the instrument. Some instruments, such as the moving-iron and electro-dynamic instruments, give a true r.m.s. indication. With other instruments the indication is either scaled up from the mean value (such as with the rectified moving-coil instrument) or scaled down from the peak value.

Sometimes quantities to be measured have complex waveforms (see Section 10.15), and whenever a quantity is non-sinusoidal, errors in instrument readings can occur if the instrument has been calibrated for sine waves only. Such waveform errors can be largely eliminated by using electronic instruments.

## 10.8   The ohmmeter

An **ohmmeter** is an instrument for measuring electrical resistance. A simple ohmmeter circuit is shown in Fig. 10.7(a). Unlike the ammeter or voltmeter, the ohmmeter circuit does not receive the energy necessary for its operation from the circuit under test. In the ohmmeter this energy is supplied by a self-contained source of voltage, such as a battery. Initially, terminals XX are short-circuited and $R$ adjusted to give f.s.d. on the milliammeter. If current $I$ is at a maximum value and voltage $E$ is constant, then resistance $R = E/I$ is at a minimum value. Thus f.s.d. on the milliammeter is made zero on the resistance scale. When terminals XX are open circuited no current flows and $R (= E/O)$ is infinity, $\infty$.

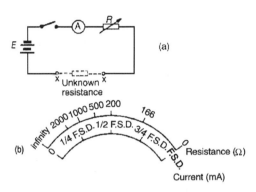

Figure 10.7

The milliammeter can thus be calibrated directly in ohms. A cramped (non-linear) scale results and is 'back to front', as shown in Fig. 10.7(b). When calibrated, an unknown resistance is placed between terminals XX and its value determined from the position of the pointer on the scale. An ohmmeter designed for measuring low values of resistance is called a **continuity tester**. An ohmmeter designed for measuring high values of resistance (i.e. megohms) is called an **insulation resistance tester** (e.g. '**Megger**').

## 10.9   Multimeters

Instruments are manufactured that combine a moving-coil meter with a number of shunts and series multipliers, to provide a range of readings on a single scale graduated to read current and voltage. If a battery is incorporated then resistance can also be measured. Such instruments are called **multimeters** or

universal instruments or **multirange instruments**. An 'Avometer' is a typical example. A particular range may be selected either by the use of separate terminals or by a selector switch. Only one measurement can be performed at a time. Often such instruments can be used in a.c. as well as d.c. circuits when a rectifier is incorporated in the instrument.

**Digital Multimeters (DMM)** are now almost universally used, the **Fluke Digital Multimeter** being an industry leader for performance, accuracy, resolution, ruggedness, reliability and safety. These instruments measure d.c. currents and voltages, resistance and continuity, a.c. (r.m.s.) currents and voltages, temperature, and much more.

## 10.10   Wattmeters

A **wattmeter** is an instrument for measuring electrical power in a circuit. Figure 10.8 shows typical connections of a wattmeter used for measuring power supplied to a load. The instrument has two coils:

(i)   a current coil, which is connected in series with the load, like an ammeter, and

(ii)   a voltage coil, which is connected in parallel with the load, like a voltmeter.

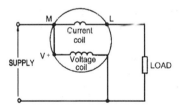

Figure 10.8

## 10.11   Instrument 'loading' effect

Some measuring instruments depend for their operation on power taken from the circuit in which measurements are being made. Depending on the 'loading' effect of the instrument (i.e. the current taken to enable it to operate), the prevailing circuit conditions may change.

The resistance of voltmeters may be calculated since each have a stated sensitivity (or 'figure of merit'), often stated in 'k$\Omega$ per volt' of f.s.d. A voltmeter should have as high a resistance as possible (− ideally infinite). In a.c. circuits the impedance of the instrument varies with frequency and thus the loading effect of the instrument can change.

**Problem 3.** Calculate the power dissipated by the voltmeter and by resistor R in Fig. 10.9 when (a) $R = 250\,\Omega$ (b) $R = 2\,M\Omega$. Assume that the voltmeter sensitivity (sometimes called figure of merit) is $10\,k\Omega/V$.

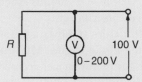

**Figure 10.9**

(a) Resistance of voltmeter, $R_v =$ sensitivity $\times$ f.s.d. Hence, $R_v = (10\ k\Omega/V) \times (200\,V) = 2000\,k\Omega = 2\,M\Omega$. Current flowing in voltmeter,

$$I_v = \frac{V}{R_v} = \frac{100}{2 \times 10^6} = 50 \times 10^{-6}\,A$$

Power dissipated by voltmeter

$$= VI_v = (100)(50 \times 10^{-6}) = 5\,mW.$$

When $R = 250\,\Omega$, current in resistor,

$$I_R = \frac{V}{R} = \frac{100}{250} = 0.4\,A$$

Power dissipated in load resistor $R = VI_R = (100)(0.4) = 40\,W$. Thus the power dissipated in the voltmeter is insignificant in comparison with the power dissipated in the load.

(b) When $R = 2\,M\Omega$, current in resistor,

$$I_R = \frac{V}{R} = \frac{100}{2 \times 10^6} = 50 \times 10^{-6}\,A$$

Power dissipated in load resistor $R = VI_R = 100 \times 50 \times 10^{-6} = 5\,mW$. In this case the higher load resistance reduced the power dissipated such that the voltmeter is using as much power as the load.

**Problem 4.** An ammeter has a f.s.d. of 100 mA and a resistance of $50\,\Omega$. The ammeter is used to measure the current in a load of resistance $500\,\Omega$ when the supply voltage is 10 V. Calculate (a) the ammeter reading expected (neglecting its resistance), (b) the actual current in the circuit, (c) the power dissipated in the ammeter, and (d) the power dissipated in the load.

From Fig. 10.10,

(a) expected ammeter reading $= V/R = 10/500 = 20\,mA$.

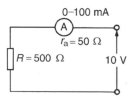

**Figure 10.10**

(b) Actual ammeter reading

$$= V/(R + r_a) = 10/(500 + 50) = 18.18\,mA.$$

Thus the ammeter itself has caused the circuit conditions to change from 20 mA to 18.18 mA.

(c) Power dissipated in the ammeter

$$= I^2 r_a = (18.18 \times 10^{-3})^2 (50) = 16.53\,mW.$$

(d) Power dissipated in the load resistor

$$= I^2 R = (18.18 \times 10^{-3})^2 (500) = 165.3\,mW.$$

**Problem 5.** A voltmeter having a f.s.d. of 100 V and a sensitivity of $1.6\,k\Omega/V$ is used to measure voltage $V_1$ in the circuit of Fig. 10.11. Determine (a) the value of voltage $V_1$ with the voltmeter not connected, and (b) the voltage indicated by the voltmeter when connected between A and B.

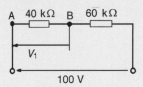

**Figure 10.11**

(a) By voltage division,

$$V_1 = \left(\frac{40}{40 + 60}\right) 100 = 40\,V$$

(b) The resistance of a voltmeter having a 100 V f.s.d. and sensitivity $1.6\,k\Omega/V$ is $100\,V \times 1.6\,k\Omega/V = 160\,k\Omega$. When the voltmeter is connected across the $40\,k\Omega$ resistor the circuit is as shown in Fig. 10.12(a) and the equivalent resistance of the parallel network is given by

$$\left(\frac{40 \times 160}{40 + 160}\right) k\Omega$$

i.e. $$\left(\frac{40 \times 160}{200}\right) k\Omega = 32\,k\Omega$$

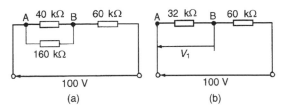

**Figure 10.12**

The circuit is now effectively as shown in Fig. 10.12(b). Thus the voltage indicated on the voltmeter is

$$\left(\frac{32}{32+60}\right)100\,V = \mathbf{34.78\,V}$$

A considerable error is thus caused by the loading effect of the voltmeter on the circuit. The error is reduced by using a voltmeter with a higher sensitivity.

---

**Problem 6.** (a) A current of 20 A flows through a load having a resistance of 2 Ω. Determine the power dissipated in the load. (b) A wattmeter, whose current coil has a resistance of 0.01 Ω is connected as shown in Fig. 10.13. Determine the wattmeter reading.

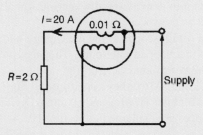

**Figure 10.13**

(a) Power dissipated in the load, $P = I^2 R = (20)^2(2)$
$$= \mathbf{800\,W}$$

(b) With the wattmeter connected in the circuit the total resistance $R_T$ is $2 + 0.01 = 2.01\,\Omega$. The wattmeter reading is thus $I^2 R_T = (20)^2(2.01)$
$$= \mathbf{804\,W}$$

---

**Now try the following exercise**

---

**Exercise 52    Further problems on instrument 'loading' effects**

1.  A 0–1 A ammeter having a resistance of 50 Ω is used to measure the current flowing in a 1 kΩ resistor when the supply voltage is 250 V. Calculate: (a) the approximate value of current (neglecting the ammeter resistance), (b) the actual current in the circuit, (c) the power dissipated in the ammeter, (d) the power dissipated in the 1 kΩ resistor.           [(a) 0.250 A
(b) 0.238 A (c) 2.832 W (d) 56.64 W]

2.  (a) A current of 15 A flows through a load having a resistance of 4 Ω. Determine the power dissipated in the load. (b) A wattmeter, whose current coil has a resistance of 0.02 Ω is connected (as shown in Fig. 10.13) to measure the power in the load. Determine the wattmeter reading assuming the current in the load is still 15 A.            [(a) 900 W (b) 904.5 W]

3.  A voltage of 240 V is applied to a circuit consisting of an 800 Ω resistor in series with a 1.6 kΩ resistor. What is the voltage across the 1.6 kΩ resistor? The p.d. across the 1.6 kΩ resistor is measured by a voltmeter of f.s.d. 250 V and sensitivity 100 Ω/V. Determine the voltage indicated.            [160 V; 156.7 V]

4.  A 240 V supply is connected across a load resistance $R$. Also connected across $R$ is a voltmeter having a f.s.d. of 300 V and a figure of merit (i.e. sensitivity) of 8 kΩ/V. Calculate the power dissipated by the voltmeter and by the load resistance if (a) $R = 100\,\Omega$ (b) $R = 1\,M\Omega$. Comment on the results obtained.
[(a) 24 mW, 576 W (b) 24 mW, 57.6 mW]

---

## 10.12    The oscilloscope

The oscilloscope is basically a graph-displaying device – it draws a graph of an electrical signal. In most applications the graph shows how signals change over time. From the graph it is possible to:

- determine the time and voltage values of a signal
- calculate the frequency of an oscillating signal
- see the 'moving parts' of a circuit represented by the signal
- tell if a malfunctioning component is distorting the signal
- find out how much of a signal is d.c. or a.c.

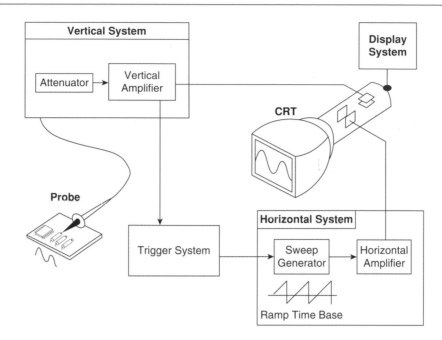

**Figure 10.14**

- tell how much of the signal is noise and whether the noise is changing with time

Oscilloscopes are used by everyone from television repair technicians to physicists. They are indispensable for anyone designing or repairing electronic equipment. The usefulness of an oscilloscope is not limited to the world of electronics. With the proper transducer (i.e. a device that creates an electrical signal in response to physical stimuli, such as sound, mechanical stress, pressure, light or heat), an oscilloscope can measure any kind of phenomena. An automobile engineer uses an oscilloscope to measure engine vibrations; a medical researcher uses an oscilloscope to measure brain waves, and so on.

Oscilloscopes are available in both analogue and digital types. An **analogue oscilloscope** works by directly applying a voltage being measured to an electron beam moving across the oscilloscope screen. The voltage deflects the beam up or down proportionally, tracing the waveform on the screen. This gives an immediate picture of the waveform.

In contrast, a **digital oscilloscope** samples the waveform and uses an analogue to digital converter (see Section 19.11, page 307) to convert the voltage being measured into digital information. It then uses this digital information to reconstruct the waveform on the screen.

For many applications either an analogue or digital oscilloscope is appropriate. However, each type does possess some unique characteristics making it more or less suitable for specific tasks.

Analogue oscilloscopes are often preferred when it is important to display rapidly varying signals in 'real time' (i.e. as they occur).

Digital oscilloscopes allow the capture and viewing of events that happen only once. They can process the digital waveform data or send the data to a computer for processing. Also, they can store the digital waveform data for later viewing and printing. Digital storage oscilloscopes are explained in Section 10.14.

## Analogue oscilloscopes

When an oscilloscope probe is connected to a circuit, the voltage signal travels through the probe to the vertical system of the oscilloscope. Figure 10.14 shows a simple block diagram that shows how an analogue oscilloscope displays a measured signal.

Depending on how the vertical scale (volts/division control) is set, an attenuator reduces the signal voltage or an amplifier increases the signal voltage. Next, the signal travels directly to the vertical deflection plates of the cathode ray tube (CRT). Voltage applied to these deflection plates causes a glowing dot to move. (An electron beam hitting phosphor inside the CRT creates the glowing dot.) A positive voltage causes the dot to move up while a negative voltage causes the dot to move down.

The signal also travels to the trigger system to start or trigger a 'horizontal sweep'. Horizontal sweep is a term referring to the action of the horizontal system causing the glowing dot to move across the screen. Triggering the horizontal system causes the horizontal time base to move the glowing dot across the screen from left to right within a specific time interval. Many sweeps in rapid sequence cause the movement of the glowing dot to blend into a solid line. At higher speeds, the dot may sweep across the screen up to 500 000 times each second.

Together, the horizontal sweeping action (i.e. the X direction) and the vertical deflection action (i.e. the Y direction), traces a graph of the signal on the screen. The trigger is necessary to stabilise a repeating signal. It ensures that the sweep begins at the same point of a repeating signal, resulting in a clear picture.

In conclusion, to use an analogue oscilloscope, three basic settings to accommodate an incoming signal need to be adjusted:

- the attenuation or amplification of the signal – use the volts/division control to adjust the amplitude of the signal before it is applied to the vertical deflection plates

- the time base – use the time/division control to set the amount of time per division represented horizontally across the screen

- the triggering of the oscilloscope – use the trigger level to stabilise a repeating signal, as well as triggering on a single event.

Also, adjusting the focus and intensity controls enable a sharp, visible display to be created.

(i)   With **direct voltage measurements**, only the Y amplifier 'volts/cm' switch on the oscilloscope is used. With no voltage applied to the Y plates the position of the spot trace on the screen is noted. When a direct voltage is applied to the Y plates the new position of the spot trace is an indication of the magnitude of the voltage. For example, in Fig. 10.15(a), with no voltage applied to the Y plates, the spot trace is in the centre of the screen (initial position) and then the spot trace moves 2.5 cm to the final position shown, on application of a d.c. voltage. With the 'volts/cm' switch on 10 volts/cm the magnitude of the direct voltage is 2.5 cm × 10 volts/cm, i.e. 25 volts.

(ii)  With **alternating voltage measurements**, let a sinusoidal waveform be displayed on an oscilloscope screen as shown in Fig. 10.15(b). If the time/cm switch is on, say, 5 ms/cm then the

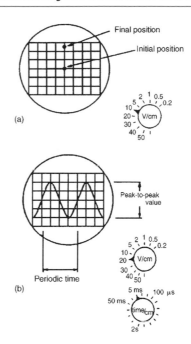

Figure 10.15

periodic time **T** of the sinewave is 5 ms/cm × 4 cm, i.e. **20 ms** or **0.02 s**. Since frequency

$$f = \frac{1}{T}, \text{ frequency} = \frac{1}{0.02} = \mathbf{50\,Hz}$$

If the 'volts/cm' switch is on, say, 20 volts/cm then the **amplitude** or **peak value** of the sinewave shown is 20 volts/cm × 2 cm, i.e. 40 V. Since

$$\text{r.m.s. voltage} = \frac{\text{peak voltage}}{\sqrt{2}} \text{ (see Chapter 14),}$$

$$\mathbf{r.m.s. \ voltage} = \frac{40}{\sqrt{2}} = \mathbf{28.28\,volts}$$

**Double beam oscilloscopes** are useful whenever two signals are to be compared simultaneously. The c.r.o. demands reasonable skill in adjustment and use. However its greatest advantage is in observing the shape of a waveform – a feature not possessed by other measuring instruments.

## Digital oscilloscopes

Some of the systems that make up digital oscilloscopes are the same as those in analogue oscilloscopes; however, digital oscilloscopes contain additional data processing systems – as shown in the block diagram of Fig. 10.16. With the added systems, the digital oscilloscope collects data for the entire waveform and then displays it.

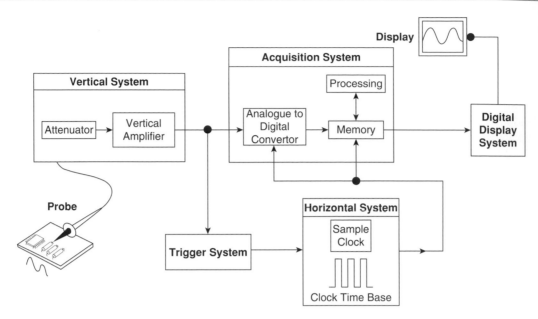

**Figure 10.16**

When a digital oscilloscope probe is attached to a circuit, the vertical system adjusts the amplitude of the signal, just as in the analogue oscilloscope. Next, the analogue to digital converter (ADC) in the acquisition system samples the signal at discrete points in time and converts the signals' voltage at these points to digital values called *sample points*. The horizontal systems' sample clock determines how often the ADC takes a sample. The rate at which the clock 'ticks' is called the sample rate and is measured in samples per second.

The sample points from the ADC are stored in memory as *waveform points*. More than one sample point may make up one waveform point.

Together, the waveform points make up one waveform *record*. The number of waveform points used to make a waveform record is called a *record length*. The trigger system determines the start and stop points of the record. The display receives these record points after being stored in memory.

Depending on the capabilities of an oscilloscope, additional processing of the sample points may take place, enhancing the display. Pre-trigger may be available, allowing events to be seen before the trigger point.

Fundamentally, with a digital oscilloscope as with an analogue oscilloscope, there is a need to adjust vertical, horizontal, and trigger settings to take a measurement.

> **Problem 7.** For the oscilloscope square voltage waveform shown in Fig. 10.17 determine (a) the

periodic time, (b) the frequency, and (c) the peak-to-peak voltage. The 'time/cm' (or timebase control) switch is on $100\,\mu s/cm$ and the 'volts/cm' (or signal amplitude control) switch is on $20\,V/cm$.

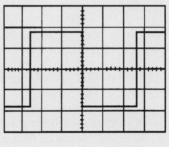

**Figure 10.17**

(*In Figs. 10.17 to 10.20 assume that the squares shown are 1 cm by 1 cm*)

(a) The width of one complete cycle is 5.2 cm. Hence the periodic time,
$$T = 5.2\,cm \times 100 \times 10^{-6}\,s/cm = \mathbf{0.52\,ms}.$$

(b) Frequency, $f = \dfrac{1}{T} = \dfrac{1}{0.52 \times 10^{-3}} = \mathbf{1.92\,kHz}$.

(c) The peak-to-peak height of the display is 3.6 cm, hence the peak-to-peak voltage

$$= 3.6\,cm \times 20\,V/cm = \mathbf{72\,V}$$

**Problem 8.** For the oscilloscope display of a pulse waveform shown in Fig. 10.18 the 'time/cm' switch is on 50 ms/cm and the 'volts/cm' switch is on 0.2 V/cm. Determine (a) the periodic time, (b) the frequency, (c) the magnitude of the pulse voltage.

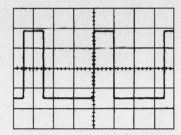

**Figure 10.18**

(a) The width of one complete cycle is 3.5 cm. Hence the periodic time,
$$T = 3.5\,\text{cm} \times 50\,\text{ms/cm} = \textbf{175 ms}.$$

(b) Frequency, $f = \dfrac{1}{T} = \dfrac{1}{0.52 \times 10^{-3}} = \textbf{5.71 Hz}.$

(c) The height of a pulse is 3.4 cm hence the magnitude of the pulse voltage
$$= 3.4\,\text{cm} \times 0.2\,\text{V/cm} = \textbf{0.68 V}$$

**Problem 9.** A sinusoidal voltage trace displayed by an oscilloscope is shown in Fig. 10.19. If the 'time/cm' switch is on 500 µs/cm and the 'volts/cm' switch is on 5 V/cm, find, for the waveform, (a) the frequency, (b) the peak-to-peak voltage, (c) the amplitude, (d) the r.m.s. value.

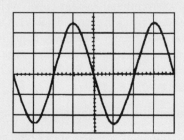

**Figure 10.19**

(a) The width of one complete cycle is 4 cm. Hence the periodic time, $T$ is 4 cm × 500 µs/cm, i.e. 2 ms.

Frequency, $f = \dfrac{1}{T} = \dfrac{1}{2 \times 10^{-3}} = \textbf{500 Hz}$

(b) The peak-to-peak height of the waveform is 5 cm. Hence the peak-to-peak voltage
$$= 5\,\text{cm} \times 5\,\text{V/cm} = \textbf{25 V}.$$

(c) Amplitude $= \frac{1}{2} \times 25\,\text{V} = \textbf{12.5 V}$

(d) The peak value of voltage is the amplitude, i.e. 12.5 V, and r.m.s.
$$\text{voltage} = \frac{\text{peak voltage}}{\sqrt{2}} = \frac{12.5}{\sqrt{2}} = \textbf{8.84 V}$$

**Problem 10.** For the double-beam oscilloscope displays shown in Fig. 10.20 determine (a) their frequency, (b) their r.m.s. values, (c) their phase difference. The 'time/cm' switch is on 100 µs/cm and the 'volts/cm' switch on 2 V/cm.

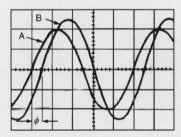

**Figure 10.20**

(a) The width of each complete cycle is 5 cm for both waveforms. Hence the periodic time, $T$, of each waveform is 5 cm × 100 µs/cm, i.e. 0.5 ms. Frequency of each waveform,
$$f = \frac{1}{T} = \frac{1}{0.5 \times 10^{-3}} = \textbf{2 kHz}$$

(b) The peak value of waveform A is 2 cm × 2 V/cm = **4 V**, hence the r.m.s. value of waveform A
$$= 4/(\sqrt{2}) = \textbf{2.83 V}$$

The peak value of waveform B is 2.5 cm × 2 V/cm = **5 V**, hence the r.m.s. value of waveform B
$$= 5/(\sqrt{2}) = \textbf{3.54 V}$$

(c)   Since 5 cm represents 1 cycle, then 5 cm represents 360°, i.e. 1 cm represents 360/5 = 72°. The phase angle $\phi = 0.5$ cm

$$= 0.5\,\text{cm} \times 72°/\text{cm} = 36°.$$

**Hence waveform A leads waveform B by 36°.**

**Now try the following exercise**

**Exercise 53   Further problems on the cathode ray oscilloscope**

1.   For the square voltage waveform displayed on an oscilloscope shown in Fig. 10.21, find (a) its frequency, (b) its peak-to-peak voltage.
[(a) 41.7 Hz (b) 176 V]

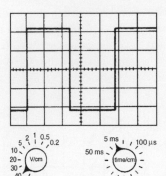

**Figure 10.21**

2.   For the pulse waveform shown in Fig. 10.22, find (a) its frequency, (b) the magnitude of the pulse voltage.   [(a) 0.56 Hz (b) 8.4 V]

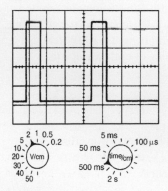

**Figure 10.22**

3.   For the sinusoidal waveform shown in Fig. 10.23, determine (a) its frequency,

(b) the peak-to-peak voltage, (c) the r.m.s. voltage.
[(a) 7.14 Hz (b) 220 V (c) 77.78 V]

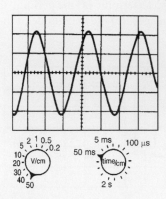

**Figure 10.23**

## 10.13   Virtual test and measuring instruments

Computer-based instruments are rapidly replacing items of conventional test equipment in many of today's test and measurement applications. Probably the most commonly available virtual test instrument is the digital storage oscilloscope (DSO). Because of the processing power available from the PC coupled with the mass storage capability, a computer-based virtual DSO is able to provide a variety of additional functions, such as spectrum analysis and digital display of both frequency and voltage. In addition, the ability to save waveforms and captured measurement data for future analysis or for comparison purposes can be extremely valuable, particularly where evidence of conformance with standards or specifications is required.

Unlike a conventional oscilloscope (which is primarily intended for waveform display) a computer-based virtual oscilloscope effectively combines several test instruments in one single package. The functions and available measurements from such an instrument usually includes:

- real time or stored waveform display

- precise time and voltage measurement (using adjustable cursors)

- digital display of voltage

- digital display of frequency and/or periodic time

- accurate measurement of phase angle

- frequency spectrum display and analysis

- data logging (stored waveform data can be exported in formats that are compatible with conventional spreadsheet packages, e.g. as .xls files)

- ability to save/print waveforms and other information in graphical format (e.g. as .jpg or .bmp files).

Virtual instruments can take various forms including:

- internal hardware in the form of a conventional PCI expansion card

- external hardware unit which is connected to the PC by means of either a conventional 25-pin parallel port connector or by means of a serial USB connector

The software (and any necessary drivers) is invariably supplied on CD-ROM or can be downloaded from the manufacturer's web site. Some manufacturers also supply software drivers together with sufficient accompanying documentation in order to allow users to control virtual test instruments from their own software developed using popular programming languages such as VisualBASIC or C++.

## 10.14  Virtual digital storage oscilloscopes

Several types of virtual DSO are currently available. These can be conveniently arranged into three different categories according to their application:

- Low-cost DSO

- High-speed DSO

- High-resolution DSO

Unfortunately, there is often some confusion between the last two categories. A high-speed DSO is designed for examining waveforms that are rapidly changing. Such an instrument does not necessarily provide high-resolution measurement. Similarly, a high-resolution DSO is useful for displaying waveforms with a high degree of precision but it may not be suitable for examining fast waveforms. The difference between these two types of DSO should become a little clearer later on.

Low-cost DSO are primarily designed for low frequency signals (typically signals up to around 20 kHz) and are usually able to sample their signals at rates of between 10K and 100K samples per second. Resolution is usually limited to either 8-bits or 12-bits (corresponding to 256 and 4096 discrete voltage levels respectively).

High-speed DSOs are rapidly replacing CRT-based oscilloscopes. They are invariably dual-channel instruments and provide all the features associated with a conventional 'scope including trigger selection, time-base and voltage ranges, and an ability to operate in X-Y mode.

Additional features available with a computer-based instrument include the ability to capture transient signals (as with a conventional digital storage 'scope') and save waveforms for future analysis. The ability to analyse a signal in terms of its frequency spectrum is yet another feature that is only possible with a DSO (see later).

### Upper frequency limit

The upper signal frequency limit of a DSO is determined primarily by the rate at which it can sample an incoming signal. Typical sampling rates for different types of virtual instrument are:

| Type of DSO | Typical sampling rate |
|---|---|
| Low-cost DSO | 20 K to 100 K per second |
| High-speed DSO | 100 M to 1000 M per second |
| High-resolution DSO | 20 M to 100 M per second |

In order to display waveforms with reasonable accuracy it is normally suggested that the sampling rate should be *at least* twice and *preferably more* than five times the highest signal frequency. Thus, in order to display a 10 MHz signal with any degree of accuracy a sampling rate of 50M samples per second will be required.

The 'five times rule' merits a little explanation. When sampling signals in a digital to analogue converter we usually apply the Nyquist criterion that the sampling frequency must be at least twice the highest analogue signal frequency. Unfortunately, this no longer applies in the case of a DSO where we need to sample at an even faster rate if we are to accurately display the signal. In practice we would need a minimum of about five points within a single cycle of a sampled waveform in order to reproduce it with approximate fidelity. Hence the sampling rate should be at least five times that of highest signal frequency in order to display a waveform reasonably faithfully.

A special case exists with dual-channel DSOs. Here the sampling rate may be shared between the two channels. Thus an effective sampling rate of 20M samples per second might equate to 10M samples per second for *each* of the two channels. In such a case the upper frequency limit would not be 4 MHz but only a mere 2 MHz.

The approximate bandwidth required to display different types of signals with reasonable precision is given in the table below:

| Signal | Bandwidth required (approx) |
|---|---|
| Low-frequency and power | d.c. to 10 kHz |
| Audio frequency (general) | d.c. to 20 kHz |
| Audio frequency (high-quality) | d.c. to 50 kHz |
| Square and pulse waveforms (up to 5 kHz) | d.c. to 100 kHz |
| Fast pulses with small rise-times | d.c. to 1 MHz |
| Video | d.c. to 10 MHz |
| Radio (LF, MF and HF) | d.c. to 50 MHz |

The general rule is that, for sinusoidal signals, the bandwidth should ideally be at least double that of the highest signal frequency whilst for square wave and pulse signals, the bandwidth should be at least ten times that of the highest signal frequency.

It is worth noting that most manufacturers define the bandwidth of an instrument as the frequency at which a sine wave input signal will fall to 0.707 of its true amplitude (i.e. the −3 dB point). To put this into context, at the cut-off frequency the displayed trace will be in error by a whopping 29%!

## Resolution

The relationship between resolution and signal accuracy (not bandwidth) is simply that the more bits used in the conversion process the more discrete voltage levels

can be resolved by the DSO. The relationship is as follows:

$$x = 2^n$$

where $x$ is the number of discrete voltage levels and $n$ is the number of bits. Thus, each time we use an additional bit in the conversion process we double the resolution of the DSO, as shown in the table below:

| Number of bits, $n$ | Number of discrete voltage levels, $x$ |
|---|---|
| 8-bit | 256 |
| 10-bit | 1024 |
| 12-bit | 4096 |
| 16-bit | 65 536 |

## Buffer memory capacity

A DSO stores its captured waveform samples in a buffer memory. Hence, for a given sampling rate, the size of this memory buffer will determine for how long the DSO can capture a signal before its buffer memory becomes full.

The relationship between sampling rate and buffer memory capacity is important. A DSO with a high sampling rate but small memory will only be able to use its full sampling rate on the top few time base ranges.

To put this into context, it's worth considering a simple example. Assume that we need to display 10 000 cycles of a 10 MHz square wave. This signal will occur in a time frame of 1 ms. If applying the 'five times rule' we would need a bandwidth of at least 50 MHz to display this signal accurately.

To reconstruct the square wave we would need a minimum of about five samples per cycle so a minimum sampling rate would be $5 \times 10$ MHz $= 50$M samples per second. To capture data at the rate of 50 M samples per second for a time interval of 1 ms requires a memory that can store 50 000 samples. If each sample uses 16-bits we would require 100 kbyte of extremely fast memory.

## Accuracy

The measurement resolution or measurement accuracy of a DSO (in terms of the smallest voltage change that can be measured) depends on the actual range that is selected. So, for example, on the 1 V range an 8-bit DSO is able to detect a voltage change of one two hundred and fifty sixth of a volt or (1/256) V or about 4 mV. For most measurement applications this will prove to be perfectly adequate as it amounts to an accuracy of about 0.4% of full-scale.

Figure 10.24 depicts a PicoScope software display showing multiple windows providing conventional oscilloscope waveform display, spectrum analyser display, frequency display, and voltmeter display.

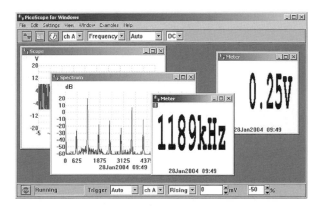

**Figure 10.24**

Adjustable cursors make it possible to carry out extremely accurate measurements. In Fig. 10.25, the peak value of the (nominal 10 V peak) waveform is measured at precisely 9625 mV (9.625 V). The time to reach the peak value (from 0 V) is measured as 246.7 µs (0.2467 ms).

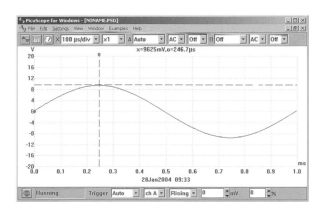

**Figure 10.25**

The addition of a second time cursor makes it possible to measure the time accurately between two events. In Fig. 10.26, event 'o' occurs 131 ns before the trigger point whilst event 'x' occurs 397 ns after the trigger point. The elapsed time between these two events is

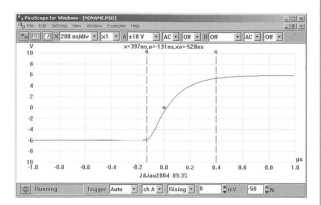

**Figure 10.26**

528 ns. The two cursors can be adjusted by means of the mouse (or other pointing device) or, more accurately, using the PC's cursor keys.

## Autoranging

Autoranging is another very useful feature that is often provided with a virtual DSO. If you regularly use a conventional 'scope for a variety of measurements you will know only too well how many times you need to make adjustments to the vertical sensitivity of the instrument.

## High-resolution DSO

High-resolution DSOs are used for precision applications where it is necessary to faithfully reproduce a waveform and also to be able to perform an accurate analysis of noise floor and harmonic content. Typical applications include small signal work and high-quality audio.

Unlike the low-cost DSO, which typically has 8-bit resolution and poor d.c. accuracy, these units are usually accurate to better than 1% and have either 12-bit or 16-bit resolution. This makes them ideal for audio, noise and vibration measurements.

The increased resolution also allows the instrument to be used as a spectrum analyser with very wide dynamic range (up to 100 dB). This feature is ideal for performing noise and distortion measurements on low-level analogue circuits.

Bandwidth alone is not enough to ensure that a DSO can accurately capture a high frequency signal. The goal of manufacturers is to achieve a flat frequency response. This response is sometimes referred to as a Maximally Flat Envelope Delay (MFED). A frequency response of this type delivers excellent pulse fidelity with minimum overshoot, undershoot and ringing.

It is important to remember that, if the input signal is not a pure sine wave it will contain a number of higher frequency harmonics. For example, a square wave will contain odd harmonics that have levels that become progressively reduced as their frequency increases. Thus, to display a 1 MHz square wave accurately you need to take into account the fact that there will be signal components present at 3 MHz, 5 MHz, 7 MHz, 9 MHz, 11 MHz, and so on.

### Spectrum analysis

The technique of Fast Fourier Transformation (FFT) calculated using software algorithms using data captured by a virtual DSO has made it possible to produce frequency spectrum displays. Such displays can be to investigate the harmonic content of waveforms as well as the relationship between several signals within a composite waveform.

Figure 10.27 shows the frequency spectrum of the 1 kHz sine wave signal from a low-distortion signal generator. Here the virtual DSO has been set to capture samples at a rate of 4096 per second within a frequency range of d.c. to 12.2 kHz. The display clearly shows the second harmonic (at a level of −50 dB or −70 dB relative to the fundamental), plus further harmonics at 3 kHz, 5 kHz and 7 kHz (all of which are greater than 75 dB down on the fundamental).

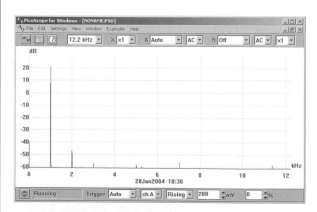

**Figure 10.27**

Problem 11. Figure 10.28 shows the frequency spectrum of a signal at 1184 kHz displayed by a high-speed virtual DSO. Determine (a) the harmonic relationship between the signals marked 'o' and 'x', (b) the difference in amplitude (expressed in dB) between the signals marked 'o'

and 'x', and (c) the amplitude of the second harmonic relative to the fundamental signal 'o'.

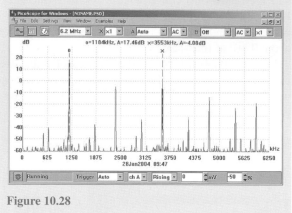

**Figure 10.28**

(a) The signal $x$ is at a frequency of 3553 kHz. This is three times the frequency of the signal at 'o' which is at 1184 kHz. Thus, $x$ **is the third harmonic of the signal 'o'**

(b) The signal at 'o' has an amplitude of +17.46 dB whilst the signal at 'x' has an amplitude of −4.08 dB. Thus, **the difference in level** $= (+17.46) − (−4.08) = \mathbf{21.54\,dB}$

(c) **The amplitude of the second harmonic** (shown at approximately 2270 kHz) $= \mathbf{-5\,dB}$

## 10.15   Waveform harmonics

(i) Let an instantaneous voltage $v$ be represented by $v = V_m \sin 2\pi ft$ volts. This is a waveform which varies sinusoidally with time $t$, has a frequency $f$, and a maximum value $V_m$. Alternating voltages are usually assumed to have wave-shapes which are sinusoidal where only one frequency is present. If the waveform is not sinusoidal it is called a **complex wave**, and, whatever its shape, it may be split up mathematically into components called the **fundamental** and a number of **harmonics**. This process is called harmonic analysis. The fundamental (or first harmonic) is sinusoidal and has the supply frequency, $f$; the other harmonics are also sine waves having frequencies which are integer multiples of $f$. Thus, if the supply frequency is 50 Hz, then the third harmonic frequency is 150 Hz, the fifth 250 Hz, and so on.

(ii) A complex waveform comprising the sum of the fundamental and a third harmonic of about half the amplitude of the fundamental is shown in Fig. 10.29(a), both waveforms being initially in phase with each other. If further odd harmonic waveforms of the appropriate amplitudes are added, a good approximation to a square wave results. In Fig. 10.29(b), the third harmonic is shown having an initial phase displacement from the fundamental. The positive and negative half cycles of each of the complex waveforms shown in Figs 10.29(a) and (b) are identical in shape, and this is a feature of waveforms containing the fundamental and only odd harmonics.

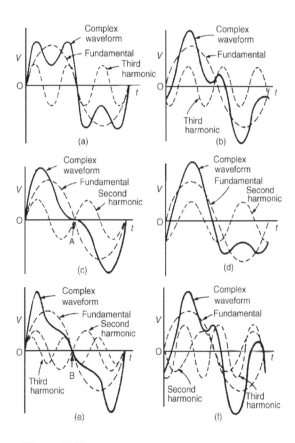

**Figure 10.29**

(iii)  A complex waveform comprising the sum of the fundamental and a second harmonic of about half the amplitude of the fundamental is shown in Fig. 10.29(c), each waveform being initially in phase with each other. If further even harmonics of appropriate amplitudes are added a good approximation to a triangular wave results. In Fig. 10.29(c), the negative cycle, if reversed, appears as a mirror image of the positive cycle about point A. In Fig. 10.29(d) the second harmonic is shown with an initial phase

displacement from the fundamental and the positive and negative half cycles are dissimilar.

(iv)  A complex waveform comprising the sum of the fundamental, a second harmonic and a third harmonic is shown in Fig. 10.29(e), each waveform being initially 'in-phase'. The negative half cycle, if reversed, appears as a mirror image of the positive cycle about point B. In Fig. 10.29(f), a complex waveform comprising the sum of the fundamental, a second harmonic and a third harmonic are shown with initial phase displacement. The positive and negative half cycles are seen to be dissimilar.

The features mentioned relative to Figs 10.29(a) to (f) make it possible to recognize the harmonics present in a complex waveform displayed on a CRO.

## 10.16    Logarithmic ratios

In electronic systems, the ratio of two similar quantities measured at different points in the system, are often expressed in logarithmic units. By definition, if the ratio of two powers $P_1$ and $P_2$ is to be expressed in **decibel (dB) units** then the number of decibels, $X$, is given by:

$$X = 10 \lg\left(\frac{P_2}{P_1}\right) \text{dB} \tag{1}$$

Thus, when the power ratio, $P_2/P_1 = 1$ then the decibel power ratio $= 10 \lg 1 = 0$, when the power ratio, $P_2/P_1 = 100$ then the decibel power ratio $= 10 \lg 100 = +20$ (i.e. a power gain), and when the power ratio, $P_2/P_1 = 1/100$ then the decibel power ratio $= 10 \lg 1/100 = -20$ (i.e. a power loss or attenuation).

Logarithmic units may also be used for voltage and current ratios. Power, $P$, is given by $P = I^2 R$ or $P = V^2/R$. Substituting in equation (1) gives:

$$X = 10 \lg\left(\frac{I_2^2 R_2}{I_1^2 R_1}\right) \text{dB}$$

or  $$X = 10 \lg\left(\frac{V_2^2/R_2}{V_1^2/R_1}\right) \text{dB}$$

If  $R_1 = R_2,$

then  $$X = 10 \lg\left(\frac{I_2^2}{I_1^2}\right) \text{dB}$$

or

$$X = 10 \lg \left( \frac{V_2^2}{V_1^2} \right) dB$$

i.e.  $X = 20 \lg \left( \dfrac{I_2}{I_1} \right) dB$

or  $X = 20 \lg \left( \dfrac{V_2}{V_1} \right) dB$

(from the laws of logarithms).

From equation (1), $X$ decibels is a logarithmic ratio of two similar quantities and is not an absolute unit of measurement. It is therefore necessary to state a **reference level** to measure a number of decibels above or below that reference. The most widely used reference level for power is 1 mW, and when power levels are expressed in decibels, above or below the 1 mW reference level, the unit given to the new power level is dBm.

A voltmeter can be re-scaled to indicate the power level directly in decibels. The scale is generally calibrated by taking a reference level of 0 dB when a power of 1 mW is dissipated in a 600 Ω resistor (this being the natural impedance of a simple transmission line). The reference voltage $V$ is then obtained from

$$P = \frac{V^2}{R},$$

i.e.  $1 \times 10^{-3} = \dfrac{V^2}{600}$

from which, $V = 0.775$ volts. In general, the number of dBm,

$$X = 20 \lg \left( \frac{V}{0.775} \right)$$

Thus $V = 0.20$ V corresponds to $20 \lg \left( \dfrac{0.2}{0.775} \right)$

$$= -11.77 \, dBm \text{ and}$$

$V = 0.90$ V corresponds to $20 \lg \left( \dfrac{0.90}{0.775} \right)$

$$= +1.3 \, dBm, \text{ and so on.}$$

A typical **decibelmeter**, or **dB meter**, scale is shown in Fig. 10.30. Errors are introduced with dB meters when the circuit impedance is not 600 Ω.

**Problem 12.**  The ratio of two powers is (a) 3 (b) 20 (c) 4 (d) 1/20. Determine the decibel power ratio in each case.

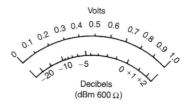

Volts

Decibels
(dBm 600 Ω)

**Figure 10.30**

From above, the power ratio in decibels, $X$, is given by:
$X = 10 \lg(P_2/P_1)$

(a)  When $\dfrac{P_2}{P_1} = 3$,

$$X = 10 \lg(3) = 10(0.477)$$
$$= \textbf{4.77 dB}$$

(b)  When $\dfrac{P_2}{P_1} = 20$,

$$X = 10 \lg(20) = 10(1.30)$$
$$= \textbf{13.0 dB}$$

(c)  When $\dfrac{P_2}{P_1} = 400$,

$$X = 10 \lg(400) = 10(2.60)$$
$$= \textbf{26.0 dB}$$

(d)  When $\dfrac{P_2}{P_1} = \dfrac{1}{20} = 0.05$,

$$X = 10 \lg(0.05) = 10(-1.30)$$
$$= \textbf{-13.0 dB}$$

(a), (b) and (c) represent power gains and (d) represents a power loss or attenuation.

**Problem 13.**  The current input to a system is 5 mA and the current output is 20 mA. Find the decibel current ratio assuming the input and load resistances of the system are equal.

From above, the decibel current ratio is

$$20 \lg \left( \frac{I_2}{I_1} \right) = 20 \lg \left( \frac{20}{5} \right)$$
$$= 20 \lg 4 = 20(0.60)$$
$$= \textbf{12 dB gain}$$

**Problem 14.** 6% of the power supplied to a cable appears at the output terminals. Determine the power loss in decibels.

If $P_1$ = input power and $P_2$ = output power then

$$\frac{P_2}{P_1} = \frac{6}{100} = 0.06$$

Decibel power ratio $= 10\lg\left(\frac{P_2}{P_1}\right) = 10\lg(0.06)$

$$= 10(-1.222) = -12.22\,\text{dB}$$

**Hence the decibel power loss, or attenuation, is 12.22 dB.**

**Problem 15.** An amplifier has a gain of 14 dB and its input power is 8 mW. Find its output power.

Decibel power ratio $= 10\lg(P_2/P_1)$ where $P_1$ = input power = 8 mW, and $P_2$ = output power. Hence

$$14 = 10\lg\left(\frac{P_2}{P_1}\right)$$

from which

$$1.4 = \lg\left(\frac{P_2}{P_1}\right)$$

and    $10^{1.4} = \dfrac{P_2}{P_1}$ from the definition of a logarithm

i.e.    $25.12 = \dfrac{P_2}{P_1}$

Output power, $P_2 = 25.12\,P_1 = (25.12)(8)$

$$= \textbf{201 mW} \text{ or } \textbf{0.201 W}$$

**Problem 16.** Determine, in decibels, the ratio of output power to input power of a 3 stage communications system, the stages having gains of 12 dB, 15 dB and −8 dB. Find also the overall power gain.

The decibel ratio may be used to find the overall power ratio of a chain simply by adding the decibel power ratios together. Hence the overall decibel

power ratio $= 12 + 15 - 8 = \textbf{19 dB gain}$.

Thus    $19 = 10\lg\left(\dfrac{P_2}{P_1}\right)$

from which    $1.9 = \lg\left(\dfrac{P_2}{P_1}\right)$

and    $10^{1.9} = \dfrac{P_2}{P_1} = 79.4$

**Thus the overall power gain, $\dfrac{P_2}{P_1} = \textbf{79.4}$**

[For the first stage,

$$12 = 10\lg\left(\frac{P_2}{P_1}\right)$$

from which

$$\frac{P_2}{P_1} = 10^{1.2} = 15.85$$

Similarly for the second stage,

$$\frac{P_2}{P_1} = 31.62$$

and for the third stage,

$$\frac{P_2}{P_1} = 0.1585$$

The overall power ratio is thus
$15.85 \times 31.62 \times 0.1585 = \textbf{79.4}$]

**Problem 17.** The output voltage from an amplifier is 4 V. If the voltage gain is 27 dB, calculate the value of the input voltage assuming that the amplifier input resistance and load resistance are equal.

Voltage gain in decibels $= 27 = 20\lg(V_2/V_1) = 20\lg(4/V_1)$. Hence

$$\frac{27}{20} = \lg\left(\frac{4}{V_1}\right)$$

i.e.    $1.35 = \lg\left(\dfrac{4}{V_1}\right)$

Thus    $10^{1.35} = \dfrac{4}{V_1}$

from which    $V_1 = \dfrac{4}{10^{1.35}}$

$$= \frac{4}{22.39}$$

$$= 0.179\,\text{V}$$

**Hence the input voltage $V_1$ is 0.179 V**

**Now try the following exercise**

---

**Exercise 54    Further problems on logarithmic ratios**

1. The ratio of two powers is (a) 3 (b) 10 (c) 20 (d) 10 000. Determine the decibel power ratio for each.            [(a) 4.77 dB (b) 10 dB (c) 13 dB (d) 40 dB]

2. The ratio of two powers is (a) $\frac{1}{10}$ (b) $\frac{1}{3}$ (c) $\frac{1}{40}$ (d) $\frac{1}{100}$. Determine the decibel power ratio for each.            [(a) −10 dB (b) −4.77 dB (c) −16.02 dB (d) −20 dB]

3. The input and output currents of a system are 2 mA and 10 mA respectively. Determine the decibel current ratio of output to input current assuming input and output resistances of the system are equal.            [13.98 dB]

4. 5% of the power supplied to a cable appears at the output terminals. Determine the power loss in decibels.            [13 dB]

5. An amplifier has a gain of 24 dB and its input power is 10 mW. Find its output power.            [2.51 W]

6. Determine, in decibels, the ratio of the output power to input power of a four stage system, the stages having gains of 10 dB, 8 dB, −5 dB and 7 dB. Find also the overall power gain.            [20 dB, 100]

7. The output voltage from an amplifier is 7 mV. If the voltage gain is 25 dB calculate the value of the input voltage assuming that the amplifier input resistance and load resistance are equal.            [0.39 mV]

8. The voltage gain of a number of cascaded amplifiers are 23 dB, −5.8 dB, −12.5 dB and 3.8 dB. Calculate the overall gain in decibels assuming that input and load resistances for each stage are equal. If a voltage of 15 mV is applied to the input of the system, determine the value of the output voltage.            [8.5 dB, 39.91 mV]

9. The scale of a voltmeter has a decibel scale added to it, which is calibrated by taking a reference level of 0 dB when a power of 1 mW is dissipated in a 600 Ω resistor. Determine the voltage at (a) 0 dB (b) 1.5 dB (c) −15 dB (d) What decibel reading corresponds to 0.5 V?            [(a) 0.775 V (b) 0.921 V (c) 0.138 V (d) −3.807 dB]

## 10.17   Null method of measurement

A **null method of measurement** is a simple, accurate and widely used method which depends on an instrument reading being adjusted to read zero current only. The method assumes:

(i) if there is any deflection at all, then some current is flowing;

(ii) if there is no deflection, then no current flows (i.e. a null condition).

Hence it is unnecessary for a meter sensing current flow to be calibrated when used in this way. A sensitive milliammeter or microammeter with centre zero position setting is called a **galvanometer**. Examples where the method is used are in the Wheatstone bridge (see Section 10.18), in the d.c. potentiometer (see Section 10.19) and with a.c. bridges (see Section 10.20).

## 10.18   Wheatstone bridge

Figure 10.31 shows a **Wheatstone bridge** circuit which compares an unknown resistance $R_x$ with others of known values, i.e. $R_1$ and $R_2$, which have fixed values, and $R_3$, which is variable. $R_3$ is varied until zero deflection is obtained on the galvanometer G. No current then

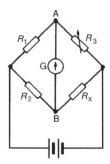

**Figure 10.31**

flows through the meter, $V_A = V_B$, and the bridge is said to be 'balanced'. At balance,

$$R_1 R_x = R_2 R_3 \text{ i.e. } R_x = \frac{R_2 R_3}{R_1} \text{ ohms}$$

**Problem 18.** In a Wheatstone bridge ABCD, a galvanometer is connected between A and C, and a battery between B and D. A resistor of unknown value is connected between A and B. When the bridge is balanced, the resistance between B and C is 100 Ω, that between C and D is 10 Ω and that between D and A is 400 Ω. Calculate the value of the unknown resistance.

The Wheatstone bridge is shown in Fig. 10.32 where $R_x$ is the unknown resistance. At balance, equating the products of opposite ratio arms, gives:

$$(R_x)(10) = (100)(400)$$

and 
$$R_x = \frac{(100)(400)}{10} = 4000 \, \Omega$$

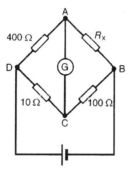

Figure 10.32

**Hence, the unknown resistance, $R_x = 4\,k\Omega$.**

### 10.19  D.C. potentiometer

The **d.c. potentiometer** is a null-balance instrument used for determining values of e.m.f.'s and p.d.s. by comparison with a known e.m.f. or p.d. In Fig. 10.33(a), using a standard cell of known e.m.f. $E_1$, the slider $S$ is moved along the slide wire until balance is obtained

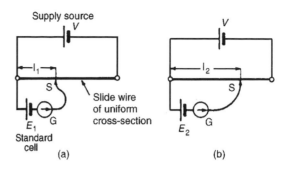

Figure 10.33

(i.e. the galvanometer deflection is zero), shown as length $l_1$.

The standard cell is now replaced by a cell of unknown e.m.f. $E_2$ (see Fig. 10.33(b)) and again balance is obtained (shown as $l_2$). Since $E_1 \propto l_1$ and $E_2 \propto l_2$ then

$$\frac{E_1}{E_2} = \frac{l_1}{l_2}$$

and 
$$E_2 = E_1 \left(\frac{l_2}{l_1}\right) \text{volts}$$

A potentiometer may be arranged as a resistive two-element potential divider in which the division ratio is adjustable to give a simple variable d.c. supply. Such devices may be constructed in the form of a resistive element carrying a sliding contact which is adjusted by a rotary or linear movement of the control knob.

**Problem 19.** In a d.c. potentiometer, balance is obtained at a length of 400 mm when using a standard cell of 1.0186 volts. Determine the e.m.f. of a dry cell if balance is obtained with a length of 650 mm.

$E_1 = 1.0186\,\text{V}$, $l_1 = 400\,\text{mm}$ and $l_2 = 650\,\text{mm}$
With reference to Fig. 10.33,

$$\frac{E_1}{E_2} = \frac{l_1}{l_2}$$

from which,

$$E_2 = E_1 \left(\frac{l_2}{l_1}\right) = (1.0186)\left(\frac{650}{400}\right)$$

$$= \mathbf{1.655\ volts}$$

**Now try the following exercise**

**Exercise 55  Further problems on the Wheatstone bridge and d.c. potentiometer**

1. In a Wheatstone bridge PQRS, a galvanometer is connected between Q and S and a voltage source between P and R. An unknown resistor $R_x$ is connected between P and Q. When the bridge is balanced, the resistance between Q and R is $200\,\Omega$, that between R and S is $10\,\Omega$ and that between S and P is $150\,\Omega$. Calculate the value of $R_x$. [$3\,\text{k}\Omega$]

2. Balance is obtained in a d.c. potentiometer at a length of 31.2 cm when using a standard cell of 1.0186 volts. Calculate the e.m.f. of a dry cell if balance is obtained with a length of 46.7 cm. [1.525 V]

## 10.20  A.C. bridges

A Wheatstone bridge type circuit, shown in Fig. 10.34, may be used in a.c. circuits to determine unknown values of inductance and capacitance, as well as resistance.

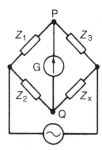

**Figure 10.34**

When the potential differences across $Z_3$ and $Z_x$ (or across $Z_1$ and $Z_2$) are equal in magnitude and phase, then the current flowing through the galvanometer, G, is zero. At balance, $Z_1 Z_x = Z_2 Z_3$ from which

$$Z_x = \frac{Z_2 Z_3}{Z_1}\ \Omega$$

There are many forms of a.c. bridge, and these include: the Maxwell, Hay, Owen and Heaviside bridges for measuring inductance, and the De Sauty, Schering and Wien bridges for measuring capacitance. A **commercial or universal bridge** is one which can be used to measure resistance, inductance or capacitance. A.c. bridges require a knowledge of complex numbers (i.e. $j$ notation, where $j = \sqrt{-1}$).

A Maxwell-Wien bridge for measuring the inductance $L$ and resistance $r$ of an inductor is shown in Fig. 10.35.

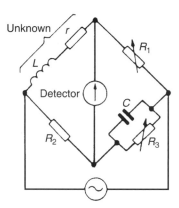

**Figure 10.35**

At balance the products of diagonally opposite impedances are equal. Thus

$$Z_1 Z_2 = Z_3 Z_4$$

Using complex quantities, $Z_1 = R_1$, $Z_2 = R_2$,

$$Z_3 = \frac{R_3(-jX_C)}{R_3 - jX_C} \left(\text{i.e. } \frac{\text{product}}{\text{sum}}\right)$$

and $Z_4 = r + jX_L$. Hence

$$R_1 R_2 = \frac{R_3(-jX_C)}{R_3 - jX_C}(r + jX_L)$$

i.e.   $R_1 R_2 (R_3 - jX_C) = (-jR_3 X_C)(r + jX_L)$

$R_1 R_2 R_3 - j R_1 R_2 X_C = -jrR_3 X_C - j^2 R_3 X_C X_L$

i.e.   $R_1 R_2 R_3 - j R_1 R_2 X_C = -jrR_3 X_C + R_3 X_C X_L$

(since $j^2 = -1$).

Equating the real parts gives:

$$R_1 R_2 R_3 = R_3 X_C X_L$$

from which,     $$X_L = \frac{R_1 R_2}{X_C}$$

i.e.     $$2\pi f L = \frac{R_1 R_2}{\frac{1}{2\pi f C}} = R_1 R_2 (2\pi f C)$$

**Hence inductance,**

$$L = R_1 R_2 C \text{ henry} \qquad (2)$$

Equating the imaginary parts gives:

$$-R_1 R_2 X_C = -r R_3 X_C$$

from which, resistance,

$$r = \frac{R_1 R_2}{R_3} \text{ ohms} \qquad (3)$$

---

**Problem 20.** For the a.c. bridge shown in Fig. 10.35 determine the values of the inductance and resistance of the coil when $R_1 = R_2 = 400\,\Omega$, $R_3 = 5\,\text{k}\Omega$ and $C = 7.5\,\mu\text{F}$.

---

From equation (2) above, inductance

$$L = R_1 R_2 C = (400)(400)(7.5 \times 10^{-6})$$

$$= \mathbf{1.2\,H}$$

From equation (3) above, resistance,

$$r = \frac{R_1 R_2}{R_3} = \frac{(400)(400)}{5000} = \mathbf{32\,\Omega}$$

---

From equation (2),

$$R_2 = \frac{L}{R_1 C}$$

and from equation (3),

$$R_3 = \frac{R_1}{r} R_2$$

Hence     $$R_3 = \frac{R_1}{r} \frac{L}{R_1 C} = \frac{L}{Cr}$$

If the frequency is constant then $R_3 \propto L/r \propto \omega L/r \propto$ Q-factor (see Chapters 15 and 16). Thus the bridge

can be adjusted to give a direct indication of Q-factor. A Q-meter is described in Section 10.21 following.

---

**Now try the following exercise**

---

**Exercise 56    Further problem on a.c. bridges**

1.  A Maxwell bridge circuit ABCD has the following arm impedances: AB, $250\,\Omega$ resistance; BC, $15\,\mu\text{F}$ capacitor in parallel with a $10\,\text{k}\Omega$ resistor; CD, $400\,\Omega$ resistor; DA, unknown inductor having inductance $L$ and resistance $R$. Determine the values of $L$ and $R$ assuming the bridge is balanced.

[1.5 H, 10 $\Omega$]

---

## 10.21   Q-meter

The **Q-factor** for a series L–C–R circuit is the voltage magnification at resonance, i.e.

$$\text{Q-factor} = \frac{\textbf{voltage across capacitor}}{\textbf{supply voltage}}$$

$$= \frac{V_c}{V} \quad (\text{see Chapter 15}).$$

The simplified circuit of a **Q-meter**, used for measuring Q-factor, is shown in Fig. 10.36. Current from a variable frequency oscillator flowing through a very low resistance $r$ develops a variable frequency voltage, $V_r$, which is applied to a series L–R–C circuit. The frequency is then varied until resonance causes voltage $V_c$ to reach a maximum value. At resonance $V_r$ and $V_c$ are noted. Then

$$\text{Q-factor} = \frac{V_c}{V_r} = \frac{V_c}{Ir}$$

In a practical Q-meter, $V_r$ is maintained constant and the electronic voltmeter can be calibrated to indicate the Q-factor directly. If a variable capacitor $C$ is used and the oscillator is set to a given frequency, then $C$ can be adjusted to give resonance. In this way inductance $L$ may be calculated using

$$f_r = \frac{1}{2\pi \sqrt{LC}}$$

Since     $$Q = \frac{2\pi f L}{R}$$

then $R$ may be calculated.

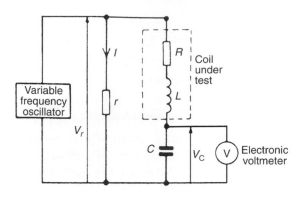

**Figure 10.36**

Q-meters operate at various frequencies and instruments exist with frequency ranges from 1 kHz to 50 MHz. Errors in measurement can exist with Q-meters since the coil has an effective parallel self capacitance due to capacitance between turns. The accuracy of a Q-meter is approximately ±5%.

---

**Problem 21.** When connected to a Q-meter an inductor is made to resonate at 400 kHz. The Q-factor of the circuit is found to be 100 and the capacitance of the Q-meter capacitor is set to 400 pF. Determine (a) the inductance, and (b) the resistance of the inductor.

---

Resonant frequency, $f_r = 400\,\text{kHz} = 400 \times 10^3\,\text{Hz}$, Q-factor $= 100$ and capacitance, $C = 400\,\text{pF} = 400 \times 10^{-12}\,\text{F}$. The circuit diagram of a Q-meter is shown in Fig. 10.36.

(a) At resonance,

$$f_r = \frac{1}{2\pi\sqrt{LC}}$$

for a series L–C–R circuit.

Hence $$2\pi f_r = \frac{1}{\sqrt{LC}}$$

from which

$$(2\pi f_r)^2 = \frac{1}{LC}$$

and **inductance**,

$$L = \frac{1}{(2\pi f_r)^2 C}$$

$$= \frac{1}{(2\pi \times 400 \times 10^3)^2 (400 \times 10^{-12})}\,\text{H}$$

$$= \mathbf{396\,\mu H \ or \ 0.396\,mH}$$

(b) Q-factor at resonance $= 2\pi f_r L/R$ from which resistance

$$R = \frac{2\pi f_r L}{Q}$$

$$= \frac{2\pi (400 \times 10^3)(0.396 \times 10^{-3})}{100}$$

$$= \mathbf{9.95\,\Omega}$$

---

**Now try the following exercise**

---

**Exercise 57  Further problem on the Q-meter**

1. A Q-meter measures the Q-factor of a series L-C-R circuit to be 200 at a resonant frequency of 250 kHz. If the capacitance of the Q-meter capacitor is set to 300 pF determine (a) the inductance $L$, and (b) the resistance $R$ of the inductor.     [(a) 1.351 mH (b) 10.61 Ω]

---

## 10.22  Measurement errors

Errors are always introduced when using instruments to measure electrical quantities. The errors most likely to occur in measurements are those due to:

(i)   the limitations of the instrument;

(ii)  the operator;

(iii) the instrument disturbing the circuit.

**(i)  Errors in the limitations of the instrument**

The **calibration accuracy** of an instrument depends on the precision with which it is constructed. Every instrument has a margin of error which is expressed as a percentage of the instruments full-scale deflection. For example, industrial grade instruments have an accuracy of ±2% of f.s.d. Thus if a voltmeter has a f.s.d. of 100 V and it indicates 40 V say, then the actual voltage may be anywhere between 40 ± (2% of 100), or 40 ±2, i.e. between 38 V and 42 V.

When an instrument is calibrated, it is compared against a standard instrument and a graph is drawn of 'error' against 'meter deflection'. A typical graph is shown in Fig. 10.37 where it is seen that the accuracy varies over the scale length. Thus a meter with a ±2% f.s.d. accuracy would tend to have an accuracy which is much better than ±2% f.s.d. over much of the range.

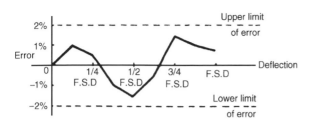

**Figure 10.37**

### (ii)  Errors by the operator

It is easy for an operator to misread an instrument. With linear scales the values of the sub-divisions are reasonably easy to determine; non-linear scale graduations are more difficult to estimate. Also, scales differ from instrument to instrument and some meters have more than one scale (as with multimeters) and mistakes in reading indications are easily made. When reading a meter scale it should be viewed from an angle perpendicular to the surface of the scale at the location of the pointer; a meter scale should not be viewed 'at an angle'. Errors by the operator are eliminated with digital instruments.

### (iii)  Errors due to the instrument disturbing the circuit

Any instrument connected into a circuit will affect that circuit to some extent. Meters require some power to operate, but provided this power is small compared with the power in the measured circuit, then little error will result. Incorrect positioning of instruments in a circuit can be a source of errors. For example, let a resistance be measured by the voltmeter-ammeter method as shown in Fig. 10.38. Assuming 'perfect' instruments, the resistance should be given by the voltmeter reading divided by the ammeter reading (i.e. $R = V/I$). However, in Fig. 10.38(a), $V/I = R + r_a$ and in Fig. 10.38(b) the current through the ammeter is that through the resistor plus that through the voltmeter. Hence the voltmeter reading divided by the ammeter reading will not give the true value of the resistance $R$ for either method of connection.

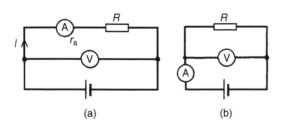

(a)                    (b)

**Figure 10.38**

---

**Problem 22.**   The current flowing through a resistor of $5\,k\Omega \pm 0.4\%$ is measured as $2.5\,mA$ with an accuracy of measurement of $\pm 0.5\%$. Determine the nominal value of the voltage across the resistor and its accuracy.

Voltage, $V = IR = (2.5 \times 10^{-3})(5 \times 10^{3}) = 12.5\,V$. The maximum possible error is $0.4\% + 0.5\% = 0.9\%$.
Hence the voltage, $V = 12.5\,V \pm 0.9\%$ of $12.5\,V$
$0.9\%$ of $12.5 = 0.9/100 \times 12.5 = 0.1125\,V = 0.11\,V$ correct to 2 significant figures.

Hence the voltage $V$ may also be expressed as **$12.5 \pm 0.11$ volts** (i.e. a voltage lying between $12.39\,V$ and $12.61\,V$).

---

**Problem 23.**   The current $I$ flowing in a resistor $R$ is measured by a $0$–$10\,A$ ammeter which gives an indication of $6.25\,A$. The voltage $V$ across the resistor is measured by a $0$–$50\,V$ voltmeter, which gives an indication of $36.5\,V$. Determine the resistance of the resistor, and its accuracy of measurement if both instruments have a limit of error of $2\%$ of f.s.d. Neglect any loading effects of the instruments.

Resistance,

$$R = \frac{V}{I} = \frac{36.5}{6.25} = 5.84\,\Omega$$

Voltage error is $\pm 2\%$ of $50\,V = \pm 1.0\,V$ and expressed as a percentage of the voltmeter reading gives

$$\frac{\pm 1}{36.5} \times 100\% = \pm 2.74\%$$

Current error is $\pm 2\%$ of $10\,A = \pm 0.2\,A$ and expressed as a percentage of the ammeter reading gives

$$\frac{\pm 0.2}{6.25} \times 100\% = \pm 3.2\%$$

Maximum relative error $=$ sum of errors $= 2.74\% + 3.2\% = \pm 5.94\%$ and $5.94\%$ of $5.84\,\Omega = 0.347\,\Omega$. Hence the resistance of the resistor may be expressed as:

**$5.84\,\Omega \pm 5.94\%$ or $5.84 \pm 0.35\,\Omega$**

(rounding off )

**Problem 24.** The arms of a Wheatstone bridge ABCD have the following resistances: AB: $R_1 = 1000\,\Omega \pm 1.0\%$; BC: $R_2 = 100\,\Omega \pm 0.5\%$; CD: unknown resistance $R_x$; DA: $R_3 = 432.5\,\Omega \pm 0.2\%$. Determine the value of the unknown resistance and its accuracy of measurement.

The Wheatstone bridge network is shown in Fig. 10.39 and at balance:

$$R_1 R_x = R_2 R_3,$$

i.e. $\quad R_x = \dfrac{R_2 R_3}{R_1} = \dfrac{(100)(432.5)}{1000} = 43.25\,\Omega$

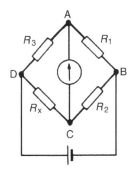

**Figure 10.39**

The maximum relative error of $R_x$ is given by the sum of the three individual errors,
i.e. $1.0\% + 0.5\% + 0.2\% = 1.7\%$.

**Hence** $\qquad\qquad$ $\mathbf{R_x = 43.25\,\Omega \pm 1.7\%}$

1.7% of $43.25\,\Omega = 0.74\,\Omega$ (rounding off). Thus $R_x$ may also be expressed as

$$\mathbf{R_x = 43.25 \pm 0.74\,\Omega}$$

**Now try the following exercises**

**Exercise 58** **Further problems on measurement errors**

1. The p.d. across a resistor is measured as 37.5 V with an accuracy of $\pm 0.5\%$. The value of the

resistor is $6\,\mathrm{k\Omega} \pm 0.8\%$. Determine the current flowing in the resistor and its accuracy of measurement.
$$[6.25\,\mathrm{mA} \pm 1.3\% \text{ or } 6.25 \pm 0.08\,\mathrm{mA}]$$

2. The voltage across a resistor is measured by a 75 V f.s.d. voltmeter which gives an indication of 52 V. The current flowing in the resistor is measured by a 20 A f.s.d. ammeter which gives an indication of 12.5 A. Determine the resistance of the resistor and its accuracy if both instruments have an accuracy of $\pm 2\%$ of f.s.d.
$$[4.16\,\Omega \pm 6.08\% \text{ or } 4.16 \pm 0.25\,\Omega]$$

3. A Wheatstone bridge PQRS has the following arm resistances: PQ, $1\,\mathrm{k\Omega} \pm 2\%$; QR, $100\,\Omega \pm 0.5\%$; RS, unknown resistance; SP, $273.6\,\Omega \pm 0.1\%$. Determine the value of the unknown resistance, and its accuracy of measurement.
$$[27.36\,\Omega \pm 2.6\% \text{ or } 27.36\,\Omega \pm 0.71\,\Omega]$$

**Exercise 59** **Short answer questions on electrical measuring instruments and measurements**

1. What is the main difference between an analogue and a digital type of measuring instrument?

2. Name the three essential devices for all analogue electrical indicating instruments

3. Complete the following statements:
   (a) An ammeter has a ...... resistance and is connected ...... with the circuit
   (b) A voltmeter has a ...... resistance and is connected ...... with the circuit

4. State two advantages and two disadvantages of a moving-coil instrument

5. What effect does the connection of (a) a shunt (b) a multiplier have on a milliammeter?

6. State two advantages and two disadvantages of a moving-coil instrument

7. Name two advantages of electronic measuring instruments compared with moving-coil or moving-iron instruments

8. Briefly explain the principle of operation of an ohmmeter

9. Name a type of ohmmeter used for measuring (a) low resistance values (b) high resistance values

10. What is a multimeter?

11. When may a rectifier instrument be used in preference to either a moving-coil or moving-iron instrument?

12. Name five quantities that a c.r.o. is capable of measuring

13. What is harmonic analysis?

14. What is a feature of waveforms containing the fundamental and odd harmonics?

15. Express the ratio of two powers $P_1$ and $P_2$ in decibel units

16. What does a power level unit of dBm indicate?

17. What is meant by a null method of measurement?

18. Sketch a Wheatstone bridge circuit used for measuring an unknown resistance in a d.c. circuit and state the balance condition

19. How may a d.c. potentiometer be used to measure p.d.'s

20. Name five types of a.c. bridge used for measuring unknown inductance, capacitance or resistance

21. What is a universal bridge?

22. State the name of an a.c. bridge used for measuring inductance

23. Briefly describe how the measurement of Q-factor may be achieved

24. Why do instrument errors occur when measuring complex waveforms?

25. Define 'calibration accuracy' as applied to a measuring instrument

26. State three main areas where errors are most likely to occur in measurements

**Exercise 60    Multi-choice questions on electrical measuring instruments and measurements**
**(Answers on page 420)**

1. Which of the following would apply to a moving coil instrument?
   (a) An uneven scale, measuring d.c.
   (b) An even scale, measuring a.c.
   (c) An uneven scale, measuring a.c.
   (d) An even scale, measuring d.c.

2. In question 1, which would refer to a moving iron instrument?

3. In question 1, which would refer to a moving coil rectifier instrument?

4. Which of the following is needed to extend the range of a milliammeter to read voltages of the order of 100 V?
   (a) a parallel high-value resistance
   (b) a series high-value resistance
   (c) a parallel low-value resistance
   (d) a series low-value resistance

5. Fig. 10.40 shows a scale of a multi-range ammeter. What is the current indicated when switched to a 25 A scale?
   (a) 84 A          (b) 5.6 A
   (c) 14 A          (d) 8.4 A

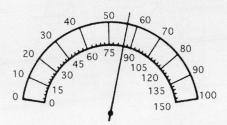

**Figure 10.40**

A sinusoidal waveform is displayed on a c.r.o. screen. The peak-to-peak distance is 5 cm and the distance between cycles is 4 cm. The 'variable' switch is on 100 μs/cm and the 'volts/cm' switch is on 10 V/cm. In questions 6 to 10, select the correct answer from the following:

(a) 25 V        (b) 5 V        (c) 0.4 ms
(d) 35.4 V      (e) 4 ms       (f) 50 V
(g) 250 Hz      (h) 2.5 V      (i) 2.5 kHz
(j) 17.7 V

6. Determine the peak-to-peak voltage

7. Determine the periodic time of the waveform

8. Determine the maximum value of the voltage

9. Determine the frequency of the waveform

10. Determine the r.m.s. value of the waveform

Figure 10.41 shows double-beam c.r.o. waveform traces. For the quantities stated in questions 11 to 17, select the correct answer from the following:

(a) 30 V   (b) 0.2 s   (c) 50 V

(d) $\dfrac{15}{\sqrt{2}}$   (e) 54° leading   (f) $\dfrac{250}{\sqrt{2}}$V

(g) 15 V   (h) 100 μs   (i) $\dfrac{50}{\sqrt{2}}$V

(j) 250 V   (k) 10 kHz   (l) 75 V

(m) 40 μs   (n) $\dfrac{3\pi}{10}$ rads lagging

(o) $\dfrac{25}{\sqrt{2}}$V   (p) 5 Hz   (q) $\dfrac{30}{\sqrt{2}}$V

(r) 25 kHz   (s) $\dfrac{75}{\sqrt{2}}$V

(t) $\dfrac{3\pi}{10}$ rads leading

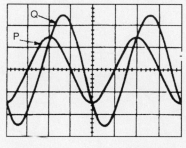

Figure 10.41

11. Amplitude of waveform P

12. Peak-to-peak value of waveform Q

13. Periodic time of both waveforms

14. Frequency of both waveforms

15. R.m.s. value of waveform P

16. R.m.s. value of waveform Q

17. Phase displacement of waveform Q relative to waveform P

18. The input and output powers of a system are 2 mW and 18 mW respectively. The decibel power ratio of output power to input power is:
(a) 9   (b) 9.54
(c) 1.9   (d) 19.08

19. The input and output voltages of a system are 500 μV and 500 mV respectively. The decibel voltage ratio of output to input voltage (assuming input resistance equals load resistance) is:
(a) 1000   (b) 30
(c) 0   (d) 60

20. The input and output currents of a system are 3 mA and 18 mA respectively. The decibel ratio of output to input current (assuming the input and load resistances are equal) is:
(a) 15.56   (b) 6
(c) 1.6   (d) 7.78

21. Which of the following statements is false?
(a) The Schering bridge is normally used for measuring unknown capacitances
(b) A.C. electronic measuring instruments can handle a much wider range of frequency than the moving coil instrument
(c) A complex waveform is one which is non-sinusoidal
(d) A square wave normally contains the fundamental and even harmonics

22. A voltmeter has a f.s.d. of 100 V, a sensitivity of 1 kΩ/V and an accuracy of ±2% of f.s.d. When the voltmeter is connected into a circuit it indicates 50 V. Which of the following statements is false?
(a) Voltage reading is 50±2 V
(b) Voltmeter resistance is 100 kΩ
(c) Voltage reading is 50 V±2%
(d) Voltage reading is 50 V±4%

23. A potentiometer is used to:
(a) compare voltages
(b) measure power factor
(c) compare currents
(d) measure phase sequence

# Chapter 11

# Semiconductor diodes

At the end of this chapter you should be able to:

- classify materials as conductors, semiconductors or insulators
- appreciate the importance of silicon and germanium
- understand n-type and p-type materials
- understand the p-n junction
- appreciate forward and reverse bias of p-n junctions
- recognise the symbols used to represent diodes in circuit diagrams
- understand the importance of diode characteristics and maximum ratings
- know the characteristics and applications of various types of diode – signal diodes, rectifiers, Zener diodes, silicon controlled rectifiers, light emitting diodes, varactor diodes and Schottky diodes.

## 11.1 Types of material

Materials may be classified as conductors, semiconductors or insulators. The classification depends on the value of resistivity of the material. Good conductors are usually metals and have resistivities in the order of $10^{-7}$ to $10^{-8}\,\Omega$m, semiconductors have resistivities in the order of $10^{-3}$ to $3 \times 10^{3}\,\Omega$m, and the resistivities of insulators are in the order of $10^{4}$ to $10^{14}\,\Omega$m. Some typical approximate values at normal room temperatures are:

### Conductors:

| | |
|---|---|
| Aluminium | $2.7 \times 10^{-8}\,\Omega$m |
| Brass (70 Cu/30 Zn) | $8 \times 10^{-8}\,\Omega$m |
| Copper (pure annealed) | $1.7 \times 10^{-8}\,\Omega$m |
| Steel (mild) | $15 \times 10^{-8}\,\Omega$m |

### Semiconductors: (at 27°C)

| | |
|---|---|
| Silicon | $2.3 \times 10^{3}\,\Omega$m |
| Germanium | $0.45\,\Omega$m |

### Insulators:

| | |
|---|---|
| Glass | $\geq 10^{10}\,\Omega$m |
| Mica | $\geq 10^{11}\,\Omega$m |
| PVC | $\geq 10^{13}\,\Omega$m |
| Rubber (pure) | $10^{12}$ to $10^{14}\,\Omega$m |

In general, over a limited range of temperatures, the resistance of a conductor increases with temperature increase, the resistance of insulators remains approximately constant with variation of temperature and the resistance of semiconductor materials decreases as the temperature increases. For a specimen of each of these materials, having the same resistance (and thus completely different dimensions), at say, 15°C, the variation for a small increase in temperature to $t°$C is as shown in Fig. 11.1.

As the temperature of semiconductor materials is raised above room temperature, the resistivity is reduced and ultimately a point is reached where they effectively become conductors. For this reason, silicon should not operate at a working temperature in excess of 150°C to 200°C, depending on its purity, and germanium should not operate at a working temperature

DOI: 10.1016/B978-0-08-089056-2.00011-5

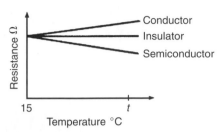

Figure 11.1

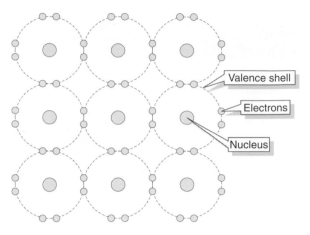

Figure 11.2

in excess of 75°C to 90°C, depending on its purity. As the temperature of a semiconductor is reduced below normal room temperature, the resistivity increases until, at very low temperatures the semiconductor becomes an insulator.

## 11.2   Semiconductor materials

From Chapter 2, it was stated that an atom contains both negative charge carriers (**electrons**) and positive charge carriers (**protons**). Electrons each carry a single unit of negative electric charge while protons each exhibit a single unit of positive charge. Since atoms normally contain an equal number of electrons and protons, the net charge present will be zero. For example, if an atom has eleven electrons, it will also contain eleven protons. The end result is that the negative charge of the electrons will be exactly balanced by the positive charge of the protons.

Electrons are in constant motion as they orbit around the nucleus of the atom. Electron orbits are organised into **shells**. The maximum number of electrons present in the first shell is two, in the second shell eight, and in the third, fourth and fifth shells it is 18, 32 and 50, respectively. In electronics, only the electron shell furthermost from the nucleus of an atom is important. It is important to note that the movement of electrons between atoms only involves those present in the outer **valence shell**.

If the valence shell contains the maximum number of electrons possible the electrons are rigidly bonded together and the material has the properties of an insulator (see Fig. 11.2). If, however, the valence shell does not have its full complement of electrons, the electrons can be easily detached from their orbital bonds, and the material has the properties associated with an electrical conductor.

In its pure state, silicon is an insulator because the covalent bonding rigidly holds all of the electrons leaving no free (easily loosened) electrons to conduct

current. If, however, an atom of a different element (i.e. an **impurity**) is introduced that has five electrons in its valence shell, a surplus electron will be present (see Fig. 11.3). These free electrons become available for use as charge carriers and they can be made to move through the lattice by applying an external potential difference to the material.

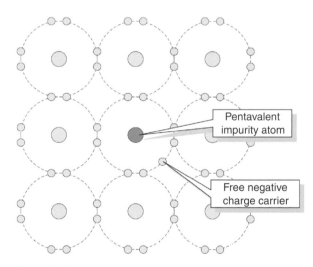

Figure 11.3

Similarly, if the impurity element introduced into the pure silicon lattice has three electrons in its valence shell, the absence of the fourth electron needed for proper covalent bonding will produce a number of spaces into which electrons can fit (see Fig. 11.4). These spaces are referred to as **holes**. Once again, current will flow when an external potential difference is applied to the material.

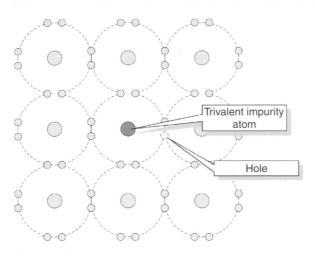

**Figure 11.4**

Regardless of whether the impurity element produces surplus electrons or holes, the material will no longer behave as an insulator, neither will it have the properties that we normally associate with a metallic conductor. Instead, we call the material a **semiconductor** – the term simply serves to indicate that the material is no longer a good insulator nor is it a good conductor but is somewhere in between. Examples of semiconductor materials include **silicon (Si), germanium (Ge), gallium arsenide (GaAs),** and **indium arsenide (InAs).**

**Antimony, arsenic** and **phosphorus** are **n-type impurities** and form an n-type material when any of these impurities are added to pure semiconductor material such as silicon or germanium. The amount of impurity added usually varies from 1 part impurity in $10^5$ parts semiconductor material to 1 part impurity to $10^8$ parts semiconductor material, depending on the resistivity required. **Indium, aluminium** and **boron** are all **p-type impurities** and form a p-type material when any of these impurities are added to a pure semiconductor.

The process of introducing an atom of another (impurity) element into the lattice of an otherwise pure material is called **doping**. When the pure material is doped with an impurity with five electrons in its valence shell (i.e. a **pentavalent impurity**) it will become an **n-type** (i.e. negative type) semiconductor material. If, however, the pure material is doped with an impurity having three electrons in its valence shell (i.e. a **trivalent impurity**) it will become a **p-type** (i.e. positive type) semiconductor material. Note that n-type semiconductor material contains an excess of negative charge carriers, and p-type material contains an excess of positive charge carriers.

In semiconductor materials, there are very few charge carriers per unit volume free to conduct. This is because the 'four electron structure' in the outer shell of the atoms (called **valency electrons**), form strong **covalent bonds** with neighbouring atoms, resulting in a tetrahedral (i.e. four-sided) structure with the electrons held fairly rigidly in place.

## 11.3 Conduction in semiconductor materials

Arsenic, antimony and phosphorus have five valency electrons and when a semiconductor is doped with one of these substances, some impurity atoms are incorporated in the tetrahedral structure. The 'fifth' valency electron is not rigidly bonded and is free to conduct, the impurity atom donating a charge carrier.

Indium, aluminium and boron have three valency electrons and when a semiconductor is doped with one of these substances, some of the semiconductor atoms are replaced by impurity atoms. One of the four bonds associated with the semiconductor material is deficient by one electron and this deficiency is called a **hole**. Holes give rise to conduction when a potential difference exists across the semiconductor material due to movement of electrons from one hole to another, as shown in Fig. 11.5. In this diagram, an electron moves from A to B, giving the appearance that the hole moves from B to A. Then electron C moves to A, giving the appearance that the hole moves to C, and so on.

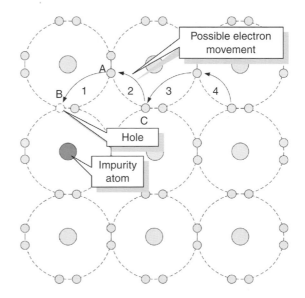

**Figure 11.5**

## 11.4    The p-n junction

A p-n junction is a piece of semiconductor material in which part of the material is p-type and part is n-type. In order to examine the charge situation, assume that separate blocks of p-type and n-type materials are pushed together. Also assume that a hole is a positive charge carrier and that an electron is a negative charge carrier.

At the junction, the donated electrons in the n-type material, called **majority carriers**, diffuse into the p-type material (diffusion is from an area of high density to an area of lower density) and the acceptor holes in the p-type material diffuse into the n-type material as shown by the arrows in Fig. 11.6. Because the n-type material has lost electrons, it acquires a positive potential with respect to the p-type material and thus tends to prevent further movement of electrons. The p-type material has gained electrons and becomes negatively charged with respect to the n-type material and hence tends to retain holes. Thus after a short while, the movement of electrons and holes stops due to the potential difference across the junction, called the **contact potential**. The area in the region of the junction becomes depleted of holes and electrons due to electron-hole recombination, and is called a **depletion layer**, as shown in Fig. 11.7.

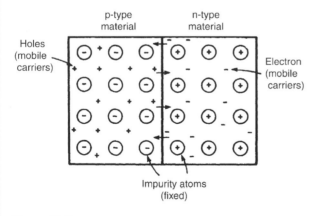

**Figure 11.6**

Problem 1.    Explain briefly the terms given below when they are associated with a p-n junction:
(a) conduction in intrinsic semiconductors,
(b) majority and minority carriers, and (c) diffusion.

(a)    Silicon or germanium with no doping atoms added are called **intrinsic semiconductors**. At room temperature, some of the electrons acquire sufficient energy for them to break the covalent bond between

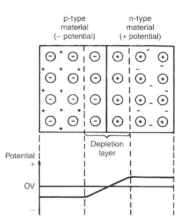

**Figure 11.7**

atoms and become free mobile electrons. This is called **thermal generation of electron-hole pairs**. Electrons generated thermally create a gap in the crystal structure called a hole, the atom associated with the hole being positively charged, since it has lost an electron. This positive charge may attract another electron released from another atom, creating a hole elsewhere. When a potential is applied across the semiconductor material, holes drift towards the negative terminal (unlike charges attract), and electrons towards the positive terminal, and hence a small current flows.

(b)    When additional mobile electrons are introduced by doping a semiconductor material with pentavalent atoms (atoms having five valency electrons), these mobile electrons are called **majority carriers**. The relatively few holes in the n-type material produced by intrinsic action are called **minority carriers**.

For p-type materials, the additional holes are introduced by doping with trivalent atoms (atoms having three valency electrons). The holes are apparently positive mobile charges and are majority carriers in the p-type material. The relatively few mobile electrons in the p-type material produced by intrinsic action are called minority carriers.

(c)    Mobile holes and electrons wander freely within the crystal lattice of a semiconductor material. There are more free electrons in n-type material than holes and more holes in p-type material than electrons. Thus, in their random wanderings, on average, holes pass into the n-type material and electrons into the p-type material. This process is called **diffusion**.

**Problem 2.** Explain briefly why a junction between p-type and n-type materials creates a contact potential.

Intrinsic semiconductors have resistive properties, in that when an applied voltage across the material is reversed in polarity, a current of the same magnitude flows in the opposite direction. When a p-n junction is formed, the resistive property is replaced by a rectifying property, that is, current passes more easily in one direction than the other.

An n-type material can be considered to be a stationary crystal matrix of fixed positive charges together with a number of mobile negative charge carriers (electrons). The total number of positive and negative charges are equal. A p-type material can be considered to be a number of stationary negative charges together with mobile positive charge carriers (holes).

Again, the total number of positive and negative charges are equal and the material is neither positively nor negatively charged. When the materials are brought together, some of the mobile electrons in the n-type material diffuse into the p-type material. Also, some of the mobile holes in the p-type material diffuse into the n-type material.

Many of the majority carriers in the region of the junction combine with the opposite carriers to complete covalent bonds and create a region on either side of the junction with very few carriers. This region, called the **depletion layer**, acts as an insulator and is in the order of 0.5 μm thick. Since the n-type material has lost electrons, it becomes positively charged. Also, the p-type material has lost holes and becomes negatively charged, creating a potential across the junction, called the **barrier** or **contact potential**.

## 11.5    Forward and reverse bias

When an external voltage is applied to a p-n junction making the p-type material positive with respect to the n-type material, as shown in Fig. 11.8, the p-n junction is **forward biased**. The applied voltage opposes the contact potential, and, in effect, closes the depletion layer. Holes and electrons can now cross the junction and a current flows. An increase in the applied voltage above that required to narrow the depletion layer (about 0.2 V for germanium and 0.6 V for silicon), results in a rapid rise in the current flow.

When an external voltage is applied to a p-n junction making the p-type material negative with respect to the

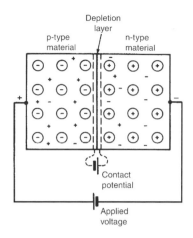

**Figure 11.8**

n-type material as is shown in Fig. 11.9, the p-n junction is **reverse biased**. The applied voltage is now in the same sense as the contact potential and opposes the movement of holes and electrons due to opening up the depletion layer. Thus, in theory, no current flows. However, at normal room temperature certain electrons in the covalent bond lattice acquire sufficient energy from the heat available to leave the lattice, generating mobile electrons and holes. This process is called **electron-hole generation by thermal excitation**.

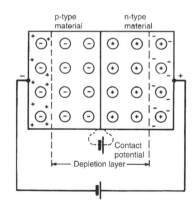

**Figure 11.9**

The electrons in the p-type material and holes in the n-type material caused by thermal excitation, are called minority carriers and these will be attracted by the applied voltage. Thus, in practice, a small current of a few microamperes for germanium and less than one microampere for silicon, at normal room temperature, flows under reverse bias conditions.

Graphs depicting the current-voltage relationship for forward and reverse biased p-n junctions, for both germanium and silicon, are shown in Fig. 11.10.

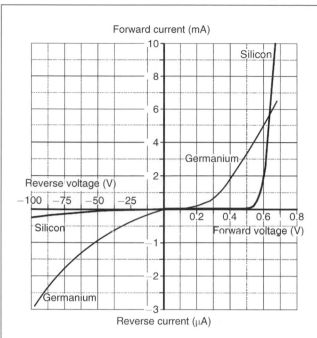

**Figure 11.10**

---

**Problem 3.** Sketch the forward and reverse characteristics of a silicon p-n junction diode and describe the shapes of the characteristics drawn.

---

A typical characteristic for a silicon p-n junction is shown in Fig. 11.10. When the positive terminal of the battery is connected to the p-type material and the negative terminal to the n-type material, the diode is forward biased. Due to like charges repelling, the holes in the p-type material drift towards the junction. Similarly the electrons in the n-type material are repelled by the negative bias voltage and also drift towards the junction. The width of the depletion layer and size of the contact potential are reduced. For applied voltages from 0 to about 0.6 V, very little current flows. At about 0.6 V, majority carriers begin to cross the junction in large numbers and current starts to flow. As the applied voltage is raised above 0.6 V, the current increases exponentially (see Fig. 11.10).

When the negative terminal of the battery is connected to the p-type material and the positive terminal to the n-type material the diode is reverse biased. The holes in the p-type material are attracted towards the negative terminal and the electrons in the n-type material are attracted towards the positive terminal (unlike charges attract). This drift increases the magnitude of both the contact potential and the thickness of the depletion layer, so that only very few majority carriers have sufficient energy to surmount the junction.

The thermally excited minority carriers, however, can cross the junction since it is, in effect, forward biased for these carriers. The movement of minority carriers results in a small constant current flowing. As the magnitude of the reverse voltage is increased a point will be reached where a large current suddenly starts to flow. The voltage at which this occurs is called the **breakdown voltage**. This current is due to two effects:

(i) the **Zener effect**, resulting from the applied voltage being sufficient to break some of the covalent bonds, and

(ii) the **avalanche effect**, resulting from the charge carriers moving at sufficient speed to break covalent bonds by collision.

---

**Problem 4.** The forward characteristic of a diode is shown in Fig. 11.11. Use the characteristic to determine (a) the current flowing in the diode when a forward voltage of 0.4 V is applied, (b) the voltage dropped across the diode when a forward current of 9 mA is flowing in it, (c) the resistance of the diode when the forward voltage is 0.6 V, and (d) whether the diode is a Ge or Si type.

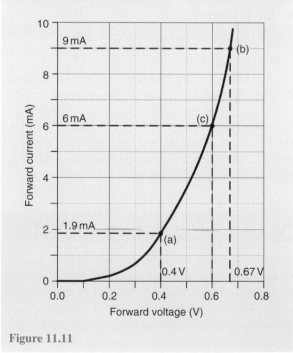

**Figure 11.11**

---

(a) From Fig. 11.11, when $V = 0.4$ V, **current flowing, $I = 1.9$ mA**

(b) When $I = 9$ mA, **the voltage dropped across the diode, $V = 0.67$ V**

(c)    From the graph, when $V = 0.6\,V$, $I = 6\,mA$.

Thus, **resistance of the diode**,

$$R = \frac{V}{I} = \frac{0.6}{6 \times 10^{-3}} = 0.1 \times 10^3 = 100\,\Omega$$

(d)    The onset of conduction occurs at approximately 0.2 V. This suggests that the diode is a **Ge type**.

---

**Problem 5.**    Corresponding readings of current, $I$, and voltage, $V$, for a semiconductor device are given in the table:

| $V_f$ (V) | 0 | 0.1 | 0.2 | 0.3 | 0.4 | 0.5 | 0.6 | 0.7 | 0.8 |
|-----------|---|-----|-----|-----|-----|-----|-----|-----|-----|
| $I_f$ (mA) | 0 | 0 | 0 | 0 | 0 | 1 | 9 | 24 | 50 |

Plot the $I/V$ characteristic for the device and identify the type of device.

---

The $I/V$ characteristic is shown in Fig. 11.12. Since the device begins to conduct when a potential of approximately 0.6 V is applied to it we can infer that **the semiconductor material is silicon** rather than germanium.

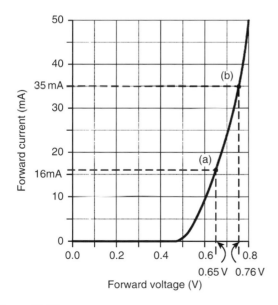

**Figure 11.12**

---

**Problem 6.**    For the characteristic of Fig. 11.12, determine for the device (a) the forward current when the forward voltage is 0.65 V, and (b) the forward voltage when the forward current is 35 mA.

---

(a)    From Fig. 11.12, when the forward voltage is 0.65 V, **the forward current = 16 mA**

(b)    When the forward current is 35 mA, **the forward voltage = 0.76 V**

---

**Now try the following exercise**

**Exercise 61    Further problems on semiconductor materials and p-n junctions**

1.    Explain what you understand by the term intrinsic semiconductor and how an intrinsic semiconductor is turned into either a p-type or an n-type material.

2.    Explain what is meant by minority and majority carriers in an n-type material and state whether the numbers of each of these carriers are affected by temperature.

3.    A piece of pure silicon is doped with (a) pentavalent impurity and (b) trivalent impurity. Explain the effect these impurities have on the form of conduction in silicon.

4.    With the aid of simple sketches, explain how pure germanium can be treated in such a way that conduction is predominantly due to (a) electrons and (b) holes.

5.    Explain the terms given below when used in semiconductor terminology: (a) covalent bond, (b) trivalent impurity, (c) pentavalent impurity, (d) electron-hole pair generation.

6.    Explain briefly why although both p-type and n-type materials have resistive properties when separate, they have rectifying properties when a junction between them exists.

7.    The application of an external voltage to a junction diode can influence the drift of holes and electrons. With the aid of diagrams explain this statement and also how the direction and magnitude of the applied voltage affects the depletion layer.

8.    State briefly what you understand by the terms: (a) reverse bias, (b) forward bias, (c) contact potential, (d) diffusion, (e) minority carrier conduction.

9. Explain briefly the action of a p-n junction diode: (a) on open-circuit, (b) when provided with a forward bias, and (c) when provided with a reverse bias. Sketch the characteristic curves for both forward and reverse bias conditions.

10. Draw a diagram illustrating the charge situation for an unbiased p-n junction. Explain the change in the charge situation when compared with that in isolated p-type and n-type materials. Mark on the diagram the depletion layer and the majority carriers in each region.

11. The graph shown in Fig. 11.13 was obtained during an experiment on a diode. (a) What type of diode is this? Give reasons. (b) Determine the forward current for a forward voltage of 0.5 V. (c) Determine the forward voltage for a forward current of 30 mA. (d) Determine the resistance of the diode when the forward voltage is 0.4 V.
[(a) Ge (b) 17 mA (c) 0.625 V (d) 50 Ω]

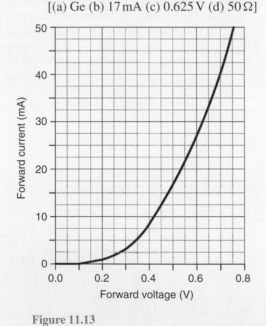

**Figure 11.13**

## 11.6 Semiconductor diodes

When a junction is formed between p-type and n-type semiconductor materials, the resulting device is called a **semiconductor diode**. This component offers an extremely low resistance to current flow in one direction and an extremely high resistance to current flow in the other. This property allows diodes to be used in applications that require a circuit to behave differently according to the direction of current flowing in it. Note that an ideal diode would pass an infinite current in one direction and no current at all in the other direction.

A semiconductor diode is an encapsulated p-n junction fitted with connecting leads or tags for connection to external circuitry. Where an appreciable current is present (as is the case with many rectifier circuits) the diode may be mounted in a metal package designed to conduct heat away from the junction. The connection to the p-type material is referred to as the **anode** while that to the n-type material is called the **cathode**.

Various different types of diode are available for different applications. These include **rectifier diodes** for use in power supplies, **Zener diodes** for use as voltage reference sources, **light emitting diodes**, and **varactor diodes**. Figure 11.14 shows the symbols used to represent diodes in electronic circuit diagrams, where 'a' is the anode and 'k' the cathode.

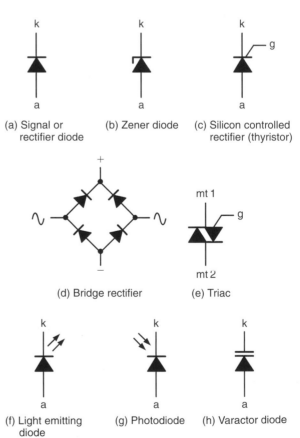

(a) Signal or rectifier diode    (b) Zener diode    (c) Silicon controlled rectifier (thyristor)

(d) Bridge rectifier    (e) Triac

(f) Light emitting diode    (g) Photodiode    (h) Varactor diode

**Figure 11.14**

## 11.7 Characteristics and maximum ratings

Signal diodes require consistent forward characteristics with low forward voltage drop. Rectifier diodes need to be able to cope with high values of reverse voltage and large values of forward current, and consistency of characteristics is of secondary importance in such applications. Table 11.1 summarises the characteristics of some common semiconductor diodes. It is worth noting that diodes are limited by the amount of forward current and reverse voltage they can withstand. This limit is based on the physical size and construction of the diode.

A typical general-purpose diode may be specified as having a forward threshold voltage of 0.6 V and a reverse breakdown voltage of 200 V. If the latter is exceeded, the diode may suffer irreversible damage. Typical values of **maximum repetitive reverse voltage** ($V_{RRM}$) or **peak inverse voltage** (PIV) range from about 50 V to over 500 V. The reverse voltage may be increased until the maximum reverse voltage for which the diode is rated is reached. If this voltage is exceeded the junction may break down and the diode may suffer permanent damage.

## 11.8 Rectification

The process of obtaining unidirectional currents and voltages from alternating currents and voltages is called **rectification**. Semiconductor diodes are commonly used to convert alternating current (a.c.) to direct current (d.c.), in which case they are referred to as **rectifiers**. The simplest form of rectifier circuit makes use of a single diode and, since it operates on only either positive or negative half-cycles of the supply, it is known as a **half-wave rectifier**. Four diodes are connected as a **bridge rectifier** – see Fig. 11.14(d) – and are often used as a **full-wave rectifier**. Note that in both cases, automatic switching of the current is carried out by the diode(s). For methods of half-wave and full-wave rectification, see Section 14.7, page 221.

## 11.9 Zener diodes

Zener diodes are heavily doped silicon diodes that, unlike normal diodes, exhibit an abrupt reverse breakdown at relatively low voltages (typically less than 6 V). A similar effect, called **avalanche breakdown**, occurs in less heavily doped diodes. These avalanche diodes also exhibit a rapid breakdown with negligible current flowing below the avalanche voltage and a relatively large current flowing once the avalanche voltage has been reached. For avalanche diodes, this breakdown voltage usually occurs at voltages above 6 V. In practice, however, both types of diode are referred to as **Zener diodes**. The symbol for a Zener diode is shown in Fig. 11.14(b) whilst a typical Zener diode characteristic is shown in Fig. 11.15.

**Table 11.1** Characteristics of some typical signal and rectifier diodes

| Device code | Material | Max repetitive reverse voltage ($V_{RRM}$) | Max forward current ($I_{F(max)}$) | Max reverse current ($I_{R(max)}$) | Application |
|---|---|---|---|---|---|
| 1N4148 | Silicon | 100 V | 75 mA | 25 nA | General purpose |
| 1N914 | Silicon | 100 V | 75 mA | 25 nA | General purpose |
| AA113 | Germanium | 60 V | 10 mA | 200 μA | RF detector |
| OA47 | Germanium | 25 V | 110 mA | 100 μA | Signal detector |
| OA91 | Germanium | 115 V | 50 mA | 275 μA | General purpose |
| 1N4001 | Silicon | 50 V | 1 A | 10 μA | Low voltage rectifier |
| 1N5404 | Silicon | 400 V | 3 A | 10 μA | High voltage rectifier |
| BY127 | Silicon | 1250 V | 1 A | 10 μA | High voltage rectifier |

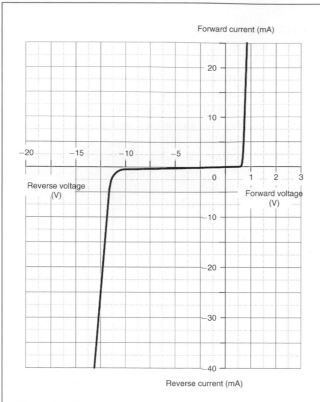

Figure 11.15

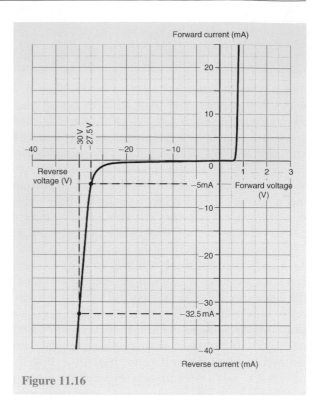

Figure 11.16

Whereas reverse breakdown is a highly undesirable effect in circuits that use conventional diodes, it can be extremely useful in the case of Zener diodes where the breakdown voltage is precisely known. When a diode is undergoing reverse breakdown and provided its maximum ratings are not exceeded, the voltage appearing across it will remain substantially constant (equal to the nominal Zener voltage) regardless of the current flowing. This property makes the Zener diode ideal for use as a **voltage regulator**.

Zener diodes are available in various families (according to their general characteristics, encapsulations and power ratings) with reverse breakdown (Zener) voltages in the range 2.4 V to 91 V.

> **Problem 7.** The characteristic of a Zener diode is shown in Fig. 11.16. Use the characteristic to determine (a) the current flowing in the diode when a reverse voltage of 30 V is applied, (b) the voltage dropped across the diode when a reverse current of 5 mA is flowing in it, (c) the voltage rating for the Zener diode, and (d) the power dissipated in the Zener diode when a reverse voltage of 30 V appears across it.

(a) When $V = -30$ V, **the current flowing in the diode, $I = -32.5$ mA**

(b) When $I = -5$ mA, **the voltage dropped across the diode, $V = -27.5$ V**

(c) The characteristic shows the onset of Zener action at 27 V; this would suggest a **Zener voltage rating of 27 V**

(d) Power, $P = V \times I$, from which, **power dissipated when the reverse voltage is 30 V,**
$P = 30 \times (32.5 \times 10^{-3}) = 0.975$ W $= $ **975 mW**

## 11.10 Silicon controlled rectifiers

Silicon controlled rectifiers (or **thyristors**) are three-terminal devices which can be used for switching and a.c. power control. Silicon controlled rectifiers can switch very rapidly from conducting to a non-conducting state. In the off state, the silicon controlled rectifier exhibits negligible leakage current, while in the on state the device exhibits very low resistance. This results in very little power loss within the silicon controlled rectifier even when appreciable power levels are being controlled.

Once switched into the conducting state, the silicon controlled rectifier will remain conducting (i.e. it is latched in the on state) until the forward current is removed from the device. In d.c. applications this necessitates the interruption (or disconnection) of the supply before the device can be reset into its non-conducting state. Where the device is used with an alternating supply, the device will automatically become reset whenever the main supply reverses. The device can then be triggered on the next half-cycle having correct polarity to permit conduction.

Like their conventional silicon diode counterparts, silicon controlled rectifiers have anode and cathode connections; control is applied by means of a gate terminal, g. The symbol for a silicon controlled rectifier is shown in Fig. 11.14(c).

In normal use, a silicon controlled rectifier (SCR) is triggered into the conducting (on) state by means of the application of a current pulse to the gate terminal – see Fig. 11.17. The effective triggering of a silicon controlled rectifier requires a gate trigger pulse having a fast rise time derived from a low-resistance source. Triggering can become erratic when insufficient gate current is available or when the gate current changes slowly.

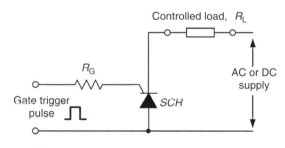

**Figure 11.17**

A typical silicon controlled rectifier for mains switching applications will require a gate trigger pulse of about 30 mA at 2.5 V to control a current of up to 5 A.

## 11.11    Light emitting diodes

Light emitting diodes (LED) can be used as general-purpose indicators and, compared with conventional filament lamps, operate from significantly smaller voltages and currents. LEDs are also very much more reliable than filament lamps. Most LEDs will provide a reasonable level of light output when a forward current of between 5 mA and 20 mA is applied.

Light emitting diodes are available in various formats with the round types being most popular. Round LEDs are commonly available in the 3 mm and 5 mm (0.2 inch)

diameter plastic packages and also in a 5 mm × 2 mm rectangular format. The viewing angle for round LEDs tends to be in the region of 20° to 40°, whereas for rectangular types this is increased to around 100°. The peak wavelength of emission depends on the type of semiconductor employed but usually lies in the range 630 to 690 nm. The symbol for an LED is shown in Fig. 11.14(f).

## 11.12    Varactor diodes

It was shown earlier that when a diode is operated in the reverse biased condition, the width of the depletion region increases as the applied voltage increases. Varying the width of the depletion region is equivalent to varying the plate separation of a very small capacitor such that the relationship between junction capacitance and applied reverse voltage will look something like that shown in Fig. 11.18. The typical variation of capacitance provided by a varactor is from about 50 pF to 10 pF as the reverse voltage is increased from 2 V to 20 V. The symbol for a varactor diode is shown in Fig. 11.14(h).

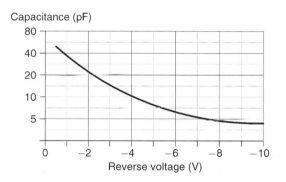

**Figure 11.18**

## 11.13    Schottky diodes

The conventional p-n junction diode explained in Section 11.4 operates well as a rectifier and switching device at relatively low frequencies (i.e. 50 Hz to 400 Hz) but its performance as a rectifier becomes seriously impaired at high frequencies due to the presence of stored charge carriers in the junction. These have the effect of momentarily allowing current to flow in the reverse direction when reverse voltage is applied. This problem becomes increasingly more problematic as the frequency of the a.c. supply is increased and the periodic time of the applied voltage becomes smaller.

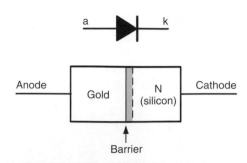

**Figure 11.19**

To avoid these problems a diode that uses a metal-semiconductor contact rather than a p-n junction (see Fig. 11.19) is employed. When compared with conventional silicon junction diodes, these **Schottky diodes** have a lower forward voltage (typically 0.35 V) and a slightly reduced maximum reverse voltage rating (typically 50 V to 200 V). Their main advantage, however, is that they operate with high efficiency in **switched-mode power supplies** (SMPS) at frequencies of up to 1 MHz. Schottky diodes are also extensively used in the construction of **integrated circuits** designed for high-speed digital logic applications.

---

**Now try the following exercises**

**Exercise 62    Further problems on semiconductor diodes**

1. Identify the types of diodes shown in Fig. 11.20.

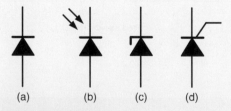

**Figure 11.20**

2. Sketch a circuit to show how a thyristor can be used as a controlled rectifier.

3. Sketch a graph showing how the capacitance of a varactor diode varies with applied reverse voltage.

4. State TWO advantages of light emitting diodes when compared with conventional filament indicating lamps.

5. State TWO applications for Schottky diodes.

6. The graph shown in Fig. 11.21 was obtained during an experiment on a Zener diode. (a) Estimate the Zener voltage for the diode. (b) Determine the reverse voltage for a reverse current of $-20$ mA. (c) Determine the reverse current for a reverse voltage of $-5.5$ V. (d) Determine the power dissipated by the diode when the reverse voltage is $-6$ V.
[(a) 5.6 (b) $-5.8$ V (c) $-5$ mA (d) 195 mW]

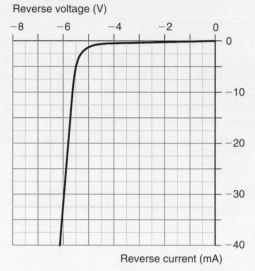

**Figure 11.21**

**Exercise 63    Short answer problems on semiconductor diodes**

1. A good conductor has a resistivity in the order of ...... to ...... $\Omega$m

2. A semiconductor has a resistivity in the order of ...... to ...... $\Omega$m

3. An insulator has a resistivity in the order of ...... to ...... $\Omega$m

4. Over a limited range, the resistance of an insulator ...... with increase in temperature

5. Over a limited range, the resistance of a semiconductor ...... with increase in temperature

6. Over a limited range, the resistance of a conductor ...... with increase in temperature

7. The working temperature of germanium should not exceed ...... °C to ...... °C, depending on its ......

8. The working temperature of silicon should not exceed ...... °C to ...... °C, depending on its ......

9. Name four semiconductor materials used in the electronics industry

10. Name two n-type impurities

11. Name two p-type impurities

12. Antimony is called ...... impurity

13. Arsenic has ...... valency electrons

14. When phosphorus is introduced into a semiconductor material, mobile ...... result

15. Boron is called a ...... impurity

16. Indium has ...... valency electrons

17. When aluminium is introduced into a semiconductor material, mobile ...... result

18. When a p-n junction is formed, the n-type material acquires a ...... charge due to losing ......

19. When a p-n junction is formed, the p-type material acquires a ...... charge due to losing ......

20. What is meant by contact potential in a p-n junction?

21. With a diagram, briefly explain what a depletion layer is in a p-n junction

22. In a p-n junction, what is diffusion?

23. To forward bias a p-n junction, the ...... terminal of the battery is connected to the p-type material

24. To reverse bias a p-n junction, the positive terminal of the battery is connected to the ...... material

25. When a germanium p-n junction is forward biased, approximately ...... mV must be applied before an appreciable current starts to flow

26. When a silicon p-n junction is forward biased, approximately ...... mV must be applied before an appreciable current starts to flow

27. When a p-n junction is reversed biased, the thickness or width of the depletion layer ......

28. If the thickness or width of a depletion layer decreases, then the p-n junction is ...... biased

29. Name five types of diodes

30. What is meant by rectification?

31. What is a zener diode? State a typical practical application and sketch its circuit diagram symbol

32. What is a thyristor? State a typical practical application and sketch its circuit diagram symbol

33. What is an LED? Sketch its circuit diagram symbol

34. What is a varactor diode? Sketch its circuit diagram symbol

35. What is a Schottky diode? State a typical practical application and sketch its circuit diagram symbol

Exercise 64   Multi-choice questions on semiconductor diodes
(Answers on page 420)

In questions 1 to 5, select which statements are true.

1. In pure silicon:
   (a) the holes are the majority carriers
   (b) the electrons are the majority carriers
   (c) the holes and electrons exist in equal numbers

(d) conduction is due to there being more electrons than holes

2. Intrinsic semiconductor materials have:
   (a) covalent bonds forming a tetrahedral structure
   (b) pentavalent atoms added
   (c) conduction by means of doping
   (d) a resistance which increases with increase of temperature

3. Pentavalent impurities:
   (a) have three valency electrons
   (b) introduce holes when added to a semi-conductor material
   (c) are introduced by adding aluminium atoms to a semiconductor material
   (d) increase the conduction of a semiconductor material

4. Free electrons in a p-type material:
   (a) are majority carriers
   (b) take no part in conduction
   (c) are minority carriers
   (d) exist in the same numbers as holes

5. When an unbiased p-n junction is formed:
   (a) the p-side is positive with respect to the n-side
   (b) a contact potential exists
   (c) electrons diffuse from the p-type material to the n-type material
   (d) conduction is by means of majority carriers

In questions 6 to 10, select which statements are false.

6. (a) The resistance of an insulator remains approximately constant with increase of temperature
   (b) The resistivity of a good conductor is about $10^7$ to $10^8$ ohm metres
   (c) The resistivity of a conductor increases with increase of temperature

(d) The resistance of a semiconductor decreases with increase of temperature

7. Trivalent impurities:
   (a) have three valeney electrons
   (b) introduce holes when added to a semi-conductor material
   (c) can be introduced to a semiconductor material by adding antimony atoms to it
   (d) increase the conductivity of a semi-conductor material when added to it

8. Free electrons in an n-type material:
   (a) are majority carriers
   (b) diffuse into the p-type material when a p-n junction is formed
   (c) as a result of the diffusion process leave the n-type material positively charged
   (d) exist in the same numbers as the holes in the n-type material

9. When a germanium p-n junction diode is forward biased:
   (a) current starts to flow in an appreciable amount when the applied voltage is about 600 mV
   (b) the thickness or width of the depletion layer is reduced
   (c) the curve representing the current flow is exponential
   (d) the positive terminal of the battery is connected to the p-type material

10. When a silicon p-n junction diode is reverse biased:
    (a) a constant current flows over a large range of voltages
    (b) current flow is due to electrons in the n-type material
    (c) current type is due to minority carriers
    (d) the magnitude of the reverse current flow is usually less than $1\,\mu A$

# Chapter 12

# Transistors

At the end of this chapter you should be able to:

- understand the structure of bipolar junction transistors (BJT) and junction gate field effect transistors (JFET)
- understand the action of BJT and JFET devices
- appreciate different classes and applications for BJT and JFET devices
- draw the circuit symbols for BJT and JFET devices
- appreciate common base, common emitter and common collector connections
- appreciate common gate, common source and common drain connections
- interpret characteristics for BJT and JFET devices
- appreciate how transistors are used as Class-A amplifiers
- use a load line to determine the performance of a transistor amplifier
- estimate quiescent operating conditions and gain from transistor characteristics and other data

## 12.1 Transistor classification

Transistors fall into **two main classes – bipolar** and **field effect**. They are also classified according to the semiconductor material employed – silicon or germanium, and to their field of application (for example, general purpose, switching, high frequency, and so on). Transistors are also classified according to the application that they are designed for, as shown in Table 12.1.

Table 12.1 Transistor classification

| | |
|---|---|
| Low-frequency | Transistors designed specifically for audio low-frequency applications (below 100 kHz) |
| High-frequency | Transistors designed specifically for high radio-frequency applications (100 kHz and above) |
| Switching | Transistors designed for switching applications |
| Low-noise | Transistors that have low-noise characteristics and which are intended primarily for the amplification of low-amplitude signals |
| High-voltage | Transistors designed specifically to handle high voltages |
| Driver | Transistors that operate at medium power and voltage levels and which are often used to precede a final (power) stage which operates at an appreciable power level |
| Small-signal | Transistors designed for amplifying small voltages in amplifiers and radio receivers |
| Power | Transistor designed to handle high currents and voltages |

DOI: 10.1016/B978-0-08-089056-2.00012-7

Note that these classifications can be combined so that it is possible, for example, to classify a transistor as a 'low-frequency power transistor' or as a 'low-noise high-frequency transistor'.

## 12.2 Bipolar junction transistors (BJT)

Bipolar transistors generally comprise n-p-n or p-n-p junctions of either silicon (Si) or germanium (Ge) material. The junctions are, in fact, produced in a single slice of silicon by diffusing impurities through a photographically reduced mask. Silicon transistors are superior when compared with germanium transistors in the vast majority of applications (particularly at high temperatures) and thus germanium devices are very rarely encountered in modern electronic equipment.

The construction of typical n-p-n and p-n-p transistors is shown in Figs12.1 and 12.2. In order to conduct the heat away from the junction (important in medium- and high-power applications) the collector is connected to the metal case of the transistor.

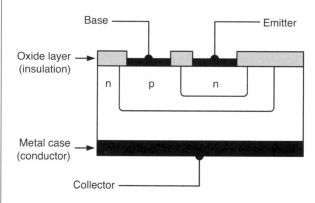

Figure 12.1

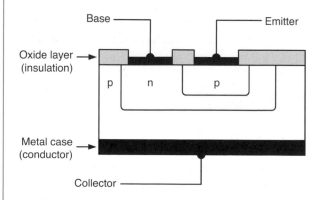

Figure 12.2

The **symbols** and simplified junction models for n-p-n and p-n-p transistors are shown in Fig. 12.3. It is important to note that the base region (p-type material in the case of an n-p-n transistor or n-type material in the case of a p-n-p transistor) is extremely narrow.

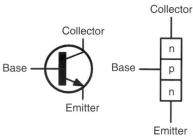

(a) n-p-n bipolar junction transistor (BJT)

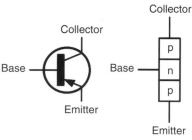

(b) p-n-p bipolar junction transistor (BJT)

Figure 12.3

## 12.3 Transistor action

In the **n-p-n transistor**, connected as shown in Fig. 12.4(a), transistor action is accounted for as follows:

(a) the majority carriers in the n-type emitter material are electrons

(b) the base-emitter junction is forward biased to these majority carriers and electrons cross the junction and appear in the base region

(c) the base region is very thin and only lightly doped with holes, so some recombination with holes occurs but many electrons are left in the base region

(d) the base-collector junction is reverse biased to holes in the base region and electrons in the collector region, but is forward biased to electrons in the base region; these electrons are attracted by the positive potential at the collector terminal

(e)  a large proportion of the electrons in the base region cross the base-collector junction into the collector region, creating a collector current

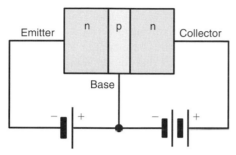

(a) n-p-n bipolar junction transistor

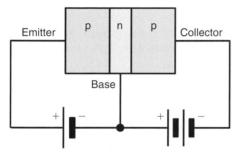

(b) p-n-p bipolar junction transistor

**Figure 12.4**

The **transistor action** for an n-p-n device is shown diagrammatically in Fig. 12.5(a). Conventional current

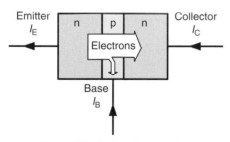

(a) n-p-n bipolar junction transistor

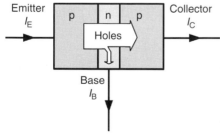

(b) p-n-p bipolar junction transistor

**Figure 12.5**

flow is taken to be in the direction of the motion of holes, that is, in the opposite direction to electron flow. Around 99.5% of the electrons leaving the emitter will cross the base-collector junction and only 0.5% of the electrons will recombine with holes in the narrow base region.

In the **p-n-p transistor**, connected as shown in Fig. 12.4(b), transistor action is accounted for as follows:

(a)  the majority carriers in the emitter p-type material are holes

(b)  the base-emitter junction is forward biased to the majority carriers and the holes cross the junction and appear in the base region

(c)  the base region is very thin and is only lightly doped with electrons so although some electron-hole pairs are formed, many holes are left in the base region

(d)  the base-collector junction is reverse biased to electrons in the base region and holes in the collector region, but forward biased to holes in the base region; these holes are attracted by the negative potential at the collector terminal

(e)  a large proportion of the holes in the base region cross the base-collector junction into the collector region, creating a collector current; conventional current flow is in the direction of hole movement

The **transistor action** for a p-n-p device is shown diagrammatically in Fig. 12.5(b). Around 99.5% of the holes leaving the emitter will cross the base-collector junction and only 0.5% of the holes will recombine with electrons in the narrow base region.

## 12.4  Leakage current

For an **n-p-n transistor**, the base-collector junction is reverse biased for majority carriers, but a small leakage current, $I_{CBO}$, flows from the collector to the base due to thermally generated minority carriers (holes in the collector and electrons in the base), being present. The base-collector junction is forward biased to these minority carriers.

Similarly, for a **p-n-p transistor**, the base-collector junction is reverse biased for majority carriers. However, a small leakage current, $I_{CBO}$, flows from the base to the collector due to thermally generated minority carriers (electrons in the collector and holes in the base), being present. Once again, the base-collector junction is forward biased to these minority carriers.

With modern transistors, leakage current is usually very small (typically less than 100 nA) and in most applications it can be ignored.

> **Problem 1.** With reference to a p-n-p transistor, explain briefly what is meant by the term 'transistor action' and why a bipolar junction transistor is so named.

For the transistor as depicted in Fig. 12.4(b), the emitter is relatively heavily doped with acceptor atoms (holes). When the emitter terminal is made sufficiently positive with respect to the base, the base-emitter junction is forward biased to the majority carriers. The majority carriers are holes in the emitter and these drift from the emitter to the base.

The base region is relatively lightly doped with donor atoms (electrons) and although some electron-hole recombinations take place, perhaps 0.5%, most of the holes entering the base, do not combine with electrons.

The base-collector junction is reverse biased to electrons in the base region, but forward biased to holes in the base region. Since the base is very thin and now is packed with holes, these holes pass the base-emitter

junction towards the negative potential of the collector terminal. The control of current from emitter to collector is largely independent of the collector-base voltage and almost wholly governed by the emitter-base voltage.

The essence of transistor action is this current control by means of the base-emitter voltage. In a p-n-p transistor, holes in the emitter and collector regions are majority carriers, but are minority carriers when in the base region. Also thermally generated electrons in the emitter and collector regions are minority carriers as are holes in the base region. However, both majority and minority carriers contribute towards the total current flow (see Fig. 12.6). It is because a transistor makes use of both types of charge carriers (holes and electrons) that they are called **bipolar**. The transistor also comprises two p-n junctions and for this reason it is a **junction transistor**; hence the name – **bipolar junction transistor**.

## 12.5 Bias and current flow

In normal operation (i.e. for operation as a linear amplifier) the base-emitter junction of a transistor is forward biased and the collector-base junction is reverse biased. The base region is, however, made very narrow so that carriers are swept across it from emitter to collector so that only a relatively small current flows in the base. To put this into context, the current flowing in the emitter circuit is typically 100 times greater than that flowing in the base. The direction of conventional current flow is from emitter to collector in the case of a p-n-p transistor, and collector to emitter in the case of an n-p-n device, as shown in Fig. 12.7.

The equation that relates current flow in the collector, base, and emitter circuits (see Fig. 12.7) is:

$$I_E = I_B + I_C$$

where $I_E$ is the emitter current, $I_B$ is the base current, and $I_C$ is the collector current (all expressed in the same units).

> **Problem 2.** A transistor operates with a collector current of 100 mA and an emitter current of 102 mA. Determine the value of base current.

Emitter current,       $I_E = I_B + I_C$
from which, base current, $I_B = I_E - I_C$
Hence, **base current,**    $I_B = 102 - 100 = \mathbf{2\,mA}$

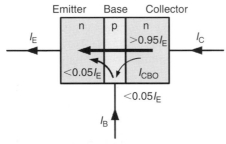

(a) n-p-n bipolar junction transistor

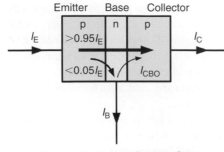

(b) p-n-p bipolar junction transistor

Figure 12.6

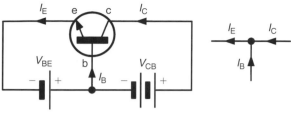

(a) n-p-n bipolar junction transistor (BJT)

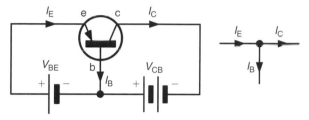

(b) p-n-p bipolar junction transistor (BJT)

**Figure 12.7**

## 12.6 Transistor operating configurations

Three basic circuit configurations are used for transistor amplifiers. These three circuit configurations depend upon which one of the three transistor connections is made common to both the input and the output. In the case of bipolar junction transistors, the configurations are known as **common-emitter**, **common-collector** (or **emitter-follower**), and **common-base**, as shown in Fig. 12.8.

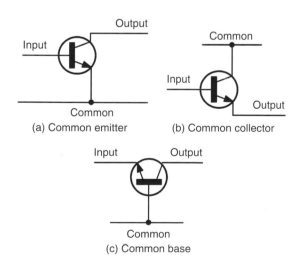

(a) Common emitter

(b) Common collector

(c) Common base

**Figure 12.8**

## 12.7 Bipolar transistor characteristics

The characteristics of a bipolar junction transistor are usually presented in the form of a set of graphs relating voltage and current present at the transistors terminals. Fig. 12.9 shows a typical **input characteristic** ($I_B$ plotted against $V_{BE}$) for an n-p-n bipolar junction transistor operating in common-emitter mode. In this mode, the input current is applied to the base and the output current appears in the collector (the emitter is effectively **common** to both the input and output circuits as shown in Fig. 12.8(a)).

The input characteristic shows that very little base current flows until the base-emitter voltage $V_{BE}$ exceeds 0.6 V. Thereafter, the base current increases rapidly – this characteristic bears a close resemblance to the forward part of the characteristic for a silicon diode.

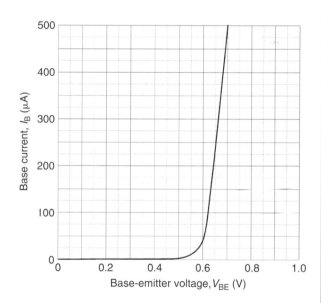

**Figure 12.9**

Figure 12.10 shows a typical set of **output (collector) characteristics** ($I_C$ plotted against $V_{CE}$) for an n-p-n bipolar transistor. Each curve corresponds to a different value of base current. Note the 'knee' in the characteristic below $V_{CE} = 2$ V. Also note that the curves are quite flat. For this reason (i.e. since the collector current does not change very much as the collector-emitter voltage changes) we often refer to this as a **constant current characteristic**.

Figure 12.11 shows a typical **transfer characteristic** for an n-p-n bipolar junction transistor. Here $I_C$ is plotted against $I_B$ for a small-signal general-purpose transistor.

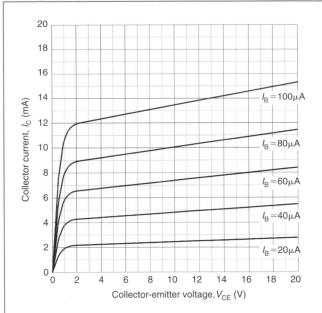

**Figure 12.10**

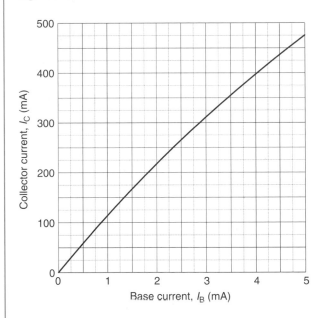

**Figure 12.11**

The slope of this curve (i.e. the ratio of $I_C$ to $I_B$) is the common-emitter current gain of the transistor which is explored further in Section 12.9.

A circuit that can be used for obtaining the common-emitter characteristics of an n-p-n BJT is shown in Fig. 12.12. For the input characteristic, VR1 is set at a particular value and the corresponding values of $V_{BE}$ and $I_B$ are noted. This is repeated for various settings of VR1 and plotting the values gives the typical input characteristic of Fig. 12.9.

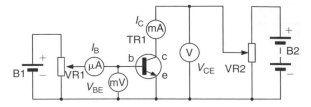

**Figure 12.12**

For the output characteristics, VR1 is varied so that $I_B$ is, say, $20\,\mu A$. Then VR2 is set at various values and corresponding values of $V_{CE}$ and $I_C$ are noted. The graph of $V_{CE}/I_C$ is then plotted for $I_B = 20\,\mu A$. This is repeated for, say, $I_B = 40\,\mu A$, $I_B = 60\,\mu A$, and so on. Plotting the values gives the typical output characteristics of Fig. 12.10.

## 12.8   Transistor parameters

The transistor characteristics met in the previous section provide us with some useful information that can help us to model the behaviour of a transistor. In particular, the three characteristic graphs can be used to determine the following parameters for operation in common-emitter mode:

**Input resistance** (from the input characteristic, Fig. 12.9)

$$\text{Static (or d.c.) input resistance} = \frac{V_{BE}}{I_B}$$

(from corresponding points on the graph)

$$\text{Dynamic (or a.c.) input resistance} = \frac{\Delta V_{BE}}{\Delta I_B}$$

(from the slope of the graph)

(Note that $\Delta V_{BE}$ means 'change of $V_{BE}$' and $\Delta I_B$ means 'change of $I_B$')

**Output resistance** (from the output characteristic, Fig. 12.10)

$$\text{Static (or d.c.) output resistance} = \frac{V_{CE}}{I_C}$$

(from corresponding points on the graph)

$$\text{Dynamic (or a.c.) output resistance} = \frac{\Delta V_{CE}}{\Delta I_C}$$

(from the slope of the graph)

(Note that $\Delta V_{CE}$ means 'change of $V_{CE}$' and $\Delta I_C$ means 'change of $I_C$')

## Current gain (from the transfer characteristic, Fig. 12.11)

Static (or d.c.) current gain $= \dfrac{I_C}{I_B}$

(from corresponding points on the graph)

Dynamic (or a.c.) current gain $= \dfrac{\Delta I_C}{\Delta I_B}$

(from the slope of the graph)

(Note that $\Delta I_C$ means 'change of $I_C$' and $\Delta I_B$ means 'change of $I_B$')

The method for determining these parameters from the relevant characteristic is illustrated in the following worked problems.

> **Problem 3.** Figure 12.13 shows the input characteristic for an n-p-n silicon transistor. When the base-emitter voltage is 0.65 V, determine (a) the value of base current, (b) the static value of input resistance, and (c) the dynamic value of input resistance.

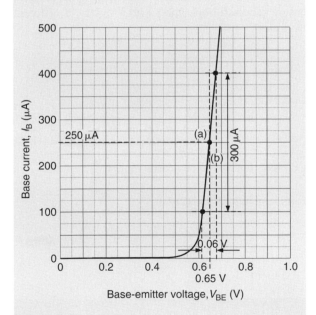

Figure 12.13

(a) From Fig. 12.13, when $V_{BE} = 0.65$ V, **base current, $I_B = 250\,\mu A$** (shown as (a) on the graph)

(b) When $V_{BE} = 0.65$ V, $I_B = 250\,\mu A$, hence, **the static value of input resistance**

$$= \frac{V_{BE}}{I_B} = \frac{0.65}{250 \times 10^{-6}} = \mathbf{2.6\,k\Omega}$$

(c) From Fig. 12.13, $V_{BE}$ changes by 0.06 V when $I_B$ changes by $300\,\mu A$ (as shown by (b) on the graph). Hence,

**dynamic value of input resistance**

$$= \frac{\Delta V_{BE}}{\Delta I_B} = \frac{0.06}{300 \times 10^{-6}} = \mathbf{200\,\Omega}$$

> **Problem 4.** Figure 12.14 shows the output characteristic for an n-p-n silicon transistor. When the collector-emitter voltage is 10 V and the base current is $80\,\mu A$, determine (a) the value of collector current, (b) the static value of output resistance, and (c) the dynamic value of output resistance.

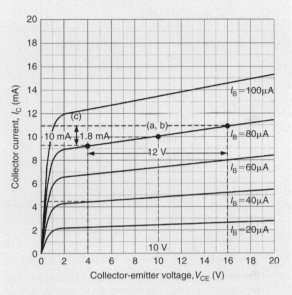

Figure 12.14

(a) From Fig. 12.14, when $V_{CE} = 10$ V and $I_B = 80\,\mu A$, (i.e. point (a, b) on the graph), the **collector current, $I_C = 10\,mA$**

(b) When $V_{CE} = 10$ V and $I_B = 80\,\mu A$ then $I_C = 10\,mA$ from part (a).
Hence,
**the static value of output resistance**

$$= \frac{V_{CE}}{I_C} = \frac{10}{10 \times 10^{-3}} = \mathbf{1\,k\Omega}$$

(c) When the change in $V_{CE}$ is 12 V, the change in $I_C$ is 1.8 mA (shown as point (c) on the graph)
Hence,
**the dynamic value of output resistance**

$$= \frac{\Delta V_{CE}}{\Delta I_C} = \frac{12}{1.8 \times 10^{-3}} = \mathbf{6.67\,k\Omega}$$

**Problem 5.** Figure 12.15 shows the transfer characteristic for an n-p-n silicon transistor. When the base current is 2.5 mA, determine (a) the value of collector current, (b) the static value of current gain, and (c) the dynamic value of current gain.

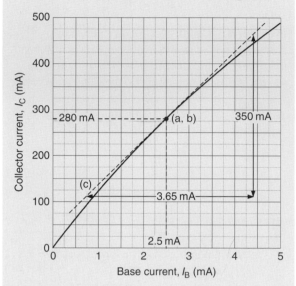

**Figure 12.15**

(a) From Fig. 12.15, when $I_B = 2.5$ mA, **collector current, $I_C = 280$ mA** (see point (a, b) on the graph)

(b) From part (a), when $I_B = 2.5$ mA, $I_C = 280$ mA hence,
**the static value of current gain**

$$= \frac{I_C}{I_B} = \frac{280 \times 10^{-3}}{2.5 \times 10^{-3}} = \mathbf{112}$$

(c) In Fig. 12.15, the tangent through the point (a, b) is shown by the broken straight line (c).
Hence,
**the dynamic value of current gain**

$$= \frac{\Delta I_C}{\Delta I_B} = \frac{(460 - 110) \times 10^{-3}}{(4.4 - 0.75) \times 10^{-3}} = \frac{350}{3.65} = \mathbf{96}$$

## 12.9 Current gain

As stated earlier, the common-emitter current gain is given by the ratio of collector current, $I_C$, to base current, $I_B$. We use the symbol $h_{FE}$ to represent the static value of common-emitter current gain, thus:

$$h_{FE} = \frac{I_C}{I_B}$$

Similarly, we use $h_{fe}$ to represent the dynamic value of common emitter current gain, thus:

$$h_{fe} = \frac{\Delta I_C}{\Delta I_B}$$

As we showed earlier, values of $h_{FE}$ and $h_{fe}$ can be obtained from the transfer characteristic ($I_C$ plotted against $I_B$). Note that $h_{FE}$ is found from corresponding static values while $h_{fe}$ is found by measuring the slope of the graph. Also note that, if the transfer characteristic is linear, there is little (if any) difference between $h_{FE}$ and $h_{fe}$.

It is worth noting that current gain ($h_{fe}$) varies with collector current. For most small-signal transistors, $h_{fe}$ is a maximum at a collector current in the range 1 mA and 10 mA. Current gain also falls to very low values for power transistors when operating at very high values of collector current. Furthermore, most transistor parameters (particularly common-emitter current gain, $h_{fe}$) are liable to wide variation from one device to the next. It is, therefore, important to design circuits on the basis of the minimum value for $h_{fe}$ in order to ensure successful operation with a variety of different devices.

**Problem 6.** A bipolar transistor has a common-emitter current gain of 125. If the transistor operates with a collector current of 50 mA, determine the value of base current.

Common-emitter current gain, $h_{FE} = \dfrac{I_C}{I_B}$

from which, **base current,**

$$I_B = \frac{I_C}{h_{FE}} = \frac{50 \times 10^{-3}}{125} = \mathbf{400\ \mu A}$$

## 12.10 Typical BJT characteristics and maximum ratings

Table 12.2 summarises the characteristics of some typical bipolar junction transistors for different applications, where $I_C$ max is the maximum collector current, $V_{CE}$ max is the maximum collector-emitter voltage, $P_{TOT}$ max is the maximum device power dissipation, and $h_{fe}$ is the typical value of common-emitter current gain.

Table 12.2   Transistor characteristics and maximum ratings

| Device | Type | $I_C$ max. | $V_{CE}$ max. | $P_{TOT}$ max. | $h_{FE}$ typical | Application |
|--------|------|------------|---------------|----------------|-------------------|-------------|
| BC108 | n-p-n | 100 mA | 20 V | 300 mW | 125 | General-purpose small-signal amplifier |
| BCY70 | n-p-n | 200 mA | −40 V | 360 mW | 150 | General-purpose small-signal amplifier |
| 2N3904 | n-p-n | 200 mA | 40 V | 310 mW | 150 | Switching |
| BF180 | n-p-n | 20 mA | 20 V | 150 mW | 100 | RF amplifier |
| 2N3053 | n-p-n | 700 mA | 40 V | 800 mW | 150 | Low-frequency amplifier/driver |
| 2N3055 | n-p-n | 15 A | 60 V | 115 W | 50 | Low-frequency power |

**Problem 7.** Which of the bipolar transistors listed in Table 12.2 would be most suitable for each of the following applications: (a) the input stage of a radio receiver, (b) the output stage of an audio amplifier, and (c) generating a 5 V square wave pulse.

(a)   **BF180**, since this transistor is designed for use in radio frequency (RF) applications

(b)   **2N3055**, since this is the only device in the list that can operate at a sufficiently high power level

(c)   **2N3904**, since switching transistors are designed for use in pulse and square wave applications

---

**Now try the following exercise**

**Exercise 65    Further problems on bipolar junction transistors**

1.   Explain, with the aid of sketches, the operation of an n-p-n transistor and also explain why the collector current is very nearly equal to the emitter current.

2.   Describe the basic principle of operation of a bipolar junction transistor, including why majority carriers crossing into the base from the emitter pass to the collector and why the collector current is almost unaffected by the collector potential.

3.   Explain what is meant by 'leakage current' in a bipolar junction transistor and why this can usually be ignored.

4.   For a transistor connected in common-emitter configuration, sketch the typical output characteristics relating collector current and the collector-emitter voltage, for various values of base current. Explain the shape of the characteristics.

5.   Sketch the typical input characteristic relating base current and the base-emitter voltage for a transistor connected in common-emitter configuration and explain its shape.

6.   With the aid of a circuit diagram, explain how the input and output characteristic of a common-emitter n-p-n transistor may be produced.

7.   Define the term 'current gain' for a bipolar junction transistor operating in common-emitter mode.

8.   A bipolar junction transistor operates with a collector current of 1.2 A and a base current

of 50 mA. What will the value of emitter current be? [1.25 A]

9. What is the value of common-emitter current gain for the transistor in problem 8? [24]

10. Corresponding readings of base current, $I_B$, and base-emitter voltage, $V_{BE}$, for a bipolar junction transistor are given in the table below:

| $V_{BE}$ (V) | 0 | 0.1 | 0.2 | 0.3 | 0.4 | 0.5 | 0.6 | 0.7 | 0.8 |
|---|---|---|---|---|---|---|---|---|---|
| $I_B$ (μA) | 0 | 0 | 0 | 0 | 1 | 3 | 19 | 57 | 130 |

Plot the $I_B/V_{BE}$ characteristic for the device and use it to determine (a) the value of $I_B$ when $V_{BE} = 0.65$ V, (b) the static value of input resistance when $V_{BE} = 0.65$ V, and (c) the dynamic value of input resistance when $V_{BE} = 0.65$ V.

[(a) 32.5 μA (b) 20 kΩ (c) 3 kΩ]

11. Corresponding readings of base current, $I_B$, and collector current, $I_C$, for a bipolar junction transistor are given in the table below:

| $I_B$ (μA) | 0 | 10 | 20 | 30 | 40 | 50 | 60 | 70 | 80 |
|---|---|---|---|---|---|---|---|---|---|
| $I_C$ (mA) | 0 | 1.1 | 2.1 | 3.1 | 4.0 | 4.9 | 5.8 | 6.7 | 7.6 |

Plot the $I_C/I_B$ characteristic for the device and use it to determine the static value of common-emitter current gain when $I_B = 45$ μA. [98]

## 12.11 Field effect transistors

Field effect transistors are available in two basic forms; junction gate and insulated gate. The gate-source junction of a **junction gate field effect transistor (JFET)** is effectively a reverse-biased p-n junction. The gate connection of an **insulated gate field effect transistor (IGFET)**, on the other hand, is insulated from the channel and charge is capacitively coupled to the channel. To keep things simple, we will consider only JFET devices. Figure 12.16 shows the basic construction of an n-channel JFET.

JFET transistors comprise a channel of p-type or n-type material surrounded by material of the opposite polarity. The ends of the channel (in which conduction takes place) form electrodes known as the source

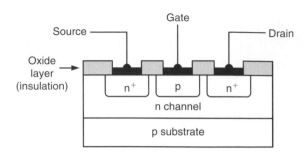

**Figure 12.16**

and drain. The effective width of the channel (in which conduction takes place) is controlled by a charge placed on the third (gate) electrode. The effective resistance between the source and drain is thus determined by the voltage present at the gate. (The + signs in Fig. 12.16 is used to indicate a region of heavy doping thus $n^+$ simply indicates a heavily doped n-type region.)

JFETs offer a very much higher input resistance when compared with bipolar transistors. For example, the input resistance of a bipolar transistor operating in common-emitter mode is usually around 2.5 kΩ. A JFET transistor operating in equivalent common-source mode would typically exhibit an input resistance of 100 MΩ! This feature makes JFET devices ideal for use in applications where a very high input resistance is desirable.

As with bipolar transistors, the characteristics of a FET are often presented in the form of a set of graphs relating voltage and current present at the transistors, terminals.

## 12.12 Field effect transistor characteristics

A typical **mutual characteristic** ($I_D$ plotted against $V_{GS}$) for a small-signal general-purpose n-channel field effect transistor operating in common-source mode is shown in Fig. 12.17. This characteristic shows that the drain current is progressively reduced as the gate-source voltage is made more negative. At a certain value of $V_{GS}$ the drain current falls to zero and the device is said to be cut-off.

Figure 12.18 shows a typical family of **output characteristics** ($I_D$ plotted against $V_{DS}$) for a small-signal general-purpose n-channel FET operating in common-source mode. This characteristic comprises a family of curves, each relating to a different value of gate-source voltage $V_{GS}$. You might also like to compare this characteristic with the output characteristic for a transistor

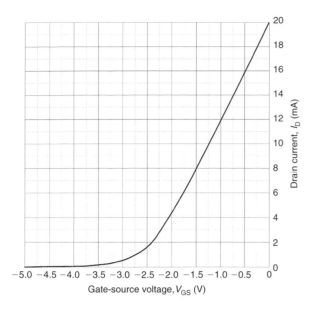

**Figure 12.17**

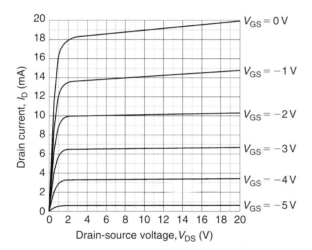

**Figure 12.18**

operating in common-emitter mode that you met earlier in Fig. 12.10.

As in the case of the bipolar junction transistor, the output characteristic curves for an n-channel FET have a 'knee' that occurs at low values of $V_{DS}$. Also, note how the curves become flattened above this value with the drain current $I_D$ not changing very significantly for a comparatively large change in drain-source voltage $V_{DS}$. These characteristics are, in fact, even flatter than those for a bipolar transistor. Because of their flatness, they are often said to represent a constant current characteristic.

The gain offered by a field effect transistor is normally expressed in terms of its **forward transconductance**

($g_{fs}$ or $Y_{fs}$,) in common-source mode. In this mode, the input voltage is applied to the gate and the output current appears in the drain (the source is effectively common to both the input and output circuits).

In common-source mode, **the static (or d.c.) forward transfer conductance** is given by:

$$g_{FS} = \frac{I_D}{V_{GS}}$$

(from corresponding points on the graph)

whilst **the dynamic (or a.c.) forward transfer conductance** is given by:

$$g_{fs} = \frac{\Delta I_D}{\Delta V_{GS}}$$

(from the slope of the graph)

(Note that $\Delta I_D$ means 'change of $I_D$' and $\Delta V_{GS}$ means 'change of $V_{GS}$')

The method for determining these parameters from the relevant characteristic is illustrated in worked problem 8 below.

Forward transfer conductance ($g_{fs}$) varies with drain current. For most small-signal devices, $g_{fs}$, is quoted for values of drain current between 1 mA and 10 mA. Most FET parameters (particularly forward transfer conductance) are liable to wide variation from one device to the next. It is, therefore, important to design circuits on the basis of the minimum value for $g_{fs}$, in order to ensure successful operation with a variety of different devices. The experimental circuit for obtaining the common-source characteristics of an n-channel JFET transistor is shown in Fig. 12.19.

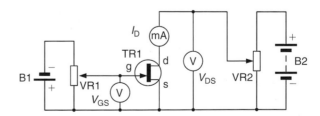

**Figure 12.19**

**Problem 8.** Figure 12.20 shows the mutual characteristic for a junction gate field effect transistor. When the gate-source voltage is −2.5 V, determine (a) the value of drain current, (b) the dynamic value of forward transconductance.

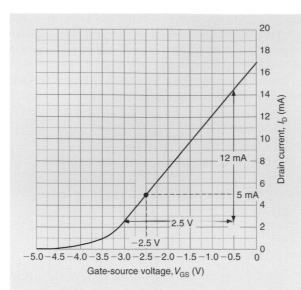

**Figure 12.20**

(a) From Fig. 12.20, when $V_{GS} = -2.5\,V$, the **drain current, $I_D = 5\,mA$**

(b) From Fig. 12.20

$$g_{fs} = \frac{\Delta I_D}{\Delta V_{GS}} = \frac{(14.5 - 2.5) \times 10^{-3}}{2.5}$$

i.e. **the dynamic value of forward transconductance** $= \dfrac{12 \times 10^{-3}}{2.5} = 4.8\,mS$

(note the unit – **siemens, S**)

---

**Problem 9.** A field effect transistor operates with a drain current of 100 mA and a gate source bias of $-1\,V$. The device has a $g_{fs}$ value of 0.25. If the bias voltage decreases to $-1.1\,V$, determine (a) the change in drain current, and (b) the new value of drain current.

---

(a) The change in gate-source voltage ($V_{GS}$) is $-0.1\,V$ and the resulting change in drain current can be determined from:

$$g_{fs} = \frac{\Delta I_D}{\Delta V_{GS}}$$

Hence, **the change in drain current**,

$$\Delta I_D = g_{fs} \times \Delta V_{GS}$$

$$= 0.25 \times -0.1$$

$$= -0.025\,A = \mathbf{-25\,mA}$$

(b) The **new value of drain current** $= (100 - 25)$

$$= \mathbf{75\,mA}$$

## 12.13 Typical FET characteristics and maximum ratings

Table 12.3 summarises the characteristics of some typical field effect transistors for different applications, where $I_D$ max is the maximum drain current, $V_{DS}$ max is the maximum drain-source voltage, $P_D$ max is the maximum drain power dissipation, and $g_{fs}$ typ is the typical value of forward transconductance for the transistor. The list includes both depletion and enhancement types as well as junction and insulated gate types.

---

**Problem 10.** Which of the field effect transistors listed in Table 12.3 would be most suitable for each of the following applications: (a) the input stage of a radio receiver, (b) the output stage of a transmitter, and (c) switching a load connected to a high-voltage supply.

---

(a) **BF244A**, since this transistor is designed for use in radio frequency (RF) applications

(b) **MRF171A**, since this device is designed for RF power applications

(c) **IRF830**, since this device is intended for switching applications and can operate at up to 500 V

## 12.14 Transistor amplifiers

Three basic circuit arrangements are used for transistor amplifiers and these are based on the three circuit configurations that we met earlier (i.e. they depend upon which one of the three transistor connections is made common to both the input and the output). In the case of **bipolar transistors**, the configurations are known as **common emitter, common collector** (or emitter follower) and **common base**.

Where **field effect transistors** are used, the corresponding configurations are **common source, common drain** (or source follower) and **common gate**.

These basic circuit configurations depicted in Figs 12.21 and 12.22 exhibit quite different performance characteristics, as shown in Tables 12.4 and 12.5 respectively.

**Table 12.3** FET characteristics and maximum ratings

| Device | Type | $I_D$ max. | $V_{DS}$ max. | $P_D$ max. | $g_{fs}$ typ. | Application |
|--------|------|-----------|---------------|-----------|---------------|-------------|
| 2N2819 | n-chan. | 10 mA | 25 V | 200 mW | 4.5 mS | General purpose |
| 2N5457 | n-chan. | 10 mA | 25 V | 310 mW | 1.2 mS | General purpose |
| 2N7000 | n-chan. | 200 mA | 60 V | 400 mW | 0.32 S | Low-power switching |
| BF244A | n-chan. | 100 mA | 30 V | 360 mW | 3.3 mS | RF amplifier |
| BSS84 | p-chan. | −130 mA | −50 V | 360 mW | 0.27 S | Low-power switching |
| IRF830 | n-chan. | 4.5 A | 500 V | 75 W | 3.0 S | Power switching |
| MRF171A | n-chan. | 4.5 A | 65 V | 115 W | 1.8 S | RF power amplifier |

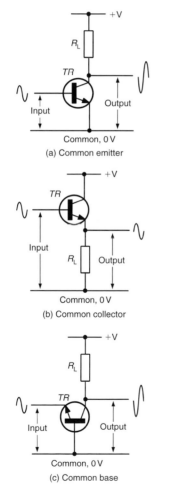

(a) Common emitter

(b) Common collector

(c) Common base

Bipolar transistor amplifier circuit configurations

(a) Common source

(b) Common drain

(c) Common gate

Field effect transistor amplifier circuit configurations

**Figure 12.21**                    **Figure 12.22**

Table 12.4  Characteristics of BJT amplifiers

| | Bipolar transistor amplifiers (see Figure 12.21) | | |
|---|---|---|---|
| Parameter | Common emitter | Common collector | Common base |
| Voltage gain | medium/high (40) | unity (1) | high (200) |
| Current gain | high (200) | high (200) | unity (1) |
| Power gain | very high (8000) | high (200) | high (200) |
| Input resistance | medium (2.5 k$\Omega$) | high (100 k$\Omega$) | low (200 $\Omega$) |
| Output resistance | medium/high (20 k$\Omega$) | low (100 $\Omega$) | high (100 k$\Omega$) |
| Phase shift | 180° | 0° | 0° |
| Typical applications | General purpose, AF and RF amplifiers | Impedance matching, input and output stages | RF and VHF amplifiers |

Table 12.5  Characteristics of FET amplifiers

| | Field effect transistor amplifiers (see Figure 12.22) | | |
|---|---|---|---|
| Parameter | Common source | Common drain | Common gate |
| Voltage gain | medium/high (40) | unity (1) | high (250) |
| Current gain | very high (200 000) | very high (200 000) | unity (1) |
| Power gain | very high (8 000 000) | very high (200 000) | high (250) |
| Input resistance | very high (1 M$\Omega$) | very high (1 M$\Omega$) | low (500 $\Omega$) |
| Output resistance | medium/high (50 k$\Omega$) | low (200 $\Omega$) | high (150 k$\Omega$) |
| Phase shift | 180° | 0° | 0° |
| Typical applications | General purpose, AF and RF amplifiers | Impedance matching stages | RF and VHF amplifiers |

A requirement of most amplifiers is that the output signal should be a faithful copy of the input signal or be somewhat larger in amplitude. Other types of amplifier are 'non-linear', in which case their input and output waveforms will not necessarily be similar. In practice, the degree of linearity provided by an amplifier can be affected by a number of factors including the amount of bias applied and the amplitude of the input signal. It is also worth noting that a linear amplifier will become non-linear when the applied input signal exceeds a threshold value. Beyond this value the amplifier is said to be overdriven and the output will become increasingly distorted if the input signal is further increased.

The optimum value of bias for **linear (Class A) amplifiers** is that value which ensures that the active devices are operated at the mid-point of their characteristics. In practice, this means that a static value of collector current will flow even when there is no signal present. Furthermore, the collector current will flow throughout the complete cycle of an input signal (i.e. conduction will take place over an angle of 360°). At no stage should the transistor be **saturated** ($V_{CE} \approx 0$ V or $V_{DS} \approx 0$ V) nor should it be **cut-off** ($V_{CE} \approx V_{CC}$ or $V_{DS} \approx V_{DD}$).

In order to ensure that a static value of collector current flows in a transistor, a small current must be applied to the base of the transistor. This current can be derived from the same voltage rail that supplies the collector circuit (via the **collector load**). Figure 12.23 shows a simple Class-A common-emitter circuit in which the **base bias resistor**, R1, and **collector load**

**resistor**, R2, are connected to a common positive supply rail.

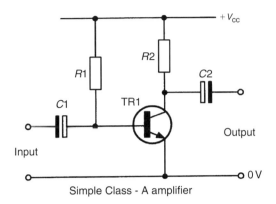

**Figure 12.23**

The a.c. signal is applied to the base terminal of the transistor via a coupling capacitor, C1. This capacitor removes the d.c. component of any signal applied to the input terminals and ensures that the base bias current delivered by R1 is unaffected by any device connected to the input. C2 couples the signal out of the stage and also prevents d.c. current flow appearing at the output terminals.

## 12.15 Load lines

The a.c. performance of a transistor amplifier stage can be predicted using a **load line** superimposed on the relevant set of output characteristics. For a bipolar transistor operating in common-emitter mode the required characteristics are $I_C$ plotted against $V_{CE}$. One end of the load line corresponds to the supply voltage ($V_{CC}$) while the other end corresponds to the value of collector or drain current that would flow with the device totally saturated ($V_{CE} = 0\,V$). In this condition:

$$I_C = \frac{V_{CC}}{R_L}$$

where $R_L$ is the value of collector or drain load resistance.

Figure 12.24 shows a load line superimposed on a set of output characteristics for a bipolar transistor operating in common-emitter mode. The quiescent point (or operating point) is the point on the load line that corresponds to the conditions that exist when no-signal is applied to the stage. In Fig. 12.24, the base bias current is set at $20\,\mu A$ so that the **quiescent point** effectively sits roughly halfway along the load line. This position

ensures that the collector voltage can swing both positively (above) and negatively (below) its quiescent value ($V_{CQ}$).

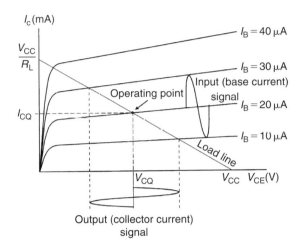

**Figure 12.24**

The effect of superimposing an alternating base current (of $20\,\mu A$ peak-peak) to the d.c. bias current (of $20\,\mu A$) can be clearly seen. The corresponding collector current signal can be determined by simply moving up and down the load line.

> **Problem 11.** The characteristic curves shown in Fig. 12.25 relate to a transistor operating in common-emitter mode. If the transistor is operated with $I_B = 30\,\mu A$, a load resistor of $1.2\,k\Omega$ and an 18 V supply, determine (a) the quiescent values of collector voltage and current ($V_{CQ}$ and $I_{CQ}$), and (b) the peak-peak output voltage that would be produced by an input signal of $40\,\mu A$ peak-peak.

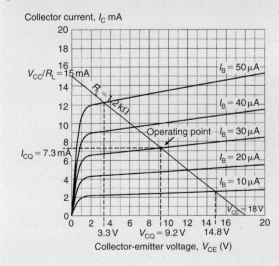

**Figure 12.25**

(a) First we need to construct the load line on Fig. 12.25. The two ends of the load line will correspond to $V_{CC}$, the 18 V supply, on the collector-emitter voltage axis and $18\,V/1.2\,k\Omega$ or 15 mA on the collector current axis.

Next we locate the **operating point** (or **quiescent point**) from the point of intersection of the $I_B = 30\,\mu A$ characteristic and the load line. Having located the operating point we can read off the **quiescent values**, i.e. the no-signal values, of collector-emitter voltage ($V_{CQ}$) and collector current ($I_{CQ}$). Hence, $V_{CQ} = 9.2\,V$ and $I_{CQ} = 7.3\,mA$

(b) Next we can determine the maximum and minimum values of collector-emitter voltage by locating the appropriate intercept points on Fig. 12.25. Note that the maximum and minimum values of base current will be $(30\,\mu A + 20\,\mu A) = 50\,\mu A$ on positive peaks of the signal and $(30\,\mu A - 20\,\mu A) = 10\,\mu A$ on negative peaks of the signal. The maximum and minimum values of $V_{CE}$ are, respectively, 14.8 V and 3.3 V. Hence,

**the output voltage swing** $= (14.8\,V - 3.3\,V)$

$= \mathbf{11.5\,V\ peak\text{-}peak}$

---

**Problem 12.** An n-p-n transistor has the following characteristics, which may be assumed to be linear between the values of collector voltage stated.

| Base current (μA) | Collector current (mA) for collector voltages of: | |
|---|---|---|
| | **1 V** | **5 V** |
| 30 | 1.4 | 1.6 |
| 50 | 3.0 | 3.5 |
| 70 | 4.6 | 5.2 |

The transistor is used as a common-emitter amplifier with load resistor $R_L = 1.2\,k\Omega$ and a collector supply of 7 V. The signal input resistance is $1\,k\Omega$. If an input current of $20\,\mu A$ peak varies sinusoidally about a mean bias of $50\,\mu A$, estimate (a) the quiescent values of collector voltage and current, (b) the output voltage swing, (c) the voltage gain, (d) the dynamic current gain, and (e) the power gain.

The characteristics are drawn as shown in Fig. 12.26.

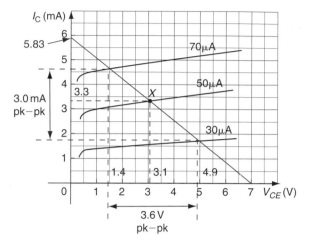

**Figure 12.26**

The two ends of the load line will correspond to $V_{CC}$, the 7 V supply, on the collector-emitter voltage axis and $7\,V/1.2\,k\Omega = 5.83\,mA$ on the collector current axis.

(a) The operating point (or quiescent point), X, is located from the point of intersection of the $I_B = 50\,\mu A$ characteristic and the load line. Having located the operating point we can read off the **quiescent values**, i.e. the no-signal values, of collector-emitter voltage ($V_{CQ}$) and collector current ($I_{CQ}$). Hence, $V_{CQ} = 3.1\,V$ and $I_{CQ} = 3.3\,mA$

(b) The maximum and minimum values of collector-emitter voltage may be determined by locating the appropriate intercept points on Fig. 12.26. Note that the maximum and minimum values of base current will be $(50\,\mu A + 20\,\mu A) = 70\,\mu A$ on positive peaks of the signal and $(50\,\mu A - 20\,\mu A) = 30\,\mu A$ on negative peaks of the signal. The maximum and minimum values of $V_{CE}$ are, respectively, 4.9 V and 1.4 V. Hence,

**the output voltage swing** $= (4.9\,V - 1.4\,V)$

$= \mathbf{3.5\,V\ peak\text{-}peak}$

(c) Voltage gain $= \dfrac{\text{change in collector voltage}}{\text{change in base voltage}}$

The change in collector voltage $= 3.5\,V$ from part (b).

The input voltage swing is given by: $i_b R_i$,

where $i_b$ is the base current swing $= (70 - 30) = 40\,\mu A$ and $R_i$ is the input resistance $= 1\,k\Omega$.

Hence,
input voltage swing $= 40 \times 10^{-6} \times 1 \times 10^{3}$

$\qquad = 40\,\text{mV}$

$\qquad = \text{change in base voltage.}$

Thus,

$$\textbf{voltage gain} = \frac{\text{change in collector voltage}}{\text{change in base voltage}}$$

$$= \frac{\Delta V_C}{\Delta V_B} = \frac{3.5}{40 \times 10^{-3}} = \textbf{87.5}$$

(d) Dynamic current gain, $h_{\text{fe}} = \dfrac{\Delta I_C}{\Delta I_B}$

From Figure 12.26, the output current swing, i.e. the change in collector current, $\Delta I_C = 3.0\,\text{mA}$ peak to peak. The input base current swing, the change in base current, $\Delta I_B = 40\,\mu\text{A}$.

Hence, **the dynamic current gain**,

$$h_{\text{fe}} = \frac{\Delta I_C}{\Delta I_B} = \frac{3.0 \times 10^{-3}}{40 \times 10^{-6}} = \textbf{75}$$

(e) For a resistive load, the power gain is given by:

$$\textbf{power gain} = \text{voltage gain} \times \text{current gain}$$

$$= 87.5 \times 75 = \textbf{6562.5}$$

---

**Now try the following exercises**

**Exercise 66    Further problems on transistors**

1. State whether the following statements are true or false:
   (a) The purpose of a transistor amplifier is to increase the frequency of the input signal.
   (b) The gain of an amplifier is the ratio of the output signal amplitude to the input signal amplitude.
   (c) The output characteristics of a transistor relate the collector current to the base current.
   (d) If the load resistor value is increased the load line gradient is reduced.
   (e) In a common-emitter amplifier, the output voltage is shifted through $180°$ with reference to the input voltage.
   (f) In a common-emitter amplifier, the input and output currents are in phase.

(g) The dynamic current gain of a transistor is always greater than the static current gain.
$\qquad$ [(a) false (b) true (c) false (d) true
$\qquad$ (e) true (f) true (g) true]

2. In relation to a simple transistor amplifier stage, explain what is meant by the terms: (a) Class-A (b) saturation (c) cut-off (d) quiescent point.

3. Sketch the circuit of a simple Class-A BJT amplifier and explain the function of the components.

4. Explain, with the aid of a labelled sketch, how a load line can be used to determine the operating point of a simple Class-A transistor amplifier.

5. Sketch circuits showing how a JFET can be connected as an amplifier in: (a) common source configuration, (b) common drain configuration, (c) common gate configuration. State typical values of voltage gain and input resistance for each circuit.

6. The output characteristics for a BJT are shown in Fig. 12.27. If this device is used in a common-emitter amplifier circuit operating from a $12\,\text{V}$ supply with a base bias of $60\,\mu\text{A}$ and a load resistor of $1\,\text{k}\Omega$, determine (a) the quiescent values of collector-emitter voltage and collector current, and (b) the peak-peak collector voltage when an $80\,\mu\text{A}$ peak-peak signal current is applied.
$\qquad$ [(a) 5 V, 7 mA  (b) 8.5 V]

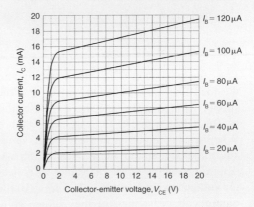

**Figure 12.27**

7. The output characteristics of a JFET are shown in Fig. 12.28. If this device is used in an

amplifier circuit operating from an 18 V supply with a gate-source bias voltage of $-3$ V and a load resistance of $900\,\Omega$, determine (a) the quiescent values of drain-source voltage and drain current, (b) the peak-peak output voltage when an input voltage of 2 V peak-peak is applied, and (c) the voltage gain of the stage.

[(a) 12.2 V, 6.1 mA  (b) 5.5 V  (c) 2.75]

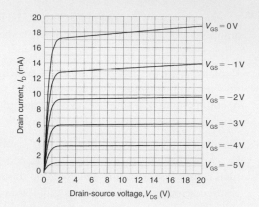

**Figure 12.28**

8. An amplifier has a current gain of 40 and a voltage gain of 30. Determine the power gain.

[1200]

9. The output characteristics of a transistor in common-emitter mode configuration can be regarded as straight lines connecting the following points.

| $I_B = 20\,\mu A$ | | $50\,\mu A$ | | $80\,\mu A$ | |
|---|---|---|---|---|---|
| $V_{CE}$ (v)  1.0 | 8.0 | 1.0 | 8.0 | 1.0 | 8.0 |
| $I_C$ (mA)  1.2 | 1.4 | 3.4 | 4.2 | 6.1 | 8.1 |

Plot the characteristics and superimpose the load line for a $1\,k\Omega$ load, given that the supply voltage is 9 V and the d.c. base bias is $50\,\mu A$. The signal input resistance is $800\,\Omega$. When a peak input current of $30\,\mu A$ varies sinusoidally about a mean bias of $50\,\mu A$, determine (a) the quiescent values of collector voltage and current, $V_{CQ}$ and $I_{CQ}$, (b) the output voltage swing, (c) the voltage gain, (d) the dynamic current gain, and (e) the power gain.

[(a) 5.2 V, 3.7 mA  (b) 5.1 V  (c) 106
(d) 87 (e) 9222]

**Exercise 67    Short answer questions on transistors**

1. In a p-n-p transistor the p-type material regions are called the ...... and ...... , and the n-type material region is called the ......

2. In an n-p-n transistor, the p-type material region is called the ...... and the n-type material regions are called the ...... and the ......

3. In a p-n-p transistor, the base-emitter junction is ......biased and the base-collector junction is ...... biased

4. In an n-p-n transistor, the base-collector junction is ...... biased and the base-emitter junction is ...... biased

5. Majority charge carriers in the emitter of a transistor pass into the base region. Most of them do not recombine because the base is ...... doped

6. Majority carriers in the emitter region of a transistor pass the base-collector junction because for these carriers it is ...... biased

7. Conventional current flow is in the direction of ...... flow

8. Leakage current flows from ...... to ...... in an n-p-n transistor

9. The input characteristic of $I_B$ against $V_{BE}$ for a transistor in common-emitter configuration is similar in shape to that of a ............

10. From a transistor input characteristic,

static input resistance $= \dfrac{......}{......}$ and

dynamic input resistance $= \dfrac{......}{......}$

11. From a transistor output characteristic,

static output resistance $= \dfrac{......}{......}$ and

dynamic output resistance $= \dfrac{......}{......}$

12. From a transistor transfer characteristic,

static current gain $= \dfrac{......}{......}$ and dynamic

current gain $= \dfrac{......}{......}$

13. Complete the following statements that refer to a transistor amplifier:
    (a) An increase in base current causes collector current to ......
    (b) When base current increases, the voltage drop across the load resistor ......
    (c) Under no-signal conditions the power supplied by the battery to an amplifier equals the power dissipated in the load plus the power dissipated in the ......
    (d) The load line has a ...... gradient
    (e) The gradient of the load line depends upon the value of ......
    (f) The position of the load line depends upon ......
    (g) The current gain of a common-emitter amplifier is always greater than ......
    (h) The operating point is generally positioned at the ...... of the load line

14. Explain, with a diagram, the construction of a junction gate field effect transistor. State the advantage of a JFET over a bipolar transistor

15. Sketch typical mutual and output characteristics for a small-signal general-purpose FET operating in common-source mode

16. Name and sketch three possible circuit arrangements used for transistor amplifiers

17. Name and sketch three possible circuit arrangements used for FETs

18. Draw a circuit diagram showing how a transistor can be used as a common-emitter amplifier. Explain briefly the purpose of all the components you show in your diagram

19. Explain how a load line is used to predict a.c. performance of a transistor amplifier

20. What is the quiescent point on a load line?

**Exercise 68    Multi-choice problems on transistors**
**(Answers on page 420)**

In Problems 1 to 10 select the correct answer from those given.

1. In normal operation, the junctions of a p-n-p transistor are:
   (a) both forward biased
   (b) base-emitter forward biased and base-collector reverse biased
   (c) both reverse biased
   (d) base-collector forward biased and base-emitter reverse biased

2. In normal operation, the junctions of an n-p-n transistor are:
   (a) both forward biased
   (b) base-emitter forward biased and base-collector reverse biased
   (c) both reverse biased
   (d) base-collector forward biased and base-emitter reverse biased

3. The current flow across the base-emitter junction of a p-n-p transistor
   (a) mainly electrons
   (b) equal numbers of holes and electrons
   (c) mainly holes
   (d) the leakage current

4. The current flow across the base-emitter junction of an n-p-n transistor consists of
   (a) mainly electrons
   (b) equal numbers of holes and electrons
   (c) mainly holes
   (d) the leakage current

5. In normal operation an n-p-n transistor connected in common-base configuration has
   (a) the emitter at a lower potential than the base
   (b) the collector at a lower potential than the base
   (c) the base at a lower potential than the emitter
   (d) the collector at a lower potential than the emitter

6. In normal operation, a p-n-p transistor connected in common-base configuration has
   (a) the emitter at a lower potential than the base
   (b) the collector at a higher potential than the base
   (c) the base at a higher potential than the emitter
   (d) the collector at a lower potential than the emitter

7. If the per unit value of electrons which leave the emitter and pass to the collector is 0.9 in an n-p-n transistor and the emitter current is 4 mA, then
   (a) the base current is approximately 4.4 mA
   (b) the collector current is approximately 3.6 mA
   (c) the collector current is approximately 4.4 mA
   (d) the base current is approximately 3.6 mA

8. The base region of a p-n-p transistor is
   (a) very thin and heavily doped with holes
   (b) very thin and heavily doped with electrons
   (c) very thin and lightly doped with holes
   (d) very thin and lightly doped with electrons

9. The voltage drop across the base-emitter junction of a p-n-p silicon transistor in normal operation is about
   (a) 200 mV (b) 600 mV (c) zero
   (d) 4.4 V

10. For a p-n-p transistor,
    (a) the number of majority carriers crossing the base-emitter junction largely depends on the collector voltage
    (b) in common-base configuration, the collector current is proportional to the collector-base voltage
    (c) in common-emitter configuration, the base current is less than the base current in common-base configuration
    (d) the collector current flow is independent of the emitter current flow for a given value of collector-base voltage

In questions 11 to 15, which refer to the amplifier shown in Fig. 12.29, select the correct answer from those given.

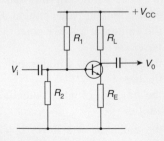

**Figure 12.29**

11. If $R_L$ short-circuited:
    (a) the amplifier signal output would fall to zero
    (b) the collector current would fall to zero
    (c) the transistor would overload

12. If $R_2$ open-circuited:
    (a) the amplifier signal output would fall to zero
    (b) the operating point would be affected and the signal would distort
    (c) the input signal would not be applied to the base

13. A voltmeter connected across $R_E$ reads zero. Most probably
    (a) the transistor base-emitter junction has short-circuited
    (b) $R_L$ has open-circuited
    (c) $R_2$ has short-circuited

14. A voltmeter connected across $R_L$ reads zero. Most probably
    (a) the $V_{CC}$ supply battery is flat
    (b) the base collector junction of the transistor has gone open circuit
    (c) $R_L$ has open-circuited

15. If $R_E$ short-circuited:
    (a) the load line would be unaffected
    (b) the load line would be affected

In questions 16 to 20, which refer to the output characteristics shown in Fig. 12.30, select the correct answer from those given.

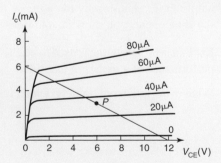

**Figure 12.30**

16. The load line represents a load resistor of
    (a) 1 kΩ    (b) 2 kΩ
    (c) 3 kΩ    (d) 0.5 kΩ

17. The no-signal collector dissipation for the operating point marked P is
    (a)  12 mW    (b)  15 mW
    (c)  18 mW    (d)  21 mW

18. The greatest permissible peak input current would be about
    (a)  30 μA    (b)  35 μA
    (c)  60 μA    (d)  80 μA

19. The greatest possible peak output voltage would then be about
    (a)  5.2 V    (b)  6.5 V
    (c)  8.8 V    (d)  13 V

20. The power dissipated in the load resistor under no-signal conditions is:
    (a)  16 mW    (b)  18 mW
    (c)  20 mW    (d)  22 mW

## Revision Test 3

This revision test covers the material contained in Chapters 8 to 12. *The marks for each question are shown in brackets at the end of each question.*

1. A conductor, 25 cm long is situated at right-angles to a magnetic field. Determine the strength of the magnetic field if a current of 12 A in the conductor produces a force on it of 4.5 N. (3)

2. An electron in a television tube has a charge of $1.5 \times 10^{-19}$ C and travels at $3 \times 10^7$ m/s perpendicular to a field of flux density $20\,\mu$T. Calculate the force exerted on the electron in the field. (3)

3. A lorry is travelling at 100 km/h. Assuming the vertical component of the earth's magnetic field is $40\,\mu$T and the back axle of the lorry is 1.98 m, find the e.m.f. generated in the axle due to motion. (4)

4. An e.m.f. of 2.5 kV is induced in a coil when a current of 2 A collapses to zero in 5 ms. Calculate the inductance of the coil. (4)

5. Two coils, $P$ and $Q$, have a mutual inductance of 100 mH. If a current of 3 A in coil $P$ is reversed in 20 ms, determine (a) the average e.m.f. induced in coil $Q$, and (b) the flux change linked with coil $Q$ if it is wound with 200 turns. (5)

6. A moving coil instrument gives a f.s.d. when the current is 50 mA and has a resistance of $40\,\Omega$. Determine the value of resistance required to enable the instrument to be used (a) as a 0–5 A ammeter, and (b) as a 0–200 V voltmeter. State the mode of connection in each case. (8)

7. An amplifier has a gain of 20 dB. Its input power is 5 mW. Calculate its output power. (3)

8. A sinusoidal voltage trace displayed on an ascilloscope is shown in Fig. RT3.1; the 'time/cm' switch is on 50 ms and the 'volts/cm' switch is on 2 V/cm. Determine for the waveform (a) the frequency (b) the peak-to-peak voltage (c) the amplitude (d) the r.m.s. value. (7)

9. With reference to a p-n junction, briefly explain the following terms: (a) majority carriers (b) contact potential (c) depletion layer (d) forward bias (e) reverse bias. (10)

10. Briefly describe each of the following, drawing their circuit diagram symbol and stating typical applications: (a) zenor diode (b) silicon controlled rectifier (c) light emitting diode (d) varactor diode (e) Schottky diode. (20)

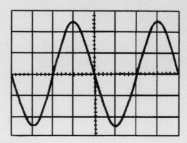

**Figure RT3.1**

11. The following values were obtained during an experiment on a varactor diode.

| Voltage, V | 5 | 10 | 15 | 20 | 25 |
|---|---|---|---|---|---|
| Capacitance, pF | 42 | 28 | 18 | 12 | 8 |

Plot a graph showing the variation of capacitance with voltage for the varactor. Label your axes clearly and use your graph to determine (a) the capacitance when the reverse voltage is $-17.5$ V, (b) the reverse voltage for a capacitance of 35 pF, and (c) the change in capacitance when the voltage changes from $-2.5$ V to $-22.5$ V. (8)

12. Briefly describe, with diagrams, the action of an n-p-n transistor. (7)

13. The output characteristics of a common-emitter transistor amplifier are given below. Assume that the characteristics are linear between the values of collector voltage stated.

| | $I_B = 10\,\mu$A | | $40\,\mu$A | | $70\,\mu$A | |
|---|---|---|---|---|---|---|
| $V_{CE}$ (V) | 1.0 | 7.0 | 1.0 | 7.0 | 1.0 | 7.0 |
| $I_C$ (mA) | 0.6 | 0.7 | 2.5 | 2.9 | 4.6 | 5.35 |

Plot the characteristics and superimpose the load line for a 1.5 k$\Omega$ load and collector supply voltage of 8 V. The signal input resistance is 1.2 k$\Omega$. When a peak input current of $30\,\mu$A varies sinusoidally about a mean bias of $40\,\mu$A, determine (a) the quiescent values of collector voltage and current, (b) the output voltage swing, (c) the voltage gain, (d) the dynamic current gain, and (e) the power gain. (18)

## General:

Charge $Q = It$   Force $F = ma$

Work $W = Fs$   Power $P = \dfrac{W}{t}$

Energy $W = Pt$

Ohm's law $V = IR$   or   $I = \dfrac{V}{R}$   or   $R = \dfrac{V}{I}$

Conductance $G = \dfrac{1}{R}$     Resistance $R = \dfrac{\rho l}{a}$

Power $P = VI = I^2 R = \dfrac{V^2}{R}$

Resistance at $\theta°C$, $R_\theta = R_0(1 + \alpha_0 \theta)$

Terminal p.d. of source, $V = E - Ir$

Series circuit $R = R_1 + R_2 + R_3 + \cdots$

Parallel network $\dfrac{1}{R} = \dfrac{1}{R_1} + \dfrac{1}{R_2} + \dfrac{1}{R_3} + \cdots$

## Capacitors and Capacitance:

$E = \dfrac{V}{d}$   $C = \dfrac{Q}{V}$   $Q = It$   $D = \dfrac{Q}{A}$

$\dfrac{D}{E} = \varepsilon_0 \varepsilon_r$   $C = \dfrac{\varepsilon_0 \varepsilon_r A(n-1)}{d}$   $W = \dfrac{1}{2}CV^2$

Capacitors in parallel $C = C_1 + C_2 + C_3 + \cdots$

Capacitors in series $\dfrac{1}{C} = \dfrac{1}{C_1} + \dfrac{1}{C_2} + \dfrac{1}{C_3} + \cdots$

## Magnetic Circuits:

$B = \dfrac{\Phi}{A}$   $F_m = NI$   $H = \dfrac{NI}{l}$   $\dfrac{B}{H} = \mu_0 \mu_r$

$S = \dfrac{mmf}{\Phi} = \dfrac{l}{\mu_0 \mu_r A}$

## Electromagnetism:

$F = BIl\sin\theta$     $F = QvB$

## Electromagnetic Induction:

$E = Blv\sin\theta$   $E = -N\dfrac{d\Phi}{dt} = -L\dfrac{dI}{dt}$

$W = \dfrac{1}{2}LI^2$   $L = \dfrac{N\Phi}{I} = \dfrac{N^2}{S}$     $E_2 = -M\dfrac{dI_1}{dt}$

$M = \dfrac{N_1 N_2}{S}$

## Measurements:

Shunt $R_s = \dfrac{I_a r_a}{I_s}$   Multiplier $R_M = \dfrac{V - I r_a}{I}$

Power in decibels $= 10\log\dfrac{P_2}{P_1}$

$= 20\log\dfrac{I_2}{I_1}$

$= 20\log\dfrac{V_2}{V_1}$

Wheatstone bridge $R_X = \dfrac{R_2 R_3}{R_1}$

Potentiometer $E_2 = E_1\left(\dfrac{l_2}{l_1}\right)$

# Further Electrical and Electronic Principles

# Chapter 13

# D.C. circuit theory

At the end of this chapter you should be able to:

- state and use Kirchhoff's laws to determine unknown currents and voltages in d.c. circuits

- understand the superposition theorem and apply it to find currents in d.c. circuits

- understand general d.c. circuit theory

- understand Thévenin's theorem and apply a procedure to determine unknown currents in d.c. circuits

- recognize the circuit diagram symbols for ideal voltage and current sources

- understand Norton's theorem and apply a procedure to determine unknown currents in d.c. circuits

- appreciate and use the equivalence of the Thévenin and Norton equivalent networks

- state the maximum power transfer theorem and use it to determine maximum power in a d.c. circuit

## 13.1 Introduction

The laws which determine the currents and voltage drops in d.c. networks are: (a) Ohm's law (see Chapter 2), (b) the laws for resistors in series and in parallel (see Chapter 5), and (c) Kirchhoff's laws (see Section 13.2 following). In addition, there are a number of circuit theorems which have been developed for solving problems in electrical networks. These include:

(i)   the superposition theorem (see Section 13.3),

(ii)  Thévenin's theorem (see Section 13.5),

(iii) Norton's theorem (see Section 13.7), and

(iv)  the maximum power transfer theorem (see Section 13.8)

## 13.2 Kirchhoff's laws

**Kirchhoff's laws state:**

(a) **Current Law.** *At any junction in an electric circuit the total current flowing towards that junction is equal to the total current flowing away from the junction, i.e. $\Sigma I = 0$*

Thus, referring to Fig. 13.1:

$$I_1 + I_2 = I_3 + I_4 + I_5$$

or     $$I_1 + I_2 - I_3 - I_4 - I_5 = 0$$

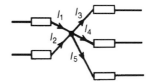

Figure 13.1

(b) **Voltage Law.** *In any closed loop in a network, the algebraic sum of the voltage drops (i.e. products of current and resistance) taken around the loop is equal to the resultant e.m.f. acting in that loop.*

Thus, referring to Fig. 13.2:

$$E_1 - E_2 = IR_1 + IR_2 + IR_3$$

DOI: 10.1016/B978-0-08-089056-2.00013-9

(Note that if current flows away from the positive terminal of a source, that source is considered by convention to be positive. Thus moving anticlockwise around the loop of Fig. 13.2, $E_1$ is positive and $E_2$ is negative.)

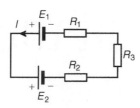

**Figure 13.2**

**Problem 1.** (a) Find the unknown currents marked in Fig. 13.3(a). (b) Determine the value of e.m.f. $E$ in Fig. 13.3(b).

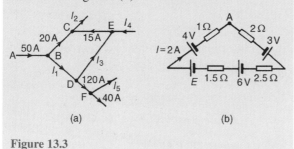

(a)                        (b)

**Figure 13.3**

(a)   Applying Kirchhoff's current law:

For junction B: $50 = 20 + I_1$

Hence          **$I_1 = 30\,\text{A}$**

For junction C: $20 + 15 = I_2$

Hence          **$I_2 = 35\,\text{A}$**

For junction D: $I_1 = I_3 + 120$

i.e.          $30 = I_3 + 120$

Hence         **$I_3 = -90\,\text{A}$**

(i.c. in the opposite direction to that shown in Fig. 13.3(a))

For junction E: $I_4 + I_3 = 15$

i.e.          $I_4 = 15 - (-90)$

Hence         **$I_4 = 105\,\text{A}$**

For junction F: $120 = I_5 + 40$

Hence         **$I_5 = 80\,\text{A}$**

(b)   Applying Kirchhoff's voltage law and moving clockwise around the loop of Fig. 13.3(b) starting at point A:

$$3 + 6 + E - 4 = (I)(2) + (I)(2.5)$$
$$+ (I)(1.5) + (I)(1)$$
$$= I(2 + 2.5 + 1.5 + 1)$$

i.e.       $5 + E = 2(7)$, since $I = 2\,\text{A}$

Hence      $E = 14 - 5 = \mathbf{9\,V}$

**Problem 2.** Use Kirchhoff's laws to determine the currents flowing in each branch of the network shown in Fig. 13.4.

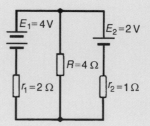

**Figure 13.4**

## Procedure

1.   Use Kirchhoff's current law and label current directions on the original circuit diagram. The directions chosen are arbitrary, but it is usual, as a starting point, to assume that current flows from the positive terminals of the batteries. This is shown in Fig. 13.5 where the three branch currents are expressed in terms of $I_1$ and $I_2$ only, since the current through $R$ is $(I_1 + I_2)$

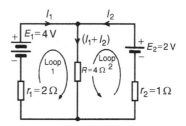

**Figure 13.5**

2.   Divide the circuit into two loops and apply Kirchhoff's voltage law to each. From loop 1 of Fig. 13.5, and moving in a clockwise direction as indicated (the direction chosen does not matter), gives

$$E_1 = I_1 r_1 + (I_1 + I_2)R$$

i.e.       $4 = 2I_1 + 4(I_1 + I_2)$,

i.e.   $6I_1 + 4I_2 = 4$            (1)

From loop 2 of Fig. 13.5, and moving in an anticlockwise direction as indicated (once again, the choice of direction does not matter; it does not have to be in the same direction as that chosen for the first loop), gives:

$$E_2 = I_2r_2 + (I_1 + I_2)R$$

i.e.        $2 = I_2 + 4(I_1 + I_2)$

i.e.   $4I_1 + 5I_2 = 2$              (2)

3. Solve Equations (1) and (2) for $I_1$ and $I_2$

$2 \times$ (1) gives:    $12I_1 + 8I_2 = 8$              (3)

$3 \times$ (2) gives:    $12I_1 + 15I_2 = 6$              (4)

(3) − (4) gives: $-7I_2 = 2$

hence $I_2 = -2/7 = \mathbf{-0.286\,A}$

(i.e. $I_2$ is flowing in the opposite direction to that shown in Fig. 13.5)

From (1)   $6I_1 + 4(-0.286) = 4$

$$6I_1 = 4 + 1.144$$

Hence                    $I_1 = \dfrac{5.144}{6} = \mathbf{0.857\,A}$

Current flowing through resistance $R$ is

$$(I_1 + I_2) = 0.857 + (-0.286)$$

$$= \mathbf{0.571\,A}$$

Note that a third loop is possible, as shown in Fig. 13.6, giving a third equation which can be used as a check:

$$E_1 - E_2 = I_1r_1 - I_2r_2$$

$$4 - 2 = 2I_1 - I_2$$

$$2 = 2I_1 - I_2$$

[Check: $2I_1 - I_2 = 2(0.857) - (-0.286) = 2$]

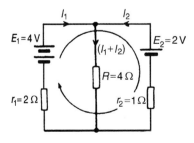

**Figure 13.6**

**Problem 3.**   Determine, using Kirchhoff's laws, each branch current for the network shown in Fig. 13.7.

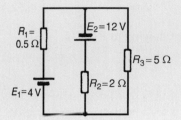

**Figure 13.7**

1. Currents, and their directions are shown labelled in Fig. 13.8 following Kirchhoff's current law. It is usual, although not essential, to follow conventional current flow with current flowing from the positive terminal of the source

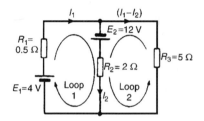

**Figure 13.8**

2. The network is divided into two loops as shown in Fig. 13.8. Applying Kirchhoff's voltage law gives:
   For loop 1:

$$E_1 + E_2 = I_1R_1 + I_2R_2$$

i.e.                    $16 = 0.5I_1 + 2I_2$              (1)

   For loop 2:

$$E_2 = I_2R_2 - (I_1 - I_2)R_3$$

Note that since loop 2 is in the opposite direction to current $(I_1 - I_2)$, the volt drop across $R_3$ (i.e. $(I_1 - I_2)(R_3)$) is by convention negative.

Thus        $12 = 2I_2 - 5(I_1 - I_2)$

i.e.        $12 = -5I_1 + 7I_2$              (2)

3. Solving Equations (1) and (2) to find $I_1$ and $I_2$:

$10 \times$ (1) gives:        $160 = 5I_1 + 20I_2$              (3)

(2) + (3) gives:        $172 = 27I_2$

hence $\qquad I_2 = \dfrac{172}{27} = \mathbf{6.37\,A}$

From (1): $\quad 16 = 0.5I_1 + 2(6.37)$

$$I_1 = \frac{16 - 2(6.37)}{0.5} = \mathbf{6.52\,A}$$

**Current flowing in $R_3 = (I_1 - I_2)$**

$$= 6.52 - 6.37 = \mathbf{0.15\,A}$$

---

**Problem 4.** For the bridge network shown in Fig. 13.9 determine the currents in each of the resistors.

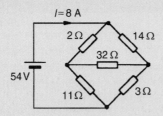

**Figure 13.9**

Let the current in the $2\,\Omega$ resistor be $I_1$, then by Kirchhoff's current law, the current in the $14\,\Omega$ resistor is $(I - I_1)$. Let the current in the $32\,\Omega$ resistor be $I_2$ as shown in Fig. 13.10. Then the current in the $11\,\Omega$ resistor is $(I_1 - I_2)$ and that in the $3\,\Omega$ resistor is $(I - I_1 + I_2)$. Applying Kirchhoff's voltage law to loop 1 and moving in a clockwise direction as shown in Fig. 13.10 gives:

$$54 = 2I_1 + 11(I_1 - I_2)$$

i.e. $\quad 13I_1 - 11I_2 = 54 \qquad\qquad (1)$

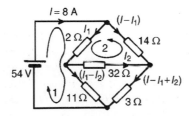

**Figure 13.10**

Applying Kirchhoff's voltage law to loop 2 and moving in a anticlockwise direction as shown in Fig. 13.10 gives:

$$0 = 2I_1 + 32I_2 - 14(I - I_1)$$

However $\qquad I = 8\,A$

Hence $\qquad 0 = 2I_1 + 32I_2 - 14(8 - I_1)$

i.e. $\quad 16I_1 + 32I_2 = 112 \qquad\qquad (2)$

Equations (1) and (2) are simultaneous equations with two unknowns, $I_1$ and $I_2$.

$16 \times (1)$ gives: $\qquad 208I_1 - 176I_2 = 864 \qquad (3)$

$13 \times (2)$ gives: $\qquad 208I_1 + 416I_2 = 1456 \qquad (4)$

$(4) - (3)$ gives: $\qquad\qquad\quad 592I_2 = 592$

$$I_2 = 1\,A$$

Substituting for $I_2$ in (1) gives:

$$13I_1 - 11 = 54$$

$$I_1 = \frac{65}{13} = 5\,A$$

Hence, the current flowing in the $2\,\Omega$ resistor

$$= I_1 = \mathbf{5\,A}$$

the current flowing in the $14\,\Omega$ resistor

$$= (I - I_1) = 8 - 5 = \mathbf{3\,A}$$

the current flowing in the $32\,\Omega$ resistor

$$= I_2 = \mathbf{1\,A}$$

the current flowing in the $11\,\Omega$ resistor

$$= (I_1 - I_2) = 5 - 1 = \mathbf{4\,A}$$

and the current flowing in the $3\,\Omega$ resistor

$$= I - I_1 + I_2 = 8 - 5 + 1 = \mathbf{4\,A}$$

---

**Now try the following exercise**

**Exercise 69 Further problems on Kirchhoff's laws**

1. Find currents $I_3$, $I_4$ and $I_6$ in Fig. 13.11

$$[I_3 = 2\,A,\ I_4 = -1\,A,\ I_6 = 3\,A]$$

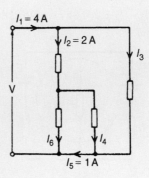

**Figure 13.11**

2. For the networks shown in Fig. 13.12, find the values of the currents marked.

[(a) $I_1 = 4$ A, $I_2 = -1$ A, $I_3 = 13$ A
(b) $I_1 = 40$ A, $I_2 = 60$ A, $I_3 = 120$ A
$I_4 = 100$ A, $I_5 = -80$ A]

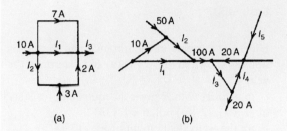

(a)　　　　　(b)

Figure 13.12

3. Calculate the currents $I_1$ and $I_2$ in Fig. 13.13.
[$I_1 = 0.8$ A, $I_2 = 0.5$ A]

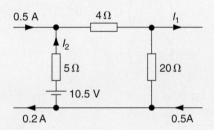

Figure 13.13

4. Use Kirchhoff's laws to find the current flowing in the 6 Ω resistor of Fig. 13.14 and the power dissipated in the 4 Ω resistor.
[2.162 A, 42.07 W]

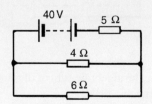

Figure 13.14

5. Find the current flowing in the 3 Ω resistor for the network shown in Fig. 13.15(a). Find also the p.d. across the 10 Ω and 2 Ω resistors.
[2.715 A, 7.410 V, 3.948 V]

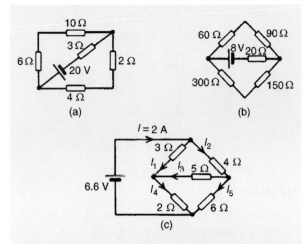

(a)　　　　　(b)

Figure 13.15

6. For the network shown in Fig. 13.15(b) find: (a) the current in the battery, (b) the current in the 300 Ω resistor, (c) the current in the 90 Ω resistor, and (d) the power dissipated in the 150 Ω resistor.

[(a) 60.38 mA (b) 15.10 mA
(c) 45.28 mA (d) 34.20 mW]

7. For the bridge network shown in Fig. 13.15(c), find the currents $I_1$ to $I_5$
[$I_1 = 1.26$ A, $I_2 = 0.74$ A, $I_3 = 0.16$ A,
$I_4 = 1.42$ A, $I_5 = 0.58$ A]

## 13.3 The superposition theorem

**The superposition theorem states:**

*In any network made up of linear resistances and containing more than one source of e.m.f., the resultant current flowing in any branch is the algebraic sum of the currents that would flow in that branch if each source was considered separately, all other sources being replaced at that time by their respective internal resistances.*

The superposition theorem is demonstrated in the following worked problems

**Problem 5.** Figure 13.16 shows a circuit containing two sources of e.m.f., each with their internal resistance. Determine the current in each branch of the network by using the superposition theorem.

Section 2

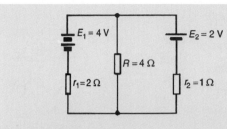

Figure 13.16

## Procedure:

1. Redraw the original circuit with source $E_2$ removed, being replaced by $r_2$ only, as shown in Fig. 13.17(a)

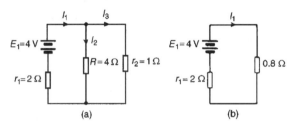

Figure 13.17

2. Label the currents in each branch and their directions as shown in Fig. 13.17(a) and determine their values. (Note that the choice of current directions depends on the battery polarity, which, by convention is taken as flowing from the positive battery terminal as shown)

   $R$ in parallel with $r_2$ gives an equivalent resistance of $(4 \times 1)/(4+1)=0.8\,\Omega$

   From the equivalent circuit of Fig. 13.17(b),

   $$I_1 = \frac{E_1}{r_1 + 0.8} = \frac{4}{2+0.8}$$

   $$= 1.429\,\text{A}$$

   From Fig. 13.17(a),

   $$I_2 = \left(\frac{1}{4+1}\right)I_1 = \frac{1}{5}(1.429) = 0.286\,\text{A}$$

   and $\quad I_3 = \left(\frac{4}{4+1}\right)I_1 = \frac{4}{5}(1.429) = 1.143\,\text{A}$

   by current division

3. Redraw the original circuit with source $E_1$ removed, being replaced by $r_1$ only, as shown in Fig. 13.18(a)

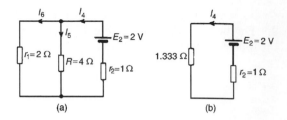

Figure 13.18

4. Label the currents in each branch and their directions as shown in Fig. 13.18(a) and determine their values.

   $r_1$ in parallel with $R$ gives an equivalent resistance of $(2 \times 4)/(2+4)=8/6=1.333\,\Omega$

   From the equivalent circuit of Fig. 13.18(b)

   $$I_4 = \frac{E_2}{1.333 + r_2} = \frac{2}{1.333 + 1} = 0.857\,\text{A}$$

   From Fig. 13.18(a),

   $$I_5 = \left(\frac{2}{2+4}\right)I_4 = \frac{2}{6}(0.857) = 0.286\,\text{A}$$

   $$I_6 = \left(\frac{4}{2+4}\right)I_4 = \frac{4}{6}(0.857) = 0.571\,\text{A}$$

5. Superimpose Fig. 13.18(a) on to Fig. 13.17(a) as shown in Fig. 13.19

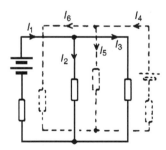

Figure 13.19

6. Determine the algebraic sum of the currents flowing in each branch.

   Resultant current flowing through source 1, i.e.

   $$I_1 - I_6 = 1.429 - 0.571$$

   $$= \mathbf{0.858\,A \;(discharging)}$$

   Resultant current flowing through source 2, i.e.

   $$I_4 - I_3 = 0.857 - 1.143$$

   $$= \mathbf{-0.286\,A \;(charging)}$$

Resultant current flowing through resistor $R$, i.e.

$$I_2 + I_5 = 0.286 + 0.286$$

$$= \mathbf{0.572\,A}$$

The resultant currents with their directions are shown in Fig. 13.20

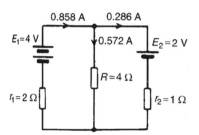

**Figure 13.20**

**Problem 6.** For the circuit shown in Fig. 13.21, find, using the superposition theorem, (a) the current flowing in and the p.d. across the $18\,\Omega$ resistor, (b) the current in the 8 V battery and (c) the current in the 3 V battery.

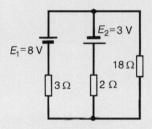

**Figure 13.21**

1. Removing source $E_2$ gives the circuit of Fig. 13.22(a)

2. The current directions are labelled as shown in Fig. 13.22(a), $I_1$ flowing from the positive terminal of $E_1$

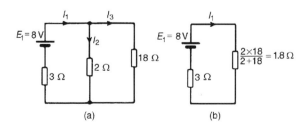

**Figure 13.22**

From Fig. 13.22(b),

$$I_1 = \frac{E_1}{3+1.8} = \frac{8}{4.8} = 1.667\,A$$

From Fig. 13.22(a),

$$I_2 = \left(\frac{18}{2+18}\right)I_1 = \frac{18}{20}(1.667) = 1.500\,A$$

and $\quad I_3 = \left(\frac{2}{2+18}\right)I_1 = \frac{2}{20}(1.667) = 0.167\,A$

3. Removing source $E_1$ gives the circuit of Fig. 13.23(a) (which is the same as Fig. 13.23(b))

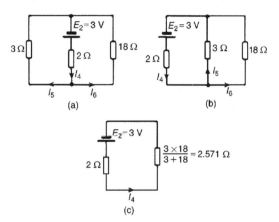

**Figure 13.23**

4. The current directions are labelled as shown in Figures 13.23(a) and 13.23(b), $I_4$ flowing from the positive terminal of $E_2$
   From Fig. 13.23(c),

$$I_4 = \frac{E_2}{2+2.571} = \frac{3}{4.571} = 0.656\,A$$

From Fig. 13.23(b),

$$I_5 = \left(\frac{18}{3+18}\right)I_4 = \frac{18}{21}(0.656) = 0.562\,A$$

$$I_6 = \left(\frac{3}{3+18}\right)I_4 = \frac{3}{21}(0.656) = 0.094\,A$$

5. Superimposing Fig. 13.23(a) on to Fig. 13.22(a) gives the circuit in Fig. 13.24

6. (a) Resultant current in the $18\,\Omega$ resistor

$$= I_3 - I_6$$

$$= 0.167 - 0.094 = \mathbf{0.073\,A}$$

P.d. across the $18\,\Omega$ resistor

$$= 0.073 \times 18 = \mathbf{1.314\,V}$$

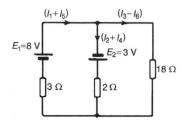

**Figure 13.24**

(b)  Resultant current in the 8 V battery
$$= I_1 + I_5 = 1.667 + 0.562$$
$$= \mathbf{2.229\,A}\ \textbf{(discharging)}$$

(c)  Resultant current in the 3 V battery
$$= I_2 + I_4 = 1.500 + 0.656$$
$$= \mathbf{2.156\,A}\ \textbf{(discharging)}$$

*For a practical laboratory experiment on the superposition theorem, see Chapter 24, page 408.*

**Now try the following exercise**

**Exercise 70   Further problems on the superposition theorem**

1.  Use the superposition theorem to find currents $I_1$, $I_2$ and $I_3$ of Fig. 13.25.
$$[I_1 = 2\,A,\ I_2 = 3\,A,\ I_3 = 5\,A]$$

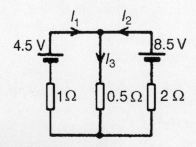

**Figure 13.25**

2.  Use the superposition theorem to find the current in the 8 Ω resistor of Fig. 13.26.
$$[0.385\,A]$$

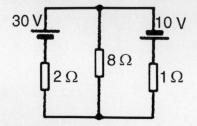

**Figure 13.26**

3.  Use the superposition theorem to find the current in each branch of the network shown in Fig. 13.27.
[10 V battery discharges at 1.429 A
4 V battery charges at 0.857 A
Current through 10 Ω resistor is 0.571 A]

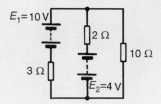

**Figure 13.27**

4.  Use the superposition theorem to determine the current in each branch of the arrangement shown in Fig. 13.28.
[24 V battery charges at 1.664 A
52 V battery discharges at 3.280 A
Current in 20 Ω resistor is 1.616 A]

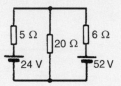

**Figure 13.28**

## 13.4   General d.c. circuit theory

The following points involving d.c. circuit analysis need to be appreciated before proceeding with problems using Thévenin's and Norton's theorems:

(i)  The open-circuit voltage, $E$, across terminals AB in Fig. 13.29 is equal to 10 V, since no current flows through the 2 Ω resistor and hence no voltage drop occurs.

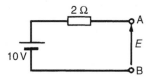

**Figure 13.29**

(ii)  The open-circuit voltage, $E$, across terminals AB in Fig. 13.30(a) is the same as the voltage across the 6 Ω resistor. The circuit may be redrawn as shown in Fig. 13.30(b)

$$E = \left(\frac{6}{6+4}\right)(50)$$

by voltage division in a series circuit,
i.e. **E = 30 V**

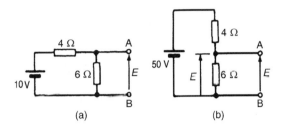

Figure 13.30

(iii) For the circuit shown in Fig. 13.31(a) representing a practical source supplying energy, $V = E - Ir$, where $E$ is the battery e.m.f., $V$ is the battery terminal voltage and $r$ is the internal resistance of the battery (as shown in Section 4.6). For the circuit shown in Fig. 13.31(b),

$$V = E - (-I)r, \text{ i.e. } V = E + Ir$$

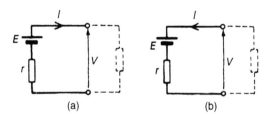

Figure 13.31

(iv) The resistance 'looking-in' at terminals AB in Fig. 13.32(a) is obtained by reducing the circuit in stages as shown in Figures 13.32(b) to (d). Hence the equivalent resistance across AB is 7 Ω.

(v) For the circuit shown in Fig. 13.33(a), the 3 Ω resistor carries no current and the p.d. across the 20 Ω resistor is 10 V. Redrawing the circuit gives Fig. 13.33(b), from which

$$E = \left(\frac{4}{4+6}\right) \times 10 = \mathbf{4\,V}$$

(vi) If the 10 V battery in Fig. 13.33(a) is removed and replaced by a short-circuit, as shown in Fig. 13.33(c), then the 20 Ω resistor may be

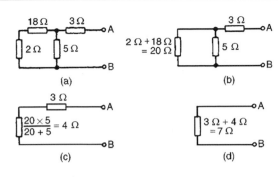

Figure 13.32

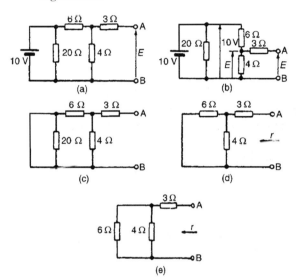

Figure 13.33

removed. The reason for this is that a short-circuit has zero resistance, and 20 Ω in parallel with zero ohms gives an equivalent resistance of $(20 \times 0)/(20+0)$ i.e. 0 Ω. The circuit is then as shown in Fig. 13.33(d), which is redrawn in Fig. 13.33(e). From Fig. 13.33(e), the equivalent resistance across AB,

$$r = \frac{6 \times 4}{6+4} + 3 = 2.4 + 3 = \mathbf{5.4\,\Omega}$$

(vii) To find the voltage across AB in Fig. 13.34: Since the 20 V supply is across the 5 Ω and 15 Ω resistors in series then, by voltage division, the voltage drop across AC,

$$V_{AC} = \left(\frac{5}{5+15}\right)(20) = 5\,V$$

Similarly,

$$V_{CB} = \left(\frac{12}{12+3}\right)(20) = 16\,V.$$

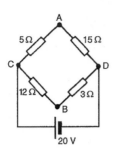

**Figure 13.34**

$V_C$ is at a potential of $+20\,\text{V}$.

$$V_A = V_C - V_{AC} = +20 - 5 = 15\,\text{V}$$

and $\quad V_B = V_C - V_{BC} = +20 - 16 = 4\,\text{V}$.

Hence the voltage between AB is $V_A - V_B = 15 - 4 = 11\,\text{V}$ and current would flow from A to B since A has a higher potential than B.

(viii) In Fig. 13.35(a), to find the equivalent resistance across AB the circuit may be redrawn as in Figs 13.35(b) and (c). From Fig. 13.27(c), the equivalent resistance across AB

$$= \frac{5 \times 15}{5 + 15} + \frac{12 \times 3}{12 + 3}$$

$$= 3.75 + 2.4 = \mathbf{6.15\,\Omega}$$

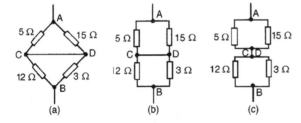

**Figure 13.35**

(ix) In the worked problems in Sections 13.5 and 13.7 following, it may be considered that Thévenin's and Norton's theorems have no obvious advantages compared with, say, Kirchhoff's laws. However, these theorems can be used to analyse part of a circuit and in much more complicated networks the principle of replacing the supply by a constant voltage source in series with a resistance (or impedance) is very useful.

## 13.5  Thévenin's theorem

**Thévenin's theorem states:**

*The current in any branch of a network is that which would result if an e.m.f. equal to the p.d. across a break*

*made in the branch, were introduced into the branch, all other e.m.f.'s being removed and represented by the internal resistances of the sources.*

The procedure adopted when using Thévenin's theorem is summarised below. To determine the current in any branch of an active network (i.e. one containing a source of e.m.f.):

(i) remove the resistance $R$ from that branch,

(ii) determine the open-circuit voltage, $E$, across the break,

(iii) remove each source of e.m.f. and replace them by their internal resistances and then determine the resistance, $r$, 'looking-in' at the break,

(iv) determine the value of the current from the equivalent circuit shown in Fig. 13.36, i.e.

$$I = \frac{E}{R + r}$$

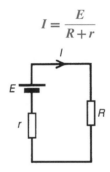

**Figure 13.36**

**Problem 7.** Use Thévenin's theorem to find the current flowing in the $10\,\Omega$ resistor for the circuit shown in Fig. 13.37.

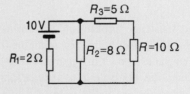

**Figure 13.37**

Following the above procedure:

(i) The $10\,\Omega$ resistance is removed from the circuit as shown in Fig. 13.38(a)

(ii) There is no current flowing in the $5\,\Omega$ resistor and current $I_1$ is given by

$$I_1 = \frac{10}{R_1 + R_2} = \frac{10}{2 + 8} = 1\,\text{A}$$

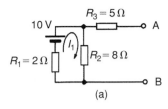

(a)

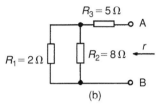

(b)

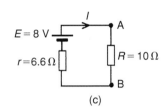

(c)

**Figure 13.38**

P.d. across $R_2 = I_1 R_2 = 1 \times 8 = 8$ V. Hence p.d. across AB, i.e. the open-circuit voltage across the break, $E = 8$ V

(iii)  Removing the source of e.m.f. gives the circuit of Fig. 13.38(b). Resistance,

$$r = R_3 + \frac{R_1 R_2}{R_1 + R_2} = 5 + \frac{2 \times 8}{2 + 8}$$

$$= 5 + 1.6 = 6.6 \,\Omega$$

(iv)  The equivalent Thévenin's circuit is shown in Fig. 13.38(c)

$$\text{Current } I = \frac{E}{R + r} = \frac{8}{10 + 6.6} = \frac{8}{16.6}$$

$$= 0.482 \,\text{A}$$

Hence the current flowing in the $10 \,\Omega$ resistor of Fig. 13.37 is **0.482 A**.

---

**Problem 8.**   For the network shown in Fig. 13.39 determine the current in the $0.8 \,\Omega$ resistor using Thévenin's theorem.

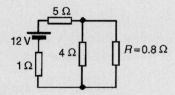

**Figure 13.39**

---

Following the procedure:

(i)   The $0.8 \,\Omega$ resistor is removed from the circuit as shown in Fig. 13.40(a).

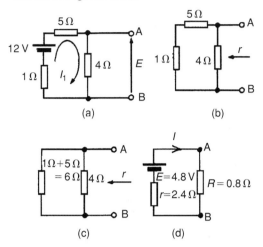

**Figure 13.40**

(ii)  Current $I_1 = \dfrac{12}{1 + 5 + 4} = \dfrac{12}{10} = 1.2 \,\text{A}$

P.d. across $4 \,\Omega$ resistor $= 4 I_1 = (4)(1.2) = 4.8$ V. Hence p.d. across AB, i.e. the open-circuit voltage across AB, $E = 4.8$ V

(iii)  Removing the source of e.m.f. gives the circuit shown in Fig. 13.40(b). The equivalent circuit of Fig. 13.40(b) is shown in Fig. 13.40(c), from which, resistance

$$r = \frac{4 \times 6}{4 + 6} = \frac{24}{10} = 2.4 \,\Omega$$

(iv)  The equivalent Thévenin's circuit is shown in Fig. 13.40(d), from which, current

$$I = \frac{E}{r + R} = \frac{4.8}{2.4 + 0.8} = \frac{4.8}{3.2}$$

$$= \mathbf{1.5 \,A = current \ in \ the \ 0.8 \,\Omega \ resistor}$$

---

**Problem 9.**   Use Thévenin's theorem to determine the current $I$ flowing in the $4 \,\Omega$ resistor shown in Fig. 13.41. Find also the power dissipated in the $4 \,\Omega$ resistor.

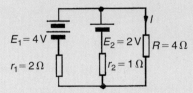

**Figure 13.41**

Following the procedure:

(i) The $4\,\Omega$ resistor is removed from the circuit as shown in Fig. 13.42(a)

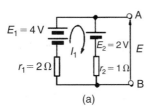

(a)

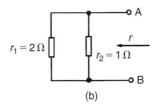

(b)

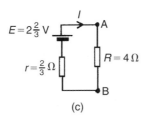

(c)

**Figure 13.42**

(ii) Current $I_1 = \dfrac{E_1 - E_2}{r_1 + r_2} = \dfrac{4-2}{2+1} = \dfrac{2}{3}\,\text{A}$

P.d. across AB,

$$E = E_1 - I_1 r_1 = 4 - \frac{2}{3}(2) = 2\frac{2}{3}\,\text{V}$$

(see Section 13.4(iii)). (Alternatively, p.d. across AB, $E = E_2 + I_1 r_2 = 2 + \frac{2}{3}(1) = 2\frac{2}{3}\,\text{V}$)

(iii) Removing the sources of e.m.f. gives the circuit shown in Fig. 13.42(b), from which, resistance

$$r = \frac{2 \times 1}{2+1} = \frac{2}{3}\,\Omega$$

(iv) The equivalent Thévenin's circuit is shown in Fig. 13.42(c), from which, current,

$$I = \frac{E}{r+R} = \frac{2\frac{2}{3}}{\frac{2}{3}+4} = \frac{8/3}{14/3} = \frac{8}{14}$$

$$= \mathbf{0.571\,A}$$

$$= \textbf{current in the 4}\,\Omega\ \textbf{resistor}$$

Power dissipated in the $4\,\Omega$ resistor,
$P = I^2 R = (0.571)^2 (4) = \mathbf{1.304\,W}$

---

**Problem 10.** Determine the current in the $5\,\Omega$ resistance of the network shown in Fig. 13.43 using Thévenin's theorem. Hence find the currents flowing in the other two branches.

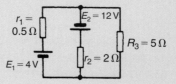

**Figure 13.43**

Following the procedure:

(i) The $5\,\Omega$ resistance is removed from the circuit as shown in Fig. 13.44(a)

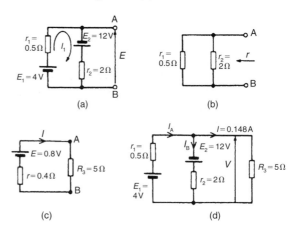

(a)           (b)

(c)           (d)

**Figure 13.44**

(ii) Current $I_1 = \dfrac{12+4}{0.5+2} = \dfrac{16}{2.5} = 6.4\,\text{A}$

P.d. across AB,

$$E = E_1 - I_1 r_1 = 4 - (6.4)(0.5) = 0.8\,\text{V}$$

(see Section 13.4(iii)). (Alternatively, $E = -E_2 + I_1 r_1 = -12 + (6.4)(2) = 0.8\,\text{V}$)

(iii) Removing the sources of e.m.f. gives the circuit shown in Fig. 13.44(b), from which resistance

$$r = \frac{0.5 \times 2}{0.5+2} = \frac{1}{2.5} = 0.4\,\Omega$$

(iv) The equivalent Thévenin's circuit is shown in Fig. 13.44(c), from which, current

$$I = \frac{E}{r+R} = \frac{0.8}{0.4+5} = \frac{0.8}{5.4} = \mathbf{0.148\,A}$$

$$= \textbf{current in the 5}\,\Omega\ \textbf{resistor}$$

From Fig. 13.44(d),

$$\text{voltage } V = IR_3 = (0.148)(5) = 0.74\,\text{V}$$

From Section 13.4(iii),

$$V = E_1 - I_A r_1$$

i.e. $\quad\quad 0.74 = 4 - (I_A)(0.5)$

Hence current, $I_A = \dfrac{4-0.74}{0.5} = \dfrac{3.26}{0.5} = \mathbf{6.52\,A}$

Also from Fig. 13.44(d),

$$V = -E_2 + I_B r_2$$

i.e. $\quad\quad 0.74 = -12 + (I_B)(2)$

Hence current $I_B = \dfrac{12+0.74}{2} = \dfrac{12.74}{2} = \mathbf{6.37\,A}$

[Check, from Fig. 13.44(d), $I_A = I_B + I$, correct to 2 significant figures by Kirchhoff's current law]

---

**Problem 11.** Use Thévenin's theorem to determine the current flowing in the $3\,\Omega$ resistance of the network shown in Fig. 13.45. The voltage source has negligible internal resistance.

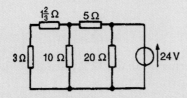

Figure 13.45

---

(Note the symbol for an ideal voltage source in Fig. 13.45 – from BS EN 60617-2: 1996, which superseded BS 3939-2: 1985 – and may be used as an alternative to the battery symbol.)

Following the procedure

(i)  The $3\,\Omega$ resistance is removed from the circuit as shown in Fig. 13.46(a).

(ii)  The $1\frac{2}{3}\,\Omega$ resistance now carries no current. P.d. across $10\,\Omega$ resistor

$$= \left(\frac{10}{10+5}\right)(24) = \mathbf{16\,V}$$

(see Section 13.4(v)). Hence p.d. across AB, $E = 16\,\text{V}$.

(iii)  Removing the source of e.m.f. and replacing it by its internal resistance means that the $20\,\Omega$ resistance is short-circuited as shown in Fig. 13.46(b)

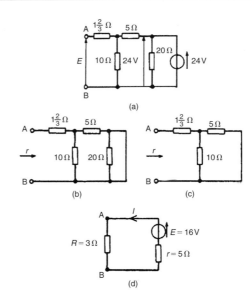

Figure 13.46

since its internal resistance is zero. The $20\,\Omega$ resistance may thus be removed as shown in Fig. 13.46(c) (see Section 13.4 (vi)).

From Fig. 13.46(c), resistance,

$$r = 1\frac{2}{3} + \frac{10\times 5}{10+5} = 1\frac{2}{3} + \frac{50}{15} = 5\,\Omega$$

(iv)  The equivalent Thévenin's circuit is shown in Fig. 13.46(d), from which, current,

$$I = \frac{E}{r+R} = \frac{16}{3+5} = \frac{16}{8} = \mathbf{2\,A}$$

$$= \textbf{current in the } 3\,\Omega \textbf{ resistance}$$

---

**Problem 12.** A Wheatstone Bridge network is shown in Fig. 13.47. Calculate the current flowing in the $32\,\Omega$ resistor, and its direction, using Thévenin's theorem. Assume the source of e.m.f. to have negligible resistance.

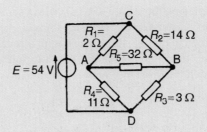

Figure 13.47

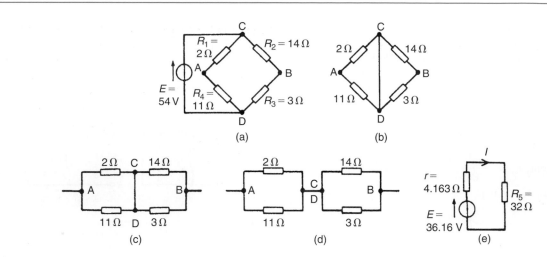

**Figure 13.48**

Following the procedure:

(i) The $32\,\Omega$ resistor is removed from the circuit as shown in Fig. 13.48(a)

(ii) The p.d. between A and C,

$$V_{AC} = \left(\frac{R_1}{R_1 + R_4}\right)(E) = \left(\frac{2}{2 + 11}\right)(54)$$

$$= 8.31\,\text{V}$$

The p.d. between B and C,

$$V_{BC} = \left(\frac{R_2}{R_2 + R_3}\right)(E) = \left(\frac{14}{14 + 3}\right)(54)$$

$$= 44.47\,\text{V}$$

Hence the p.d. between A and B $=44.47 - 8.31 = \mathbf{36.16\,V}$

Point C is at a potential of $+54\,\text{V}$. Between C and A is a voltage drop of $8.31\,\text{V}$. Hence the voltage at point A is $54 - 8.31 = 45.69\,\text{V}$. Between C and B is a voltage drop of $44.47\,\text{V}$. Hence the voltage at point B is $54 - 44.47 = 9.53\,\text{V}$. Since the voltage at A is greater than at B, current must flow in the direction A to B. (See Section 13.4 (vii))

(iii) Replacing the source of e.m.f. with a short-circuit (i.e. zero internal resistance) gives the circuit shown in Fig. 13.48(b). The circuit is redrawn and simplified as shown in Fig. 13.48(c) and (d), from which the resistance between terminals A and B,

$$r = \frac{2 \times 11}{2 + 11} + \frac{14 \times 3}{14 + 3}$$

$$= \frac{22}{13} + \frac{42}{17}$$

$$= 1.692 + 2.471$$

$$= \mathbf{4.163\,\Omega}$$

(iv) The equivalent Thévenin's circuit is shown in Fig. 13.48(e), from which, current

$$I = \frac{E}{r + R_5}$$

$$= \frac{36.16}{4.163 + 32} = 1\,\text{A}$$

**Hence the current in the $32\,\Omega$ resistor of Fig. 13.47 is 1A, flowing from A to B**
*For a practical laboratory experiment on Thévenin's theorem, see Chapter 24, page 410.*

---

**Now try the following exercise**

**Exercise 71    Further problems on Thévenin's theorem**

1. Use Thévenin's theorem to find the current flowing in the $14\,\Omega$ resistor of the network shown in Fig. 13.49. Find also the power dissipated in the $14\,\Omega$ resistor.
   
   [0.434 A, 2.64 W]

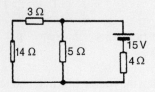

**Figure 13.49**

2. Use Thévenin's theorem to find the current flowing in the 6 Ω resistor shown in Fig. 13.50 and the power dissipated in the 4 Ω resistor.

[2.162 A, 42.07 W]

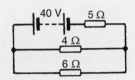

**Figure 13.50**

3. Repeat problems 1 to 4 of Exercise 70, page 190, using Thévenin's theorem.

4. In the network shown in Fig. 13.51, the battery has negligible internal resistance. Find, using Thévenin's theorem, the current flowing in the 4 Ω resistor.       [0.918 A]

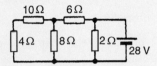

**Figure 13.51**

5. For the bridge network shown in Fig. 13.52, find the current in the 5 Ω resistor, and its direction, by using Thévenin's theorem.

[0.153 A from B to A]

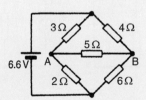

**Figure 13.52**

## 13.6    Constant-current source

A source of electrical energy can be represented by a source of e.m.f. in series with a resistance. In Section 13.5, the Thévenin constant-voltage source consisted of a constant e.m.f. $E$ in series with an internal resistance $r$. However this is not the only form of representation. A source of electrical energy can also be represented by a constant-current source in parallel with a resistance. It may be shown that the two forms

are equivalent. An **ideal constant-voltage generator** is one with zero internal resistance so that it supplies the same voltage to all loads. An **ideal constant-current generator** is one with infinite internal resistance so that it supplies the same current to all loads.

Note the symbol for an ideal current source (from BS EN 60617-2: 1996, which superseded BS 3939-2: 1985), shown in Fig. 13.53.

## 13.7    Norton's theorem

**Norton's theorem** **states:**

*The current that flows in any branch of a network is the same as that which would flow in the branch if it were connected across a source of electrical energy, the short-circuit current of which is equal to the current that would flow in a short-circuit across the branch, and the internal resistance of which is equal to the resistance which appears across the open-circuited branch terminals.*

The procedure adopted when using Norton's theorem is summarised below. To determine the current flowing in a resistance $R$ of a branch AB of an active network:

(i)    short-circuit branch AB

(ii)    determine the short-circuit current $I_{SC}$ flowing in the branch

(iii)    remove all sources of e.m.f. and replace them by their internal resistance (or, if a current source exists, replace with an open-circuit), then determine the resistance $r$, 'looking-in' at a break made between A and B

(iv)    determine the current $I$ flowing in resistance $R$ from the Norton equivalent network shown in Fig. 13.53, i.e.

$$I = \left( \frac{r}{r+R} \right) I_{SC}$$

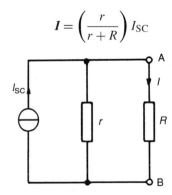

**Figure 13.53**

**Problem 13.** Use Norton's theorem to determine the current flowing in the $10\,\Omega$ resistance for the circuit shown in Fig. 13.54.

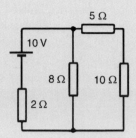

**Figure 13.54**

Following the above procedure:

(i) The branch containing the $10\,\Omega$ resistance is short-circuited as shown in Fig. 13.55(a)

(ii) Fig. 13.55(b) is equivalent to Fig. 13.55(a).

Hence $I_{SC} = \dfrac{10}{2} = 5\,A$

(iii) If the $10\,V$ source of e.m.f. is removed from Fig. 13.55(a) the resistance 'looking-in' at a break made between A and B is given by:

$$r = \frac{2 \times 8}{2+8} = 1.6\,\Omega$$

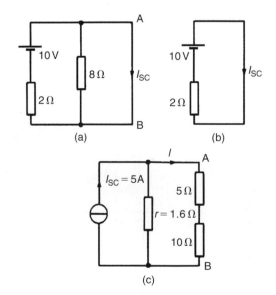

(a)

(b)

(c)

**Figure 13.55**

(iv) From the Norton equivalent network shown in Fig. 13.55(c) the current in the $10\,\Omega$ resistance,

by current division, is given by:

$$I = \left(\frac{1.6}{1.6+5+10}\right)(5) = \mathbf{0.482\,A}$$

as obtained previously in Problem 7 using Thévenin's theorem.

**Problem 14.** Use Norton's theorem to determine the current $I$ flowing in the $4\,\Omega$ resistance shown in Fig. 13.56.

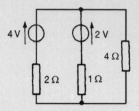

**Figure 13.56**

Following the procedure:

(i) The $4\,\Omega$ branch is short-circuited as shown in Fig. 13.57(a)

(ii) From Fig. 13.57(a),

$$I_{SC} = I_1 + I_2 = \frac{4}{2} + \frac{2}{1} = 4\,A$$

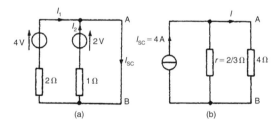

(a)

(b)

**Figure 13.57**

(iii) If the sources of e.m.f. are removed the resistance 'looking-in' at a break made between A and B is given by:

$$r = \frac{2 \times 1}{2+1} = \frac{2}{3}\,\Omega$$

(iv) From the Norton equivalent network shown in Fig. 13.57(b) the current in the $4\,\Omega$ resistance is given by:

$$I = \left(\frac{\frac{2}{3}}{\frac{2}{3}+4}\right)(4) = \mathbf{0.571\,A},$$

as obtained previously in problems 2, 5 and 9 using Kirchhoff's laws and the theorems of superposition and Thévenin

**Problem 15.** Determine the current in the 5 Ω resistance of the network shown in Fig. 13.58 using Norton's theorem. Hence find the currents flowing in the other two branches.

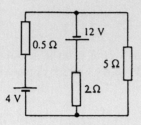

Figure 13.58

Following the procedure:

(i) The 5 Ω branch is short-circuited as shown in Fig. 13.59(a)

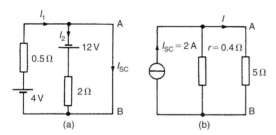

Figure 13.59

(ii) From Fig. 13.59(a),

$$I_{SC} = I_1 - I_2 = \frac{4}{0.5} - \frac{12}{2} = 8 - 6 = \mathbf{2\,A}$$

(iii) If each source of e.m.f. is removed the resistance 'looking-in' at a break made between A and B is given by:

$$r = \frac{0.5 \times 2}{0.5 + 2} = 0.4\,\Omega$$

(iv) From the Norton equivalent network shown in Fig. 13.59(b) the current in the 5 Ω resistance is given by:

$$I = \left(\frac{0.4}{0.4 + 5}\right)(2) = \mathbf{0.148\,A},$$

as obtained previously in Problem 10 using Thévenin's theorem.

The currents flowing in the other two branches are obtained in the same way as in Problem 10. Hence the current flowing from the 4 V source is **6.52 A** and the current flowing from the 12 V source is **6.37 A**.

**Problem 16.** Use Norton's theorem to determine the current flowing in the 3 Ω resistance of the network shown in Fig. 13.60. The voltage source has negligible internal resistance.

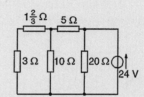

Figure 13.60

Following the procedure:

(i) The branch containing the 3 Ω resistance is short-circuited as shown in Fig. 13.61(a)

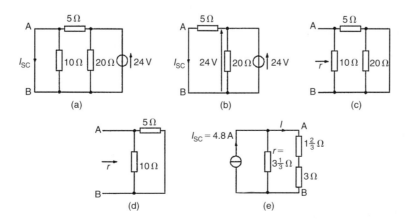

Figure 13.61

(ii)  From the equivalent circuit shown in Fig. 13.61(b),

$$I_{SC} = \frac{24}{5} = 4.8\,\text{A}$$

(iii)  If the 24 V source of e.m.f. is removed the resistance 'looking-in' at a break made between A and B is obtained from Fig. 13.61(c) and its equivalent circuit shown in Fig. 13.61(d) and is given by:

$$r = \frac{10 \times 5}{10 + 5} = \frac{50}{15} = 3\frac{1}{3}\,\Omega$$

(iv)  From the Norton equivalent network shown in Fig. 13.61(e) the current in the 3 Ω resistance is given by:

$$I = \left( \frac{3\frac{1}{3}}{3\frac{1}{3} + 1\frac{2}{3} + 3} \right)(4.8) = \mathbf{2\,A},$$

as obtained previously in Problem 11 using Thévenin's theorem.

**Problem 17.**  Determine the current flowing in the 2 Ω resistance in the network shown in Fig. 13.62.

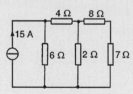

**Figure 13.62**

Following the procedure:

(i)  The 2 Ω resistance branch is short-circuited as shown in Fig. 13.63(a)

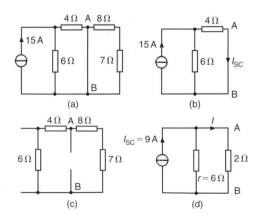

**Figure 13.63**

(ii)  Fig. 13.63(b) is equivalent to Fig. 13.63(a). Hence

$$I_{SC} = \frac{6}{6 + 4}(15) = \mathbf{9\,A} \text{ by current division.}$$

(iii)  If the 15 A current source is replaced by an open-circuit then from Fig. 13.63(c) the resistance 'looking-in' at a break made between A and B is given by $(6+4)\,\Omega$ in parallel with $(8+7)\,\Omega$, i.e.

$$r = \frac{(10)(15)}{10 + 15} = \frac{150}{25} = 6\,\Omega$$

(iv)  From the Norton equivalent network shown in Fig. 13.63(d) the current in the 2 Ω resistance is given by:

$$I = \left( \frac{6}{6 + 2} \right)(9) = \mathbf{6.75\,A}$$

**Now try the following exercise**

**Exercise 72    Further problems on Norton's theorem**

1.  Repeat Problems 1–4 of Exercise 70, page 186, by using Norton's theorem.

2.  Repeat Problems 1, 2, 4 and 5 of Exercise 71, page 196, by using Norton's theorem.

3.  Determine the current flowing in the 6 Ω resistance of the network shown in Fig. 13.64 by using Norton's theorem.  [2.5 mA]

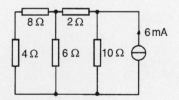

**Figure 13.64**

## 13.8 Thévenin and Norton equivalent networks

The Thévenin and Norton networks shown in Fig. 13.65 are equivalent to each other. The resistance 'looking-in' at terminals AB is the same in each of the networks, i.e. $r$.

If terminals AB in Fig. 13.65(a) are short-circuited, the short-circuit current is given by $E/r$. If terminals

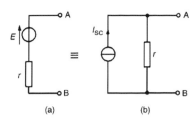

**Figure 13.65**

AB in Fig. 13.65(b) are short-circuited, the short-circuit current is $I_{SC}$. For the circuit shown in Fig. 13.65(a) to be equivalent to the circuit in Fig. 13.65(b) the same short-circuit current must flow. Thus $I_{SC} = E/r$.

Figure 13.66 shows a source of e.m.f. $E$ in series with a resistance $r$ feeding a load resistance $R$

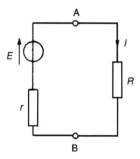

**Figure 13.66**

From Fig. 13.66,

$$I = \frac{E}{r+R} = \frac{E/r}{(r+R)/r} = \left(\frac{r}{r+R}\right)\frac{E}{r}$$

i.e.    $I = \left(\dfrac{r}{r+R}\right)I_{SC}$

From Fig. 13.67 it can be seen that, when viewed from the load, the source appears as a source of current $I_{SC}$ which is divided between $r$ and $R$ connected in parallel.

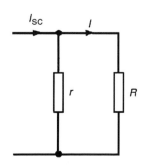

**Figure 13.67**

Thus the two representations shown in Fig. 13.65 are equivalent.

**Problem 18.**   Convert the circuit shown in Fig. 13.68 to an equivalent Norton network.

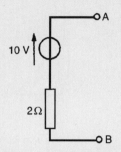

**Figure 13.68**

If terminals AB in Fig. 13.68 are short-circuited, the short-circuit current $I_{SC} = 10/2 = 5$ A.

The resistance 'looking-in' at terminals AB is $2\,\Omega$. Hence the equivalent Norton network is as shown in Fig. 13.69

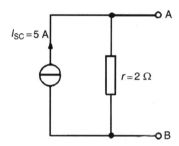

**Figure 13.69**

**Problem 19.**   Convert the network shown in Fig. 13.70 to an equivalent Thévenin circuit.

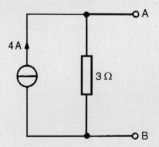

**Figure 13.70**

The open-circuit voltage $E$ across terminals AB in Fig. 13.70 is given by:

$$E = (I_{SC})(r) = (4)(3) = 12\,\text{V}.$$

Section 2

The resistance 'looking-in' at terminals AB is $3\,\Omega$. Hence the equivalent Thévenin circuit is as shown in Fig. 13.71

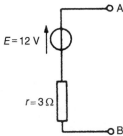

**Figure 13.71**

**Problem 20.** (a) Convert the circuit to the left of terminals AB in Fig. 13.72 to an equivalent Thévenin circuit by initially converting to a Norton equivalent circuit. (b) Determine the current flowing in the $1.8\,\Omega$ resistor.

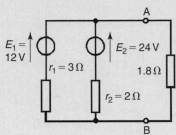

**Figure 13.72**

(a) For the branch containing the $12\,V$ source, converting to a Norton equivalent circuit gives $I_{SC}=12/3=4\,A$ and $r_1=3\,\Omega$. For the branch containing the $24\,V$ source, converting to a Norton equivalent circuit gives $I_{SC2}=24/2=12\,A$ and $r_2=2\,\Omega$. Thus Fig. 13.73(a) shows a network equivalent to Fig. 13.72

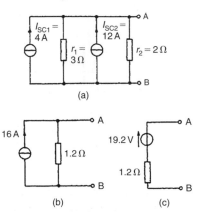

**Figure 13.73**

From Fig. 13.73(a) the total short-circuit current is $4+12=16\,A$ and the total resistance is given by $(3\times2)/(3+2)=\textbf{1.2}\,\boldsymbol{\Omega}$. Thus Fig. 13.73(a) simplifies to Fig. 13.73(b). The open-circuit voltage across AB of Fig. 13.73(b), $E=(16)(1.2)=\textbf{19.2\,V}$, and the resistance 'looking-in' at AB is $1.2\,\Omega$. Hence the Thévenin equivalent circuit is as shown in Fig. 13.73(c).

(b) When the $1.8\,\Omega$ resistance is connected between terminals A and B of Fig. 13.73(c) the current $I$ flowing is given by

$$I=\left(\frac{19.2}{1.2+1.8}\right)=\textbf{6.4\,A}$$

**Problem 21.** Determine by successive conversions between Thévenin and Norton equivalent networks a Thévenin equivalent circuit for terminals AB of Fig. 13.74. Hence determine the current flowing in the $200\,\Omega$ resistance.

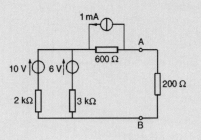

**Figure 13.74**

For the branch containing the $10\,V$ source, converting to a Norton equivalent network gives $I_{SC}=10/2000=5\,mA$ and $r_1=2\,k\Omega$
For the branch containing the $6\,V$ source, converting to a Norton equivalent network gives $I_{SC}=6/3000=2\,mA$ and $r_2=3\,k\Omega$

Thus the network of Fig. 13.74 converts to Fig. 13.75(a). Combining the $5\,mA$ and $2\,mA$ current sources gives the equivalent network of Fig. 13.75(b) where the short-circuit current for the original two branches considered is $7\,mA$ and the resistance is $(2\times3)/(2+3)=1.2\,k\Omega$

Both of the Norton equivalent networks shown in Fig. 13.75(b) may be converted to Thévenin equivalent circuits. The open-circuit voltage across CD is $(7\times10^{-3})(1.2\times10^{3})=8.4\,V$ and the resistance 'looking-in' at CD is $1.2\,k\Omega$. The open-circuit voltage across EF is $(1\times10^{-3})(600)=0.6\,V$ and the resistance 'looking-in' at EF is $0.6\,k\Omega$. Thus Fig. 13.75(b)

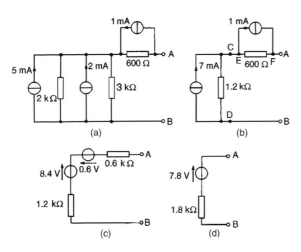

Figure 13.75

converts to Fig. 13.75(c). Combining the two Thévenin circuits gives $E = 8.4 - 0.6 = 7.8\,\text{V}$ and the resistance $r = (1.2 + 0.6)\,\text{k}\Omega = \mathbf{1.8\,k\Omega}$

Thus the Thévenin equivalent circuit for terminals AB of Fig. 13.74 is as shown in Fig. 13.75(d)

Hence the current $I$ flowing in a $200\,\Omega$ resistance connected between A and B is given by

$$I = \frac{7.8}{1800 + 200}$$

$$= \frac{7.8}{2000} = \mathbf{3.9\,mA}$$

**Now try the following exercise**

**Exercise 73    Further problems on Thévenin and Norton equivalent networks**

1.  Convert the circuits shown in Fig. 13.76 to Norton equivalent networks.

$[(a)\ I_{SC} = 25\,\text{A},\ r = 2\,\Omega$
$(b)\ I_{SC} = 2\,\text{mA},\ r = 5\,\Omega]$

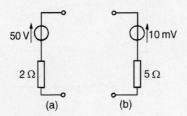

Figure 13.76

2.  Convert the networks shown in Fig. 13.77 to Thévenin equivalent circuits.

$[(a)\ E = 20\,\text{V},\ r = 4\,\Omega$
$(b)\ E = 12\,\text{mV},\ r = 3\,\Omega]$

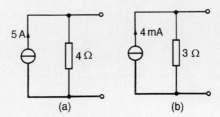

Figure 13.77

3.  (a) Convert the network to the left of terminals AB in Fig. 13.78 to an equivalent Thévenin circuit by initially converting to a Norton equivalent network.

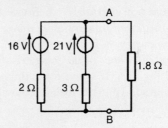

Figure 13.78

(b) Determine the current flowing in the $1.8\,\Omega$ resistance connected between A and B in Fig. 13.78.

$[(a)\ E = 18\,\text{V},\ r = 1.2\,\Omega\ (b)\ 6\,\text{A}]$

4.  Determine, by successive conversions between Thévenin and Norton equivalent networks, a Thévenin equivalent circuit for terminals AB of Fig. 13.79. Hence determine the current flowing in a $6\,\Omega$ resistor connected between A and B.

$[E = 9\frac{1}{3}\,\text{V},\ r = 1\,\Omega,\ 1\frac{1}{3}\,\text{A}]$

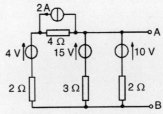

Figure 13.79

5.  For the network shown in Fig. 13.80, convert each branch containing a voltage source to its

Norton equivalent and hence determine the current flowing in the 5 Ω resistance.

[1.22 A]

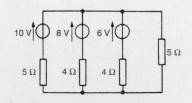

Figure 13.80

## 13.9 Maximum power transfer theorem

**The maximum power transfer theorem states:**
*The power transferred from a supply source to a load is at its maximum when the resistance of the load is equal to the internal resistance of the source.*

Hence, in Fig. 13.81, when $R = r$ the power transferred from the source to the load is a maximum.

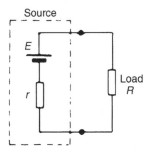

Figure 13.81

Typical practical applications of the maximum power transfer theorem are found in stereo amplifier design, seeking to maximise power delivered to speakers, and in electric vehicle design, seeking to maximise power delivered to drive a motor.

**Problem 22.** The circuit diagram of Fig. 13.82 shows dry cells of source e.m.f. 6 V, and internal resistance 2.5 Ω. If the load resistance $R_L$ is varied from 0 to 5 Ω in 0.5 Ω steps, calculate the power dissipated by the load in each case. Plot a graph of $R_L$ (horizontally) against power (vertically) and determine the maximum power dissipated.

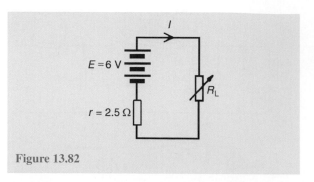

Figure 13.82

When $R_L = 0$, current $I = E/(r + R_L) = 6/2.5 = 2.4$ A and power dissipated in $R_L$, $P = I^2 R_L$ i.e. $P = (2.4)^2(0) = 0$ W.
When $R_L = 0.5$ Ω,
current $I = E/(r + R_L) = 6/(2.5 + 0.5) = 2$ A and $P = I^2 R_L = (2)^2(0.5) = 2.00$ W.
When $R_L = 1.0$ Ω, current $I = 6/(2.5 + 1.0) = 1.714$ A and $P = (1.714)^2(1.0) = 2.94$ W.

With similar calculations the following table is produced:

| $R_L$ (Ω) | 0 | 0.5 | 1.0 | 1.5 | 2.0 | 2.5 |
|---|---|---|---|---|---|---|
| $I = \dfrac{E}{r + R_L}$ | 2.4 | 2.0 | 1.714 | 1.5 | 1.333 | 1.2 |
| $P = I^2 R_L$ (W) | 0 | 2.00 | 2.94 | 3.38 | 3.56 | 3.60 |
| $R_L$ (Ω) | | 3.0 | 3.5 | 4.0 | 4.5 | 5.0 |
| $I = \dfrac{E}{r + R_L}$ | | 1.091 | 1.0 | 0.923 | 0.857 | 0.8 |
| $P = I^2 R_L$ (W) | | 3.57 | 3.50 | 3.41 | 3.31 | 3.20 |

A graph of $R_L$ against $P$ is shown in Fig. 13.83. **The maximum value of power is 3.60 W** which occurs when $R_L$ is 2.5 Ω, i.e. **maximum power occurs when $R_L = r$**, which is what the maximum power transfer theorem states.

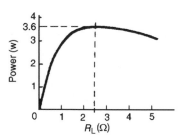

Figure 13.83

**Problem 23.** A d.c. source has an open-circuit voltage of 30 V and an internal resistance of 1.5 Ω. State the value of load resistance that gives maximum power dissipation and determine the value of this power.

The circuit diagram is shown in Fig. 13.84. From the maximum power transfer theorem, for maximum power dissipation, $R_L = r = \mathbf{1.5\,\Omega}$

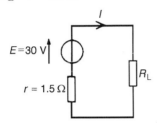

**Figure 13.84**

From Fig. 13.84, current $I = E/(r + R_L)$

$$= 30/(1.5 + 1.5) = 10\,\text{A}$$

Power $P = I^2 R_L = (10)^2(1.5) = \mathbf{150\,W} =$ maximum power dissipated

**Problem 24.** Find the value of the load resistor $R_L$ shown in Fig. 13.85 that gives maximum power dissipation and determine the value of this power.

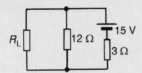

**Figure 13.85**

Using the procedure for Thévenin's theorem:
(i) Resistance $R_L$ is removed from the circuit as shown in Fig. 13.86(a)

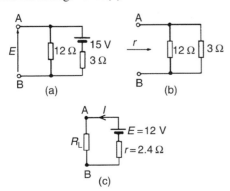

**Figure 13.86**

(ii) The p.d. across AB is the same as the p.d. across the 12 Ω resistor. Hence

$$E = \left(\frac{12}{12 + 3}\right)(15) = 12\,\text{V}$$

(iii) Removing the source of e.m.f. gives the circuit of Fig. 13.86(b), from which, resistance,

$$r = \frac{12 \times 3}{12 + 3} = \frac{36}{15} = 2.4\,\Omega$$

(iv) The equivalent Thévenin's circuit supplying terminals AB is shown in Fig. 13.86(c), from which,

$$\text{current, } I = \frac{E}{r + R_L}$$

For maximum power, $R_L = r = \mathbf{2.4\,\Omega}$

$$\text{Thus current, } I = \frac{12}{2.4 + 2.4} = 2.5\,\text{A}$$

Power, $P$, dissipated in load $R_L$,
$P = I^2 R_L = (2.5)^2(2.4) = \mathbf{15\,W}.$

**Now try the following exercises**

**Exercise 74    Further problems on the maximum power transfer theorem**

1. A d.c. source has an open-circuit voltage of 20 V and an internal resistance of 2 Ω. Determine the value of the load resistance that gives maximum power dissipation. Find the value of this power.                [2 Ω, 50 W]

2. Determine the value of the load resistance $R_L$ shown in Fig. 13.87 that gives maximum power dissipation and find the value of the power.

    [$R_L = 1.6\,\Omega$, $P = 57.6\,\text{W}$]

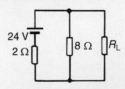

**Figure 13.87**

3. A d.c. source having an open-circuit voltage of 42 V and an internal resistance of 3 Ω is connected to a load of resistance $R_L$. Determine the maximum power dissipated by the load.                [147 W]

4. A voltage source comprising six 2 V cells, each having an internal resistance of $0.2\,\Omega$, is connected to a load resistance $R$. Determine the maximum power transferred to the load.
[30 W]

5. The maximum power dissipated in a $4\,\Omega$ load is 100 W when connected to a d.c. voltage $V$ and internal resistance $r$. Calculate (a) the current in the load, (b) internal resistance $r$, and (c) voltage $V$. [(a) 5 A (b) $4\,\Omega$ (c) 40 V]

### Exercise 75 Short answer questions on d.c. circuit theory

1. Name two laws and three theorems which may be used to find unknown currents and p.d.'s in electrical circuits

2. State Kirchhoff's current law

3. State Kirchhoff's voltage law

4. State, in your own words, the superposition theorem

5. State, in your own words, Thévenin's theorem

6. State, in your own words, Norton's theorem

7. State the maximum power transfer theorem for a d.c. circuit

### Exercise 76 Multi-choice questions on d.c. circuit theory
### (Answers on page 420)

1. Which of the following statements is true: For the junction in the network shown in Fig. 13.88:
   (a) $I_5 - I_4 = I_3 - I_2 + I_1$
   (b) $I_1 + I_2 + I_3 = I_4 + I_5$
   (c) $I_2 + I_3 + I_5 = I_1 + I_4$
   (d) $I_1 - I_2 - I_3 - I_4 + I_5 = 0$

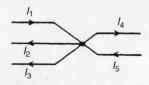

Figure 13.88

2. Which of the following statements is true? For the circuit shown in Fig. 13.89:
   (a) $E_1 + E_2 + E_3 = Ir_1 + Ir_2 + I_3 r_3$
   (b) $E_2 + E_3 - E_1 - I(r_1 + r_2 + r_3) = 0$
   (c) $I(r_1 + r_2 + r_3) = E_1 - E_2 - E_3$
   (d) $E_2 + E_3 - E_1 = Ir_1 + Ir_2 + Ir_3$

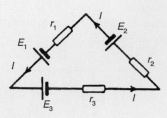

Figure 13.89

3. For the circuit shown in Fig. 13.90, the internal resistance $r$ is given by:
   (a) $\dfrac{I}{V-E}$      (b) $\dfrac{V-E}{I}$
   (c) $\dfrac{I}{E-V}$      (d) $\dfrac{E-V}{I}$

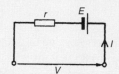

Figure 13.90

4. For the circuit shown in Fig. 13.91, voltage V is:

   (a) 12 V   (b) 2 V   (c) 10 V   (d) 0 V

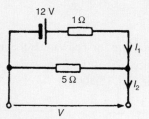

Figure 13.91

5. For the circuit shown in Fig. 13.91, current $I_1$ is:
   (a) 2 A      (b) 14.4 A
   (c) 0.5 A     (d) 0 A

6. For the circuit shown in Fig. 13.91, current $I_2$ is:
   (a) 2 A      (b) 14.4 A
   (c) 0.5 A     (d) 0 A

7.  The equivalent resistance across terminals AB of Fig. 13.92 is:
    (a)  $9.31\,\Omega$  (b)  $7.24\,\Omega$
    (c)  $10.0\,\Omega$  (d)  $6.75\,\Omega$

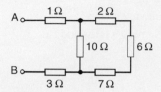

Figure 13.92

8.  With reference to Fig. 13.93, which of the following statements is correct?
    (a)  $V_{PQ}=2\,V$
    (b)  $V_{PQ}=15\,V$
    (c)  When a load is connected between P and Q, current would flow from Q to P
    (d)  $V_{PQ}=20\,V$

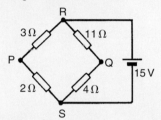

Figure 13.93

9.  In Fig. 13.93, if the 15 V battery is replaced by a short-circuit, the equivalent resistance across terminals PQ is:
    (a)  $20\,\Omega$  (b)  $4.20\,\Omega$
    (c)  $4.13\,\Omega$  (d)  $4.29\,\Omega$

10.  For the circuit shown in Fig. 13.94, maximum power transfer from the source is required. For this to be so, which of the following statements is true?
    (a)  $R_2=10\,\Omega$  (b)  $R_2=30\,\Omega$
    (c)  $R_2=7.5\,\Omega$  (d)  $R_2=15\,\Omega$

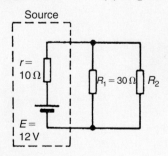

Figure 13.94

11.  The open-circuit voltage $E$ across terminals XY of Fig. 13.95 is:
    (a)  $0\,V$  (b)  $20\,V$
    (c)  $4\,V$  (d)  $16\,V$

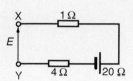

Figure 13.95

12.  The maximum power transferred by the source in Fig. 13.96 is:
    (a)  $5\,W$  (b)  $200\,W$
    (c)  $40\,W$  (d)  $50\,W$

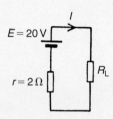

Figure 13.96

13.  For the circuit shown in Fig. 13.97, voltage $V$ is:
    (a)  $0\,V$  (b)  $20\,V$
    (c)  $4\,V$  (d)  $16\,V$

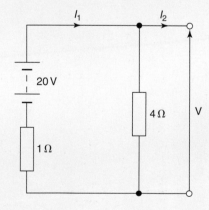

Figure 13.97

14.  For the circuit shown in Fig. 13.97, current $I_1$ is:

Section 2

(a) 25 A      (b) 4 A

(c) 0 A      (d) 20 A

15. For the circuit shown in Fig. 13.97, current $I_2$ is:

     (a) 25 A      (b) 4 A

     (c) 0 A      (d) 20 A

16. The current flowing in the branches of a d.c. circuit may be determined using:

     (a) Kirchhoff's laws

     (b) Lenz's law

     (c) Faraday's laws

     (d) Fleming's left-hand rule

# Chapter 14

# Alternating voltages and currents

At the end of this chapter you should be able to:

- appreciate why a.c. is used in preference to d.c.
- describe the principle of operation of an a.c. generator
- distinguish between unidirectional and alternating waveforms
- define cycle, period or periodic time $T$ and frequency $f$ of a waveform
- perform calculations involving $T = 1/f$
- define instantaneous, peak, mean and r.m.s. values, and form and peak factors for a sine wave
- calculate mean and r.m.s. values and form and peak factors for given waveforms
- understand and perform calculations on the general sinusoidal equation $v = V_m \sin(\omega t \pm \phi)$
- understand lagging and leading angles
- combine two sinusoidal waveforms (a) by plotting graphically, (b) by drawing phasors to scale and (c) by calculation
- understand rectification, and describe methods of obtaining half-wave and full-wave rectification
- appreciate methods of smoothing a rectified output waveform

## 14.1 Introduction

Electricity is produced by generators at power stations and then distributed by a vast network of transmission lines (called the National Grid system) to industry and for domestic use. It is easier and cheaper to generate alternating current (a.c.) than direct current (d.c.) and a.c. is more conveniently distributed than d.c. since its voltage can be readily altered using transformers. Whenever d.c. is needed in preference to a.c., devices called rectifiers are used for conversion (see Section 14.7).

## 14.2 The a.c. generator

Let a single turn coil be free to rotate at constant angular velocity symmetrically between the poles of a magnet system as shown in Fig. 14.1.

An e.m.f. is generated in the coil (from Faraday's laws) which varies in magnitude and reverses its direction at regular intervals. The reason for this is shown in Fig. 14.2. In positions (a), (e) and (i) the conductors of the loop are effectively moving along the magnetic field, no flux is cut and hence no e.m.f. is induced. In position (c) maximum flux is cut and hence maximum e.m.f. is induced. In position (g), maximum flux is cut

DOI: 10.1016/B978-0-08-089056-2.00014-0

**Section 2**

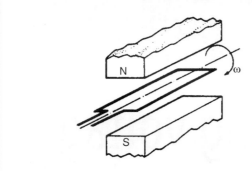

**Figure 14.1**

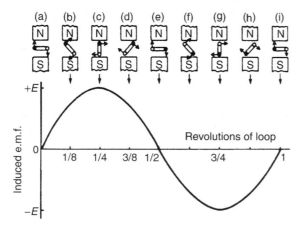

**Figure 14.2**

and hence maximum e.m.f. is again induced. However, using Fleming's right-hand rule, the induced e.m.f. is in the opposite direction to that in position (c) and is thus shown as $-E$. In positions (b), (d), (f) and (h) some flux is cut and hence some e.m.f. is induced. If all such

positions of the coil are considered, in one revolution of the coil, one cycle of alternating e.m.f. is produced as shown. This is the principle of operation of the a.c. generator (i.e. the alternator).

## 14.3 Waveforms

If values of quantities which vary with time $t$ are plotted to a base of time, the resulting graph is called a **waveform**. Some typical waveforms are shown in Fig. 14.3. Waveforms (a) and (b) are **unidirectional waveforms**, for, although they vary considerably with time, they flow in one direction only (i.e. they do not cross the time axis and become negative). Waveforms (c) to (g) are called **alternating waveforms** since their quantities are continually changing in direction (i.e. alternately positive and negative).

A waveform of the type shown in Fig. 14.3(g) is called a **sine wave**. It is the shape of the waveform of e.m.f. produced by an alternator and thus the mains electricity supply is of 'sinusoidal' form.

One complete series of values is called a **cycle** (i.e. from O to P in Fig. 14.3(g)).

The time taken for an alternating quantity to complete one cycle is called the **period** or the **periodic time, $T$,** of the waveform.

The number of cycles completed in one second is called the **frequency, $f$,** of the supply and is measured in **hertz, Hz**. The standard frequency of the electricity supply in Great Britain is 50 Hz

$$T = \frac{1}{f} \quad \text{or} \quad f = \frac{1}{T}$$

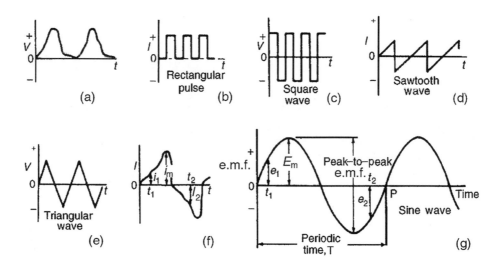

**Figure 14.3**

**Problem 1.**  Determine the periodic time for frequencies of (a) 50 Hz and (b) 20 kHz.

(a)  Periodic time $T = \dfrac{1}{f} = \dfrac{1}{50} = 0.02$ s or **20 ms**

(b)  Periodic time $T = \dfrac{1}{f} = \dfrac{1}{20\,000}$

$$= 0.00005 \text{ s or } \mathbf{50\,\mu s}$$

**Problem 2.**  Determine the frequencies for periodic times of (a) 4 ms (b) 4 μs.

(a)  Frequency $f = \dfrac{1}{T} = \dfrac{1}{4 \times 10^{-3}}$

$$= \dfrac{1000}{4} = \mathbf{250\,Hz}$$

(b)  Frequency $f = \dfrac{1}{T} = \dfrac{1}{4 \times 10^{-6}} = \dfrac{1\,000\,000}{4}$

$$= \mathbf{250\,000\,Hz}$$

$$\text{or } \mathbf{250\,kHz} \text{ or } \mathbf{0.25\,MHz}$$

**Problem 3.**  An alternating current completes 5 cycles in 8 ms. What is its frequency?

Time for 1 cycle $= (8/5)$ ms $= 1.6$ ms $=$ periodic time $T$.

$$\text{Frequency } f = \dfrac{1}{T} = \dfrac{1}{1.6 \times 10^{-3}} = \dfrac{1000}{1.6}$$

$$= \dfrac{10\,000}{16} = \mathbf{625\,Hz}$$

**Now try the following exercise**

**Exercise 77    Further problems on frequency and periodic time**

1.  Determine the periodic time for the following frequencies:
    (a)  2.5 Hz    (b)  100 Hz    (c)  40 kHz
    [(a) 0.4 s (b) 10 ms (c) 25 μs]

2.  Calculate the frequency for the following periodic times:
    (a)  5 ms    (b)  50 μs    (c)  0.2 s
    [(a) 200 Hz (b) 20 kHz (c) 5 Hz]

3.  An alternating current completes 4 cycles in 5 ms. What is its frequency?        [800 Hz]

## 14.4    A.C. values

**Instantaneous values** are the values of the alternating quantities at any instant of time. They are represented by small letters, $i$, $v$, $e$, etc., (see Fig. 14.3(f) and (g)).

The largest value reached in a half-cycle is called the **peak value** or the **maximum value** or the **amplitude** of the waveform. Such values are represented by $V_m$, $I_m$, $E_m$, etc. (see Fig. 14.3(f) and (g)). A **peak-to-peak** value of e.m.f. is shown in Fig. 14.3(g) and is the difference between the maximum and minimum values in a cycle.

The **average** or **mean value** of a symmetrical alternating quantity, (such as a sine wave), is the average value measured over a half-cycle, (since over a complete cycle the average value is zero).

$$\text{Average or mean value} = \frac{\text{area under the curve}}{\text{length of base}}$$

The area under the curve is found by approximate methods such as the trapezoidal rule, the mid-ordinate rule or Simpson's rule. Average values are represented by $V_{AV}$, $I_{AV}$, $E_{AV}$, etc.

For a sine wave:

$$\text{average value} = \mathbf{0.637 \times maximum\ value}$$

$$\text{(i.e. } \mathbf{2/\pi \times maximum\ value)}$$

The **effective value** of an alternating current is that current which will produce the same heating effect as an equivalent direct current. The effective value is called the **root mean square (r.m.s.) value** and whenever an alternating quantity is given, it is assumed to be the r.m.s. value. For example, the domestic mains supply in Great Britain is 240 V and is assumed to mean '240 V r.m.s.'. The symbols used for r.m.s. values are I, V, E, etc. For a non-sinusoidal waveform as shown in Fig. 14.4 the r.m.s. value is given by:

$$I = \sqrt{\frac{i_1^2 + i_2^2 + \cdots + i_n^2}{n}}$$

where $n$ is the number of intervals used.

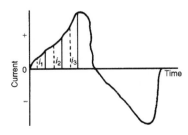

**Figure 14.4**

For a sine wave:

$$\text{r.m.s. value} = 0.707 \times \text{maximum value}$$

$$(\text{i.e.} 1/\sqrt{2} \times \text{maximum value})$$

$$\text{Form factor} = \frac{\text{r.m.s. value}}{\text{average value}}$$

For a sine wave, form factor $= 1.11$

$$\text{Peak factor} = \frac{\text{maximum value}}{\text{r.m.s. value}}$$

For a sine wave, peak factor $= 1.41$.
    The values of form and peak factors give an indication of the shape of waveforms.

---

**Problem 4.**   For the periodic waveforms shown in Fig. 14.5 determine for each: (i) frequency (ii) average value over half a cycle (iii) r.m.s. value (iv) form factor and (v) peak factor.

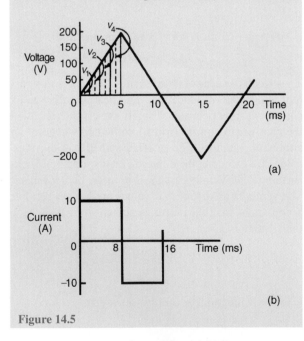

**Figure 14.5**

---

(a)   **Triangular waveform** (Fig. 14.5(a)).

(i)   Time for 1 complete cycle $= 20\,\text{ms} =$ periodic time, $T$. Hence

$$\text{frequency } f = \frac{1}{T} = \frac{1}{20 \times 10^{-3}}$$

$$= \frac{1000}{20} = \mathbf{50\,Hz}$$

(ii)   Area under the triangular waveform for a half-cycle $= \frac{1}{2} \times$ base $\times$ height

$$= \frac{1}{2} \times (10 \times 10^{-3}) \times 200 = 1 \text{ volt second}$$

$$\left.\begin{array}{c}\text{Average value}\\\text{of waveform}\end{array}\right\} = \frac{\text{area under curve}}{\text{length of base}}$$

$$= \frac{1 \text{ volt second}}{10 \times 10^{-3}\text{second}}$$

$$= \frac{1000}{10} = \mathbf{100\,V}$$

(iii)   In Fig. 14.5(a), the first 1/4 cycle is divided into 4 intervals. Thus

$$\text{r.m.s. value} = \sqrt{\frac{v_1^2 + v_2^2 + v_3^2 + v_4^2}{4}}$$

$$= \sqrt{\frac{25^2 + 75^2 + 125^2 + 175^2}{4}}$$

$$= \mathbf{114.6\,V}$$

(Note that the greater the number of intervals chosen, the greater the accuracy of the result. For example, if twice the number of ordinates as that chosen above are used, the r.m.s. value is found to be 115.6 V)

(iv)   Form factor $= \dfrac{\text{r.m.s. value}}{\text{average value}}$

$$= \frac{114.6}{100} = \mathbf{1.15}$$

(v)   Peak factor $= \dfrac{\text{maximum value}}{\text{r.m.s. value}}$

$$= \frac{200}{114.6} = \mathbf{1.75}$$

(b)   **Rectangular waveform** (Fig. 14.5(b)).

(i)   Time for 1 complete cycle $= 16\,\text{ms} =$ periodic time, $T$. Hence

$$\text{frequency, } f = \frac{1}{T} = \frac{1}{16 \times 10^{-3}} = \frac{1000}{16}$$

$$= \mathbf{62.5\,Hz}$$

(ii)   $\left.\begin{array}{c}\text{Average value over}\\\text{half a cycle}\end{array}\right\} = \dfrac{\text{area under curve}}{\text{length of base}}$

$$= \frac{10 \times (8 \times 10^{-3})}{8 \times 10^{-3}}$$

$$= \mathbf{10\,A}$$

(iii)   The r.m.s. value $= \sqrt{\dfrac{i_1^2 + i_2^2 + i_3^2 + i_4^2}{4}} = \mathbf{10\,A}$,
however many intervals are chosen, since the waveform is rectangular.

(iv)   Form factor $= \dfrac{\text{r.m.s. value}}{\text{average value}} = \dfrac{10}{10} = \mathbf{1}$

(v)    Peak factor $= \dfrac{\text{maximum value}}{\text{r.m.s. value}} = \dfrac{10}{10} = \mathbf{1}$

---

**Problem 5.**   The following table gives the corresponding values of current and time for a half-cycle of alternating current.

| time $t$ (ms) | 0 | 0.5 | 1.0 | 1.5 | 2.0 | 2.5 |
|---|---|---|---|---|---|---|
| current $i$ (A) | 0 | 7 | 14 | 23 | 40 | 56 |

| time $t$ (ms) | 3.0 | 3.5 | 4.0 | 4.5 | 5.0 |
|---|---|---|---|---|---|
| current $i$ (A) | 68 | 76 | 60 | 5 | 0 |

Assuming the negative half-cycle is identical in shape to the positive half-cycle, plot the waveform and find (a) the frequency of the supply, (b) the instantaneous values of current after 1.25 ms and 3.8 ms, (c) the peak or maximum value, (d) the mean or average value, and (e) the r.m.s. value of the waveform.

---

The half-cycle of alternating current is shown plotted in Fig. 14.6

(a)   Time for a half-cycle $= 5$ ms; hence the time for 1 cycle, i.e. the periodic time, $T = 10$ ms or 0.01 s

$$\text{Frequency, } f = \frac{1}{T} = \frac{1}{0.01} = \mathbf{100\,Hz}$$

(b)   Instantaneous value of current after 1.25 ms is **19 A**, from Fig. 14.6. Instantaneous value of current after 3.8 ms is **70 A**, from Fig. 14.6

(c)   Peak or maximum value $= \mathbf{76\,A}$

(d)   Mean or average value $= \dfrac{\text{area under curve}}{\text{length of base}}$

Using the mid-ordinate rule with 10 intervals, each of width 0.5 ms gives:

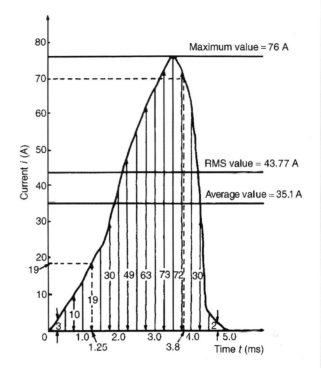

**Figure 14.6**

$$\left.\begin{array}{r}\text{area under}\\ \text{curve}\end{array}\right\} = (0.5 \times 10^{-3})[3 + 10 + 19 + 30 \\ + 49 + 63 + 73 \\ + 72 + 30 + 2]$$

(see Fig. 14.6)

$$= (0.5 \times 10^{-3})(351)$$

$$\left.\begin{array}{r}\text{Hence mean or}\\ \text{average value}\end{array}\right\} = \frac{(0.5 \times 10^{-3})(351)}{5 \times 10^{-3}}$$

$$= \mathbf{35.1\,A}$$

(e)   R.m.s. value $= \sqrt{\dfrac{\begin{array}{c}3^2 + 10^2 + 19^2 + 30^2 + 49^2 \\ + 63^2 + 73^2 + 72^2 + 30^2 + 2^2\end{array}}{10}}$

$$= \sqrt{\frac{19\,157}{10}} = \mathbf{43.8\,A}$$

---

**Problem 6.**   Calculate the r.m.s. value of a sinusoidal current of maximum value 20 A.

For a sine wave,

r.m.s. value $= 0.707 \times$ maximum value

$$= 0.707 \times 20 = \mathbf{14.14\,A}$$

---

**Problem 7.**   Determine the peak and mean values for a 240 V mains supply.

For a sine wave, r.m.s. value of voltage $V = 0.707 \times V_m$. A 240 V mains supply means that 240 V is the r.m.s. value, hence

$$V_m = \frac{V}{0.707} = \frac{240}{0.707} = \mathbf{339.5\,V}$$

$$= \mathbf{peak\ value}$$

Mean value

$$V_{AV} = 0.637V_m = 0.637 \times 339.5 = \mathbf{216.3\,V}$$

**Problem 8.**   A supply voltage has a mean value of 150 V. Determine its maximum value and its r.m.s. value.

For a sine wave, mean value $= 0.637 \times$ maximum value. Hence

$$\mathbf{maximum\ value} = \frac{\text{mean value}}{0.637} = \frac{150}{0.637}$$

$$= \mathbf{235.5\,V}$$

**R.m.s. value** $= 0.707 \times$ maximum value
$$= 0.707 \times 235.5 = \mathbf{166.5\,V}$$

---

**Now try the following exercise**

**Exercise 78   Further problems on a.c. values of waveforms**

1.  An alternating current varies with time over half a cycle as follows:

| Current (A) | 0 | 0.7 | 2.0 | 4.2 | 8.4 |
|---|---|---|---|---|---|
| time (ms) | 0 | 1 | 2 | 3 | 4 |

| Current (A) | 8.2 | 2.5 | 1.0 | 0.4 | 0.2 | 0 |
|---|---|---|---|---|---|---|
| time (ms) | 5 | 6 | 7 | 8 | 9 | 10 |

The negative half-cycle is similar. Plot the curve and determine:
(a) the frequency (b) the instantaneous values at 3.4 ms and 5.8 ms (c) its mean value and (d) its r.m.s. value.

[(a) 50 Hz (b) 5.5 A, 3.1 A
(c) 2.8 A (d) 4.0 A]

2.  For the waveforms shown in Fig. 14.7 determine for each (i) the frequency (ii) the average value over half a cycle (iii) the r.m.s. value (iv) the form factor (v) the peak factor.

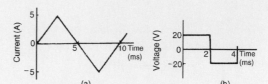

[(a)   (i) 100 Hz (ii) 2.50 A (iii) 2.87 A
        (iv) 1.15    (v) 1.74
 (b)   (i) 250 Hz (ii) 20 V   (iii) 20 V
        (iv) 1.0     (v) 1.0
 (c)   (i) 125 Hz (ii) 18 A   (iii) 19.56 A
        (iv) 1.09    (v) 1.23
 (d)   (i) 250 Hz (ii) 25 V   (iii) 50 V
        (iv) 2.0     (v) 2.0]

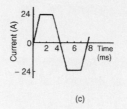
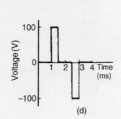

**Figure 14.7**

3.  An alternating voltage is triangular in shape, rising at a constant rate to a maximum of 300 V in 8 ms and then falling to zero at a constant rate in 4 ms. The negative half-cycle is identical in shape to the positive half-cycle. Calculate (a) the mean voltage over half a cycle, and (b) the r.m.s. voltage.
[(a) 150 V (b) 170 V]

4.  An alternating e.m.f. varies with time over half a cycle as follows:

| E.m.f. (V) | 0 | 45 | 80 | 155 | 215 |
|---|---|---|---|---|---|
| time (ms) | 0 | 1.5 | 3.0 | 4.5 | 6.0 |

| E.m.f. (V) | 320 | 210 | 95 | 0 |
|---|---|---|---|---|
| time (ms) | 7.5 | 9.0 | 10.5 | 12.0 |

The negative half-cycle is identical in shape to the positive half-cycle. Plot the waveform and determine (a) the periodic time and frequency (b) the instantaneous value of voltage at 3.75 ms (c) the times when the

voltage is 125 V (d) the mean value, and (e) the
r.m.s. value

[(a) 24 ms, 41.67 Hz   (b) 115 V
(c) 4 ms and 10.1 ms    (d) 142 V
(e) 171 V]

5.  Calculate the r.m.s. value of a sinusoidal curve
    of maximum value 300 V         [212.1 V]

6.  Find the peak and mean values for a 200 V
    mains supply              [282.9 V, 180.2 V]

7.  Plot a sine wave of peak value 10.0 A. Show
    that the average value of the waveform is
    6.37 A over half a cycle, and that the r.m.s.
    value is 7.07 A

8.  A sinusoidal voltage has a maximum value of
    120 V. Calculate its r.m.s. and average values.
                              [84.8 V, 76.4 V]

9.  A sinusoidal current has a mean value of
    15.0 A. Determine its maximum and r.m.s.
    values.                  [23.55 A, 16.65 A]

## 14.5   Electrical safety – insulation and fuses

**Insulation** is used to prevent 'leakage', and when determining what type of insulation should be used, the
maximum voltage present must be taken into account.
For this reason, **peak values are always considered
when choosing insulation materials**.

**Fuses** are the weak link in a circuit and are used to
break the circuit if excessive current is drawn. Excessive
current could lead to a fire. Fuses rely on the heating effect of the current, and for this reason, **r.m.s.
values must always be used when calculating the
appropriate fuse size**.

## 14.6   The equation of a sinusoidal waveform

In Fig. 14.8, 0A represents a vector that is free to rotate
anticlockwise about 0 at an angular velocity of $\omega$ rad/s.
A rotating vector is known as a **phasor**.

After time $t$ seconds the vector 0A has turned through
an angle $\omega t$. If the line BC is constructed perpendicular
to 0A as shown, then

$$\sin \omega t = \frac{BC}{0B} \quad \text{i.e. } BC = 0B \sin \omega t$$

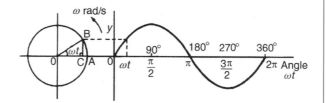

Figure 14.8

If all such vertical components are projected on to a
graph of $y$ against angle $\omega t$ (in radians), a sine curve
results of maximum value 0A. Any quantity which
varies sinusoidally can thus be represented as a phasor.

A sine curve may not always start at $0°$. To show this
a periodic function is represented by $y = \sin(\omega t \pm \phi)$,
where $\phi$ is the phase (or angle) difference compared
with $y = \sin \omega t$. In Fig. 14.9(a), $y_2 = \sin(\omega t + \phi)$ starts
$\phi$ radians earlier than $y_1 = \sin \omega t$ and is thus said to
**lead** $y_1$ by $\phi$ radians. Phasors $y_1$ and $y_2$ are shown in
Fig. 14.9(b) at the time when $t = 0$.

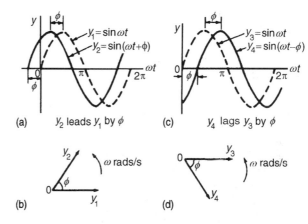

Figure 14.9

In Fig. 14.9(c), $y_4 = \sin(\omega t - \phi)$ starts $\phi$ radians later
than $y_3 = \sin \omega t$ and is thus said to **lag** $y_3$ by $\phi$ radians.
Phasors $y_3$ and $y_4$ are shown in Fig. 14.9(d) at the time
when $t = 0$.

**Given the general sinusoidal voltage,**
$v = V_m \sin(\omega t \pm \phi)$, **then**

(i)   **Amplitude or maximum value** $= V_m$

(ii)  **Peak to peak value** $= 2V_m$

(iii) **Angular velocity** $= \omega$ **rad/s**

(iv)  **Periodic time,** $T = 2\pi/\omega$ **seconds**

(v)   **Frequency,** $f = \omega/2\pi$ **Hz (since** $\omega = 2\pi f$)

(vi)  $\phi =$ **angle of lag or lead (compared with**
      $v = V_m \sin \omega t$)

**Problem 9.** An alternating voltage is given by $v = 282.8\sin 314t$ volts. Find (a) the r.m.s. voltage, (b) the frequency, and (c) the instantaneous value of voltage when $t = 4$ ms.

(a) The general expression for an alternating voltage is $v = V_m\sin(\omega t \pm \phi)$. Comparing $v = 282.8\sin 314t$ with this general expression gives the peak voltage as 282.8 V. Hence the r.m.s. voltage $= 0.707 \times$ maximum value
$$= 0.707 \times 282.8 = \mathbf{200\,V}$$

(b) Angular velocity, $\omega = 314$ rad/s, i.e. $2\pi f = 314$. Hence frequency,
$$f = \frac{314}{2\pi} = \mathbf{50\,Hz}$$

(c) When $t = 4$ ms,
$$v = 282.8\sin(314 \times 4 \times 10^{-3})$$
$$= 282.8\sin(1.256) = \mathbf{268.9\,V}$$

Note that 1.256 radians $= \left[1.256 \times \dfrac{180°}{\pi}\right]$
$$= 71.96°$$

Hence $v = 282.8\sin 71.96° = \mathbf{268.9\,V}$, as above.

**Problem 10.** An alternating voltage is given by $v = 75\sin(200\pi t - 0.25)$ volts. Find (a) the amplitude, (b) the peak-to-peak value, (c) the r.m.s. value, (d) the periodic time, (e) the frequency, and (f) the phase angle (in degrees and minutes) relative to $75\sin 200\pi t$.

Comparing $v = 75\sin(200\pi t - 0.25)$ with the general expression $v = V_m\sin(\omega t \pm \phi)$ gives:

(a) Amplitude, or peak value $= \mathbf{75\,V}$

(b) Peak-to-peak value $= 2 \times 75 = \mathbf{150\,V}$

(c) The r.m.s. value $= 0.707 \times$ maximum value
$$= 0.707 \times 75 = \mathbf{53\,V}$$

(d) Angular velocity, $\omega = 200\pi$ rad/s. Hence periodic time,
$$T = \frac{2\pi}{\omega} = \frac{2\pi}{200\pi} = \frac{1}{100} = \mathbf{0.01\,s}\text{ or }\mathbf{10\,ms}$$

(e) Frequency, $f = \dfrac{1}{T} = \dfrac{1}{0.01} = \mathbf{100\,Hz}$

(f) Phase angle, $\phi = 0.25$ radians lagging $75\sin 200\pi t$
$$0.25\text{ rads} = 0.25 \times \frac{180°}{\pi} = 14.32°$$
Hence phase angle $= \mathbf{14.32°\ lagging}$

**Problem 11.** An alternating voltage, $v$, has a periodic time of 0.01 s and a peak value of 40 V. When time $t$ is zero, $v = -20$ V. Express the instantaneous voltage in the form $v = V_m\sin(\omega t \pm \phi)$.

Amplitude, $V_m = 40$ V.

Periodic time $T = \dfrac{2\pi}{\omega}$ hence angular velocity,
$$\omega = \frac{2\pi}{T} = \frac{2\pi}{0.01} = 200\pi\text{ rad/s.}$$
$v = V_m\sin(\omega t + \phi)$ thus becomes
$$v = 40\sin(200\pi t + \phi)\text{ volts.}$$

When time $t = 0$, $v = -20$ V
i.e. $-20 = 40\sin\phi$
so that $\sin\phi = -20/40 = -0.5$

Hence $\phi = \sin^{-1}(-0.5) = -30°$
$$= \left(-30 \times \frac{\pi}{180}\right)\text{ rads} = -\frac{\pi}{6}\text{ rads}$$

Thus $v = \mathbf{40\sin\left(200\pi t - \dfrac{\pi}{6}\right)\,V}$

**Problem 12.** The current in an a.c. circuit at any time $t$ seconds is given by: $i = 120\sin(100\pi t + 0.36)$ amperes. Find (a) the peak value, the periodic time, the frequency and phase angle relative to $120\sin 100\pi t$, (b) the value of the current when $t = 0$, (c) the value of the current when $t = 8$ ms, (d) the time when the current first reaches 60 A, and (e) the time when the current is first a maximum.

(a) Peak value $= \mathbf{120\,A}$

Periodic time $T = \dfrac{2\pi}{\omega}$
$$= \frac{2\pi}{100\pi}\quad(\text{since }\omega = 100\pi)$$
$$= \frac{1}{50} = \mathbf{0.02\,s}\text{ or }\mathbf{20\,ms}$$

Frequency, $f = \dfrac{1}{T} = \dfrac{1}{0.02} = \mathbf{50\,Hz}$

Phase angle $= 0.36$ rads

$$= 0.36 \times \frac{180°}{\pi} = \mathbf{20.63°\ leading}$$

(b)  When $t = 0$,

$$i = 120\sin(0 + 0.36)$$

$$= 120\sin 20.63° = \mathbf{42.3\,A}$$

(c)  When $t = 8$ ms,

$$i = 120\sin\left[100\pi\left(\frac{8}{10^3}\right) + 0.36\right]$$

$$= 120\sin 2.8733$$

$$= \mathbf{31.8\,A}$$

(d)  When $i = 60$ A, $60 = 120\sin(100\pi t + 0.36)$  thus
$(60/120) = \sin(100\pi t + 0.36)$ so that
$(100\pi t + 0.36) = \sin^{-1} 0.5 = 30°$
$= \pi/6$ rads $= 0.5236$ rads.
Hence time,

$$t = \frac{0.5236 - 0.36}{100\pi} = \mathbf{0.521\ ms}$$

(e)  When the current is a maximum, $i = 120$ A.

Thus $\qquad 120 = 120\sin(100\pi t + 0.36)$

$$1 = \sin(100\pi t + 0.36)$$

$$(100\pi t + 0.36) = \sin^{-1} 1 = 90°$$

$$= (\pi/2)\,\text{rads}$$

$$= 1.5708\,\text{rads.}$$

Hence time, $\qquad t = \dfrac{1.5708 - 0.36}{100\pi} = \mathbf{3.85\ ms}$

*For a practical laboratory experiment on the use of the CRO to measure voltage, frequency and phase, see Chapter 24, page 412.*

---

**Now try the following exercise**

**Exercise 79    Further problems on**
$$v = V_m\sin(\omega t \pm \phi)$$

1.  An alternating voltage is represented by $v = 20\sin 157.1\,t$ volts. Find (a) the maximum value (b) the frequency (c) the periodic time. (d) What is the angular velocity of the phasor representing this waveform?

[(a) 20 V (b) 25 Hz (c) 0.04 s
(d) 157.1 rads/s]

2.  Find the peak value, the r.m.s. value, the frequency, the periodic time and the phase angle (in degrees) of the following alternating quantities:

(a)  $v = 90\sin 400\pi t$ volts
[90 V, 63.63 V, 200 Hz, 5 ms, 0°]

(b)  $i = 50\sin(100\pi t + 0.30)$ amperes
[50 A, 35.35 A, 50 Hz, 0.02 s, 17.19° lead]

(c)  $e = 200\sin(628.4\,t - 0.41)$ volts
[200 V, 141.4 V, 100 Hz, 0.01 s, 23.49° lag]

3.  A sinusoidal current has a peak value of 30 A and a frequency of 60 Hz. At time $t = 0$, the current is zero. Express the instantaneous current $i$ in the form $i = I_m\sin\omega t$.

[$i = 30\sin 120\pi t$ A]

4.  An alternating voltage $v$ has a periodic time of 20 ms and a maximum value of 200 V. When time $t = 0$, $v = -75$ volts. Deduce a sinusoidal expression for $v$ and sketch one cycle of the voltage showing important points.

[$v = 200\sin(100\pi t - 0.384)$ V]

5.  The voltage in an alternating current circuit at any time $t$ seconds is given by $v = 60\sin 40t$ volts. Find the first time when the voltage is (a) 20 V (b) $-30$ V.

[(a) 8.496 ms (b) 91.63 ms]

6.  The instantaneous value of voltage in an a.c. circuit at any time $t$ seconds is given by $v = 100\sin(50\pi t - 0.523)$ V. Find:

(a)  the peak-to-peak voltage, the frequency, the periodic time and the phase angle

(b)  the voltage when $t = 0$

(c)  the voltage when $t = 8$ ms

(d)  the times in the first cycle when the voltage is 60 V

(e)  the times in the first cycle when the voltage is $-40$ V

(f)  the first time when the voltage is a maximum.

Sketch the curve for one cycle showing relevant points.

[(a) 200 V, 25 Hz, 0.04 s, 29.97° lagging
(b) $-49.95$ V (c) 66.96 V (d) 7.426 ms, 19.23 ms (e) 25.95 ms, 40.71 ms
(f) 13.33 ms]

## 14.7 Combination of waveforms

The resultant of the addition (or subtraction) of two sinusoidal quantities may be determined either:

(a) by plotting the periodic functions graphically (see worked Problems 13 and 16), or

(b) by resolution of phasors by drawing or calculation (see worked Problems 14 and 15)

**Problem 13.** The instantaneous values of two alternating currents are given by $i_1 = 20\sin\omega t$ amperes and $i_2 = 10\sin(\omega t + \pi/3)$ amperes. By plotting $i_1$ and $i_2$ on the same axes, using the same scale, over one cycle, and adding ordinates at intervals, obtain a sinusoidal expression for $i_1 + i_2$

$i_1 = 20\sin\omega t$ and $i_2 = 10\sin(\omega t + \pi/3)$ are shown plotted in Fig. 14.10. Ordinates of $i_1$ and $i_2$ are added at, say, 15° intervals (a pair of dividers are useful for this). For example,

at 30°, $i_1 + i_2 = 10 + 10 = 20$ A

at 60°, $i_1 + i_2 = 17.3 + 8.7 = 26$ A

at 150°, $i_1 + i_2 = 10 + (-5) = 5$ A, and so on.

| $\omega t$ (degrees) | 0 | 30 | 60 | 90 |
|---|---|---|---|---|
| $\sin\omega t$ | 0 | 0.5 | 0.866 | 1 |
| $i_1 = 20\sin\omega t$ | 0 | 10 | 17.32 | 20 |

| $(\omega t + 60)$ | 60 | 90 | 120 | 150 |
|---|---|---|---|---|
| $\sin(\omega t + \frac{\pi}{3})$ | 0.866 | 1 | 0.866 | 0.5 |
| $i_2 = 10\sin(\omega t + \frac{\pi}{3})$ | 8.66 | 10 | 8.66 | 5 |

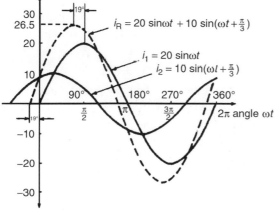

**Figure 14.10**

The resultant waveform for $i_1 + i_2$ is shown by the broken line in Fig. 14.10. It has the same period, and hence frequency, as $i_1$ and $i_2$. The amplitude or peak value is 26.5 A

The resultant waveform leads the curve $i_1 = 20\sin\omega t$ by 19° i.e. $(19 \times \pi/180)$ rads $= 0.332$ rads

Hence the sinusoidal expression for the resultant $i_1 + i_2$ is given by:

$$i_R = i_1 + i_2 = 26.5\sin(\omega t + 0.332)\,A$$

**Problem 14.** Two alternating voltages are represented by $v_1 = 50\sin\omega t$ volts and $v_2 = 100\sin(\omega t - \pi/6)$ V. Draw the phasor diagram and find, by calculation, a sinusoidal expression to represent $v_1 + v_2$

Phasors are usually drawn at the instant when time $t = 0$. Thus $v_1$ is drawn horizontally 50 units long and $v_2$ is drawn 100 units long lagging $v_1$ by $\pi/6$ rads, i.e. 30°. This is shown in Fig. 14.11(a) where 0 is the point of rotation of the phasors.

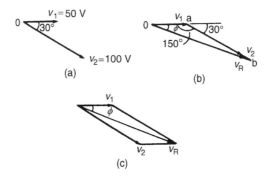

**Figure 14.11**

Procedure to draw phasor diagram to represent $v_1 + v_2$:

(i) Draw $v_1$ horizontal 50 units long, i.e. oa of Fig. 14.11(b)

(ii) Join $v_2$ to the end of $v_1$ at the appropriate angle, i.e. ab of Fig. 14.11(b)

(iii) The resultant $v_R = v_1 + v_2$ is given by the length ob and its phase angle may be measured with respect to $v_1$

Alternatively, when two phasors are being added the resultant is always the diagonal of the parallelogram, as shown in Fig. 14.11(c).

From the drawing, by measurement, $v_R = 145$ V and angle $\phi = 20°$ lagging $v_1$.

A more accurate solution is obtained by calculation, using the cosine and sine rules. Using the cosine rule on triangle 0ab of Fig. 14.11(b) gives:

$$v_R^2 = v_1^2 + v_2^2 - 2v_1v_2\cos 150°$$

$$= 50^2 + 100^2 - 2(50)(100)\cos 150°$$

$$= 2500 + 10000 - (-8660)$$

$$v_R = \sqrt{21\,160} = 145.5\,\text{V}$$

Using the sine rule,

$$\frac{100}{\sin\phi} = \frac{145.5}{\sin 150°}$$

from which   $\sin\phi = \dfrac{100\sin 150°}{145.5}$

$$= 0.3436$$

and $\phi = \sin^{-1}0.3436 = 0.35$ radians, and lags $v_1$. Hence

$$v_\mathbf{R} = v_1 + v_2 = \mathbf{145.5\sin(\omega t - 0.35)\,V}$$

---

**Problem 15.** Find a sinusoidal expression for $(i_1 + i_2)$ of Problem 13, (a) by drawing phasors, (b) by calculation.

(a) The relative positions of $i_1$ and $i_2$ at time $t = 0$ are shown as phasors in Fig. 14.12(a). The phasor diagram in Fig. 14.12(b) shows the resultant $i_R$, and $i_R$ is measured as 26 A and angle $\phi$ as 19° or 0.33 rads leading $i_1$.

**Hence, by drawing, $i_R = 26\sin(\omega t + 0.33)\,A$**

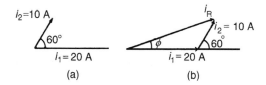

**Figure 14.12**

(b) From Fig. 14.12(b), by the cosine rule:

$$i_R^2 = 20^2 + 10^2 - 2(20)(10)(\cos 120°)$$

from which $i_R = \mathbf{26.46\,A}$

By the sine rule:

$$\frac{10}{\sin\phi} = \frac{26.46}{\sin 120°}$$

from which $\phi = 19.10°$   (i.e. 0.333 rads)

**Hence, by calculation,**

$$i_\mathbf{R} = \mathbf{26.46\sin(\omega t + 0.333)\,A}$$

An alternative method of calculation is to use **complex numbers** (see '*Engineering Mathematics*').

Then $i_1 + i_2 = 20\sin\omega t + 10\sin\left(\omega t + \dfrac{\pi}{3}\right)$

$$\equiv 20\angle 0 + 10\angle\frac{\pi}{3}\text{rad or}$$

$$20\angle 0° + 10\angle 60°$$

$$= (20 + j0) + (5 + j8.66)$$

$$= (25 + j8.66)$$

$$= 26.46\angle 19.106° \text{ or } 26.46\angle 0.333\,\text{rad}$$

$$\equiv \mathbf{26.46\sin(\omega t + 0.333)\,A}$$

---

**Problem 16.** Two alternating voltages are given by $v_1 = 120\sin\omega t$ volts and $v_2 = 200\sin(\omega t - \pi/4)$ volts. Obtain sinusoidal expressions for $v_1 - v_2$ (a) by plotting waveforms, and (b) by resolution of phasors.

---

(a) $v_1 = 120\sin\omega t$ and $v_2 = 200\sin(\omega t - \pi/4)$ are shown plotted in Fig. 14.13. Care must be taken when subtracting values of ordinates especially when at least one of the ordinates is negative. For example

at 30°, $v_1 - v_2 = 60 - (-52) = 112\,\text{V}$
at 60°, $v_1 - v_2 = 104 - 52 = 52\,\text{V}$
at 150°, $v_1 - v_2 = 60 - 193 = -133\,\text{V}$ and so on.

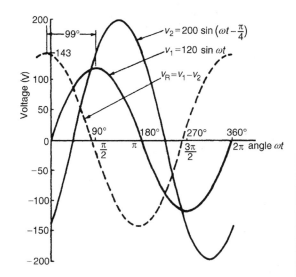

**Figure 14.13**

The resultant waveform, $v_R = v_1 - v_2$, is shown by the broken line in Fig. 14.13 The maximum value of $v_R$ is 143 V and the waveform is seen to lead $v_1$ by 99° (i.e. 1.73 radians)

**Hence, by drawing,**

$$v_R = v_1 - v_2 = 143 \sin(\omega t + 1.73) \text{ volts}$$

(b) The relative positions of $v_1$ and $v_2$ are shown at time $t = 0$ as phasors in Fig. 14.14(a). Since the resultant of $v_1 - v_2$ is required, $-v_2$ is drawn in the opposite direction to $+v_2$ and is shown by the broken line in Fig. 14.14(a). The phasor diagram with the resultant is shown in Fig. 14.14(b) where $-v_2$ is added phasorially to $v_1$.

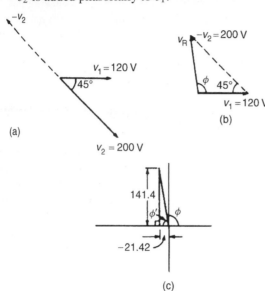

(a)

(b)

(c)

**Figure 14.14**

By resolution:

Sum of horizontal components of $v_1$ and $v_2 = 120\cos 0° - 200\cos 45° = -21.42$

Sum of vertical components of $v_1$ and $v_2 = 120\sin 0° + 200\sin 45° = 141.4$

From Fig. 14.14(c), resultant

$$v_R = \sqrt{(-21.42)^2 + (141.4)^2}$$
$$= 143.0$$

and $\tan\phi' = \dfrac{141.4}{21.42}$

$$= \tan 6.6013$$

from which, $\phi' = \tan^{-1} 6.6013$

$$= 81.39°$$

and $\phi = 98.61°$ or 1.721 radians

**Hence, by resolution of phasors,**

$$v_R = v_1 - v_2 = 143.0 \sin(\omega t + 1.721) \text{ volts}$$

**(By complex numbers:**

$$v_R = v_1 - v_2 = 120\angle 0 - 200\angle -\frac{\pi}{4}$$
$$= (120 + j0) - (141.42 - j141.42)$$
$$= -21.42 + j141.42$$
$$= 143.0\angle 98.61° \quad \text{or} \quad 143.9\angle 1.721 \text{ rad}$$

Hence, $\quad v_R = v_1 - v_2 = 143.0 \sin(\omega t + 1.721) \text{ volts})$

Now try the following exercise

**Exercise 80    Further problems on the combination of periodic functions**

1. The instantaneous values of two alternating voltages are given by $v_1 = 5\sin\omega t$ and $v_2 = 8\sin(\omega t - \pi/6)$. By plotting $v_1$ and $v_2$ on the same axes, using the same scale, over one cycle, obtain expressions for (a) $v_1 + v_2$ and (b) $v_1 - v_2$

    [(a) $v_1 + v_2 = 12.6\sin(\omega t - 0.32)$ V
    (b) $v_1 - v_2 = 4.4\sin(\omega t + 2)$ V]

2. Repeat Problem 1 using calculation
    [(a) $12.58\sin(\omega t - 0.324)$
    (b) $4.44\sin(\omega t + 2.02)$]

3. Construct a phasor diagram to represent $i_1 + i_2$ where $i_1 = 12\sin\omega t$ and $i_2 = 15\sin(\omega t + \pi/3)$. By measurement, or by calculation, find a sinusoidal expression to represent $i_1 + i_2$
    [$23.43\sin(\omega t + 0.588)$]

    Determine, either by plotting graphs and adding ordinates at intervals, or by calculation, the following periodic functions in the form $v = V_m \sin(\omega t \pm \phi)$

4. $10\sin\omega t + 4\sin(\omega t + \pi/4)$
    [$13.14\sin(\omega t + 0.217)$]

5. $80\sin(\omega t + \pi/3) + 50\sin(\omega t - \pi/6)$
    [$94.34\sin(\omega t + 0.489)$]

6. $100\sin\omega t - 70\sin(\omega t - \pi/3)$
    [$88.88\sin(\omega t + 0.751)$]

7. The voltage drops across two components when connected in series across

an a.c. supply are $v_1 = 150\sin 314.2t$ and $v_2 = 90\sin(314.2t - \pi/5)$ volts respectively. Determine (a) the voltage of the supply, in trigonometric form, (b) the r.m.s. value of the supply voltage, and (c) the frequency of the supply.

$$[(a) \ 229\sin(314.2t - 0.233)\,V$$
$$(b) \ 161.9\,V \ (c) \ 50\,Hz]$$

8. If the supply to a circuit is $25\sin 628.3t$ volts and the voltage drop across one of the components is $18\sin(628.3t - 0.52)$ volts, calculate (a) the voltage drop across the remainder of the circuit, (b) the supply frequency, and (c) the periodic time of the supply.

$$[(a) \ 12.96\sin(628.3t + 0.762)\,V$$
$$(b) \ 100\,Hz \ (c) \ 10\,ms]$$

9. The voltages across three components in a series circuit when connected across an a.c. supply are:

$$v_1 = 30\sin\left(300\pi t - \frac{\pi}{6}\right) \text{ volts},$$

$$v_2 = 40\sin\left(300\pi t + \frac{\pi}{4}\right) \text{ volts and}$$

$$v_3 = 50\sin\left(300\pi t + \frac{\pi}{3}\right) \text{ volts}.$$

Calculate (a) the supply voltage, in sinusoidal form, (b) the frequency of the supply, (c) the periodic time, and (d) the r.m.s. value of the supply.

$$[(a) \ 97.39\sin(300\pi t + 0.620)\,V$$
$$(b) \ 150\,Hz \ (c) \ 6.67\,ms \ (d) \ 68.85\,V]$$

## 14.8 Rectification

The process of obtaining unidirectional currents and voltages from alternating currents and voltages is called rectification. Automatic switching in circuits is achieved using diodes (see Chapter 11).

### Half-wave rectification

Using a single diode, D, as shown in Fig. 14.15, **half-wave rectification** is obtained. When P is sufficiently positive with respect to Q, diode D is switched on and current i flows. When P is negative with respect to Q, diode D is switched off. Transformer T isolates the equipment from direct connection with the mains supply and enables the mains voltage to be changed.

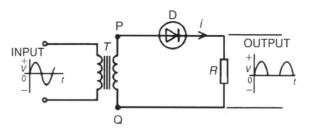

Figure 14.15

Thus, an alternating, sinusoidal waveform applied to the transformer primary is rectified into a unidirectional waveform. Unfortunately, the output waveform shown in Fig. 14.15 is not constant (i.e. steady), and as such, would be unsuitable as a d.c. power supply for electronic equipment. It would, however, be satisfactory as a battery charger. In section 14.8, methods of smoothing the output waveform are discussed.

### Full-wave rectification using a centre-tapped transformer

Two diodes may be used as shown in Fig. 14.16 to obtain **full-wave rectification** where a centre-tapped transformer T is used. When P is sufficiently positive with respect to Q, diode $D_1$ conducts and current flows (shown by the broken line in Fig. 14.16). When S is positive with respect to Q, diode $D_2$ conducts and current flows (shown by the continuous line in Fig. 14.16).

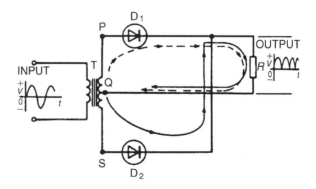

Figure 14.16

The current flowing in the load $R$ is in the same direction for both half-cycles of the input. The output waveform is thus as shown in Fig. 14.16. The output is unidirectional, but is not constant; however, it is better than the output waveform produced with a half-wave rectifier. Section 14.8 explains how the waveform may be improved so as to be of more use.

A **disadvantage** of this type of rectifier is that centre-tapped transformers are expensive.

## Full-wave bridge rectification

Four diodes may be used in a **bridge rectifier** circuit, as shown in Fig. 14.17 to obtain **full-wave rectification**. (Note, the term 'bridge' means a network of four elements connected to form a square, the input being applied to two opposite corners and the output being taken from the remaining two corners.) As for the rectifier shown in Fig. 14.16, the current flowing in load $R$ is in the same direction for both half-cycles of the input giving the output waveform shown.

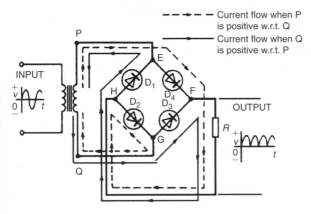

Figure 14.17

*Following the broken line in Fig. 14.17:*
When P is positive with respect to Q, current flows from the transformer to point E, through diode $D_4$ to point F, then through load $R$ to point H, through $D_2$ to point G, and back to the transformer.

*Following the full line in Fig. 14.17:*
When Q is positive with respect to P, current flows from the transformer to point G, through diode $D_3$ to point F, then through load R to point H, through $D_1$ to point E, and back to the transformer. The output waveform is not steady and needs improving; a method of smoothing is explained in the next section.

## 14.9 Smoothing of the rectified output waveform

The pulsating outputs obtained from the half- and full-wave rectifier circuits are not suitable for the operation of equipment that requires a steady d.c. output, such as would be obtained from batteries. For example, for applications such as audio equipment, a supply with a large variation is unacceptable since it produces 'hum' in the output. **Smoothing** is the process of removing the worst of the output waveform variations.

To smooth out the pulsations a large capacitor, C, is connected across the output of the rectifier, as shown in Fig. 14.18; the effect of this is to maintain the output voltage at a level which is very near to the peak of the output waveform. The improved waveforms for half-wave and full-wave rectifiers are shown in more detail in Fig. 14.19.

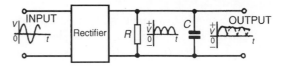

**Figure 14.18**

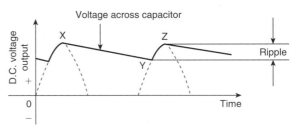

(a) Half-wave rectifier

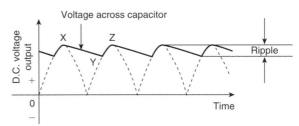

(b) Full-wave rectifier

**Figure 14.19**

During each pulse of output voltage, the capacitor C charges to the same potential as the peak of the waveform, as shown as point X in Fig. 14.19. As the waveform dies away, the capacitor discharges across the load, as shown by XY. The output voltage is then restored to the peak value the next time the rectifier conducts, as shown by YZ. This process continues as shown in Fig. 14.19.

Capacitor C is called a **reservoir capacitor** since it stores and releases charge between the peaks of the rectified waveform.

The variation in potential between points X and Y is called **ripple**, as shown in Fig. 14.19; the object is to reduce ripple to a minimum. Ripple may be reduced even further by the addition of inductance and another

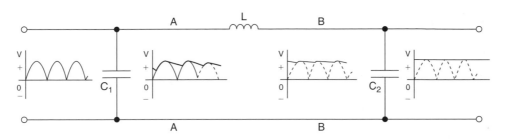

**Figure 14.20**

capacitor in a '**filter**' circuit arrangement, as shown in Fig. 14.20.

The output voltage from the rectifier is applied to capacitor $C_1$ and the voltage across points AA is shown in Fig. 14.20, similar to the waveforms of Fig. 14.19. The load current flows through the inductance L; when current is changing, e.m.f.'s are induced, as explained in Chapter 9. By Lenz's law, the induced voltages will oppose those causing the current changes.

As the ripple voltage increases and the load current increases, the induced e.m.f. in the inductor will oppose the increase. As the ripple voltage falls and the load current falls, the induced e.m.f. will try to maintain the current flow.

The voltage across points BB in Fig. 14.20 and the current in the inductance are almost ripple-free. A further capacitor, $C_2$, completes the process.

*For a practical laboratory experiment on the use of the CRO with a bridge rectifier circuit, see Chapter 24, page 413.*

---

**Now try the following exercises**

**Exercise 81  Short answer questions on alternating voltages and currents**

1. Briefly explain the principle of operation of the simple alternator

2. What is meant by (a) waveform (b) cycle

3. What is the difference between an alternating and a unidirectional waveform?

4. The time to complete one cycle of a waveform is called the ......

5. What is frequency? Name its unit

6. The mains supply voltage has a special shape of waveform called a ......

7. Define peak value

8. What is meant by the r.m.s. value?

9. The domestic mains electricity voltage in Great Britain is ......

10. What is the mean value of a sinusoidal alternating e.m.f. which has a maximum value of 100 V?

11. The effective value of a sinusoidal waveform is ...... × maximum value

12. What is a phasor quantity?

13. Complete the statement:
Form factor = ...... ÷ ......, and for a sine wave, form factor = ......

14. Complete the statement:
Peak factor = ...... ÷ ......, and for a sine wave, peak factor = ......

15. A sinusoidal current is given by $i = I_m \sin(\omega t \pm \alpha)$. What do the symbols $I_m$, $\omega$ and $\alpha$ represent?

16. How is switching obtained when converting a.c. to d.c.?

17. Draw an appropriate circuit diagram suitable for half-wave rectifications and explain its operation

18. Explain, with a diagram, how full-wave rectification is obtained using a centre-tapped transformer

19. Explain, with a diagram, how full-wave rectification is obtained using a bridge rectifier circuit

20. Explain a simple method of smoothing the output of a rectifier

Section 2

**Exercise 82   Multi-choice questions on alternating voltages and currents**
**(Answers on page 420)**

1.  The value of an alternating current at any given instant is:
    (a) a maximum value
    (b) a peak value
    (c) an instantaneous value
    (d) an r.m.s. value

2.  An alternating current completes 100 cycles in 0.1 s. Its frequency is:
    (a)  20 Hz          (b)  100 Hz
    (c)  0.002 Hz       (d)  1 kHz

3.  In Fig. 14.21, at the instant shown, the generated e.m.f. will be:
    (a)  zero
    (b)  an r.m.s. value
    (c)  an average value
    (d)  a maximum value

**Figure 14.21**

4.  The supply of electrical energy for a consumer is usually by a.c. because:
    (a) transmission and distribution are more easily effected
    (b) it is most suitable for variable speed motors
    (c) the volt drop in cables is minimal
    (d) cable power losses are negligible

5.  Which of the following statements is false?
    (a) It is cheaper to use a.c. than d.c.
    (b) Distribution of a.c. is more convenient than with d.c. since voltages may be readily altered using transformers
    (c) An alternator is an a.c. generator
    (d) A rectifier changes d.c. to a.c.

6.  An alternating voltage of maximum value 100 V is applied to a lamp. Which of the following direct voltages, if applied to the lamp, would cause the lamp to light with the same brilliance?
    (a)  100 V          (b)  63.7 V
    (c)  70.7 V         (d)  141.4 V

7.  The value normally stated when referring to alternating currents and voltages is the:
    (a) instantaneous value
    (b) r.m.s. value
    (c) average value
    (d) peak value

8.  State which of the following is false. For a sine wave:
    (a) the peak factor is 1.414
    (b) the r.m.s. value is 0.707 × peak value
    (c) the average value is 0.637 × r.m.s. value
    (d) the form factor is 1.11

9.  An a.c. supply is 70.7 V, 50 Hz. Which of the following statements is false?
    (a) The periodic time is 20 ms
    (b) The peak value of the voltage is 70.7 V
    (c) The r.m.s. value of the voltage is 70.7 V
    (d) The peak value of the voltage is 100 V

10. An alternating voltage is given by $v = 100\sin(50\pi t - 0.30)$ V.
    Which of the following statements is true?
    (a) The r.m.s. voltage is 100 V
    (b) The periodic time is 20 ms
    (c) The frequency is 25 Hz
    (d) The voltage is leading $v = 100\sin 50\pi t$ by 0.30 radians

11. The number of complete cycles of an alternating current occurring in one second is known as:
    (a) the maximum value of the alternating current
    (b) the frequency of the alternating current
    (c) the peak value of the alternating current
    (d) the r.m.s. or effective value

12. A rectifier conducts:
    (a) direct currents in one direction
    (b) alternating currents in one direction
    (c) direct currents in both directions
    (d) alternating currents in both directions

This revision test covers the material contained in Chapter 13 to 14. *The marks for each question are shown in brackets at the end of each question.*

1. Find the current flowing in the $5\,\Omega$ resistor of the circuit shown in Fig. RT4.1 using (a) Kirchhoff's laws, (b) the superposition theorem, (c) Thévenin's theorem, (d) Norton's theorem.
   Demonstrate that the same answer results from each method.

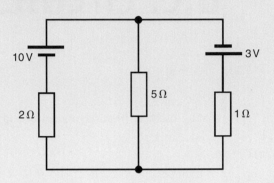

   **Figure RT4.1**

   Find also the current flowing in each of the other two branches of the circuit. (27)

2. A d.c. voltage source has an internal resistance of $2\,\Omega$ and an open-circuit voltage of 24 V. State the value of load resistance that gives maximum power dissipation and determine the value of this power. (5)

3. A sinusoidal voltage has a mean value of 3.0 A. Determine it's maximum and r.m.s. values. (4)

4. The instantaneous value of current in an a.c. circuit at any time $t$ seconds is given by: $i = 50\sin(100\pi t - 0.45)$ mA. Determine
   (a) the peak to peak current, the frequency, the periodic time and the phase angle (in degrees)
   (b) the current when $t = 0$
   (c) the current when $t = 8$ ms
   (d) the first time when the voltage is a maximum.
   Sketch the current for one cycle showing relevant points. (14)

Section 2

# Chapter 15

# Single-phase series a.c. circuits

At the end of this chapter you should be able to:

- draw phasor diagrams and current and voltage waveforms for (a) purely resistive (b) purely inductive and (c) purely capacitive a.c. circuits
- perform calculations involving $X_L = 2\pi f L$ and $X_C = 1/(2\pi f C)$
- draw circuit diagrams, phasor diagrams and voltage and impedance triangles for $R$–$L$, $R$–$C$ and $R$–$L$–$C$ series a.c. circuits and perform calculations using Pythagoras' theorem, trigonometric ratios and $Z = V/I$
- understand resonance
- derive the formula for resonant frequency and use it in calculations
- understand Q-factor and perform calculations using

$$\frac{V_L(\text{or } V_C)}{V} \quad \text{or} \quad \frac{\omega_r L}{R} \quad \text{or} \quad \frac{1}{\omega_r C R} \quad \text{or} \quad \frac{1}{R}\sqrt{\frac{L}{C}}$$

- understand bandwidth and half-power points
- perform calculations involving $(f_2 - f_1) = f_r / Q$
- understand selectivity and typical values of Q-factor
- appreciate that power $P$ in an a.c. circuit is given by $P = VI\cos\phi$ or $I_R^2 R$ and perform calculations using these formulae
- understand true, apparent and reactive power and power factor and perform calculations involving these quantities

## 15.1 Purely resistive a.c. circuit

In a purely resistive a.c. circuit, the current $I_R$ and applied voltage $V_R$ are in phase. See Fig. 15.1

## 15.2 Purely inductive a.c. circuit

In a purely inductive a.c. circuit, the current $I_L$ **lags** the applied voltage $V_L$ by 90° (i.e. $\pi/2$ rads). See Fig. 15.2

In a purely inductive circuit the opposition to the flow of alternating current is called the **inductive reactance**, $X_L$

$$X_L = \frac{V_L}{I_L} = 2\pi f L \ \Omega$$

where $f$ is the supply frequency, in hertz, and $L$ is the inductance, in henry's. $X_L$ is proportional to $f$ as shown in Fig. 15.3

DOI: 10.1016/B978-0-08-089056-2.00015-2

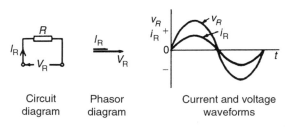

Circuit diagram　　Phasor diagram　　Current and voltage waveforms

Figure 15.1

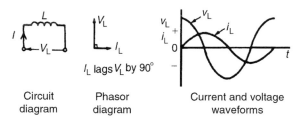

$I_L$ lags $V_L$ by 90°

Circuit diagram　　Phasor diagram　　Current and voltage waveforms

Figure 15.2

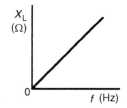

Figure 15.3

**Problem 1.** (a) Calculate the reactance of a coil of inductance 0.32 H when it is connected to a 50 Hz supply. (b) A coil has a reactance of 124 $\Omega$ in a circuit with a supply of frequency 5 kHz. Determine the inductance of the coil.

(a) Inductive reactance,

$$X_L = 2\pi fL = 2\pi(50)(0.32) = \mathbf{100.5\,\Omega}$$

(b) Since $X_L = 2\pi fL$, inductance

$$L = \frac{X_L}{2\pi f} = \frac{124}{2\pi(5000)}\,H = \mathbf{3.95\,mH}$$

**Problem 2.** A coil has an inductance of 40 mH and negligible resistance. Calculate its inductive reactance and the resulting current if connected to (a) a 240 V, 50 Hz supply, and (b) a 100 V, 1 kHz supply.

(a) Inductive reactance,

$$X_L = 2\pi fL$$
$$= 2\pi(50)(40 \times 10^{-3}) = \mathbf{12.57\,\Omega}$$

$$\text{Current, } I = \frac{V}{X_L} = \frac{240}{12.57} = \mathbf{19.09\,A}$$

(b) Inductive reactance,

$$X_L = 2\pi(1000)(40 \times 10^{-3}) = \mathbf{251.3\,\Omega}$$

$$\text{Current, } I = \frac{V}{X_L} = \frac{100}{251.3} = \mathbf{0.398\,A}$$

## 15.3　Purely capacitive a.c. circuit

In a purely capacitive a.c. circuit, the current $I_C$ **leads** the applied voltage $V_C$ by 90° (i.e. $\pi/2$ rads). See Fig. 15.4

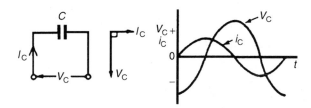

Figure 15.4

In a purely capacitive circuit the opposition to the flow of alternating current is called the **capacitive reactance**, $X_C$

$$X_C = \frac{V_C}{I_C} = \frac{1}{2\pi fC}\,\Omega$$

where $C$ is the capacitance in farads.
$X_C$ varies with frequency $f$ as shown in Fig. 15.5

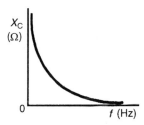

Figure 15.5

**Problem 3.** Determine the capacitive reactance of a capacitor of $10\,\mu\text{F}$ when connected to a circuit of frequency (a) $50\,\text{Hz}$ (b) $20\,\text{kHz}$

(a) Capacitive reactance

$$X_C = \frac{1}{2\pi fC}$$

$$= \frac{1}{2\pi(50)(10 \times 10^{-6})}$$

$$= \frac{10^6}{2\pi(50)(10)} = \mathbf{318.3\,\Omega}$$

(b) $X_C = \dfrac{1}{2\pi fC}$

$$= \frac{1}{2\pi(20 \times 10^3)(10 \times 10^{-6})}$$

$$= \frac{10^6}{2\pi(20 \times 10^3)(10)}$$

$$= \mathbf{0.796\,\Omega}$$

Hence as the frequency is increased from $50\,\text{Hz}$ to $20\,\text{kHz}$, $X_C$ decreases from $318.3\,\Omega$ to $0.796\,\Omega$ (see Fig. 15.5)

**Problem 4.** A capacitor has a reactance of $40\,\Omega$ when operated on a $50\,\text{Hz}$ supply. Determine the value of its capacitance.

Since

$$X_C = \frac{1}{2\pi fC}$$

capacitance

$$C = \frac{1}{2\pi fX_C}$$

$$= \frac{1}{2\pi(50)(40)}\,\text{F}$$

$$= \frac{10^6}{2\pi(50)(40)}\,\mu\text{F}$$

$$= \mathbf{79.58\,\mu F}$$

**Problem 5.** Calculate the current taken by a $23\,\mu\text{F}$ capacitor when connected to a $240\,\text{V}$, $50\,\text{Hz}$ supply.

$$\text{Current}\quad I = \frac{V}{X_C}$$

$$= \frac{V}{\left(\dfrac{1}{2\pi fC}\right)}$$

$$= 2\pi fCV$$

$$= 2\pi(50)(23 \times 10^{-6})(240)$$

$$= \mathbf{1.73\,A}$$

## CIVIL

The relationship between voltage and current for the inductive and capacitive circuits can be summarised using the word 'CIVIL', which represents the following: **In a capacitor $(C)$ the current $(I)$ is ahead of the voltage $(V)$, and the voltage $(V)$ is ahead of the current $(I)$ for the inductor $(L)$.**

**Now try the following exercise**

**Exercise 83 Further problems on purely inductive and capacitive a.c. circuits**

1. Calculate the reactance of a coil of inductance $0.2\,\text{H}$ when it is connected to (a) a $50\,\text{Hz}$, (b) a $600\,\text{Hz}$, and (c) a $40\,\text{kHz}$ supply.
   [(a) $62.83\,\Omega$ (b) $754\,\Omega$ (c) $50.27\,\text{k}\Omega$]

2. A coil has a reactance of $120\,\Omega$ in a circuit with a supply frequency of $4\,\text{kHz}$. Calculate the inductance of the coil. [$4.77\,\text{mH}$]

3. A supply of $240\,\text{V}$, $50\,\text{Hz}$ is connected across a pure inductance and the resulting current is $1.2\,\text{A}$. Calculate the inductance of the coil.
   [$0.637\,\text{H}$]

4. An e.m.f. of $200\,\text{V}$ at a frequency of $2\,\text{kHz}$ is applied to a coil of pure inductance $50\,\text{mH}$. Determine (a) the reactance of the coil, and (b) the current flowing in the coil.
   [(a) $628\,\Omega$ (b) $0.318\,\text{A}$]

5. A $120\,\text{mH}$ inductor has a $50\,\text{mA}$, $1\,\text{kHz}$ alternating current flowing through it. Find the p.d. across the inductor. [$37.7\,\text{V}$]

6. Calculate the capacitive reactance of a capacitor of $20\,\mu\text{F}$ when connected to an a.c. circuit of frequency (a) $20\,\text{Hz}$, (b) $500\,\text{Hz}$, (c) $4\,\text{kHz}$
   [(a) $397.9\,\Omega$ (b) $15.92\,\Omega$ (c) $1.989\,\Omega$]

7. A capacitor has a reactance of $80\,\Omega$ when connected to a $50\,\text{Hz}$ supply. Calculate the value of its capacitance. [$39.79\,\mu\text{F}$]

8. Calculate the current taken by a $10\,\mu\text{F}$ capacitor when connected to a 200 V, 100 Hz supply.
   [1.257 A]

9. A capacitor has a capacitive reactance of $400\,\Omega$ when connected to a 100 V, 25 Hz supply. Determine its capacitance and the current taken from the supply.    [$15.92\,\mu\text{F}$, 0.25 A]

10. Two similar capacitors are connected in parallel to a 200 V, 1 kHz supply. Find the value of each capacitor if the circuit current is 0.628 A.    [$0.25\,\mu\text{F}$]

## 15.4   *R–L* series a.c. circuit

In an a.c. circuit containing inductance $L$ and resistance $R$, the applied voltage $V$ is the phasor sum of $V_R$ and $V_L$ (see Fig. 15.6), and thus the current $I$ lags the applied voltage $V$ by an angle lying between $0°$ and $90°$ (depending on the values of $V_R$ and $V_L$), shown as angle $\phi$. In any a.c. series circuit the current is common to each component and is thus taken as the reference phasor.

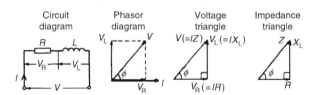

| Circuit diagram | Phasor diagram | Voltage triangle | Impedance triangle |

**Figure 15.6**

From the phasor diagram of Fig. 15.6, the **'voltage triangle'** is derived.
**For the *R–L* circuit:**

$$V = \sqrt{V_R^2 + V_L^2} \quad \text{(by Pythagoras' theorem)}$$

and

$$\tan\phi = \frac{V_L}{V_R} \quad \text{(by trigonometric ratios)}$$

In an a.c. circuit, the ratio applied voltage $V$ to current $I$ is called the **impedance**, $Z$, i.e.

$$Z = \frac{V}{I}\,\Omega$$

If each side of the voltage triangle in Fig. 15.6 is divided by current $I$ then the **'impedance triangle'** is derived.

**For the *R–L* circuit:** $Z = \sqrt{R^2 + X_L^2}$

$$\tan\phi = \frac{X_L}{R}$$

$$\sin\phi = \frac{X_L}{Z}$$

and
$$\cos\phi = \frac{R}{Z}$$

**Problem 6.**   In a series $R–L$ circuit the p.d. across the resistance $R$ is 12 V and the p.d. across the inductance $L$ is 5 V. Find the supply voltage and the phase angle between current and voltage.

From the voltage triangle of Fig. 15.6, supply voltage

$$V = \sqrt{12^2 + 5^2}$$

i.e.
$$V = \textbf{13 V}$$

(Note that in a.c. circuits, the supply voltage is **not** the arithmetic sum of the p.d.'s across components. It is, in fact, the **phasor sum**)

$$\tan\phi = \frac{V_L}{V_R} = \frac{5}{12}$$

from which, circuit phase angle

$$\phi = \tan^{-1}\left(\frac{5}{12}\right) = \textbf{22.62° lagging}$$

('Lagging' infers that the current is 'behind' the voltage, since phasors revolve anticlockwise)

**Problem 7.**   A coil has a resistance of $4\,\Omega$ and an inductance of 9.55 mH. Calculate (a) the reactance, (b) the impedance, and (c) the current taken from a 240 V, 50 Hz supply. Determine also the phase angle between the supply voltage and current.

$R = 4\,\Omega$, $L = 9.55\,\text{mH} = 9.55 \times 10^{-3}\,\text{H}$, $f = 50\,\text{Hz}$ and $V = 240\,\text{V}$

(a)   Inductive reactance,

$$X_L = 2\pi f L$$

$$= 2\pi(50)(9.55 \times 10^{-3})$$

$$= \textbf{3}\,\Omega$$

(b)   Impedance,

$$Z = \sqrt{R^2 + X_L^2} = \sqrt{4^2 + 3^2} = 5\,\Omega$$

(c)   Current,

$$I = \frac{V}{Z} = \frac{240}{5} = 48\,A$$

The circuit and phasor diagrams and the voltage and impedance triangles are as shown in Fig. 15.6

Since

$$\tan\phi = \frac{X_L}{R}$$

$$\phi = \tan^{-1}\frac{X_L}{R}$$

$$= \tan^{-1}\frac{3}{4}$$

$$= \textbf{36.87° lagging}$$

**Problem 8.**   A coil takes a current of 2 A from a 12 V d.c. supply. When connected to a 240 V, 50 Hz supply the current is 20 A. Calculate the resistance, impedance, inductive reactance and inductance of the coil.

Resistance

$$R = \frac{\text{d.c. voltage}}{\text{d.c. current}} = \frac{12}{2} = 6\,\Omega$$

Impedance

$$Z = \frac{\text{a.c. voltage}}{\text{a.c. current}} = \frac{240}{20} = 12\,\Omega$$

Since

$$Z = \sqrt{R^2 + X_L^2}$$

inductive reactance,

$$X_L = \sqrt{Z^2 - R^2} = \sqrt{12^2 - 6^2} = 10.39\,\Omega$$

Since $X_L = 2\pi fL$, inductance,

$$L = \frac{X_L}{2\pi f} = \frac{10.39}{2\pi(50)} = \textbf{33.1 mH}$$

This problem indicates a simple method for finding the inductance of a coil, i.e. firstly to measure the current when the coil is connected to a d.c. supply of known voltage, and then to repeat the process with an a.c. supply.

*For a practical laboratory experiment on the measurement of inductance of a coil, see Chapter 24, page 414.*

**Problem 9.**   A coil of inductance 318.3 mH and negligible resistance is connected in series with a 200 Ω resistor to a 240 V, 50 Hz supply. Calculate (a) the inductive reactance of the coil, (b) the impedance of the circuit, (c) the current in the circuit, (d) the p.d. across each component, and (e) the circuit phase angle.

$L = 318.3\,\text{mH} = 0.3183\,\text{H}, R = 200\,\Omega$,
$V = 240\,\text{V}$ and $f = 50\,\text{Hz}$.
The circuit diagram is as shown in Fig. 15.6

(a)   Inductive reactance

$$X_L = 2\pi fL = 2\pi(50)(0.3183) = \textbf{100}\,\Omega$$

(b)   Impedance

$$Z = \sqrt{R^2 + X_L^2}$$

$$= \sqrt{200^2 + 100^2} = \textbf{223.6}\,\Omega$$

(c)   Current

$$I = \frac{V}{Z} = \frac{240}{223.6} = \textbf{1.073 A}$$

(d)   The p.d. across the coil,

$$V_L = IX_L = 1.073 \times 100 = \textbf{107.3 V}$$

The p.d. across the resistor,

$$V_R = IR = 1.073 \times 200 = \textbf{214.6 V}$$

[Check: $\sqrt{V_R^2 + V_L^2} = \sqrt{214.6^2 + 107.3^2} = 240\,V$, the supply voltage]

(e)   From the impedance triangle, angle

$$\phi = \tan^{-1}\frac{X_L}{R} = \tan^{-1}\left(\frac{100}{200}\right)$$

**Hence the phase angle $\phi = $ 26.57° lagging.**

**Problem 10.**   A coil consists of a resistance of 100 Ω and an inductance of 200 mH. If an alternating voltage, $v$, given by $v = 200\sin 500t$ volts is applied across the coil, calculate (a) the circuit impedance, (b) the current flowing, (c) the p.d. across the resistance, (d) the p.d. across the inductance, and (e) the phase angle between voltage and current.

Since $v = 200 \sin 500t$ volts then $V_m = 200\,\text{V}$ and $\omega = 2\pi f = 500\,\text{rad/s}$

Hence r.m.s. voltage

$$V = 0.707 \times 200 = 141.4\,\text{V}$$

Inductive reactance,

$$X_L = 2\pi fL$$
$$= \omega L = 500 \times 200 \times 10^{-3} = 100\,\Omega$$

(a)  Impedance

$$Z = \sqrt{R^2 + X_L^2}$$
$$= \sqrt{100^2 + 100^2} = \mathbf{141.4\,\Omega}$$

(b)  Current

$$I = \frac{V}{Z} = \frac{141.4}{141.4} = \mathbf{1\,A}$$

(c)  P.d. across the resistance

$$V_R = IR = 1 \times 100 = \mathbf{100\,V}$$

P.d. across the inductance

$$V_L = IX_L = 1 \times 100 = \mathbf{100\,V}$$

(d)  Phase angle between voltage and current is given by:

$$\tan\phi = \frac{X_L}{R}$$

from which,

$$\phi = \tan^{-1}\left(\frac{100}{100}\right)$$

hence        $\boldsymbol{\phi = 45°}$ or $\dfrac{\pi}{4}$ **rads**

**Problem 11.**   A pure inductance of 1.273 mH is connected in series with a pure resistance of $30\,\Omega$. If the frequency of the sinusoidal supply is 5 kHz and the p.d. across the $30\,\Omega$ resistor is 6 V, determine the value of the supply voltage and the voltage across the 1.273 mH inductance. Draw the phasor diagram.

The circuit is shown in Fig. 15.7(a)

Supply voltage, $V = IZ$

Current $I = \dfrac{V_R}{R} = \dfrac{6}{30} = 0.20\,\text{A}$

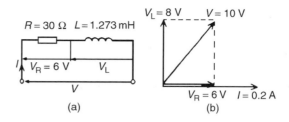

**Figure 15.7**

Inductive reactance

$$X_L = 2\pi fL$$
$$= 2\pi (5 \times 10^3)(1.273 \times 10^{-3})$$
$$= 40\,\Omega$$

Impedance,

$$Z = \sqrt{R^2 + X_L^2} = \sqrt{30^2 + 40^2} = 50\,\Omega$$

Supply voltage

$$V = IZ = (0.20)(50) = \mathbf{10\,V}$$

Voltage across the 1.273 mH inductance,

$$V_L = IX_L = (0.2)(40) = \mathbf{8\,V}$$

The phasor diagram is shown in Fig. 15.7(b)
  (Note that in a.c. circuits, the supply voltage is **not** the arithmetic sum of the p.d.'s across components but the **phasor sum**)

**Problem 12.**   A coil of inductance 159.2 mH and resistance $20\,\Omega$ is connected in series with a $60\,\Omega$ resistor to a 240 V, 50 Hz supply. Determine (a) the impedance of the circuit, (b) the current in the circuit, (c) the circuit phase angle, (d) the p.d. across the $60\,\Omega$ resistor, and (e) the p.d. across the coil. (f) Draw the circuit phasor diagram showing all voltages.

The circuit diagram is shown in Fig. 15.8(a). When impedances are connected in series the individual resistances may be added to give the total circuit resistance. The equivalent circuit is thus shown in Fig. 15.8(b).
Inductive reactance $X_L = 2\pi fL$

$$= 2\pi (50)(159.2 \times 10^{-3}) = 50\,\Omega$$

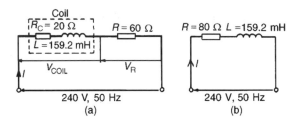

**Figure 15.8**

(a)   Circuit impedance,

$$Z = \sqrt{R^2 + X_L^2} = \sqrt{80^2 + 50^2} = 94.34\,\Omega$$

(b)   Circuit current, $I = \dfrac{V}{Z} = \dfrac{240}{94.34} = 2.544\,\text{A}$.

(c)   Circuit phase angle $\phi = \tan^{-1} X_L/R$
$$= \tan^{-1}(50/80)$$
$$= 32° \text{ lagging}$$

From Fig. 15.8(a):

(d)   $V_R = IR = (2.544)(60) = 152.6\,\text{V}$

(e)   $V_{\text{COIL}} = IZ_{\text{COIL}}$, where

$$Z_{\text{COIL}} = \sqrt{R_C^2 + X_L^2} = \sqrt{20^2 + 50^2} = 53.85\,\Omega.$$

Hence $V_{\text{COIL}} = (2.544)(53.85) = 137.0\,\text{V}$

(f)   For the phasor diagram, shown in Fig. 15.9,
$V_L = IX_L = (2.544)(50) = 127.2\,\text{V}$.
$V_{R\text{COIL}} = IR_C = (2.544)(20) = 50.88\,\text{V}$

The 240 V supply voltage is the phasor sum of $V_{\text{COIL}}$ and $V_R$ as shown in the phasor diagram in Fig. 15.9

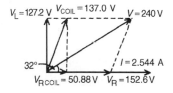

**Figure 15.9**

**Now try the following exercise**

**Exercise 84   Further problems on R–L a.c. series circuits**

1.   Determine the impedance of a coil which has a resistance of $12\,\Omega$ and a reactance of $16\,\Omega$.
[$20\,\Omega$]

2.   A coil of inductance 80 mH and resistance $60\,\Omega$ is connected to a 200 V, 100 Hz supply. Calculate the circuit impedance and the current taken from the supply. Find also the phase angle between the current and the supply voltage.
[$78.27\,\Omega$, 2.555 A, 39.95° lagging]

3.   An alternating voltage given by $v = 100 \sin 240t$ volts is applied across a coil of resistance $32\,\Omega$ and inductance 100 mH. Determine (a) the circuit impedance, (b) the current flowing, (c) the p.d. across the resistance, and (d) the p.d. across the inductance.
[(a) $40\,\Omega$ (b) 1.77 A
(c) 56.64 V (d) 42.48 V]

4.   A coil takes a current of 5 A from a 20 V d.c. supply. When connected to a 200 V, 50 Hz a.c. supply the current is 25 A. Calculate the (a) resistance, (b) impedance, and (c) inductance of the coil.
[(a) $4\,\Omega$ (b) $8\,\Omega$ (c) 22.05 mH]

5.   A resistor and an inductor of negligible resistance are connected in series to an a.c. supply. The p.d. across the resistor is 18 V and the p.d. across the inductor is 24 V. Calculate the supply voltage and the phase angle between voltage and current.   [30 V, 53.13° lagging]

6.   A coil of inductance 636.6 mH and negligible resistance is connected in series with a $100\,\Omega$ resistor to a 250 V, 50 Hz supply. Calculate (a) the inductive reactance of the coil, (b) the impedance of the circuit, (c) the current in the circuit, (d) the p.d. across each component, and (e) the circuit phase angle.
[(a) $200\,\Omega$ (b) $223.6\,\Omega$ (c) 1.118 A
(d) 223.6 V, 111.8 V (e) 63.43° lagging]

## 15.5   R–C series a.c. circuit

In an a.c. series circuit containing capacitance $C$ and resistance $R$, the applied voltage $V$ is the phasor sum of $V_R$ and $V_C$ (see Fig. 15.10) and thus the current $I$ leads the applied voltage $V$ by an angle lying between 0° and 90° (depending on the values of $V_R$ and $V_C$), shown as angle $\alpha$.

From the phasor diagram of Fig. 15.10, the '**voltage triangle**' is derived.

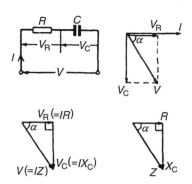

**Figure 15.10**

**For the $R-C$ circuit:**

$$V = \sqrt{V_R^2 + V_C^2} \quad \text{(by Pythagoras' theorem)}$$

and

$$\tan\alpha = \frac{V_C}{V_R} \quad \text{(by trigonometric ratios)}$$

As stated in Section 15.4, in an a.c. circuit, the ratio applied voltage $V$ to current $I$ is called the **impedance** $Z$, i.e. $Z = V/I\,\Omega$

If each side of the voltage triangle in Fig. 15.10 is divided by current $I$ then the **'impedance triangle'** is derived.

**For the $R-C$ circuit: $Z = \sqrt{R^2 + X_C^2}$**

$$\tan\alpha = \frac{X_C}{R} \quad \sin\alpha = \frac{X_C}{Z} \quad \text{and} \quad \cos\alpha = \frac{R}{Z}$$

**Problem 13.** A resistor of $25\,\Omega$ is connected in series with a capacitor of $45\,\mu F$. Calculate (a) the impedance, and (b) the current taken from a 240 V, 50 Hz supply. Find also the phase angle between the supply voltage and the current.

$R = 25\,\Omega$, $C = 45\,\mu F = 45 \times 10^{-6}\,F$, $V = 240\,V$ and $f = 50\,Hz$. The circuit diagram is as shown in Fig. 15.10

Capacitive reactance,

$$X_C = \frac{1}{2\pi fC}$$

$$= \frac{1}{2\pi(50)(45 \times 10^{-6})} = 70.74\,\Omega$$

(a) Impedance $Z = \sqrt{R^2 + X_C^2} = \sqrt{25^2 + 70.74^2}$

$$= \mathbf{75.03\,\Omega}$$

(b) Current $I = V/Z = 240/75.03 = \mathbf{3.20\,A}$

Phase angle between the supply voltage and current, $\alpha = \tan^{-1}(X_C/R)$ hence

$$\alpha = \tan^{-1}\left(\frac{70.74}{25}\right) = \mathbf{70.54°\ leading}$$

('Leading' infers that the current is 'ahead' of the voltage, since phasors revolve anticlockwise)

**Problem 14.** A capacitor $C$ is connected in series with a $40\,\Omega$ resistor across a supply of frequency 60 Hz. A current of 3 A flows and the circuit impedance is $50\,\Omega$. Calculate (a) the value of capacitance, $C$, (b) the supply voltage, (c) the phase angle between the supply voltage and current, (d) the p.d. across the resistor, and (e) the p.d. across the capacitor. Draw the phasor diagram.

(a) Impedance $Z = \sqrt{R^2 + X_C^2}$

Hence $X_C = \sqrt{Z^2 - R^2} = \sqrt{50^2 - 40^2} = 30\,\Omega$

$X_C = \dfrac{1}{2\pi fC}$ hence,

$$C = \frac{1}{2\pi fX_C} = \frac{1}{2\pi(60)(30)} \, F = \mathbf{88.42\,\mu F}$$

(b) Since $Z = V/I$ then $V = IZ = (3)(50) = \mathbf{150\,V}$

(c) Phase angle, $\alpha = \tan^{-1}X_C/R = \tan^{-1}(30/40) = \mathbf{36.87°\ leading}$.

(d) P.d. across resistor, $V_R = IR = (3)(40) = \mathbf{120\,V}$

(e) P.d. across capacitor, $V_C = IX_C = (3)(30) = \mathbf{90\,V}$

The phasor diagram is shown in Fig. 15.11, where the supply voltage $V$ is the phasor sum of $V_R$ and $V_C$.

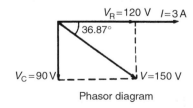

Phasor diagram

**Figure 15.11**

**Now try the following exercise**

**Exercise 85   Further problems on *R–C* a.c. circuits**

1. A voltage of 35 V is applied across a *C–R* series circuit. If the voltage across the resistor is 21 V, find the voltage across the capacitor.
   [28 V]

2. A resistance of 50 Ω is connected in series with a capacitance of 20 μF. If a supply of 200 V, 100 Hz is connected across the arrangement find (a) the circuit impedance, (b) the current flowing, and (c) the phase angle between voltage and current.
   [(a) 93.98 Ω (b) 2.128 A (c) 57.86° leading]

3. A 24.87 μF capacitor and a 30 Ω resistor are connected in series across a 150 V supply. If the current flowing is 3 A find (a) the frequency of the supply, (b) the p.d. across the resistor and (c) the p.d. across the capacitor.
   [(a) 160 Hz (b) 90 V (c) 120 V]

4. An alternating voltage $v = 250 \sin 800t$ volts is applied across a series circuit containing a 30 Ω resistor and 50 μF capacitor. Calculate

(a) the circuit impedance, (b) the current flowing, (c) the p.d. across the resistor, (d) the p.d. across the capacitor, and (e) the phase angle between voltage and current.
   [(a) 39.05 Ω (b) 4.526 A (c) 135.8 V (d) 113.2 V (e) 39.81° leading]

5. A 400 Ω resistor is connected in series with a 2358 pF capacitor across a 12 V a.c. supply. Determine the supply frequency if the current flowing in the circuit is 24 mA   [225 kHz]

## 15.6   *R–L–C* series a.c. circuit

In an a.c. series circuit containing resistance $R$, inductance $L$ and capacitance $C$, the applied voltage $V$ is the phasor sum of $V_R$, $V_L$ and $V_C$ (see Fig. 15.12). $V_L$ and $V_C$ are anti-phase, i.e. displaced by 180°, and there are three phasor diagrams possible – each depending on the relative values of $V_L$ and $V_C$.

**When** $X_L > X_C$ (Fig. 15.12(b)):

$$Z = \sqrt{R^2 + (X_L - X_C)^2}$$

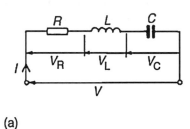

(a)

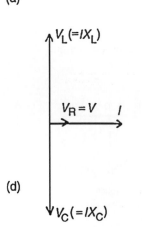

(d)

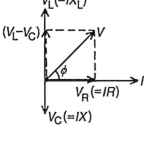

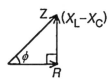

IMPEDANCE TRIANGLE

(b)

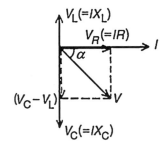

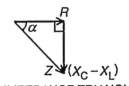

IMPEDANCE TRIANGLE

(c)

Figure 15.12

and $\tan\phi = \dfrac{X_L - X_C}{R}$

When $X_C > X_L$ (Fig. 15.12(c)):

$$Z = \sqrt{R^2 + (X_C - X_L)^2}$$

and $\tan\alpha = \dfrac{X_C - X_L}{R}$

When $X_L = X_C$ (Fig. 15.12(d)), the applied voltage $V$ and the current $I$ are in phase. This effect is called **series resonance** (see Section 15.7).

> **Problem 15.** A coil of resistance $5\,\Omega$ and inductance $120\,\text{mH}$ in series with a $100\,\mu\text{F}$ capacitor, is connected to a $300\,\text{V}$, $50\,\text{Hz}$ supply. Calculate (a) the current flowing, (b) the phase difference between the supply voltage and current, (c) the voltage across the coil, and (d) the voltage across the capacitor.

The circuit diagram is shown in Fig. 15.13

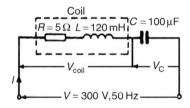

**Figure 15.13**

$$X_L = 2\pi f L = 2\pi(50)(120 \times 10^{-3}) = \mathbf{37.70\,\Omega}$$

$$X_C = \frac{1}{2\pi f C} = \frac{1}{2\pi(50)(100 \times 10^{-6})} = \mathbf{31.83\,\Omega}$$

Since $X_L$ is greater than $X_C$ the circuit is inductive.

$$X_L - X_C = 37.70 - 31.83 = 5.87\,\Omega$$

Impedance

$$Z = \sqrt{R^2 + (X_L - X_C)^2}$$
$$= \sqrt{5^2 + 5.87^2} = 7.71\,\Omega$$

(a) Current $I = \dfrac{V}{Z} = \dfrac{300}{7.71} = \mathbf{38.91\,A}$

(b) Phase angle

$$\phi = \tan^{-1}\left(\frac{X_L - X_C}{R}\right)$$
$$= \tan^{-1}\left(\frac{5.87}{5}\right) = \mathbf{49.58°}$$

(c) Impedance of coil

$$Z_{COIL} = \sqrt{R^2 + X_L^2}$$
$$= \sqrt{5^2 + 37.7^2} = 38.03\,\Omega$$

Voltage across coil

$$V_{COIL} = I Z_{COIL}$$
$$= (38.91)(38.03) = \mathbf{1480\,V}$$

Phase angle of coil

$$= \tan^{-1}\frac{X_L}{R}$$
$$= \tan^{-1}\left(\frac{37.7}{5}\right) = \mathbf{82.45°\ lagging}$$

(d) Voltage across capacitor

$$V_C = I X_C = (38.91)(31.83) = \mathbf{1239\,V}$$

The phasor diagram is shown in Fig. 15.14. The supply voltage $V$ is the phasor sum of $V_{COIL}$ and $V_C$.

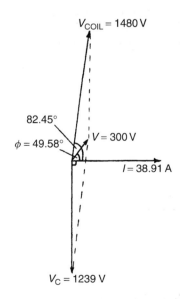

**Figure 15.14**

## Series-connected impedances

For series-connected impedances the total circuit impedance can be represented as a single $L$–$C$–$R$ circuit by combining all values of resistance together, all values of inductance together and all values of capacitance together, (remembering that for series connected capacitors

$$\frac{1}{C} = \frac{1}{C_1} + \frac{1}{C_2} + \cdots \Big)$$

For example, the circuit of Fig. 15.15(a) showing three impedances has an equivalent circuit of Fig. 15.15(b).

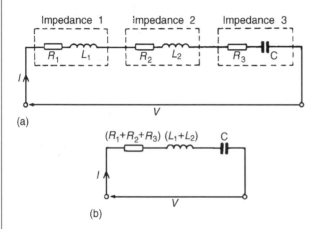

(a)

(b)

**Figure 15.15**

> **Problem 16.** The following three impedances are connected in series across a 40 V, 20 kHz supply:
> (i) a resistance of 8 Ω, (ii) a coil of inductance 130 μH and 5 Ω resistance, and (iii) a 10 Ω resistor in series with a 0.25 μF capacitor. Calculate (a) the circuit current, (b) the circuit phase angle, and (c) the voltage drop across each impedance.

The circuit diagram is shown in Fig. 15.16(a). Since the total circuit resistance is $8+5+10$, i.e. 23 Ω, an equivalent circuit diagram may be drawn as shown in Fig. 15.16(b).

Inductive reactance,

$$X_L = 2\pi fL = 2\pi(20 \times 10^3)(130 \times 10^{-6}) = 16.34\,\Omega$$

Capacitive reactance,

$$X_C = \frac{1}{2\pi fC} = \frac{1}{2\pi(20 \times 10^3)(0.25 \times 10^{-6})}$$

$$= 31.83\,\Omega$$

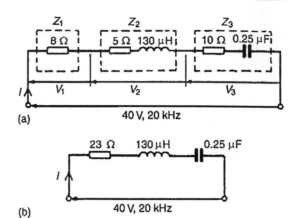

**Figure 15.16**

Since $X_C > X_L$, the circuit is capacitive (see phasor diagram in Fig. 15.12(c)).

$$X_C - X_L = 31.83 - 16.34 = 15.49\,\Omega$$

(a) Circuit impedance, $Z = \sqrt{R^2 + (X_C - X_L)^2}$

$$= \sqrt{23^2 + 15.49^2} = 27.73\,\Omega$$

Circuit current, $I = V/Z = 40/27.73 = \mathbf{1.442\,A}$

From Fig. 15.12(c), circuit phase angle

$$\phi = \tan^{-1}\left(\frac{X_C - X_L}{R}\right)$$

i.e.

$$\phi = \tan^{-1}\left(\frac{15.49}{23}\right) = \mathbf{33.96°\ leading}$$

(b) From Fig. 15.16(a),

$$V_1 = IR_1 = (1.442)(8) = \mathbf{11.54\,V}$$

$$V_2 = IZ_2 = I\sqrt{5^2 + 16.34^2}$$

$$= (1.442)(17.09) = \mathbf{24.64\,V}$$

$$V_3 = IZ_3 = I\sqrt{10^2 + 31.83^2}$$

$$= (1.442)(33.36) = \mathbf{48.11\,V}$$

The 40 V supply voltage is the phasor sum of $V_1$, $V_2$ and $V_3$

> **Problem 17.** Determine the p.d.'s $V_1$ and $V_2$ for the circuit shown in Fig. 15.17 if the frequency of the supply is 5 kHz. Draw the phasor diagram and hence determine the supply voltage $V$ and the circuit phase angle.

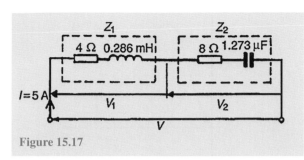

Figure 15.17

**For impedance $Z_1$:** $R_1 = 4\,\Omega$ and

$$X_L = 2\pi fL$$

$$= 2\pi(5 \times 10^3)(0.286 \times 10^{-3})$$

$$= 8.985\,\Omega$$

$$V_1 = IZ_1 = I\sqrt{R^2 + X_L^2}$$

$$= 5\sqrt{4^2 + 8.985^2} = 49.18\,\text{V}$$

Phase angle $\psi_1 = \tan^{-1}\left(\dfrac{X_L}{R}\right) = \tan^{-1}\left(\dfrac{8.985}{4}\right)$

$$= 66.0°\,\text{lagging}$$

**For impedance $Z_2$:** $R_2 = 8\,\Omega$ and

$$X_C = \frac{1}{2\pi fC} = \frac{1}{2\pi(5 \times 10^3)(1.273 \times 10^{-6})}$$

$$= 25.0\,\Omega$$

$$V_2 = IZ_2 = I\sqrt{R^2 + X_C^2} = 5\sqrt{8^2 + 25.0^2}$$

$$= 131.2\,\text{V}.$$

Phase angle $\phi_2 = \tan^{-1}\left(\dfrac{X_C}{R}\right) = \tan^{-1}\left(\dfrac{25.0}{8}\right)$

$$= 72.26°\,\text{leading}$$

The phasor diagram is shown in Fig. 15.18.

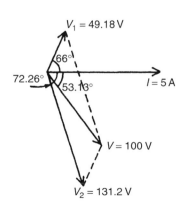

Figure 15.18

The phasor sum of $V_1$ and $V_2$ gives the supply voltage $V$ of 100 V at a phase angle of **53.13° leading**. These values may be determined by drawing or by calculation – either by resolving into horizontal and vertical components or by the cosine and sine rules.

---

**Now try the following exercise**

**Exercise 86    Further problems on $R-L-C$ a.c. circuits**

1. A 40 μF capacitor in series with a coil of resistance 8 Ω and inductance 80 mH is connected to a 200 V, 100 Hz supply. Calculate (a) the circuit impedance, (b) the current flowing, (c) the phase angle between voltage and current, (d) the voltage across the coil, and (e) the voltage across the capacitor.

   [(a) 13.18 Ω (b) 15.17 A (c) 52.63° lagging (d) 772.1 V (e) 603.6 V]

2. Find the values of resistance $R$ and inductance $L$ in the circuit of Fig. 15.19.

   [$R=131\,\Omega$, $L=0.545\,\text{H}$]

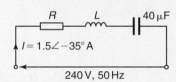

Figure 15.19

3. Three impedances are connected in series across a 100 V, 2 kHz supply. The impedances comprise:
   (i) an inductance of 0.45 mH and 2 Ω resistance,
   (ii) an inductance of 570 μH and 5 Ω resistance, and
   (iii) a capacitor of capacitance 10 μF and resistance 3 Ω.

   Assuming no mutual inductive effects between the two inductances calculate (a) the circuit impedance, (b) the circuit current, (c) the circuit phase angle, and (d) the voltage across each impedance. Draw the phasor diagram.

   [(a) 11.12 Ω (b) 8.99 A (c) 25.92° lagging (d) 53.92 V, 78.53 V, 76.46 V]

4. For the circuit shown in Fig. 15.20 determine the voltages $V_1$ and $V_2$ if the supply frequency

is 1 kHz. Draw the phasor diagram and hence determine the supply voltage $V$ and the circuit phase angle.

$$[V_1 = 26.0\,\text{V}, V_2 = 67.05\,\text{V},$$
$$V = 50\,\text{V}, 53.14° \text{ leading}]$$

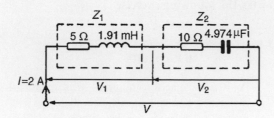

Figure 15.20

## 15.7 Series resonance

As stated in Section 15.6, for an $R-L-C$ series circuit, when $X_L = X_C$ (Fig. 15.12(d)), the applied voltage $V$ and the current $I$ are in phase. This effect is called **series resonance**. At resonance:

(i)  $V_L = V_C$

(ii)  $Z = R$ (i.e. the minimum circuit impedance possible in an $L-C-R$ circuit)

(iii)  $I = V/R$ (i.e. the maximum current possible in an $L-C-R$ circuit)

(iv)  Since $X_L = X_C$, then $2\pi f_r L = 1/2\pi f_r C$ from which,

$$f_r^2 = \frac{1}{(2\pi)^2 LC}$$

and

$$f_r = \frac{1}{2\pi \sqrt{LC}}\,\text{Hz}$$

where $f_r$ is the resonant frequency.

(v)  The series resonant circuit is often described as an **acceptor circuit** since it has its minimum impedance, and thus maximum current, at the resonant frequency.

(vi)  Typical graphs of current $I$ and impedance $Z$ against frequency are shown in Fig. 15.21

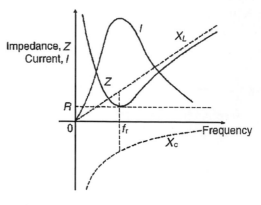

Figure 15.21

**Problem 18.** A coil having a resistance of $10\,\Omega$ and an inductance of $125\,\text{mH}$ is connected in series with a $60\,\mu\text{F}$ capacitor across a $120\,\text{V}$ supply. At what frequency does resonance occur? Find the current flowing at the resonant frequency.

Resonant frequency,

$$f_r = \frac{1}{2\pi\sqrt{LC}}\,\text{Hz} = \frac{1}{2\pi\sqrt{\left[\left(\dfrac{125}{10^3}\right)\left(\dfrac{60}{10^6}\right)\right]}}$$

$$= \frac{1}{2\pi\sqrt{\left(\dfrac{125 \times 6}{10^8}\right)}} = \frac{1}{2\pi\left(\dfrac{\sqrt{(125)(6)}}{10^4}\right)}$$

$$= \frac{10^4}{2\pi\sqrt{(125)(6)}} = \textbf{58.12\,Hz}$$

At resonance, $X_L = X_C$ and impedance $Z = R$. Hence current, $I = V/R = 120/10 = \textbf{12\,A}$

**Problem 19.** The current at resonance in a series $L-C-R$ circuit is $100\,\mu\text{A}$. If the applied voltage is $2\,\text{mV}$ at a frequency of $200\,\text{kHz}$, and the circuit inductance is $50\,\mu\text{H}$, find (a) the circuit resistance, and (b) the circuit capacitance.

(a)  $I = 100\,\mu\text{A} = 100 \times 10^{-6}\,\text{A}$ and $V = 2\,\text{mV} = 2 \times 10^{-3}\,\text{V}$.

At resonance, impedance $Z$ = resistance $R$. Hence

$$R = \frac{V}{I} = \frac{2 \times 10^{-3}}{100 \times 10^{-6}} = \frac{2 \times 10^6}{100 \times 10^3} = \textbf{20\,\Omega}$$

(b)  At resonance $X_L = X_C$ i.e.

$$2\pi fL = \frac{1}{2\pi f C}$$

Hence capacitance

$$C = \frac{1}{(2\pi f)^2 L}$$

$$= \frac{1}{(2\pi \times 200 \times 10^3)^2 (50 \times 10^{-6})} \, \text{F}$$

$$= \frac{(10^6)(10^6)}{(4\pi)^2 (10^{10})(50)} \, \mu\text{F}$$

$$= \mathbf{0.0127\, \mu F} \text{ or } \mathbf{12.7\, nF}$$

## 15.8  Q-factor

At resonance, if $R$ is small compared with $X_L$ and $X_C$, it is possible for $V_L$ and $V_C$ to have voltages many times greater than the supply voltage (see Fig. 15.12(d), page 234)

**Voltage magnification at resonance**

$$= \frac{\textbf{voltage across } L \textbf{ (or } C\textbf{)}}{\textbf{supply voltage } V}$$

This ratio is a measure of the quality of a circuit (as a resonator or tuning device) and is called the **Q-factor**. Hence

$$\text{Q-factor} = \frac{V_L}{V} = \frac{IX_L}{IR} = \frac{X_L}{R} = \frac{2\pi f_r L}{R}$$

Alternatively,

$$\text{Q-factor} = \frac{V_C}{V} = \frac{IX_C}{IR} = \frac{X_C}{R} = \frac{1}{2\pi f_r C R}$$

At resonance

$$f_r = \frac{1}{2\pi \sqrt{LC}}$$

i.e.  $$2\pi f_r = \frac{1}{\sqrt{LC}}$$

Hence

$$\text{Q-factor} = \frac{2\pi f_r L}{R} = \frac{1}{\sqrt{LC}}\left(\frac{L}{R}\right) = \frac{1}{R}\sqrt{\frac{L}{C}}$$

**Problem 20.** A coil of inductance 80 mH and negligible resistance is connected in series with a capacitance of 0.25 μF and a resistor of resistance 12.5 Ω across a 100 V, variable frequency supply. Determine (a) the resonant frequency, and (b) the current at resonance. How many times greater than the supply voltage is the voltage across the reactances at resonance?

(a)  Resonant frequency

$$f_r = \frac{1}{2\pi \sqrt{\left(\dfrac{80}{10^3}\right)\left(\dfrac{0.25}{10^6}\right)}}$$

$$= \frac{1}{2\pi \sqrt{\dfrac{(8)(0.25)}{10^8}}} = \frac{10^4}{2\pi \sqrt{2}}$$

$$= \mathbf{1125.4\, Hz} \text{ or } \mathbf{1.1254\, kHz}$$

(b)  Current at resonance $I = V/R = 100/12.5 = \mathbf{8\,A}$
Voltage across inductance, at resonance,

$$V_L = IX_L = (I)(2\pi fL)$$

$$= (8)(2\pi)(1125.4)(80 \times 10^{-3})$$

$$= 4525.5\,\text{V}$$

(Also, voltage across capacitor,

$$V_C = IX_C = \frac{I}{2\pi f C}$$

$$= \frac{8}{2\pi (1125.4)(0.25 \times 10^{-6})}$$

$$= 4525.5\,\text{V})$$

Voltage magnification at resonance $= V_L/V$ or $V_C/V = 4525.5/100 = \mathbf{45.255}$ i.e. at resonance, the voltage across the reactances are 45.255 times greater than the supply voltage. Hence **the Q-factor of the circuit is 45.255**

**Problem 21.** A series circuit comprises a coil of resistance 2 Ω and inductance 60 mH, and a 30 μF capacitor. Determine the Q-factor of the circuit at resonance.

At resonance,

$$\text{Q-factor} = \frac{1}{R}\sqrt{\frac{L}{C}} = \frac{1}{2}\sqrt{\frac{60 \times 10^{-3}}{30 \times 10^{-6}}}$$

$$= \frac{1}{2}\sqrt{\frac{60 \times 10^6}{30 \times 10^3}}$$

$$= \frac{1}{2}\sqrt{2000} = \textbf{22.36}$$

**Problem 22.** A coil of negligible resistance and inductance 100 mH is connected in series with a capacitance of 2 μF and a resistance of 10 Ω across a 50 V, variable frequency supply. Determine (a) the resonant frequency, (b) the current at resonance, (c) the voltages across the coil and the capacitor at resonance, and (d) the Q-factor of the circuit.

(a)  Resonant frequency,

$$f_r = \frac{1}{2\pi\sqrt{LC}} = \frac{1}{2\pi\sqrt{\left(\dfrac{100}{10^3}\right)\left(\dfrac{2}{10^6}\right)}}$$

$$= \frac{1}{2\pi\sqrt{\dfrac{20}{10^8}}} = \frac{1}{\dfrac{2\pi\sqrt{20}}{10^4}}$$

$$= \frac{10^4}{2\pi\sqrt{20}} = \textbf{355.9 Hz}$$

(b)  Current at resonance $I = V/R = 50/10 = \textbf{5 A}$

(c)  Voltage across coil at resonance,

$$V_L = IX_L = I(2\pi f_r L)$$

$$= (5)(2\pi \times 355.9 \times 100 \times 10^{-3}) = \textbf{1118 V}$$

Voltage across capacitance at resonance,

$$V_C = IX_C = \frac{I}{2\pi f_r C}$$

$$= \frac{5}{2\pi(355.9)(2 \times 10^{-6})} = \textbf{1118 V}$$

(d)  Q-factor (i.e. voltage magnification at resonance)

$$= \frac{V_L}{V} \text{ or } \frac{V_C}{V} = \frac{1118}{50} = \textbf{22.36}$$

Q-factor may also have been determined by

$$\frac{2\pi f_r L}{R} \quad \text{or} \quad \frac{1}{2\pi f_r CR} \quad \text{or} \quad \frac{1}{R}\sqrt{\frac{L}{C}}$$

**Now try the following exercise**

**Exercise 87   Further problems on series resonance and Q-factor**

1.  Find the resonant frequency of a series a.c. circuit consisting of a coil of resistance 10 Ω and inductance 50 mH and capacitance 0.05 μF. Find also the current flowing at resonance if the supply voltage is 100 V.

    [3.183 kHz, 10 A]

2.  The current at resonance in a series $L$–$C$–$R$ circuit is 0.2 mA. If the applied voltage is 250 mV at a frequency of 100 kHz and the circuit capacitance is 0.04 μF, find the circuit resistance and inductance.

    [1.25 kΩ, 63.3 μH]

3.  A coil of resistance 25 Ω and inductance 100 mH is connected in series with a capacitance of 0.12 μF across a 200 V, variable frequency supply. Calculate (a) the resonant frequency, (b) the current at resonance, and (c) the factor by which the voltage across the reactance is greater than the supply voltage.

    [(a) 1.453 kHz (b) 8 A (c) 36.51]

4.  A coil of 0.5 H inductance and 8 Ω resistance is connected in series with a capacitor across a 200 V, 50 Hz supply. If the current is in phase with the supply voltage, determine the capacitance of the capacitor and the p.d. across its terminals.          [20.26 μF, 3.928 kV]

5.  Calculate the inductance which must be connected in series with a 1000 pF capacitor to give a resonant frequency of 400 kHz.

    [0.158 mH]

6.  A series circuit comprises a coil of resistance 20 Ω and inductance 2 mH and a 500 pF capacitor. Determine the Q-factor of the circuit at resonance. If the supply voltage is 1.5 V, what is the voltage across the capacitor?

    [100, 150 V]

## 15.9  Bandwidth and selectivity

Figure 15.22 shows how current $I$ varies with frequency in an $R–L–C$ series circuit. At the resonant frequency $f_r$, current is a maximum value, shown as $I_r$. Also shown are the points A and B where the current is 0.707 of the maximum value at frequencies $f_1$ and $f_2$. The power delivered to the circuit is $I^2R$. At $I = 0.707\, I_r$, the power is $(0.707\, I_r)^2 R = 0.5\, I_r^2 R$, i.e. half the power that occurs at frequency $f_r$. The points corresponding to $f_1$ and $f_2$ are called the **half-power points**. The distance between these points, i.e. $(f_2 - f_1)$, is called the **bandwidth**.

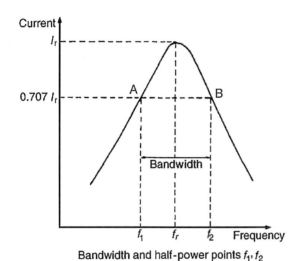

**Figure 15.22**   Bandwidth and half-power points $f_1$, $f_2$

It may be shown that

$$Q = \frac{f_r}{(f_2 - f_1)}$$

or

$$(f_2 - f_1) = \frac{f_r}{Q}$$

**Problem 23.**   A filter in the form of a series $L–R–C$ circuit is designed to operate at a resonant frequency of 5 kHz. Included within the filter is a 20 mH inductance and 10 Ω resistance. Determine the bandwidth of the filter.

Q-factor at resonance is given by:

$$Q_r = \frac{\omega_r L}{R} = \frac{(2\pi \times 5000)(20 \times 10^{-3})}{10}$$

$$= 62.83$$

Since $Q_r = f_r/(f_2 - f_1)$, **bandwidth,**

$$(f_2 - f_1) = \frac{f_r}{Q} = \frac{5000}{62.83} = \textbf{79.6 Hz}$$

**Selectivity** is the ability of a circuit to respond more readily to signals of a particular frequency to which it is tuned than to signals of other frequencies. The response becomes progressively weaker as the frequency departs from the resonant frequency. The higher the Q-factor, the narrower the bandwidth and the more selective is the circuit. Circuits having high Q-factors (say, in the order of 100 to 300) are therefore useful in communications engineering. A high Q-factor in a series power circuit has disadvantages in that it can lead to dangerously high voltages across the insulation and may result in electrical breakdown.

*For a practical laboratory experiment on series a.c. circuits and resonance, see Chapter 24, page 415.*

## 15.10  Power in a.c. circuits

In Figs 15.23(a)–(c), the value of power at any instant is given by the product of the voltage and current at that instant, i.e. the instantaneous power, $p = vi$, as shown by the broken lines.

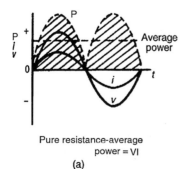

Pure resistance-average power = VI    (a)

Pure inductance-average power = 0    (b)

Pure capacitance-average power = 0    (c)

**Figure 15.23**

(a)   For a purely resistive a.c. circuit, the average power dissipated, $P$, is given by: $P = VI = I^2R = V^2/R$ **watts** ($V$ and $I$ being r.m.s. values) See Fig. 15.23(a)

(b)   For a purely inductive a.c. circuit, the average power is zero. See Fig. 15.23(b)

(c)   For a purely capacitive a.c. circuit, the average power is zero. See Fig. 15.23(c)

Figure 15.24 shows current and voltage waveforms for an $R$–$L$ circuit where the current lags the voltage by angle $\phi$. The waveform for power (where $p = vi$) is shown by the broken line, and its shape, and hence average power, depends on the value of angle $\phi$.

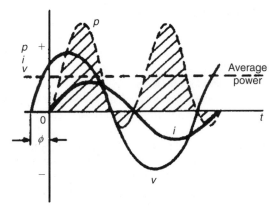

**Figure 15.24**

For an $R$–$L$, $R$–$C$ or $R$–$L$–$C$ series a.c. circuit, the average power $P$ is given by:

$$P = VI\cos\phi \text{ watts}$$

or   $$P = I^2R \text{ watts}$$

($V$ and $I$ being r.m.s. values)

> **Problem 24.**   An instantaneous current, $i = 250\sin\omega t$ mA flows through a pure resistance of $5\,\text{k}\Omega$. Find the power dissipated in the resistor.

Power dissipated, $P = I^2R$ where $I$ is the r.m.s. value of current. If $i = 250\sin\omega t$ mA, then $I_m = 0.250$ A and r.m.s. current, $I = (0.707 \times 0.250)$ A. Hence **power** $P = (0.707 \times 0.250)^2(5000) = \textbf{156.2 watts}$.

> **Problem 25.**   A series circuit of resistance $60\,\Omega$ and inductance $75\,\text{mH}$ is connected to a $110\,\text{V}$, $60\,\text{Hz}$ supply. Calculate the power dissipated.

Inductive reactance, $X_L = 2\pi fL$

$$= 2\pi(60)(75 \times 10^{-3})$$

$$= 28.27\,\Omega$$

Impedance, $Z = \sqrt{R^2 + X_L^2}$

$$= \sqrt{60^2 + 28.27^2}$$

$$= 66.33\,\Omega$$

Current, $I = V/Z = 110/66.33 = 1.658$ A.

To calculate power dissipation in an a.c. circuit two formulae may be used:

(i) $P = I^2R = (1.658)^2(60) = \textbf{165 W}$

or

(ii) $P = VI\cos\phi$ where $\cos\phi = \dfrac{R}{Z} = \dfrac{60}{66.33}$
$$= 0.9046$$

Hence $P = (110)(1.658)(0.9046) = \textbf{165 W}$

## 15.11   Power triangle and power factor

Figure 15.25(a) shows a phasor diagram in which the current $I$ lags the applied voltage $V$ by angle $\phi$. The horizontal component of $V$ is $V\cos\phi$ and the vertical component of $V$ is $V\sin\phi$. If each of the voltage phasors is multiplied by $I$, Fig. 15.25(b) is obtained and is known as the **'power triangle'**.

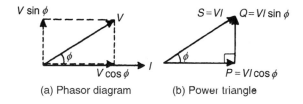

(a) Phasor diagram     (b) Power triangle

**Figure 15.25**

**Apparent power,**

$$S = VI \text{ voltamperes (VA)}$$

**True or active power,**

$$P = VI\cos\phi \text{ watts (W)}$$

**Reactive power,**

$$Q = VI\sin\phi \text{ reactive}$$

$$\text{voltamperes (var)}$$

$$\text{Power factor} = \frac{\text{True power } P}{\text{Apparent power } S}$$

For sinusoidal voltages and currents,

$$\text{power factor} = \frac{P}{S} = \frac{VI\cos\phi}{VI}$$

i.e.  $\text{p.f.} = \cos\phi = \dfrac{R}{Z}$   (from Fig. 15.6)

The relationships stated above are also true when current $I$ leads voltage $V$.

**Problem 26.** A pure inductance is connected to a 150 V, 50 Hz supply, and the apparent power of the circuit is 300 VA. Find the value of the inductance.

Apparent power $S = VI$.
Hence current $I = S/V = 300/150 = 2$ A.
Inductive reactance $X_L = V/I = 150/2 = 75\,\Omega$.
Since $X_L = 2\pi fL$,

$$\text{inductance } L = \frac{X_L}{2\pi f} = \frac{75}{2\pi(50)} = \mathbf{0.239\,H}$$

**Problem 27.** A transformer has a rated output of 200 kVA at a power factor of 0.8. Determine the rated power output and the corresponding reactive power.

$VI = 200\,\text{kVA} = 200 \times 10^3$ and p.f. $= 0.8 = \cos\phi$.
**Power output,** $P = VI\cos\phi - (200 \times 10^3)(0.8)$
$$= \mathbf{160\,kW}.$$
Reactive power, $Q = VI\sin\phi$. If $\cos\phi = 0.8$, then $\phi = \cos^{-1}0.8 = 36.87°$.
Hence $\sin\phi = \sin 36.87° = 0.6$
Hence **reactive power,** $Q = (200 \times 10^3)(0.6)$
$$= \mathbf{120\,kvar}.$$

**Problem 28.** A load takes 90 kW at a power factor of 0.5 lagging. Calculate the apparent power and the reactive power.

True power $P = 90\,\text{kW} = VI\cos\phi$ and
power factor $= 0.5 = \cos\phi$.

**Apparent power,** $S = VI = \dfrac{P}{\cos\phi} = \dfrac{90}{0.5} = \mathbf{180\,kVA}$

Angle $\phi = \cos^{-1}0.5 = 60°$
hence $\sin\phi = \sin 60° = 0.866$
Hence **reactive power,**
$Q = VI\sin\phi = 180 \times 10^3 \times 0.866 = \mathbf{156\,kvar}.$

**Problem 29.** The power taken by an inductive circuit when connected to a 120 V, 50 Hz supply is 400 W and the current is 8 A. Calculate (a) the resistance, (b) the impedance, (c) the reactance, (d) the power factor, and (e) the phase angle between voltage and current.

(a) Power $P = I^2 R$ hence $R = \dfrac{P}{I^2} = \dfrac{400}{8^2} = \mathbf{6.25\,\Omega}.$

(b) Impedance $Z = \dfrac{V}{I} = \dfrac{120}{8} = \mathbf{15\,\Omega}.$

(c) Since $Z = \sqrt{R^2 + X_L^2}$, then
$X_L = \sqrt{Z^2 - R^2} = \sqrt{15^2 - 6.25^2} = \mathbf{13.64\,\Omega}.$

(d) **Power factor** $= \dfrac{\text{true power}}{\text{apparent power}} = \dfrac{VI\cos\phi}{VI}$
$$= \frac{400}{(120)(8)} = \mathbf{0.4167}$$

(e) p.f. $= \cos\phi = 0.4167$ hence
phase angle, $\phi = \cos^{-1}0.4167 = \mathbf{65.37°\ lagging}.$

**Problem 30.** A circuit consisting of a resistor in series with a capacitor takes 100 watts at a power factor of 0.5 from a 100 V, 60 Hz supply. Find (a) the current flowing, (b) the phase angle, (c) the resistance, (d) the impedance, and (e) the capacitance.

(a) Power factor $= \dfrac{\text{true power}}{\text{apparent power}}$, i.e. $0.5 = \dfrac{100}{100 \times I}$
hence current,

$$I = \frac{100}{(0.5)(100)} = \mathbf{2\,A}$$

(b) Power factor $= 0.5 = \cos\phi$ hence phase angle,
$\phi = \cos^{-1}0.5 = \mathbf{60°\ leading}$

(c) Power $P = I^2 R$ hence resistance,

$$R = \frac{P}{I^2} = \frac{100}{2^2} = \mathbf{25\,\Omega}$$

(d) Impedance $Z = \dfrac{V}{I} = \dfrac{100}{2} = \mathbf{50\,\Omega}$

(e) Capacitive reactance, $X_C = \sqrt{Z^2 - R^2}$
$$= \sqrt{50^2 - 25^2} = \mathbf{43.30\,\Omega}.$$

$$X_C = 1/2\pi fC.$$

Hence capacitance $C = \dfrac{1}{2\pi f X_C} = \dfrac{1}{2\pi (60)(43.30)}$ F

$$= 61.26\,\mu\text{F}$$

**Now try the following exercises**

**Exercise 88    Further problems on power in a.c. circuits**

1. A voltage $v = 200\sin\omega t$ volts is applied across a pure resistance of $1.5\,\text{k}\Omega$. Find the power dissipated in the resistor.
   [13.33 W]

2. A $50\,\mu\text{F}$ capacitor is connected to a 100 V, 200 Hz supply. Determine the true power and the apparent power.     [0, 628.3 VA]

3. A motor takes a current of 10 A when supplied from a 250 V a.c. supply. Assuming a power factor of 0.75 lagging find the power consumed. Find also the cost of running the motor for 1 week continuously if 1 kWh of electricity costs 12.20 p.
   [1875 W, £38.43]

4. A motor takes a current of 12 A when supplied from a 240 V a.c. supply. Assuming a power factor of 0.70 lagging, find the power consumed.     [2.016 kW]

5. A transformer has a rated output of 100 kVA at a power factor of 0.6. Determine the rated power output and the corresponding reactive power.     [60 kW, 80 kvar]

6. A substation is supplying 200 kVA and 150 kvar. Calculate the corresponding power and power factor.     [132 kW, 0.66]

7. A load takes 50 kW at a power factor of 0.8 lagging. Calculate the apparent power and the reactive power.     [62.5 kVA, 37.5 kvar]

8. A coil of resistance $400\,\Omega$ and inductance 0.20 H is connected to a 75 V, 400 Hz supply. Calculate the power dissipated in the coil.
   [5.452 W]

9. An $80\,\Omega$ resistor and a $6\,\mu\text{F}$ capacitor are connected in series across a 150 V, 200 Hz supply. Calculate (a) the circuit impedance, (b) the current flowing, and (c) the power dissipated in the circuit.
   [(a) $154.9\,\Omega$ (b) 0.968 A (c) 75 W]

10. The power taken by a series circuit containing resistance and inductance is 240 W when connected to a 200 V, 50 Hz supply. If the current flowing is 2 A find the values of the resistance and inductance.     [$60\,\Omega$, 255 mH]

11. The power taken by a $C$–$R$ series circuit, when connected to a 105 V, 2.5 kHz supply, is 0.9 kW and the current is 15 A. Calculate (a) the resistance, (b) the impedance, (c) the reactance, (d) the capacitance, (e) the power factor, and (f) the phase angle between voltage and current.
    [(a) $4\,\Omega$ (b) $7\,\Omega$ (c) $5.745\,\Omega$ (d) $11.08\,\mu\text{F}$ (e) 0.571 (f) 55.18° leading]

12. A circuit consisting of a resistor in series with an inductance takes 210 W at a power factor of 0.6 from a 50 V, 100 Hz supply. Find (a) the current flowing, (b) the circuit phase angle, (c) the resistance, (d) the impedance, and (e) the inductance.
    [(a) 7 A (b) 53.13° lagging (c) $4.286\,\Omega$ (d) $7.143\,\Omega$ (e) 9.095 mH]

13. A 200 V, 60 Hz supply is applied to a capacitive circuit. The current flowing is 2 A and the power dissipated is 150 W. Calculate the values of the resistance and capacitance.
    [$37.5\,\Omega$, $28.61\,\mu\text{F}$]

**Exercise 89    Short answer questions on single-phase a.c. circuits**

1. Complete the following statements:
   (a) In a purely resistive a.c. circuit the current is ...... with the voltage
   (b) In a purely inductive a.c. circuit the current ...... the voltage by ...... degrees
   (c) In a purely capacitive a.c. circuit the current ...... the voltage by ...... degrees

2. Draw phasor diagrams to represent (a) a purely resistive a.c. circuit (b) a purely inductive a.c. circuit (c) a purely capacitive a.c. circuit

3. What is inductive reactance? State the symbol and formula for determining inductive reactance

4.  What is capacitive reactance? State the symbol and formula for determining capacitive reactance

5.  Draw phasor diagrams to represent (a) a coil (having both inductance and resistance), and (b) a series capacitive circuit containing resistance

6.  What does 'impedance' mean when referring to an a.c. circuit?

7.  Draw an impedance triangle for an $R$–$L$ circuit. Derive from the triangle an expression for (a) impedance, and (b) phase angle

8.  Draw an impedance triangle for an $R$–$C$ circuit. From the triangle derive an expression for (a) impedance, and (b) phase angle

9.  What is series resonance?

10.  Derive a formula for resonant frequency $f_r$ in terms of $L$ and $C$

11.  What does the Q-factor in a series circuit mean?

12.  State three formulae used to calculate the Q-factor of a series circuit at resonance

13.  State an advantage of a high Q-factor in a series high-frequency circuit

14.  State a disadvantage of a high Q-factor in a series power circuit

15.  State two formulae which may be used to calculate power in an a.c. circuit

16.  Show graphically that for a purely inductive or purely capacitive a.c. circuit the average power is zero

17.  Define 'power factor'

18.  Define (a) apparent power (b) reactive power

19.  Define (a) bandwidth (b) selectivity

**Exercise 90    Multi-choice questions on single-phase a.c. circuits**
**(Answers on page 420)**

1.  An inductance of 10 mH connected across a 100 V, 50 Hz supply has an inductive reactance of
    (a)  $10\pi\ \Omega$          (b)  $1000\pi\ \Omega$
    (c)  $\pi\ \Omega$          (d)  $\pi\ H$

2.  When the frequency of an a.c. circuit containing resistance and inductance is increased, the current
    (a)  decreases          (b)  increases
    (c)  stays the same

3.  In question 2, the phase angle of the circuit
    (a)  decreases          (b)  increases
    (c)  stays the same

4.  When the frequency of an a.c. circuit containing resistance and capacitance is decreased, the current
    (a)  decreases          (b)  increases
    (c)  stays the same

5.  In question 4, the phase angle of the circuit
    (a)  decreases          (b)  increases
    (c)  stays the same

6.  A capacitor of 1 μF is connected to a 50 Hz supply. The capacitive reactance is
    (a)  $50\ \text{M}\Omega$          (b)  $\dfrac{10}{\pi}\ \text{k}\Omega$
    (c)  $\dfrac{\pi}{10^4}\ \Omega$          (d)  $\dfrac{10}{\pi}\ \Omega$

7.  In a series a.c. circuit the voltage across a pure inductance is 12 V and the voltage across a pure resistance is 5 V. The supply voltage is
    (a)  13 V          (b)  17 V
    (c)  7 V          (d)  2.4 V

8.  Inductive reactance results in a current that
    (a)  leads the voltage by 90°
    (b)  is in phase with the voltage
    (c)  leads the voltage by $\pi$ rad
    (d)  lags the voltage by $\pi/2$ rad

9.  Which of the following statements is false?
    (a)  Impedance is at a minimum at resonance in an a.c. circuit
    (b)  The product of r.m.s. current and voltage gives the apparent power in an a.c. circuit
    (c)  Current is at a maximum at resonance in an a.c. circuit
    (d)  $\dfrac{\text{Apparent power}}{\text{True power}}$ gives power factor

10.  The impedance of a coil, which has a resistance of $X$ ohms and an inductance of $Y$ henrys, connected across a supply of frequency $K$ Hz, is

(a) $2\pi KY$  (b) $X+Y$
(c) $\sqrt{X^2+Y^2}$  (d) $\sqrt{X^2+(2\pi KY)^2}$

11. In question 10, the phase angle between the current and the applied voltage is given by
(a) $\tan^{-1}\dfrac{Y}{X}$  (b) $\tan^{-1}\dfrac{2\pi KY}{X}$
(c) $\tan^{-1}\dfrac{X}{2\pi KY}$  (d) $\tan\left(\dfrac{2\pi KY}{X}\right)$

12. When a capacitor is connected to an a.c. supply the current
(a) leads the voltage by $180°$
(b) is in phase with the voltage
(c) leads the voltage by $\pi/2$ rad
(d) lags the voltage by $90°$

13. When the frequency of an a.c. circuit containing resistance and capacitance is increased the impedance
(a) increases  (b) decreases
(c) stays the same

14. In an $R-L-C$ series a.c. circuit a current of $5\,A$ flows when the supply voltage is $100\,V$. The phase angle between current and voltage is $60°$ lagging. Which of the following statements is false?
(a) The circuit is effectively inductive
(b) The apparent power is $500\,VA$
(c) The equivalent circuit reactance is $20\,\Omega$
(d) The true power is $250\,W$

15. A series a.c. circuit comprising a coil of inductance $100\,mH$ and resistance $1\,\Omega$ and a $10\,\mu F$ capacitor is connected across a $10\,V$ supply. At resonance the p.d. across the capacitor is
(a) $10\,kV$  (b) $1\,kV$
(c) $100\,V$  (d) $10\,V$

16. The amplitude of the current $I$ flowing in the circuit of Fig. 15.26 is:
(a) $21\,A$  (b) $16.8\,A$
(c) $28\,A$  (d) $12\,A$

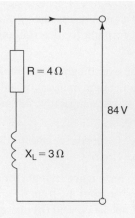

**Figure 15.26**

17. If the supply frequency is increased at resonance in a series $R-L-C$ circuit and the values of $L, C$ and $R$ are constant, the circuit will become:
(a) capacitive  (b) resistive
(c) inductive  (d) resonant

18. For the circuit shown in Fig. 15.27, the value of Q-factor is:
(a) $50$  (b) $100$
(c) $5\times10^{-4}$  (d) $40$

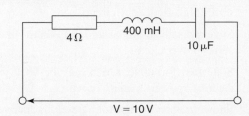

**Figure 15.27**

19. A series $R-L-C$ circuit has a resistance of $8\,\Omega$, an inductance of $100\,mH$ and a capacitance of $5\,\mu F$. If the current flowing is $2\,A$, the impedance at resonance is:
(a) $160\,\Omega$  (b) $16\,\Omega$
(c) $8\,m\Omega$  (d) $8\,\Omega$

# Chapter 16

# Single-phase parallel a.c. circuits

At the end of this chapter you should be able to:

- calculate unknown currents, impedances and circuit phase angle from phasor diagrams for (a) $R-L$ (b) $R-C$ (c) $L-C$ (d) $LR-C$ parallel a.c. circuits
- state the condition for parallel resonance in an $LR-C$ circuit
- derive the resonant frequency equation for an $LR-C$ parallel a.c. circuit
- determine the current and dynamic resistance at resonance in an $LR-C$ parallel circuit
- understand and calculate $Q$-factor in an $LR-C$ parallel circuit
- understand how power factor may be improved

## 16.1 Introduction

In parallel circuits, such as those shown in Figs 16.1 and 16.2, the voltage is common to each branch of the network and is thus taken as the reference phasor when drawing phasor diagrams.

For any parallel a.c. circuit:

**True or active power,** $\quad \mathbf{P}=VI\cos\phi \text{ watts (W)}$

or $\qquad\qquad\qquad P=I_R^2 R \text{ watts}$

**Apparent power,** $\qquad \mathbf{S}=VI \text{ voltamperes (VA)}$

**Reactive power,** $\qquad \mathbf{Q}=VI\sin\phi$ reactive
$\qquad\qquad\qquad\qquad$ voltamperes (var)

$$\text{Power factor} = \frac{\text{true power}}{\text{apparent power}} = \frac{P}{S} = \cos\phi$$

(These formulae are the same as for series a.c. circuits as used in Chapter 15.)

DOI: 10.1016/B978-0-08-089056-2.00016-4

## 16.2 $R-L$ parallel a.c. circuit

In the two branch parallel circuit containing resistance $R$ and inductance $L$ shown in Fig. 16.1, the current flowing in the resistance, $I_R$, is in-phase with the supply voltage $V$ and the current flowing in the inductance, $I_L$, lags the supply voltage by 90°. The supply current $I$ is the phasor sum of $I_R$ and $I_L$ and thus the current $I$ lags the applied voltage $V$ by an angle lying between 0° and 90° (depending on the values of $I_R$ and $I_L$), shown as angle $\phi$ in the phasor diagram.

Circuit diagram        Phasor diagram

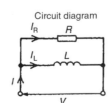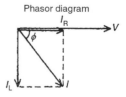

**Figure 16.1**

**Section 2**

From the phasor diagram: $I = \sqrt{I_R^2 + I_L^2}$ (by Pythagoras' theorem) where

$$I_R = \frac{V}{R} \quad \text{and} \quad I_L = \frac{V}{X_L}$$

$$\tan\phi = \frac{I_L}{I_R} \quad \sin\phi = \frac{I_L}{I} \quad \text{and} \quad \cos\phi = \frac{I_R}{I}$$

(by trigonometric ratios)

$$\text{Circuit impedance, } Z = \frac{V}{I}$$

**Problem 1.** A 20 Ω resistor is connected in parallel with an inductance of 2.387 mH across a 60 V, 1 kHz supply. Calculate (a) the current in each branch, (b) the supply current, (c) the circuit phase angle, (d) the circuit impedance, and (e) the power consumed.

The circuit and phasor diagrams are as shown in Fig. 16.1

(a)   Current flowing in the resistor,

$$I_R = \frac{V}{R} = \frac{60}{20} = 3\,\text{A}$$

Current flowing in the inductance,

$$I_L = \frac{V}{X_L} = \frac{V}{2\pi f L}$$

$$= \frac{60}{2\pi(1000)(2.387 \times 10^{-3})} = 4\,\text{A}$$

(b)   From the phasor diagram, supply current,

$$I = \sqrt{I_R^2 + I_L^2} = \sqrt{3^2 + 4^2} = 5\,\text{A}$$

(c)   Circuit phase angle,

$$\phi = \tan^{-1}\frac{I_L}{I_R} = \tan^{-1}\frac{4}{3} = \textbf{53.13° lagging}$$

(d)   Circuit impedance,

$$Z = \frac{V}{I} = \frac{60}{5} = 12\,\Omega$$

(e)   Power consumed

$$P = VI\cos\phi = (60)(5)(\cos 53.13°)$$

$$= \textbf{180 W}$$

(Alternatively, power consumed,
$P = I_R^2 R = (3)^2(20) = \textbf{180 W}$)

**Now try the following exercise**

**Exercise 91   Further problems on R–L parallel a.c. circuits**

1.   A 30 Ω resistor is connected in parallel with a pure inductance of 3 mH across a 110 V, 2 kHz supply. Calculate (a) the current in each branch, (b) the circuit current, (c) the circuit phase angle, (d) the circuit impedance, (e) the power consumed, and (f) the circuit power factor.
[(a) $I_R = 3.67$ A, $I_L = 2.92$ A (b) 4.69 A (c) 38.51° lagging (d) 23.45 Ω (e) 404 W (f) 0.782 lagging]

2.   A 40 Ω resistance is connected in parallel with a coil of inductance L and negligible resistance across a 200 V, 50 Hz supply and the supply current is found to be 8 A. Sketch the phasor diagram and determine the inductance of the coil. [102 mH]

## 16.3   R–C parallel a.c. circuit

In the two branch parallel circuit containing resistance R and capacitance C shown in Fig. 16.2, $I_R$ is in-phase with the supply voltage V and the current flowing in the capacitor, $I_C$, leads V by 90°. The supply current I is the phasor sum of $I_R$ and $I_C$ and thus the current I leads the applied voltage V by an angle lying between 0° and 90° (depending on the values of $I_R$ and $I_C$), shown as angle α in the phasor diagram.

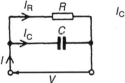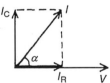

Figure 16.2

From the phasor diagram: $I = \sqrt{I_R^2 + I_C^2}$ (by Pythagoras' theorem) where

$$I_R = \frac{V}{R} \quad \text{and} \quad I_C = \frac{V}{X_C}$$

$$\tan\alpha = \frac{I_C}{I_R} \quad \sin\alpha = \frac{I_C}{I} \quad \text{and} \quad \cos\alpha = \frac{I_R}{I}$$

(by trigonometric ratios)

$$\text{Circuit impedance, } Z = \frac{V}{I}$$

**Problem 2.** A 30 μF capacitor is connected in parallel with an 80 Ω resistor across a 240 V, 50 Hz supply. Calculate (a) the current in each branch, (b) the supply current, (c) the circuit phase angle, (d) the circuit impedance, (e) the power dissipated, and (f) the apparent power.

The circuit and phasor diagrams are as shown in Fig. 16.2

(a)  Current in resistor,

$$I_R = \frac{V}{R} = \frac{240}{80} = 3\,A$$

Current in capacitor,

$$I_C = \frac{V}{X_C} = \frac{V}{\left(\dfrac{1}{2\pi f C}\right)} = 2\pi f C V$$

$$= 2\pi(50)(30 \times 10^6)(240) = 2.262\,A$$

(b)  Supply current,

$$I = \sqrt{I_R^2 + I_C^2} = \sqrt{3^2 + 2.262^2}$$

$$= 3.757\,A$$

(c)  Circuit phase angle,

$$\alpha = \tan^{-1}\frac{I_C}{I_R} = \tan^{-1}\frac{2.262}{3}$$

$$= 37.02° \text{ leading}$$

(d)  Circuit impedance,

$$Z = \frac{V}{I} = \frac{240}{3.757} = 63.88\,\Omega$$

(e)  True or active power dissipated,

$$P = VI\cos\alpha = (240)(3.757)\cos 37.02°$$

$$= 720\,W$$

(Alternatively, true power

$$P = I_R^2 R = (3)^2(80) = 720\,W)$$

(f)  Apparent power,

$$S = VI = (240)(3.757) = 901.7\,VA$$

**Problem 3.** A capacitor $C$ is connected in parallel with a resistor $R$ across a 120 V, 200 Hz supply. The supply current is 2 A at a power factor of 0.6 leading. Determine the values of $C$ and $R$.

The circuit diagram is shown in Fig. 16.3(a).

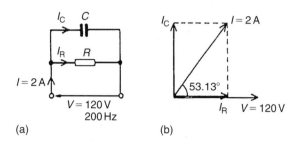

(a)                        (b)

**Figure 16.3**

Power factor $= \cos\phi = 0.6$ leading, hence

$$\phi = \cos^{-1}0.6 = 53.13° \text{ leading}.$$

From the phasor diagram shown in Fig. 16.3(b),

$$I_R = I\cos 53.13° = (2)(0.6)$$

$$= 1.2\,A$$

and    $I_C = I\sin 53.13° = (2)(0.8)$

$$= 1.6\,A$$

(Alternatively, $I_R$ and $I_C$ can be measured from the scaled phasor diagram.)
From the circuit diagram,

$$I_R = \frac{V}{R} \text{ from which}$$

$$R = \frac{V}{I_R}$$

$$= \frac{120}{1.2} = 100\,\Omega$$

and    $I_C = \dfrac{V}{X_C}$

$$= 2\pi f C V \text{ from which}$$

$$C = \frac{I_C}{2\pi f V}$$

$$= \frac{1.6}{2\pi(200)(120)}$$

$$= 10.61\,μF$$

**Now try the following exercise**

**Exercise 92   Further problems on R–C parallel a.c. circuits**

1.  A 1500 nF capacitor is connected in parallel with a $16\,\Omega$ resistor across a 10 V, 10 kHz supply. Calculate (a) the current in each branch, (b) the supply current, (c) the circuit phase angle, (d) the circuit impedance, (e) the power consumed, (f) the apparent power, and (g) the circuit power factor. Sketch the phasor diagram.

    [(a) $I_R = 0.625$ A, $I_C = 0.943$ A (b) 1.131 A (c) 56.46° leading (d) $8.84\,\Omega$ (e) 6.25 W (f) 11.31 VA (g) 0.553 leading]

2.  A capacitor $C$ is connected in parallel with a resistance $R$ across a 60 V, 100 Hz supply. The supply current is 0.6 A at a power factor of 0.8 leading. Calculate the value of $R$ and $C$.

    [$R = 125\,\Omega$, $C = 9.55\,\mu$F]

## 16.4   *L–C* parallel circuit

In the two branch parallel circuit containing inductance $L$ and capacitance $C$ shown in Fig. 16.4, $I_L$ lags $V$ by 90° and $I_C$ leads $V$ by 90°

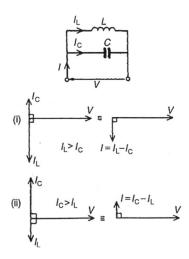

**Figure 16.4**

Theoretically there are three phasor diagrams possible – each depending on the relative values of $I_L$ and $I_C$:

(i)   $I_L > I_C$ (giving a supply current, $I = I_L - I_C$ lagging $V$ by 90°)

(ii)  $I_C > I_L$ (giving a supply current, $I = I_C - I_L$ leading $V$ by 90°)

(iii) $I_L = I_C$ (giving a supply current, $I = 0$).

The latter condition is not possible in practice due to circuit resistance inevitably being present (as in the circuit described in Section 16.5).

**For the $L - C$ parallel circuit,**

$$I_L = \frac{V}{X_L} \quad I_C = \frac{V}{X_C}$$

$I =$ **phasor difference between $I_L$ and $I_C$, and**

$$Z = \frac{V}{I}$$

**Problem 4.**   A pure inductance of 120 mH is connected in parallel with a 25 μF capacitor and the network is connected to a 100 V, 50 Hz supply. Determine (a) the branch currents, (b) the supply current and its phase angle, (c) the circuit impedance, and (d) the power consumed.

The circuit and phasor diagrams are as shown in Fig. 16.4

(a)   Inductive reactance,

$$X_L = 2\pi fL = 2\pi (50)(120 \times 10^{-3})$$
$$= 37.70\,\Omega$$

Capacitive reactance,

$$X_C = \frac{1}{2\pi fC} = \frac{1}{2\pi (50)(25 \times 10^{-6})}$$
$$= 127.3\,\Omega$$

Current flowing in inductance,

$$I_L = \frac{V}{X_L} = \frac{100}{37.70} = \mathbf{2.653\,A}$$

Current flowing in capacitor,

$$I_C = \frac{V}{X_C} = \frac{100}{127.3} = \mathbf{0.786\,A}$$

(b)   $I_L$ and $I_C$ are anti-phase, hence supply current,

$$I = I_L - I_C = 2.653 - 0.786 = \mathbf{1.867\,A}$$

and **the current lags the supply voltage $V$ by 90°** (see Fig. 16.4(i))

(c) Circuit impedance,

$$Z = \frac{V}{I} = \frac{100}{1.867} = \mathbf{53.56\,\Omega}$$

(d) Power consumed,

$$P = VI\cos\phi = (100)(1.867)\cos 90° = \mathbf{0\,W}$$

---

**Problem 5.** Repeat Problem 4 for the condition when the frequency is changed to 150 Hz.

(a) Inductive reactance,

$$X_L = 2\pi(150)(120 \times 10^{-3}) = 113.1\,\Omega$$

Capacitive reactance,

$$X_C = \frac{1}{2\pi(150)(25 \times 10^{-6})} = 42.44\,\Omega$$

Current flowing in inductance,

$$I_L = \frac{V}{X_L} = \frac{100}{113.1} = \mathbf{0.884\,A}$$

Current flowing in capacitor,

$$I_C = \frac{V}{X_C} = \frac{100}{42.44} = \mathbf{2.356\,A}$$

(b) Supply current,

$$I = I_C - I_L = 2.356 - 0.884 = \mathbf{1.472\,A}$$

**leading $V$ by 90°** (see Fig. 16.4(ii))

(c) Circuit impedance,

$$Z = \frac{V}{I} = \frac{100}{1.472} = \mathbf{67.93\,\Omega}$$

(d) Power consumed,

$$P = VI\cos\phi = \mathbf{0\,W} \text{ (since } \phi = 90°)$$

---

**From problems 4 and 5:**

(i) When $X_L < X_C$ then $I_L > I_C$ and $I$ lags $V$ by 90°

(ii) When $X_L > X_C$ then $I_L < I_C$ and $I$ leads $V$ by 90°

(iii) In a parallel circuit containing no resistance the power consumed is zero

---

**Now try the following exercise**

1. An inductance of 80 mH is connected in parallel with a capacitance of 10 μF across a 60 V, 100 Hz supply. Determine (a) the branch currents, (b) the supply current, (c) the circuit phase angle, (d) the circuit impedance, and (e) the power consumed.
   [(a) $I_C = 0.377$ A, $I_L = 1.194$ A (b) 0.817 A
   (c) 90° lagging (d) 73.44 Ω (e) 0 W]

2. Repeat Problem 1 for a supply frequency of 200 Hz.
   [(a) $I_C = 0.754$ A, $I_L = 0.597$ A (b) 0.157 A
   (c) 90° leading (d) 382.2 Ω (e) 0 W]

## 16.5    *LR–C* parallel a.c. circuit

In the two branch circuit containing capacitance $C$ in parallel with inductance $L$ and resistance $R$ in series (such as a coil) shown in Fig. 16.5(a), the phasor diagram for the $LR$ branch alone is shown in Fig. 16.5(b) and the phasor diagram for the $C$ branch is shown alone in Fig. 16.5(c). Rotating each and superimposing on one another gives the complete phasor diagram shown in Fig. 16.5(d).

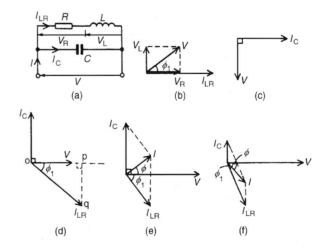

**Figure 16.5**

The current $I_{LR}$ of Fig. 16.5(d) may be resolved into horizontal and vertical components. The horizontal component, shown as $op$ is $I_{LR}\cos\phi_1$ and the vertical component, shown as $pq$ is $I_{LR} \sin \phi_1$. There are three possible conditions for this circuit:

(i) $I_C > I_{LR} \sin \phi_1$ (giving a supply current $I$ leading $V$ by angle $\phi$–as shown in Fig. 16.5(e))

(ii) $I_{LR} \sin \phi > I_C$ (giving $I$ lagging $V$ by angle $\phi$–as shown in Fig. 16.5(f))

(iii) $I_C = I_{LR} \sin \phi_1$ (this is called **parallel resonance**, see Section 16.6)

There are two methods of finding the phasor sum of currents $I_{LR}$ and $I_C$ in Fig. 16.5(e) and (f). These are: (i) by a scaled phasor diagram, or (ii) by resolving each current into their 'in-phase' (i.e. horizontal) and 'quadrature' (i.e. vertical) **components**, as demonstrated in problems 6 and 7. With reference to the phasor diagrams of Fig. 16.5:

**Impedance of *LR* branch, $Z_{LR} = \sqrt{R^2 + X_L^2}$**

**Current, $I_{LR} = \dfrac{V}{Z_{LR}}$ and $I_C = \dfrac{V}{X_C}$**

**Supply current**

$I$ = phasor sum of $I_{LR}$ and $I_C$ (by drawing)

$$= \sqrt{(I_{LR} \cos \phi_1)^2 + (I_{LR} \sin \phi_1 \sim I_C)^2}$$

**(by calculation)**

where $\sim$ means 'the difference between'.

**Circuit impedance $Z = \dfrac{V}{I}$**

$$\tan \phi_1 = \frac{V_L}{V_R} = \frac{X_L}{R}$$

$$\sin \phi_1 = \frac{X_L}{Z_{LR}} \quad \text{and} \quad \cos \phi_1 = \frac{R}{Z_{LR}}$$

$$\tan \phi = \frac{I_{LR} \sin \phi_1 \sim I_C}{I_{LR} \cos \phi_1} \quad \text{and} \quad \cos \phi = \frac{I_{LR} \cos \phi_1}{I}$$

---

**Problem 6.** A coil of inductance 159.2 mH and resistance $40\,\Omega$ is connected in parallel with a $30\,\mu$F capacitor across a 240 V, 50 Hz supply. Calculate (a) the current in the coil and its phase angle, (b) the current in the capacitor and its phase angle, (c) the supply current and its phase angle, (d) the circuit impedance, (e) the power consumed, (f) the apparent power, and (g) the reactive power. Draw the phasor diagram.

---

The circuit diagram is shown in Fig. 16.6(a).

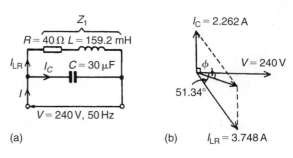

(a)  (b)

**Figure 16.6**

(a) For the coil, inductive reactance $X_L = 2\pi f L = 2\pi(50)(159.2 \times 10^{-3}) = 50\,\Omega$.

$$\text{Impedance } Z_1 = \sqrt{R^2 + X_L^2}$$
$$= \sqrt{40^2 + 50^2}$$
$$= 64.03\,\Omega$$

Current in coil

$$I_{LR} = \frac{V}{Z_1} = \frac{240}{64.03} = \mathbf{3.748\,A}$$

Branch phase angle

$$\phi_1 = \tan^{-1} \frac{X_L}{R} = \tan^{-1} \frac{50}{40}$$
$$= \tan^{-1} 1.25 = \mathbf{51.34°\ lagging}$$

(see phasor diagram in Fig. 16.6(b))

(b) Capacitive reactance,

$$X_C = \frac{1}{2\pi f C} = \frac{1}{2\pi(50)(30 \times 10^{-6})}$$
$$= 106.1\,\Omega$$

Current in capacitor,

$$I_C = \frac{V}{X_C} = \frac{240}{106.1}$$
$$= \mathbf{2.262\,A\ leading\ the\ supply}$$
$$\mathbf{voltage\ by\ 90°}$$

(see phasor diagram of Fig. 16.6(b)).

(c) The supply current $I$ is the phasor sum of $I_{LR}$ and $I_C$. This may be obtained by drawing the phasor diagram to scale and measuring the current $I$ and its phase angle relative to $V$. (Current $I$ will always be the diagonal of the parallelogram formed as in Fig. 16.6(b).)

Alternatively the current $I_{LR}$ and $I_C$ may be resolved into their horizontal (or 'in-phase') and vertical (or 'quadrature') components.

The horizontal component of $I_{LR}$ is:

$$I_{LR}\cos 51.34° = 3.748\cos 51.34° = 2.341\,A.$$

The horizontal component of $I_C$ is

$$I_C \cos 90° = 0$$

Thus the total horizontal component,

$$I_H = 2.341\,A$$

The vertical component of $I_{LR}$

$$= -I_{LR}\sin 51.34° = -3.748\sin 51.34°$$
$$= -2.927\,A$$

The vertical component of $I_C$

$$= I_C \sin 90° = 2.262\sin 90° = 2.262\,A$$

Thus the total vertical component,

$$I_V = -2.927 + 2.262 = -0.665\,A$$

$I_H$ and $I_V$ are shown in Fig. 16.7, from which,

$$I = \sqrt{2.341^2 + (-0.665)^2} = 2.434\,A$$

Angle $\phi = \tan^{-1}\dfrac{0.665}{2.341} = 15.86°$ lagging

Hence **the supply current $I = 2.434\,A$ lagging $V$ by 15.86°**

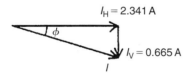

Figure 16.7

(d)  Circuit impedance,

$$Z = \frac{V}{I} = \frac{240}{2.434} = 98.60\,\Omega$$

(e)  Power consumed,

$$P = VI\cos\phi = (240)(2.434)\cos 15.86°$$
$$= 562\,W$$

(Alternatively, $P = I_R^2 R = I_{LR}^2 R$ (in this case)
$$= (3.748)^2(40) = 562\,W)$$

(f)  Apparent power,

$$S = VI = (240)(2.434) = 584.2\,VA$$

(g)  Reactive power,

$$Q = VI\sin\phi = (240)(2.434)(\sin 15.86°)$$
$$= 159.6\,var$$

**Problem 7.** A coil of inductance 0.12 H and resistance 3 kΩ is connected in parallel with a 0.02 μF capacitor and is supplied at 40 V at a frequency of 5 kHz. Determine (a) the current in the coil, and (b) the current in the capacitor. (c) Draw to scale the phasor diagram and measure the supply current and its phase angle; check the answer by calculation. Determine (d) the circuit impedance and (e) the power consumed.

The circuit diagram is shown in Fig. 16.8(a).

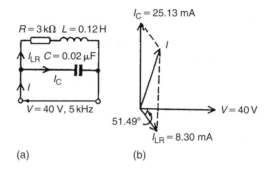

(a)    (b)

Figure 16.8

(a)  Inductive reactance,

$$X_L = 2\pi fL = 2\pi(5000)(0.12) = 3770\,\Omega$$

Impedance of coil,

$$Z_1 = \sqrt{R^2 + X_L} = \sqrt{3000^2 + 3770^2}$$
$$= 4818\,\Omega$$

Current in coil,

$$I_{LR} = \frac{V}{Z_1} = \frac{40}{4818} = 8.30\,mA$$

Branch phase angle

$$\phi = \tan^{-1}\frac{X_L}{R} = \tan^{-1}\frac{3770}{3000}$$
$$= 51.49°\ lagging$$

**Section 2**

(b) Capacitive reactance,

$$X_C = \frac{1}{2\pi fC} = \frac{1}{2\pi(5000)(0.02 \times 10^{-6})}$$

$$= 1592\,\Omega$$

Capacitor current,

$$I_C = \frac{V}{X_C} = \frac{40}{1592}$$

$$= \textbf{25.13 mA leading } V \textbf{ by } 90°$$

(c) Currents $I_{LR}$ and $I_C$ are shown in the phasor diagram of Fig. 16.8(b). The parallelogram is completed as shown and the supply current is given by the diagonal of the parallelogram. The current $I$ is measured as **19.3 mA** leading voltage $V$ by **74.5°**. By calculation,

$$I = \sqrt{(I_{LR}\cos 51.49°)^2 + (I_C - I_{LR}\sin 51.49°)^2}$$

$$= 19.34\,\text{mA}$$

and

$$\phi = \tan^{-1}\left(\frac{I_C - I_{LR}\sin 51.5°}{I_{LR}\cos 51.5°}\right) = 74.50°$$

(d) Circuit impedance,

$$Z = \frac{V}{I} = \frac{40}{19.34 \times 10^{-3}} = \textbf{2.068 k}\Omega$$

(e) Power consumed,

$$P = VI\cos\phi$$

$$= (40)(19.34 \times 10^{-3})\cos 74.50°$$

$$= \textbf{206.7 mW}$$

$$(\text{Alternatively, } P = I_R^2 R$$

$$= I_{LR}^2 R$$

$$= (8.30 \times 10^{-3})^2(3000)$$

$$= \textbf{206.7 mW})$$

**Now try the following exercise**

**Exercise 94  Further problems on LR–C parallel a.c. circuit**

1. A coil of resistance $60\,\Omega$ and inductance 318.4 mH is connected in parallel with a 15 μF

capacitor across a 200 V, 50 Hz supply. Calculate (a) the current in the coil, (b) the current in the capacitor, (c) the supply current and its phase angle, (d) the circuit impedance, (e) the power consumed, (f) the apparent power, and (g) the reactive power. Sketch the phasor diagram.
[(a) 1.715 A (b) 0.943 A (c) 1.028 A at 30.88° lagging (d) 194.6 Ω (e) 176.5 W (f) 205.6 VA (g) 105.5 var]

2. A 25 nF capacitor is connected in parallel with a coil of resistance $2\,k\Omega$ and inductance 0.20 H across a 100 V, 4 kHz supply. Determine (a) the current in the coil, (b) the current in the capacitor, (c) the supply current and its phase angle (by drawing a phasor diagram to scale, and also by calculation), (d) the circuit impedance, and (e) the power consumed.
[(a) 18.48 mA (b) 62.83 mA (c) 46.17 mA at 81.49° leading (d) 2.166 kΩ (e) 0.683 W]

## 16.6  Parallel resonance and Q-factor

**Parallel resonance**

**Resonance** occurs in the two branch network containing capacitance $C$ in parallel with inductance $L$ and resistance $R$ in series (see Fig. 16.5(a)) when the quadrature (i.e. vertical) component of current $I_{LR}$ is equal to $I_C$. At this condition the supply current $I$ is in-phase with the supply voltage $V$.

**Resonant frequency**

When the quadrature component of $I_{LR}$ is equal to $I_C$ then: $I_C = I_{LR}\sin\phi_1$ (see Fig. 16.9). Hence

$$\frac{V}{X_C} = \left(\frac{V}{Z_{LR}}\right)\left(\frac{X_L}{Z_{LR}}\right) \quad \text{(from Section 16.5)}$$

from which,

$$Z_{LR}^2 = X_L X_C = (2\pi f_r L)\left(\frac{1}{2\pi f_r C}\right) = \frac{L}{C} \quad (1)$$

Hence

$$\left[\sqrt{R^2 + X_L^2}\right]^2 = \frac{L}{C} \quad \text{and} \quad R^2 + X_L^2 = \frac{L}{C}$$

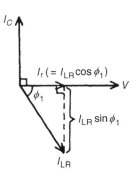

**Figure 16.9**

Thus $(2\pi f_r L)^2 = \dfrac{L}{C} - R^2$ and

$$2\pi f_r L = \sqrt{\dfrac{L}{C} - R^2}$$

and $\qquad f_r = \dfrac{1}{2\pi L}\sqrt{\dfrac{L}{C} - R^2}$

$$= \dfrac{1}{2\pi}\sqrt{\dfrac{L}{L^2 C} - \dfrac{R^2}{L^2}}$$

i.e. parallel resonant frequency,

$$f_r = \dfrac{1}{2\pi}\sqrt{\dfrac{1}{LC} - \dfrac{R^2}{L^2}}$$

(When $R$ is negligible, then $f_r = \dfrac{1}{2\pi\sqrt{LC}}$, which is the same as for series resonance)

## Current at resonance

Current at resonance,

$$I_r = I_{LR}\cos\phi_1 \quad \text{(from Fig. 16.9)}$$

$$= \left(\dfrac{V}{Z_{LR}}\right)\left(\dfrac{R}{Z_{LR}}\right) \quad \text{(from Section 16.5)}$$

$$= \dfrac{VR}{Z_{LR}^2}$$

However, from equation (1), $Z_{LR}^2 = L/C$ hence

$$I_r = \dfrac{VR}{(L/C)} = \dfrac{VRC}{L} \qquad (2)$$

The current is at a **minimum** at resonance.

## Dynamic resistance

Since the current at resonance is in-phase with the voltage the impedance of the circuit acts as a resistance. This resistance is known as the **dynamic resistance**, $R_D$ (or sometimes, the dynamic impedance).

From equation (2), impedance at resonance

$$= \dfrac{V}{I_r} = \dfrac{V}{\left(\dfrac{VRC}{L}\right)}$$

$$= \dfrac{L}{RC}$$

i.e. dynamic resistance,

$$R_D = \dfrac{L}{RC}\text{ ohms}$$

Graphs of current and impedance against frequency near to resonance for a parallel circuit are shown in Fig. 16.10, and are seen to be the reverse of those in a series circuit (from page 238).

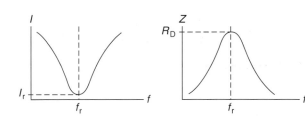

**Figure 16.10**

## Rejector circuit

The parallel resonant circuit is often described as a **rejector** circuit since it presents its maximum impedance at the resonant frequency and the resultant current is a minimum.

## Mechanical analogy

Electrical resonance for the parallel circuit can be likened to a mass hanging on a spring which, if pulled down and released, will oscillate up and down but due to friction the oscillations will slowly die. To maintain the oscillation the mass would require a small force applied each time it reaches its point of maximum travel and this is exactly what happens with the electrical circuit. A small current is required to overcome the losses and maintain the oscillations of current. Figure 16.11 shows the two cases.

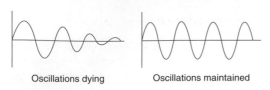

Oscillations dying        Oscillations maintained

Figure 16.11

## Applications of resonance

One use for resonance is to establish a condition of **stable frequency** in circuits designed to produce a.c. signals. Usually, a parallel circuit is used for this purpose, with the capacitor and inductor directly connected together, exchanging energy between each other. Just as a pendulum can be used to stabilise the frequency of a clock mechanism's oscillations, so can a parallel circuit be used to stabilise the electrical frequency of an a.c. oscillator circuit.

Another use for resonance is in applications where the effects of greatly increased or decreased impedance at a particular frequency is desired. A resonant circuit can be used to 'block' (i.e. present high impedance toward) a frequency or range of frequencies, thus acting as a sort of frequency **'filter'** to strain certain frequencies out of a mix of others. In fact, these particular circuits are called filters, and their design is considered in Chapter 17. In essence, this is how analogue radio receiver tuner circuits work to filter, or select, one station frequency out of the mix of different radio station frequency signals intercepted by the antenna.

## Q-factor

Currents higher than the supply current can circulate within the parallel branches of a parallel resonant circuit, the current leaving the capacitor and establishing the magnetic field of the inductor, this then collapsing and recharging the capacitor, and so on. The **Q-factor** of a parallel resonant circuit is the ratio of the current circulating in the parallel branches of the circuit to the supply current, i.e. the current magnification.

$$Q\text{-factor at resonance} = \text{current magnification}$$

$$= \frac{\text{circulating current}}{\text{supply current}}$$

$$= \frac{I_C}{I_r} = \frac{I_{LR}\sin\phi_1}{I_r}$$

$$= \frac{I_{LR}\sin\phi_1}{I_{LR}\cos\phi_1}$$

$$= \frac{\sin\phi_1}{\cos\phi_1} = \tan\phi_1$$

$$= \frac{X_L}{R}$$

i.e.      $Q\text{-factor at resonance} = \dfrac{2\pi f_r L}{R}$

(which is the same as for a series circuit).
Note that in a **parallel** circuit the $Q$-factor is a measure of **current magnification**, whereas in a **series** circuit it is a measure of **voltage magnification**.
At mains frequencies the $Q$-factor of a parallel circuit is usually low, typically less than 10, but in radio-frequency circuits the $Q$-factor can be very high.

---

**Problem 8.** A pure inductance of 150 mH is connected in parallel with a 40 μF capacitor across a 50 V, variable frequency supply. Determine (a) the resonant frequency of the circuit and (b) the current circulating in the capacitor and inductance at resonance.

---

The circuit diagram is shown in Fig. 16.12.

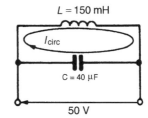

Figure 16.12

(a) Parallel resonant frequency,

$$f_r = \frac{1}{2\pi}\sqrt{\frac{1}{LC} - \frac{R^2}{L^2}}$$

However, resistance $R = 0$, hence,

$$f_r = \frac{1}{2\pi}\sqrt{\frac{1}{LC}}$$

$$= \frac{1}{2\pi}\sqrt{\frac{1}{(150 \times 10^{-3})(40 \times 10^{-6})}}$$

$$= \frac{1}{2\pi}\sqrt{\frac{10^7}{(15)(4)}} = \frac{10^3}{2\pi}\sqrt{\frac{1}{6}}$$

$$= \mathbf{64.97\,Hz}$$

(b)  Current circulating in $L$ and $C$ at resonance,

$$I_{\text{CIRC}} = \frac{V}{X_C} = \frac{V}{\left(\dfrac{1}{2\pi f_r C}\right)} = 2\pi f_r C V$$

Hence

$$I_{\text{CIRC}} = 2\pi (64.97)(40 \times 10^{-6})(50)$$
$$= \mathbf{0.816\,A}$$

(Alternatively,

$$I_{\text{CIRC}} = \frac{V}{X_L} = \frac{V}{2\pi f_r L} = \frac{50}{2\pi (64.97)(0.15)}$$
$$= \mathbf{0.817\,A})$$

---

**Problem 9.**   A coil of inductance 0.20 H and resistance 60 Ω is connected in parallel with a 20 µF capacitor across a 20 V, variable frequency supply. Calculate (a) the resonant frequency, (b) the dynamic resistance, (c) the current at resonance and (d) the circuit $Q$-factor at resonance.

(a)  Parallel resonant frequency,

$$f_r = \frac{1}{2\pi}\sqrt{\frac{1}{LC} - \frac{R^2}{L^2}}$$

$$= \frac{1}{2\pi}\sqrt{\frac{1}{(0.20)(20 \times 10^{-6})} - \frac{(60)^2}{(0.20)^2}}$$

$$= \frac{1}{2\pi}\sqrt{2\,50\,000 - 90\,000} = \frac{1}{2\pi}\sqrt{1\,60\,000}$$

$$= \frac{1}{2\pi}(400) = \mathbf{63.66\,Hz}$$

(b)  Dynamic resistance,

$$R_D = \frac{L}{RC} = \frac{0.20}{(60)(20 \times 10^{-6})} = \mathbf{166.7\,\Omega}$$

(c)  Current at resonance,

$$I_r = \frac{V}{R_D} = \frac{20}{166.7} = \mathbf{0.12\,A}$$

(d)  Circuit $Q$-factor at resonance

$$= \frac{2\pi f_r L}{R} = \frac{2\pi (63.66)(0.20)}{60} = \mathbf{1.33}$$

---

Alternatively, $Q$-factor at resonance

= current magnification (for a parallel circuit)

$$= \frac{I_C}{I_r}$$

$$I_c = \frac{V}{X_C} = \frac{V}{\left(\dfrac{1}{2\pi f_r C}\right)} = 2\pi f_r C V$$

$$= 2\pi (63.66)(20 \times 10^{-6})(20) = 0.16\,A$$

Hence   $Q$-factor $= I_C/I_r = 0.16/0.12 = \mathbf{1.33}$,   as obtained above.

---

**Problem 10.**   A coil of inductance 100 mH and resistance 800 Ω is connected in parallel with a variable capacitor across a 12 V, 5 kHz supply. Determine for the condition when the supply current is a minimum: (a) the capacitance of the capacitor, (b) the dynamic resistance, (c) the supply current, and (d) the $Q$-factor.

(a)  The supply current is a minimum when the parallel circuit is at resonance and resonant frequency,

$$f_r = \frac{1}{2\pi}\sqrt{\frac{1}{LC} - \frac{R^2}{L^2}}$$

Transposing for $C$ gives:

$$(2\pi f_r)^2 = \frac{1}{LC} - \frac{R^2}{L^2}$$

$$(2\pi f_r)^2 + \frac{R^2}{L^2} = \frac{1}{LC}$$

and $C = \dfrac{1}{L\left\{(2\pi f_r)^2 + \dfrac{R^2}{L^2}\right\}}$

When $L = 100$ mH, $R = 800$ Ω and $f_r = 5000$ Hz,

$$C = \frac{1}{100 \times 10^{-3}\left\{(2\pi (5000))^2 + \dfrac{800^2}{(100 \times 10^{-3})^2}\right\}}$$

$$= \frac{1}{0.1\{\pi^2 10^8 + (0.64)(10^8)\}}\,\text{F}$$

$$= \frac{10^6}{0.1(10.51 \times 10^8)}\,\mu\text{F}$$

$$= \mathbf{0.009515\,\mu F}\ \text{or}\ \mathbf{9.515\,nF}$$

(b) Dynamic resistance,

$$R_D = \frac{L}{CR} = \frac{100 \times 10^{-3}}{(9.515 \times 10^{-9})(800)}$$
$$= 13.14\,k\Omega$$

(c) Supply current at resonance,

$$I_r = \frac{V}{R_D} = \frac{12}{13.14 \times 10^3} = 0.913\,mA$$

(d) $Q$-factor at resonance

$$= \frac{2\pi f_r L}{R} = \frac{2\pi (5000)(100 \times 10^{-3})}{800} = 3.93$$

Alternatively, $Q$-factor at resonance

$$= \frac{I_C}{I_r} = \frac{(V/X_C)}{I_r} = \frac{2\pi f_r CV}{I_r}$$

$$= \frac{2\pi (5000)(9.515 \times 10^{-9})(12)}{0.913 \times 10^{-3}} = 3.93$$

*For a practical laboratory experiment on parallel a.c. circuits and resonance, see Chapter 24, page 417.*

**Now try the following exercise**

**Exercise 95    Further problems on parallel resonance and Q-factor**

1. A 0.15 μF capacitor and a pure inductance of 0.01 H are connected in parallel across a 10 V, variable frequency supply. Determine (a) the resonant frequency of the circuit, and (b) the current circulating in the capacitor and inductance.    [(a) 4.11 kHz (b) 38.74 mA]

2. A 30 μF capacitor is connected in parallel with a coil of inductance 50 mH and unknown resistance $R$ across a 120 V, 50 Hz supply. If the circuit has an overall power factor of 1 find (a) the value of $R$, (b) the current in the coil, and (c) the supply current.
[(a) 37.68 Ω (b) 2.94 A (c) 2.714 A]

3. A coil of resistance 25 Ω and inductance 150 mH is connected in parallel with a 10 μF capacitor across a 60 V, variable frequency supply. Calculate (a) the resonant frequency, (b) the dynamic resistance, (c) the current at resonance and (d) the $Q$-factor at resonance.
[(a) 127.2 Hz (b) 600 Ω (c) 0.10 A (d) 4.80]

4. A coil having resistance $R$ and inductance 80 mH is connected in parallel with a 5 nF capacitor across a 25 V, 3 kHz supply. Determine for the condition when the current is a minimum, (a) the resistance $R$ of the coil, (b) the dynamic resistance, (c) the supply current, and (d) the $Q$-factor.
[(a) 3.705 kΩ (b) 4.318 kΩ (c) 5.79 mA (d) 0.41]

5. A coil of resistance 1.5 kΩ and 0.25 H inductance is connected in parallel with a variable capacitance across a 10 V, 8 kHz supply. Calculate (a) the capacitance of the capacitor when the supply current is a minimum, (b) the dynamic resistance, and (c) the supply current.
[(a) 1561 pF (b) 106.8 kΩ (c) 93.66 μA]

6. A parallel circuit as shown in Fig. 16.13 is tuned to resonance by varying capacitance $C$. Resistance, $R = 30\,\Omega$, inductance, $L = 400\,\mu H$, and the supply voltage, $V = 200\,V$, 5 MHz.

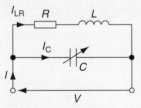

Figure 16.13

Calculate (a) the value of $C$ to give resonance at 5 MHz, (b) the dynamic impedance, (c) the $Q$-factor, (d) the bandwidth, (e) the current in each branch, (f) the supply current, and (g) the power dissipated at resonance.
[(a) 2.533 pF (b) 5.264 MΩ (c) 418.9 (d) 11.94 kHz (e) $I_C = 15.915\angle 90°$ mA, $I_{LR} = 15.915\angle -89.863°$ mA (f) 38 μA (g) 7.60 mW]

## 16.7  Power factor improvement

From page 243, in any a.c. circuit, **power factor $= \cos\phi$**, where $\phi$ is the phase angle between supply current and supply voltage.

Industrial loads such as a.c. motors are essentially inductive (i.e. R-L) and may have a low power factor. For example, let a motor take a current of 50 A at a power factor of 0.6 lagging from a 240 V, 50 Hz supply, as shown in the circuit diagram of Fig. 16.14(a).

If power factor = 0.6 lagging, then:

$$\cos\phi = 0.6 \text{ lagging}$$

Hence,

$$\text{phase angle, } \phi = \cos^{-1}0.6 = 53.13° \text{ lagging}$$

Lagging means that $I$ lags $V$ (remember CIVIL), and the phasor diagram is as shown in Fig. 16.14(b).

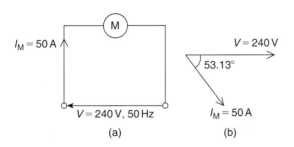

**Figure 16.14**

How can this power factor of 0.6 be 'improved' or 'corrected' to, say, unity?

Unity power factor means: $\cos\phi = 1$ from which, $\phi = 0$

So how can the circuit of Fig. 16.14(a) be modified so that the circuit phase angle is changed from 53.13° to 0°? The answer is to connect a capacitor in parallel with the motor as shown in Fig. 16.15(a).

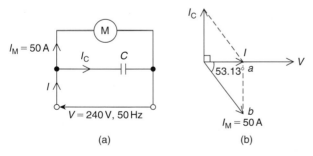

**Figure 16.15**

When a capacitor is connected in parallel with the inductive load, it takes a current shown as $I_C$. In the phasor diagram of Fig. 16.15(b), current $I_C$ is shown leading the voltage $V$ by 90° (again, remember CIVIL). The supply current in Fig. 16.15(a) is shown as $I$ and is now the phasor sum of $I_M$ and $I_C$.

In the phasor diagram of Fig. 16.15(b), current $I$ is shown as the phasor sum of $I_M$ and $I_C$ and is in phase with $V$, i.e. the circuit phase angle is 0°, which means that the power factor is $\cos 0° = 1$.

Thus, by connecting a capacitor in parallel with the motor, the power factor has been improved from 0.6 lagging to unity.

From right angle triangles, $\cos 53.13°$

$$= \frac{\text{adjacent}}{\text{hypotenuse}} = \frac{I}{50}$$

from which,    **supply current, $I = 50\cos 53.13°$**

$$= 30\,A$$

**Before the capacitor was connected, the supply current was 50 A. Now it is 30 A.**

Herein lies **the advantage of power factor improvement – the supply current has been reduced.**

When power factor is improved, **the supply current is reduced, the supply system has lower losses** (i.e. lower $I^2R$ losses) and therefore **cheaper running costs**.

**Problem 11.** In the circuit of Fig. 16.16, what value of capacitor is needed to improve the power factor from 0.6 lagging to unity?

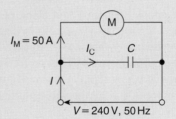

**Figure 16.16**

This is the same circuit as used above where the supply current was reduced from 50 A to 30 A by power factor improvement. In the phasor diagram of Fig. 16.17, current $I_C$ needs to equal $ab$ if $I$ is to be in phase with $V$.

From right angle triangles, $\sin 53.13° = \dfrac{\text{opposite}}{\text{hypotenuse}}$

$$= \frac{ab}{50}$$

from which,    $ab = 50\sin 53.13° = 40\,A$

Hence, **a capacitor has to be of such a value as to take 40 A for the power factor to be improved from 0.6 to 1.**

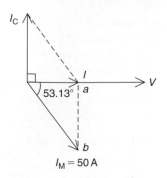

**Figure 16.17**

From a.c. theory, in the circuit of Fig. 16.16,

$$I_C = \frac{V}{X_c} = \frac{V}{\left(\dfrac{1}{2\pi f C}\right)} = 2\pi f C V$$

from which,

**capacitance**, $C = \dfrac{I_c}{2\pi f V} = \dfrac{40}{2\pi(50)(240)} = \mathbf{530.5\,\mu F}$

---

In **practical situations** a power factor of 1 is not normally required but a power factor in the region of **0.8** or better is usually aimed for. (Actually, a power factor of 1 means resonance!)

> **Problem 12.** An inductive load takes a current of 60 A at a power factor of 0.643 lagging when connected to a 240 V, 60 Hz supply. It is required to improve the power factor to 0.80 lagging by connecting a capacitor in parallel with the load. Calculate (a) the new supply current, (b) the capacitor current, and (c) the value of the power factor correction capacitor.

(a)  A power factor of 0.643 means

$$\cos\phi_1 = 0.643$$

from which,   $\phi_1 = \cos^{-1}0.643 = 50°$

A power factor of 0.80 means

$$\cos\phi_2 = 0.80$$

from which,   $\phi_2 = \cos^{-1}0.80 = 36.87°$

The phasor diagram is shown in Fig. 16.18, where the new supply current $I$ is shown by length $Ob$

From triangle $Oac$,   $\cos 50° = \dfrac{Oa}{60}$ from which,

$$Oa = 60\cos 50° = 38.57\,\text{A}$$

From triangle $Oab$,

$$\cos 36.87° = \frac{Oa}{Ob} = \frac{38.57}{I}$$

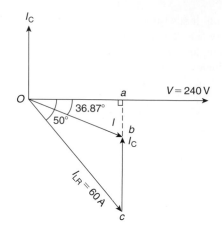

**Figure 16.18**

from which,  **new supply current**,

$$I = \frac{38.57}{\cos 36.87} = \mathbf{48.21\,A}$$

(b)  The new supply current $I$ is the phasor sum of $I_C$ and $I_{LR}$

Thus, if   $I = I_C + I_{LR}$ then $I_C = I - I_{LR}$

i.e.  **capacitor current**,

$$I_C = 48.21\angle-36.87° - 60\angle-50°$$
$$= (38.57 - j28.93) - (38.57 - j45.96)$$
$$= (0 + j17.03)\,\text{A}  \quad\text{or}\quad \mathbf{17.03\angle90°\,A}$$

(c)  Current,   $I_C = \dfrac{V}{X_c} = \dfrac{V}{\left(\dfrac{1}{2\pi f C}\right)} = 2\pi f C V$

from which, **capacitance**,

$$C = \frac{I_c}{2\pi f V} = \frac{17.03}{2\pi(60)(240)} = \mathbf{188.2\,\mu F}$$

> **Problem 13.** A 400 V alternator is supplying a load of 42 kW at a power factor of 0.7 lagging. Calculate (a) the kVA loading and (b) the current taken from the alternator. (c) If the power factor is now raised to unity find the new kVA loading.

(a)  Power $= VI\cos\phi = (VI)$ (power factor)

Hence $VI = \dfrac{\text{power}}{\text{p.f.}} = \dfrac{42 \times 10^3}{0.7} = \mathbf{60\,kVA}$

(b)  $VI = 60\,000\,\text{VA}$

hence $I = \dfrac{60\,000}{V} = \dfrac{60\,000}{400} = \mathbf{150\,A}$

(c) The kVA loading remains at **60 kVA** irrespective of changes in power factor.

> **Problem 14.** A motor has an output of 4.8 kW, an efficiency of 80% and a power factor of 0.625 lagging when operated from a 240 V, 50 Hz supply. It is required to improve the power factor to 0.95 lagging by connecting a capacitor in parallel with the motor. Determine (a) the current taken by the motor, (b) the supply current after power factor correction, (c) the current taken by the capacitor, (d) the capacitance of the capacitor, and (e) the kvar rating of the capacitor.

(a)  Efficiency $= \dfrac{\text{power output}}{\text{power input}}$

hence  $\dfrac{80}{100} = \dfrac{4800}{\text{power input}}$

and power input $= \dfrac{4800}{0.8} = 6000\,\text{W}$

Hence,  $6000 = V I_M \cos\phi = (240)(I_M)(0.625)$, since $\cos\phi = \text{p.f.} = 0.625$. Thus current taken by the motor,

$$I_M = \frac{6000}{(240)(0.625)} = \textbf{40 A}$$

The circuit diagram is shown in Fig. 16.19(a). The phase angle between $I_M$ and $V$ is given by: $\phi = \cos^{-1}0.625 = 51.32°$, hence the phasor diagram is as shown in Fig. 16.19(b).

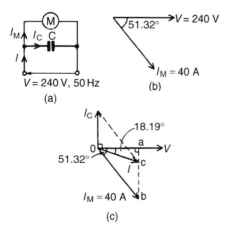

(a)

(b)

(c)

**Figure 16.19**

(b)  When a capacitor $C$ is connected in parallel with the motor a current $I_C$ flows which leads $V$ by $90°$.

The phasor sum of $I_M$ and $I_C$ gives the supply current $I$, and has to be such as to change the circuit power factor to 0.95 lagging, i.e. a phase angle of $\cos^{-1}0.95$ or $18.19°$ lagging, as shown in Fig. 16.19(c). The horizontal component of $I_M$ (shown as $oa$)

$$= I_M \cos 51.32°$$
$$= 40 \cos 51.32° = 25\,\text{A}$$

The horizontal component of $I$ (also given by $oa$)

$$= I \cos 18.19°$$
$$= 0.95\,I$$

Equating the horizontal components gives: $25 = 0.95\,I$. Hence the supply current after p.f. correction,

$$I = \frac{25}{0.95} = \textbf{26.32 A}$$

(c)  The vertical component of $I_M$ (shown as $ab$)

$$= I_M \sin 51.32°$$
$$= 40 \sin 51.32° = 31.22\,\text{A}$$

The vertical component of $I$ (shown as $ac$)

$$= I \sin 18.19°$$
$$= 26.32 \sin 18.19° = 8.22\,\text{A}$$

The magnitude of the capacitor current $I_C$ (shown as $bc$) is given by

$$ab - ac \quad \text{i.e.} \quad I_C = 31.22 - 8.22 = \textbf{23 A}$$

(d)  Current $I_C = \dfrac{V}{X_C} = \dfrac{V}{\left(\dfrac{1}{2\pi f C}\right)} = 2\pi f C V$

from which

$$C = \frac{I_C}{2\pi f V} = \frac{23}{2\pi (50)(240)}\,F = \textbf{305\,\textmu F}$$

(e)  **kvar rating** of the capacitor

$$= \frac{V I_C}{1000} = \frac{(240)(23)}{1000} = \textbf{5.52 kvar}$$

In this problem the supply current has been reduced from 40 A to 26.32 A without altering the current or power taken by the motor. This means that the $I^2R$ losses are reduced, and results in a saving of costs.

**Problem 15.** A 250 V, 50 Hz single-phase supply feeds the following loads (i) incandescent lamps taking a current of 10 A at unity power factor, (ii) fluorescent lamps taking 8 A at a power factor of 0.7 lagging, (iii) a 3 kVA motor operating at full load and at a power factor of 0.8 lagging and (iv) a static capacitor. Determine, for the lamps and motor, (a) the total current, (b) the overall power factor, and (c) the total power. (d) Find the value of the static capacitor to improve the overall power factor to 0.975 lagging.

A phasor diagram is constructed as shown in Fig. 16.20(a), where 8 A is lagging voltage $V$ by $\cos^{-1}0.7$, i.e. $45.57°$, and the motor current is $(3000/250)$, i.e. 12 A lagging $V$ by $\cos^{-1}0.8$, i.e. $36.87°$

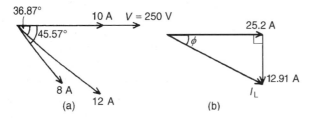

**Figure 16.20**

(a)  The horizontal component of the currents

$$= 10\cos 0° + 12\cos 36.87° + 8\cos 45.57°$$

$$= 10 + 9.6 + 5.6 = 25.2\,\text{A}$$

The vertical component of the currents

$$= 10\sin 0° + 12\sin 36.87° + 8\sin 45.57°$$

$$= 0 + 7.2 + 5.713 = 12.91\,\text{A}$$

From Fig. 16.20(b), total current,

$$I_L = \sqrt{25.2^2 + 12.91^2} = \textbf{28.31 A} \text{ at a phase angle}$$
of $\phi = \tan^{-1}(12.91/25.2)$ i.e. $27.13°$ lagging.

(b)  Power factor

$$= \cos\phi = \cos 27.13° = \textbf{0.890 lagging}$$

(c)  Total power,

$$P = V I_L \cos\phi = (250)(28.31)(0.890)$$

$$= \textbf{6.3 kW}$$

(d)  To improve the power factor, a capacitor is connected in parallel with the loads. The capacitor

takes a current $I_C$ such that the supply current falls from 28.31 A to $I$, lagging $V$ by $\cos^{-1}0.975$, i.e. $12.84°$. The phasor diagram is shown in Fig. 16.21.

$$oa = 28.31\cos 27.13° = I\cos 12.84°$$

$$\text{hence } I = \frac{28.31\cos 27.13°}{\cos 12.84°} = 25.84\,\text{A}$$

Current $I_C = bc = (ab - ac)$

$$= 28.31\sin 27.13° - 25.84\sin 12.84°$$

$$= 12.91 - 5.742 = 7.168\,\text{A}$$

$$I_C = \frac{V}{X_C} = \frac{V}{\left(\dfrac{1}{2\pi fc}\right)} = 2\pi fCV$$

Hence capacitance

$$C = \frac{I_C}{2\pi f\,V} = \frac{7.168}{2\pi(50)(250)}\,F = \textbf{91.27 μF}$$

Thus to improve the power factor from 0.890 to 0.975 lagging a 91.27 μF capacitor is connected in parallel with the loads.

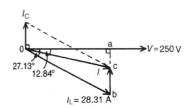

**Figure 16.21**

**Now try the following exercises**

**Exercise 96    Further problems on power factor improvement**

1.  A 415 V alternator is supplying a load of 55 kW at a power factor of 0.65 lagging. Calculate (a) the kVA loading and (b) the current taken from the alternator. (c) If the power factor is now raised to unity find the new kVA loading.

    [(a) 84.6 kVA  (b) 203.9 A  (c) 84.6 kVA]

2.  A single phase motor takes 30 A at a power factor of 0.65 lagging from a 240 V, 50 Hz

supply. Determine (a) the current taken by the capacitor connected in parallel to correct the power factor to unity, and (b) the value of the supply current after power factor correction.

[(a) 22.80 A (b) 19.50 A]

3. A $20\,\Omega$ non-reactive resistor is connected in series with a coil of inductance $80\,\text{mH}$ and negligible resistance. The combined circuit is connected to a $200\,\text{V}$, $50\,\text{Hz}$ supply. Calculate (a) the reactance of the coil, (b) the impedance of the circuit, (c) the current in the circuit, (d) the power factor of the circuit, (e) the power absorbed by the circuit, (f) the value of a power factor correction capacitor to produce a power factor of unity, and (g) the value of a power factor correction capacitor to produce a power factor of 0.9.

[(a) $25.13\,\Omega$ (b) $32.12\angle51.49°\,\Omega$
(c) $6.227\angle-51.49°$ A (d) 0.623
(e) 775.5 W (f) $77.56\,\mu\text{F}$ (g) $47.67\,\mu\text{F}$]

4. A motor has an output of $6\,\text{kW}$, an efficiency of 75% and a power factor of 0.64 lagging when operated from a $250\,\text{V}$, $60\,\text{Hz}$ supply. It is required to raise the power factor to 0.925 lagging by connecting a capacitor in parallel with the motor. Determine (a) the current taken by the motor, (b) the supply current after power factor correction, (c) the current taken by the capacitor, (d) the capacitance of the capacitor, and (e) the kvar rating of the capacitor.

[(a) 50 A (b) 34.59 A (c) 25.28 A
(d) $268.2\,\mu\text{F}$ (e) 6.32 kvar]

5. A supply of $250\,\text{V}$, $80\,\text{Hz}$ is connected across an inductive load and the power consumed is $2\,\text{kW}$, when the supply current is $10\,\text{A}$. Determine the resistance and inductance of the circuit. What value of capacitance connected in parallel with the load is needed to improve the overall power factor to unity?

[$R=20\,\Omega$, $L=29.84\,\text{mH}$, $C=47.75\,\mu\text{F}$]

6. A $200\,\text{V}$, $50\,\text{Hz}$ single-phase supply feeds the following loads: (i) fluorescent lamps taking a current of $8\,\text{A}$ at a power factor of 0.9 leading, (ii) incandescent lamps taking a current of $6\,\text{A}$ at unity power factor, (iii) a motor taking a current of $12\,\text{A}$ at a power factor of 0.65 lagging.

Determine the total current taken from the supply and the overall power factor. Find also the value of a static capacitor connected in parallel with the loads to improve the overall power factor to 0.98 lagging.

[21.74 A, 0.966 lagging, $21.74\,\mu\text{F}$]

---

**Exercise 97    Short answer questions on single-phase parallel a.c. circuits**

1. Draw a phasor diagram for a two-branch parallel circuit containing capacitance $C$ in one branch and resistance $R$ in the other, connected across a supply voltage $V$

2. Draw a phasor diagram for a two-branch parallel circuit containing inductance $L$ and resistance $R$ in one branch and capacitance $C$ in the other, connected across a supply voltage $V$

3. Draw a phasor diagram for a two-branch parallel circuit containing inductance $L$ in one branch and capacitance $C$ in the other for the condition in which inductive reactance is greater than capacitive reactance

4. State two methods of determining the phasor sum of two currents

5. State two formulae which may be used to calculate power in a parallel circuit

6. State the condition for resonance for a two-branch circuit containing capacitance $C$ in parallel with a coil of inductance $L$ and resistance $R$

7. Develop a formula for the resonant frequency in an $LR$–$C$ parallel circuit, in terms of resistance $R$, inductance $L$ and capacitance $C$

8. What does $Q$-factor of a parallel circuit mean?

9. Develop a formula for the current at resonance in an $LR$–$C$ parallel circuit in terms of resistance $R$, inductance $L$, capacitance $C$ and supply voltage $V$

10. What is dynamic resistance? State a formula for dynamic resistance

11. Explain a simple method of improving the power factor of an inductive circuit

12. Why is it advantageous to improve power factor?

**Exercise 98    Multi-choice questions on single-phase parallel a.c. circuits**
**(Answers on page 421)**

A two-branch parallel circuit containing a $10\,\Omega$ resistance in one branch and a $100\,\mu$F capacitor in the other, has a 120 V, $2/3\pi$ kHz supply connected across it. Determine the quantities stated in questions 1 to 8, selecting the correct answer from the following list:

(a) 24 A           (b) $6\,\Omega$
(c) $7.5\,\mathrm{k}\Omega$        (d) 12 A
(e) $\tan^{-1}\frac{3}{4}$ leading   (f) 0.8 leading
(g) $7.5\,\Omega$         (h) $\tan^{-1}\frac{4}{3}$ leading

(i) 16 A           (j) $\tan^{-1}\frac{5}{3}$ lagging
(k) 1.44 kW        (l) 0.6 leading
(m) $12.5\,\Omega$        (n) 2.4 kW
(o) $\tan^{-1}\frac{4}{3}$ lagging   (p) 0.6 lagging
(q) 0.8 lagging        (r) 1.92 kW
(s) 20 A

1. The current flowing in the resistance

2. The capacitive reactance of the capacitor

3. The current flowing in the capacitor

4. The supply current

5. The supply phase angle

6. The circuit impedance

7. The power consumed by the circuit

8. The power factor of the circuit

9. A two-branch parallel circuit consists of a 15 mH inductance in one branch and a $50\,\mu$F capacitor in the other across a 120 V, $1/\pi$ kHz supply. The supply current is:

(a)  8 A leading by $\dfrac{\pi}{2}$ rad

(b)  16 A lagging by $90°$

(c)  8 A lagging by $90°$

(d)  16 A leading by $\dfrac{\pi}{2}$ rad

10. The following statements, taken correct to 2 significant figures, refer to the circuit shown in Fig. 16.22. Which are false?
(a) The impedance of the $R$–$L$ branch is $5\,\Omega$
(b) $I_{LR} = 50$ A
(c) $I_C = 20$ A
(d) $L = 0.80$ H

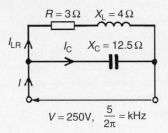

**Figure 16.22**

(e) $C = 16\,\mu$F
(f) The 'in-phase' component of the supply current is 30 A
(g) The 'quadrature' component of the supply current is 40 A
(h) $I = 36$ A
(i) Circuit phase $= 33°41'$ leading
(j) Circuit impedance $= 6.9\,\Omega$
(k) Circuit power factor $= 0.83$ lagging
(l) Power consumed $= 9.0$ kW

11. Which of the following statements is false?
(a) The supply current is a minimum at resonance in a parallel circuit
(b) The $Q$-factor at resonance in a parallel circuit is the voltage magnification
(c) Improving power factor reduces the current flowing through a system
(d) The circuit impedance is a maximum at resonance in a parallel circuit

12. An $LR$–$C$ parallel circuit has the following component values: $R = 10\,\Omega$, $L = 10$ mH, $C = 10\,\mu$F and $V = 100$ V. Which of the following statements is false?
(a) The resonant frequency $f_r$ is $1.5/\pi$ kHz
(b) The current at resonance is 1 A
(c) The dynamic resistance is $100\,\Omega$
(d) The circuit $Q$-factor at resonance is 30

13. The magnitude of the impedance of the circuit shown in Fig. 16.23 is:
    (a) $7\,\Omega$        (b) $5\,\Omega$
    (c) $2.4\,\Omega$        (d) $1.71\,\Omega$

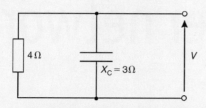

Figure 16.23

14. In the circuit shown in Fig. 16.24, the magnitude of the supply current $I$ is:
    (a) 17A        (b) 7A
    (c) 15A        (d) 23A

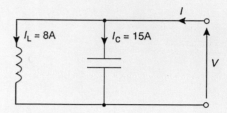

Figure 16.24

# Chapter 17

# Filter networks

At the end of this chapter you should be able to:

- appreciate the purpose of a filter network
- understand basic types of filter sections, i.e. low-pass, high-pass, band-pass and band-stop filters
- define cut-off frequency, two-port networks and characteristic impedance
- design low- and high-pass filter sections given nominal impedance and cut-off frequency
- determine the values of components comprising a band-pass filter given cut-off frequencies
- appreciate the difference between ideal and practical filter characteristics

## 17.1 Introduction

**Attenuation** is a reduction or loss in the magnitude of a voltage or current due to its transmission over a line. A **filter** is a network designed to pass signals having frequencies within certain bands (called **pass-bands**) with little attenuation, but greatly attenuates signals within other bands (called **attenuation bands** or **stop-bands**).

A filter is frequency sensitive and is thus composed of reactive elements. Since certain frequencies are to be passed with minimal loss, ideally the inductors and capacitors need to be pure components since the presence of resistance results in some attenuation at all frequencies.

Between the pass-band of a filter, where ideally the attenuation is zero, and the attenuation band, where ideally the attenuation is infinite, is the **cut-off frequency**, this being the frequency at which the attenuation changes from zero to some finite value.

A filter network containing no source of power is termed **passive**, and one containing one or more power sources is known as an **active** filter network.

Filters are used for a variety of purposes in nearly every type of electronic communications and control equipment. The bandwidths of filters used in communications systems vary from a fraction of a hertz to many megahertz, depending on the application.

There are four basic types of filter sections:

(a)  low-pass

(b)  high-pass

(c)  band-pass

(d)  band-stop

## 17.2 Two-port networks and characteristic impedance

Networks in which electrical energy is fed in at one pair of terminals and taken out at a second pair of terminals are called **two-port networks**. The network between the input port and the output port is a transmission network for which a known relationship exists between the input and output currents and voltages.

Figure 17.1(a) shows a **T-network**, which is termed **symmetrical** if $Z_A = Z_B$, and Fig. 17.1(b) shows a **$\pi$-network** which is symmetrical if $Z_E = Z_F$.

If $Z_A \neq Z_B$ in Fig. 17.1(a) and $Z_E \neq Z_F$ in Fig. 17.1(b), the sections are termed **asymmetrical**.

DOI: 10.1016/B978-0-08-089056-2.00017-6

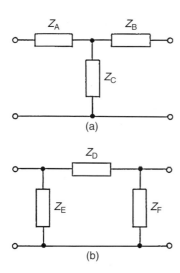

**Figure 17.1**

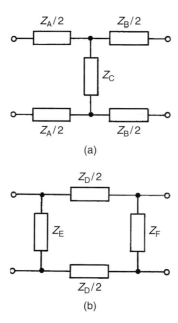

**Figure 17.2**

Both networks shown have one common terminal, which may be earthed, and are therefore said to be **unbalanced**. The **balanced** form of the T-network is shown in Fig. 17.2(a) and the balanced form of the $\pi$-network is shown in Fig. 17.2(b).

The input impedance of a network is the ratio of voltage to current at the input terminals. With a two-port network the input impedance often varies according to the load impedance across the output terminals. For any passive two-port network it is found that a particular value of load impedance can always be found which will produce an input impedance having the same

value as the load impedance. This is called the **iterative impedance** for an asymmetrical network and its value depends on which pair of terminals is taken to be the input and which the output (there are thus two values of iterative impedance, one for each direction).

For a symmetrical network there is only one value for the iterative impedance and this is called the **characteristic impedance** $Z_0$ of the symmetrical two-port network.

## 17.3    Low-pass filters

Figure 17.3 shows simple unbalanced T- and $\pi$-section filters using series inductors and shunt capacitors. If either section is connected into a network and a continuously increasing frequency is applied, each would have a frequency-attenuation characteristic as shown in Fig. 17.4. This is an ideal characteristic and assumes pure reactive elements. All frequencies are seen to be passed from zero up to a certain value without attenuation, this value being shown as $f_c$, the cut-off frequency; all values of frequency above $f_c$ are attenuated. It is for this reason that the networks shown in Fig. 17.3(a) and (b) are known as **low-pass filters**.

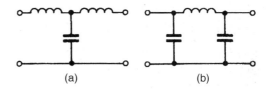

**Figure 17.3**

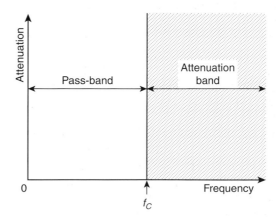

**Figure 17.4**

The electrical circuit diagram symbol for a low-pass filter is shown in Fig. 17.5.

Figure 17.5

Summarising, **a low-pass filter is one designed to pass signals at frequencies below a specified cut-off frequency.**

In practise, the characteristic curve of a low-pass prototype filter section looks more like that shown in Fig. 17.6. The characteristic may be improved somewhat closer to the ideal by connecting two or more identical sections in cascade. This produces a much sharper cut-off characteristic, although the attenuation in the pass-band is increased a little.

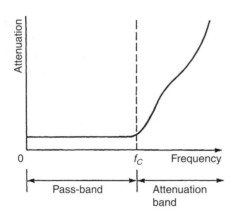

Figure 17.6

When rectifiers are used to produce the d.c. supplies of electronic systems, a large ripple introduces undesirable noise and may even mask the effect of the signal voltage. Low-pass filters are added to smooth the output voltage waveform, this being one of the most common applications of filters in electrical circuits.

Filters are employed to isolate various sections of a complete system and thus to prevent undesired interactions. For example, the insertion of low-pass decoupling filters between each of several amplifier stages and a common power supply reduces interaction due to the common power supply impedance.

### Cut-off frequency and nominal impedance calculations

A low-pass symmetrical T-network and a low-pass symmetrical $\pi$-network are shown in Fig. 17.7. It may

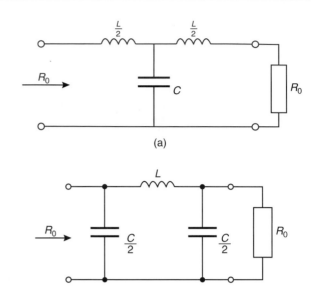

Figure 17.7

be shown that the cut-off frequency, $f_c$, for each section is the same, and is given by:

$$f_c = \frac{1}{\pi\sqrt{LC}} \tag{1}$$

When the frequency is very low, the characteristic impedance is purely resistive. This value of characteristic impedance is known as the **design impedance** or the **nominal impedance** of the section and is often given the symbol $R_0$, where

$$R_0 = \sqrt{\frac{L}{C}} \tag{2}$$

Problem 1. Determine the cut-off frequency and the nominal impedance for the low-pass T-connected section shown in Fig. 17.8.

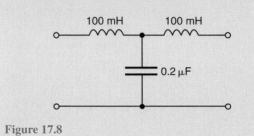

Figure 17.8

Comparing Fig. 17.8 with the low-pass section of Fig. 17.7(a), shows that:

$$\frac{L}{2} = 100\,\text{mH},$$

i.e. inductance,    $L = 200\,\text{mH} = 0.2\,\text{H},$

and capacitance    $C = 0.2\,\mu\text{F} = 0.2 \times 10^{-6}\,\text{F}.$

From equation (1), **cut-off frequency**,

$$f_c = \frac{1}{\pi\sqrt{LC}}$$

$$= \frac{1}{\pi\sqrt{(0.2 \times 0.2 \times 10^{-6})}} = \frac{10^3}{\pi(0.2)}$$

i.e.    $f_c = 1592\,\text{Hz}\quad\text{or}\quad 1.592\,\text{kHz}$

From equation (2), **nominal impedance**,

$$R_0 = \sqrt{\frac{L}{C}} = \sqrt{\frac{0.2}{0.2 \times 10^{-6}}}$$

$$= 1000\,\Omega\quad\text{or}\quad 1\,\text{k}\Omega$$

**Problem 2.** Determine the cut-off frequency and the nominal impedance for the low-pass $\pi$-connected section shown in Fig. 17.9.

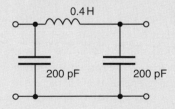

0.4 H

200 pF        200 pF

**Figure 17.9**

Comparing Fig. 17.9 with the low-pass section of Fig. 17.7(b), shows that:

$$\frac{C}{2} = 200\,\text{pF},$$

i.e. capacitance,    $C = 400\,\text{pF} = 400 \times 10^{-12}\,\text{F},$

and inductance    $L = 0.4\,\text{H}.$

From equation (1), **cut-off frequency**,

$$f_c = \frac{1}{\pi\sqrt{LC}}$$

$$= \frac{1}{\pi\sqrt{(0.4 \times 400 \times 10^{-12})}} = \frac{10^6}{\pi\sqrt{160}}$$

i.e.    $f_c = 25.16\,\text{kHz}$

From equation (2), **nominal impedance**,

$$R_0 = \sqrt{\frac{L}{C}} = \sqrt{\frac{0.4}{400 \times 10^{-12}}} = 31.62\,\text{k}\Omega$$

## To determine values of *L* and *C* given $R_0$ and $f_c$

If the values of the nominal impedance $R_0$ and the cut-off frequency $f_c$ are known for a low-pass T- or $\pi$-section, it is possible to determine the values of inductance and capacitance required to form the section. It may be shown that:

$$\text{capacitance } C = \frac{1}{\pi R_0 f_c} \tag{3}$$

and    $$\text{inductance } L = \frac{R_0}{\pi f_c} \tag{4}$$

**Problem 3.** A filter section is to have a characteristic impedance at zero frequency of $600\,\Omega$ and a cut-off frequency of 5 MHz. Design (a) a low-pass T-section filter, and (b) a low-pass $\pi$-section filter to meet these requirements.

The characteristic impedance at zero frequency is the nominal impedance $R_0$, i.e. $R_0 = 600\,\Omega$; cut-off frequency $f_c = 5\,\text{MHz} = 5 \times 10^6\,\text{Hz}.$
From equation (3), capacitance,

$$C = \frac{1}{\pi R_0 f_c} = \frac{1}{\pi(600)(5 \times 10^6)}\,\text{F}$$

$$= 1.06 \times 10^{-10}\,\text{F} = 106\,\text{pF}$$

From equation (4), inductance,

$$L = \frac{R_0}{\pi f_c} = \frac{600}{\pi(5 \times 10^6)}\,\text{H}$$

$$= 3.82 \times 10^{-5} = 38.2\,\mu\text{H}$$

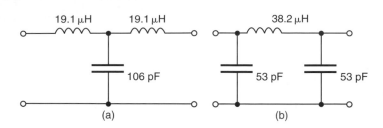

**Figure 17.10**

(a) A low-pass T-section filter is shown in Fig. 17.10(a), where the series arm inductances are each $\dfrac{L}{2}$ (see Fig. 17.7(a)), i.e. $\dfrac{38.2}{2} = 19.1\,\mu\text{H}$

(b) A low-pass $\pi$-section filter is shown in Fig. 17.10(b), where the shunt arm capacitances are each $\dfrac{C}{2}$ (see Fig. 17.7(b)), i.e. $\dfrac{106}{2} = 53\,\text{pF}$

**Now try the following exercise**

**Exercise 99   Further problems on low-pass filter sections**

1.  Determine the cut-off frequency and the nominal impedance of each of the low-pass filter sections shown in Fig. 17.11.
    [(a) 1592 Hz; 5 kΩ (b) 9545 Hz; 600 Ω]

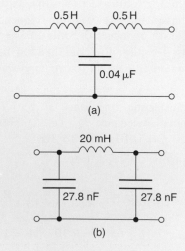

**Figure 17.11**

2.  A filter section is to have a characteristic impedance at zero frequency of 500 Ω and a cut-off frequency of 1 kHz. Design (a) a low-pass T-section filter, and (b) a low-pass $\pi$-section filter to meet these requirements.
    [(a) Each series arm 79.60 mH shunt arm
    0.6366 µF
    (b) Series arm 159.2 mH, each shunt arm
    0.3183 µF]

3.  Determine the value of capacitance required in the shunt arm of a low-pass T-section if the inductance in each of the series arms is 40 mH and the cut-off frequency of the filter is 2.5 kHz.   [0.203 µF]

4.  The nominal impedance of a low-pass $\pi$-section filter is 600 Ω. If the capacitance in each of the shunt arms is 0.1 µF determine the inductance in the series arm.   [72 mH]

## 17.4   High-pass filters

Figure 17.12 shows simple unbalanced T- and $\pi$-section filters using series capacitors and shunt inductors. If either section is connected into a network and a continuously increasing frequency is applied, each would have a frequency-attenuation characteristic as shown in Fig. 17.13.

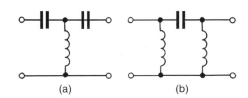

**Figure 17.12**

Once again this is an ideal characteristic assuming pure reactive elements. All frequencies below the

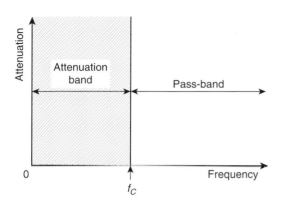

Figure 17.13

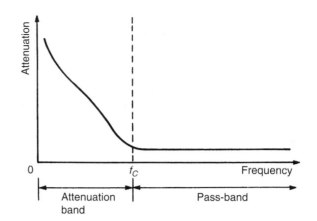

Figure 17.15

cut-off frequency $f_c$ are seen to be attenuated and all frequencies above $f_c$ are passed without loss.

It is for this reason that the networks shown in Figs 17.12(a) and (b) are known as **high-pass filters**. The electrical circuit diagram symbol for a high-pass filter is shown in Fig. 17.14.

Figure 17.14

Summarising, **a high-pass filter is one designed to pass signals at frequencies above a specified cut-off frequency**.

The characteristic shown in Fig. 17.13 is ideal in that it is assumed that there is no attenuation at all in the pass-bands and infinite attenuation in the attenuation band. Both of these conditions are impossible to achieve in practice. Due to resistance, mainly in the inductive elements the attenuation in the pass-band will not be zero, and in a practical filter section the attenuation in the attenuation band will have a finite value. In addition to the resistive loss there is often an added loss due to mismatching.

Ideally when a filter is inserted into a network it is matched to the impedance of that network. However the characteristic impedance of a filter section will vary with frequency and the termination of the section may be an impedance that does not vary with frequency in the same way.

Figure 17.13 showed an ideal high-pass filter section characteristic of attenuation against frequency. In practise, the characteristic curve of a high-pass prototype filter section would look more like that shown in Fig. 17.15.

## Cut-off frequency and nominal impedance calculations

A high-pass symmetrical T-network and a high-pass symmetrical $\pi$-network are shown in Fig. 17.16. It may be shown that the cut-off frequency, $f_c$, for each section is the same, and is given by:

$$f_c = \frac{1}{4\pi\sqrt{LC}} \qquad (5)$$

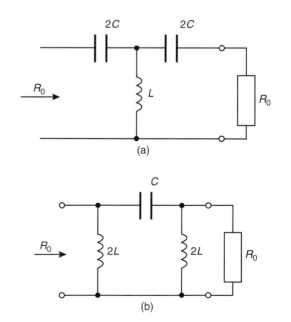

Figure 17.16

When the frequency is very high, the characteristic impedance is purely resistive. This value of characteristic impedance is then the **nominal imped-ance** of the section and is given by:

$$R_0 = \sqrt{\frac{L}{C}} \qquad (6)$$

**Problem 4.**   Determine the cut-off frequency and the nominal impedance for the high-pass T-connected section shown in Fig. 17.17.

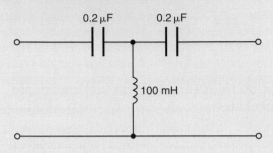

0.2 µF      0.2 µF

100 mH

**Figure 17.17**

Comparing Fig. 17.17 with the high-pass section of Fig. 17.16(a), shows that:

$$2C = 0.2\,\mu\text{F},$$

i.e. capacitance,    $C = 0.1\,\mu\text{F} = 0.1 \times 10^{-6}$,

and inductance,    $L = 100\,\text{mH} = 0.1\,\text{H}.$

From equation (5), **cut-off frequency**,

$$f_c = \frac{1}{4\pi \sqrt{LC}}$$

$$= \frac{1}{4\pi \sqrt{(0.1 \times 0.1 \times 10^{-6})}} = \frac{10^3}{4\pi (0.1)}$$

i.e.    $f_c = 796\,\text{Hz}$

From equation (6), **nominal impedance**,

$$R_0 = \sqrt{\frac{L}{C}} = \sqrt{\frac{0.1}{0.1 \times 10^{-6}}}$$

$$= 1000\,\Omega \quad \text{or} \quad 1\,\text{k}\Omega$$

**Problem 5.**   Determine the cut-off frequency and the nominal impedance for the high-pass $\pi$-connected section shown in Fig. 17.18.

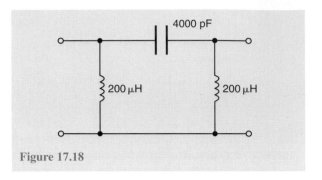

4000 pF

200 µH      200 µH

**Figure 17.18**

Comparing Fig. 17.18 with the high-pass section of Fig. 17.16(b), shows that:

$$2L = 200\,\mu\text{H},$$

i.e. inductance,    $L = 100\,\mu\text{H} = 10^{-4}\,\text{H},$

and capacitance,   $C = 4000\,\text{pF} = 4 \times 10^{-9}\,\text{F}.$

From equation (5), **cut-off frequency**,

$$f_c = \frac{1}{4\pi \sqrt{LC}}$$

$$= \frac{1}{4\pi \sqrt{(10^{-4} \times 4 \times 10^{-9})}} = 1.26 \times 10^5$$

i.e.    $f_c = 126\,\text{kHz}$

From equation (6), **nominal impedance**,

$$R_0 = \sqrt{\frac{L}{C}} = \sqrt{\frac{10^{-4}}{4 \times 10^{-9}}}$$

$$= \sqrt{\frac{10^5}{4}} = 158\,\Omega$$

### To determine values of $L$ and $C$ given $R_0$ and $f_c$

If the values of the nominal impedance $R_0$ and the cut-off frequency $f_c$ are known for a high-pass T- or $\pi$-section, it is possible to determine the values of inductance and capacitance required to form the section. It may be shown that:

$$\text{capacitance } C = \frac{1}{4\pi R_0 f_c} \qquad (7)$$

and $$\text{inductance } L = \frac{R_0}{4\pi f_c} \qquad (8)$$

> **Problem 6.** A filter section is required to pass all frequencies above 25 kHz and to have a nominal impedance of 600 Ω. Design (a) a high-pass T-section filter, and (b) a high-pass $\pi$-section filter to meet these requirements.

Cut-off frequency $f_c = 25\,\text{kHz} = 25 \times 10^3\,\text{Hz}$, and nominal impedance, $R_0 = 600\,\Omega$.

From equation (7), capacitance,

$$C = \frac{1}{4\pi R_0 f_c} = \frac{1}{4\pi(600)(25 \times 10^3)}\,\text{F}$$

$$= \frac{10^{12}}{4\pi(600)(25 \times 10^3)}\,\text{pF}$$

$$= 5305\,\text{pF} \quad \text{or} \quad 5.305\,\text{nF}$$

From equation (8), inductance,

$$L = \frac{R_0}{4\pi f_c} = \frac{600}{4\pi(25 \times 10^3)}$$

$$= 0.00191\,\text{H} = 1.91\,\text{mH}$$

(a) A high-pass T-section filter is shown in Fig. 17.19(a), where the series arm capacitances

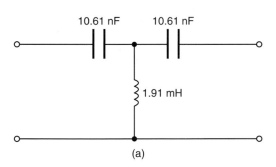

(a)

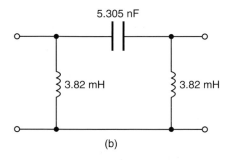

(b)

**Figure 17.19**

are each $2C$ (see Fig. 17.16(a)), i.e. $2 \times 5.305 = 10.61\,\text{nF}$

(b) A high-pass $\pi$-section filter is shown in Fig. 17.19(b), where the shunt arm inductances are each $2L$ (see Fig. 17.6(b)), i.e. $2 \times 1.91 = 3.82\,\text{mH}$.

**Now try the following exercise**

**Exercise 100   Further problems on high-pass filter sections**

1. Determine the cut-off frequency and the nominal impedance of each of the high-pass filter sections shown in Fig. 17.20.
   [(a) 22.51 kHz; 14.14 kΩ
   (b) 281.3 Hz; 1414 Ω]

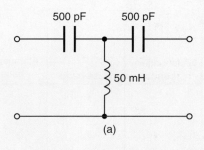

(a)

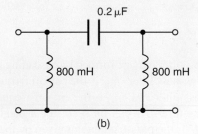

(b)

**Figure 17.20**

2. A filter section is required to pass all frequencies above 4 kHz and to have a nominal impedance 750 Ω. Design (a) an appropriate high-pass T section filter, and (b) an appropriate high-pass $\pi$-section filter to meet these requirements.
   [(a) Each series arm = 53.06 nF, shunt arm = 14.92 mH
   (b) Series arm = 26.53 nF, each shunt arm = 29.84 mH]

3. The inductance in each of the shunt arms of a high-pass $\pi$-section filter is 50 mH. If the nominal impedance of the section is 600 $\Omega$, determine the value of the capacitance in the series arm. [69.44 nF]

4. Determine the value of inductance required in the shunt arm of a high-pass T-section filter if in each series arm it contains a 0.5 $\mu$F capacitor. The cut-off frequency of the filter section is 1500 Hz. [11.26 mH]

## 17.5 Band-pass filters

A **band-pass filter is one designed to pass signals with frequencies between two specified cut-off frequencies**. The characteristic of an ideal band-pass filter is shown in Fig. 17.21.

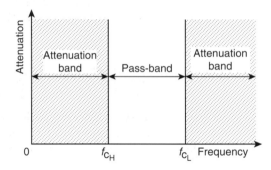

Figure 17.21

Such a filter may be formed by cascading a high-pass and a low-pass filter. $f_{C_H}$ is the cut-off frequency of the high-pass filter and $f_{C_L}$ is the cut-off frequency of the low-pass filter. As can be seen, for a band-pass filter $f_{C_L} > f_{C_H}$, the pass-band being given by the difference between these values.

The electrical circuit diagram symbol for a band-pass filter is shown in Fig. 17.22.

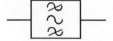

Figure 17.22

A typical practical characteristic for a band-pass filter is shown in Fig. 17.23.

Crystal and ceramic devices are used extensively as band-pass filters. They are common in the intermediate-frequency amplifiers of v.h.f. radios where a precisely

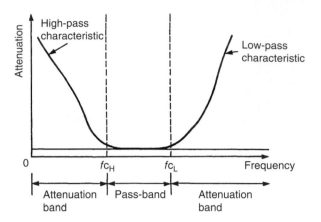

Figure 17.23

defined bandwidth must be maintained for good performance.

**Problem 7.** A band-pass filter is comprised of a low-pass T-section filter having a cut-off frequency of 15 kHz, connected in series with a high-pass T-section filter having a cut-off frequency of 10 kHz. The terminating impedance of the filter is 600 $\Omega$. Determine the values of the components comprising the composite filter.

**For the low-pass T-section filter:**

$$f_{C_L} = 15\,000\,\text{Hz}$$

From equation (3), capacitance,

$$C = \frac{1}{\pi R_0 f_c} = \frac{1}{\pi(600)(15\,000)}$$

$$= 35.4 \times 10^{-9} = 35.4\,\text{nF}$$

From equation (4), inductance,

$$L = \frac{R_0}{\pi f_c} = \frac{600}{\pi(15\,000)}$$

$$= 0.01273\,\text{H} = 12.73\,\text{mH}$$

Thus, from Fig. 17.7(a), the series arm inductances are each $\dfrac{L}{2}$ i.e.

$$\frac{12.73}{2} = \mathbf{6.37\,mH}$$

and the shunt arm capacitance is **35.4 nF**.

**For the high-pass T-section filter**:

$$f_{C_H} = 10\,000\,\text{Hz}$$

**Figure 17.24**

From equation (7), capacitance,

$$C = \frac{1}{4\pi R_0 f_c} = \frac{1}{4\pi (600)(10\,000)}$$

$$= 1.33 \times 10^{-8} = 13.3\,\text{nF}$$

From equation (8), inductance,

$$L = \frac{R_0}{4\pi f_c} = \frac{600}{4\pi (10\,000)}$$

$$= 4.77 \times 10^{-3} = 4.77\,\text{mH}.$$

Thus, from Fig. 17.16(a), the series arm capacitances are each $2C$,

i.e.        $2 \times 13.3 = \mathbf{26.6\,nF}$

and the shunt arm inductance is **4.77 mH**. The composite, band-pass filter is shown in Fig. 17.24.

The attenuation against frequency characteristic will be similar to Fig. 17.23 where $f_{C_H} = 10\,\text{kHz}$ and $f_{C_L} = 15\,\text{kHz}$.

---

**Now try the following exercise**

**Exercise 101    Further problems on band-pass filters**

1.  A band-pass filter is comprised of a low-pass T-section filter having a cut-off frequency of 20 kHz, connected in series with a high-pass T-section filter having a cut-off frequency of 8 kHz. The terminating impedance of the filter is 600 Ω. Determine the values of the components comprising the composite filter.
    [Low-pass T-section: each series arm 4.77 mH, shunt arm 26.53 nF
    High-pass T-section: each series arm 33.16 nF, shunt arm 5.97 mH]

2.  A band-pass filter is comprised of a low-pass $\pi$-section filter having a cut-off frequency of 50 kHz, connected in series with a high-pass $\pi$-section filter having a cut-off frequency of 40 kHz. The terminating impedance of the filter is 620 Ω. Determine the values of the components comprising the composite filter.
    [Low-pass $\pi$-section: series arm 3.95 mH, each shunt arm 5.13 nF
    High-pass $\pi$-section: series arm 3.21 nF, each shunt arm 2.47 mH]

## 17.6    Band-stop filters

A **band-stop filter is one designed to pass signals with all frequencies except those between two specified cut-off frequencies**. The characteristic of an ideal band-stop filter is shown in Fig. 17.25.

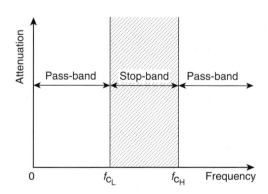

**Figure 17.25**

Such a filter may be formed by connecting a high-pass and a low-pass filter in parallel. As can be seen, for a band-stop filter $f_{C_H} > f_{C_L}$, the stop-band being given by the difference between these values.

The electrical circuit diagram symbol for a band-stop filter is shown in Fig. 17.26.

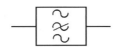

**Figure 17.26**

A typical practical characteristic for a band-stop filter is shown in Fig. 17.27.

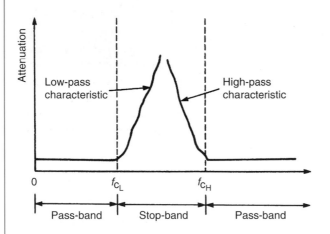

**Figure 17.27**

Sometimes, as in the case of interference from 50 Hz power lines in an audio system, the exact frequency of a spurious noise signal is known. Usually such interference is from an odd harmonic of 50 Hz, for example, 250 Hz. A sharply tuned band-stop filter, designed to attenuate the 250 Hz noise signal, is used to minimise the effect of the output. A high-pass filter with cut-off frequency greater than 250 Hz would also remove the interference, but some of the lower frequency components of the audio signal would be lost as well.

Filter design can be a complicated area. For more, see *Electrical Circuit Theory and Technology*.

**Now try the following exercises**

**Exercise 102    Short answer questions on filters**

1.  Define a filter

2.  Define the cut-off frequency for a filter

3.  Define a two-port network

4.  Define characteristic impedance for a two-port network

5.  A network designed to pass signals at frequencies below a specified cut-off frequency is called a ...... filter

6.  A network designed to pass signals with all frequencies except those between two specified cut-off frequencies is called a ...... filter

7.  A network designed to pass signals with frequencies between two specified cut-off frequencies is called a ...... filter

8.  A network designed to pass signals at frequencies above a specified cut-off frequency is called a ...... filter

9.  State one application of a low-pass filter

10.  Sketch (a) an ideal, and (b) a practical attenuation/frequency characteristic for a low-pass filter

11.  Sketch (a) an ideal, and (b) a practical attenuation/frequency characteristic for a high-pass filter

12.  Sketch (a) an ideal, and (b) a practical attenuation/frequency characteristic for a band-pass filter

14.  State one application of a band-pass filter

13.  Sketch (a) an ideal, and (b) a practical attenuation/frequency characteristic for a band-stop filter

15.  State one application of a band-stop filter

**Exercise 103    Multi-choice questions on filters**
**(Answers on page 421)**

1.  A network designed to pass signals with all frequencies except those between two specified cut-off frequencies is called a:
    (a) low-pass filter      (b) high-pass filter
    (c) band-pass filter    (d) band-stop filter

2.  A network designed to pass signals at frequencies above a specified cut-off frequency is called a:
    (a) low-pass filter      (b) high-pass filter
    (c) band-pass filter    (d) band-stop filter

3. A network designed to pass signals at frequencies below a specified cut-off frequency is called a:
   (a) low-pass filter    (b) high-pass filter
   (c) band-pass filter    (d) band-stop filter

4. A network designed to pass signals with frequencies between two specified cut-off frequencies is called a:
   (a) low-pass filter    (b) high-pass filter
   (c) band-pass filter    (d) band-stop filter

5. A low-pass T-connected symmetrical filter section has an inductance of 200 mH in each of its series arms and a capacitance of 0.5 $\mu$F in its shunt arm. The cut-off frequency of the filter is:
   (a) 1007 Hz    (b) 251.6 Hz
   (c) 711.8 Hz    (d) 177.9 Hz

6. A low-pass $\pi$-connected symmetrical filter section has an inductance of 200 mH in its series arm and capacitances of 400 pF in each of its shunt arms. The cut-off frequency of the filter is:
   (a) 25.16 kHz    (b) 6.29 kHz
   (c) 17.79 kHz    (d) 35.59 kHz

The following refers to questions 7 and 8.

A filter section is to have a nominal impedance of 620 $\Omega$ and a cut-off frequency of 2 MHz.

7. A low-pass T-connected symmetrical filter section is comprised of:
   (a) 98.68 $\mu$H in each series arm, 128.4 pF in shunt arm
   (b) 49.34 $\mu$H in each series arm, 256.7 pF in shunt arm
   (c) 98.68 $\mu$H in each series arm, 256.7 pF in shunt arm
   (d) 49.34 $\mu$H in each series arm, 128.4 pF in shunt arm

8. A low-pass $\pi$-connected symmetrical filter section is comprised of:
   (a) 98.68 $\mu$H in each series arm, 128.4 pF in shunt arm
   (b) 49.34 $\mu$H in each series arm, 256.7 pF in shunt arm

   (c) 98.68 $\mu$H in each series arm, 256.7 pF in shunt arm
   (d) 49.34 $\mu$H in each series arm, 128.4 pF in shunt arm

9. A high-pass T-connected symmetrical filter section has capacitances of 400 nF in each of its series arms and an inductance of 200 mH in its shunt arm. The cut-off frequency of the filter is:
   (a) 1592 Hz    (b) 1125 Hz
   (c) 281 Hz    (d) 398 Hz

10. A high-pass $\pi$-connected symmetrical filter section has a capacitance of 5000 pF in its series arm and inductances of 500 $\mu$H in each of its shunt arms. The cut-off frequency of the filter is:
    (a) 201.3 kHz    (b) 71.18 kHz
    (c) 50.33 kHz    (d) 284.7 kHz

The following refers to questions 11 and 12.

A filter section is required to pass all frequencies above 50 kHz and to have a nominal impedance of 650 $\Omega$.

11. A high-pass T-connected symmetrical filter section is comprised of:
    (a) Each series arm 2.45 nF, shunt arm 1.03 mH
    (b) Each series arm 4.90 nF, shunt arm 2.08 mH
    (c) Each series arm 2.45 nF, shunt arm 2.08 mH
    (d) Each series arm 4.90 nF, shunt arm 1.03 mH

12. A high-pass $\pi$-connected symmetrical filter section is comprised of:
    (a) Series arm 4.90 nF, and each shunt arm 1.04 mH
    (b) Series arm 4.90 nF, and each shunt arm 2.07 mH
    (c) Series arm 2.45 nF, and each shunt arm 2.07 mH
    (d) Series arm 2.45 nF, and each shunt arm 1.04 mH

# Chapter 18

# D.C. transients

At the end of this chapter you should be able to:

- understand the term 'transient'

- describe the transient response of capacitor and resistor voltages, and current in a series $C - R$ d.c. circuit

- define the term 'time constant'

- calculate time constant in a $C - R$ circuit

- draw transient growth and decay curves for a $C - R$ circuit

- use equations $v_C = V(1 - e^{-t/\tau})$, $v_R = Ve^{-t/\tau}$ and $i = Ie^{-t/\tau}$ for a $C - R$ circuit

- describe the transient response when discharging a capacitor

- describe the transient response of inductor and resistor voltages, and current in a series $L - R$ d.c. circuit

- calculate time constant in an $L - R$ circuit

- draw transient growth and decay curves for an $L - R$ circuit

- use equations $v_L = Ve^{-t/\tau}$, $v_R = V(1 - e^{-t/\tau})$ and $i = I(1 - e^{-t/\tau})$

- describe the transient response for current decay in an $L - R$ circuit

- understand the switching of inductive circuits

- describe the effects of time constant on a rectangular waveform via integrator and differentiator circuits

## 18.1 Introduction

When a d.c. voltage is applied to a capacitor $C$ and resistor $R$ connected in series, there is a short period of time immediately after the voltage is connected, during which the current flowing in the circuit and voltages across $C$ and $R$ are changing.

Similarly, when a d.c. voltage is connected to a circuit having inductance $L$ connected in series with resistance $R$, there is a short period of time immediately after the voltage is connected, during which the current flowing in the circuit and the voltages across $L$ and $R$ are changing. These changing values are called **transients**.

DOI: 10.1016/B978-0-08-089056-2.00018-8

## 18.2 Charging a capacitor

(a) The circuit diagram for a series connected $C - R$ circuit is shown in Fig. 18.1. When switch S is closed then by Kirchhoff's voltage law:

$$V = v_C + v_R \qquad (1)$$

(b) The battery voltage $V$ is constant. The capacitor voltage $v_C$ is given by $q/C$, where $q$ is the charge on the capacitor. The voltage drop across $R$ is given by $iR$, where $i$ is the current flowing in the circuit.

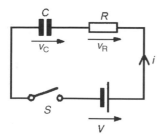

Figure 18.1

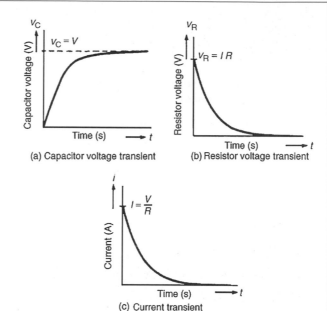

(a) Capacitor voltage transient

(b) Resistor voltage transient

(c) Current transient

Figure 18.2

Hence at all times:

$$V = \frac{q}{C} + iR \qquad (2)$$

At the instant of closing $S$, (initial circuit condition), assuming there is no initial charge on the capacitor, $q_0$ is zero, hence $v_{Co}$ is zero. Thus from Equation (1), $V = 0 + v_{Ro}$, i.e. $v_{Ro} = V$. This shows that the resistance to current is solely due to $R$, and the initial current flowing, $i_0 = I = V/R$.

(c) A short time later at time $t_1$ seconds after closing $S$, the capacitor is partly charged to, say, $q_1$ coulombs because current has been flowing. The voltage $v_{C1}$ is now $(q_1/C)$ volts. If the current flowing is $i_1$ amperes, then the voltage drop across $R$ has fallen to $i_1 R$ volts. Thus, equation (2) is now $V = (q_1/C) + i_1 R$.

(d) A short time later still, say at time $t_2$ seconds after closing the switch, the charge has increased to $q_2$ coulombs and $v_C$ has increased to $(q_2/C)$ volts. Since $V = v_C + v_R$ and $V$ is a constant, then $v_R$ decreases to $i_2 R$, Thus $v_C$ is increasing and $i$ and $v_R$ are decreasing as time increases.

(e) Ultimately, a few seconds after closing $S$, (i.e. at the final or **steady-state** condition), the capacitor is fully charged to, say, $Q$ coulombs, current no longer flows, i.e. $i = 0$, and hence $v_R = iR = 0$. It follows from equation (1) that $v_C = V$.

(f) Curves showing the changes in $v_C$, $v_R$ and $i$ with time are shown in Fig. 18.2.
The curve showing the variation of $v_C$ with time is called an **exponential growth curve** and the graph is called the 'capacitor voltage/time' characteristic. The curves showing the variation of $v_R$ and $i$ with time are called **exponential decay curves**, and the graphs are called 'resistor voltage/time' and 'current/time' characteristics respectively. (The name 'exponential' shows that

the shape can be expressed mathematically by an exponential mathematical equation, as shown in Section 18.4.)

## 18.3    Time constant for a $C - R$ circuit

(a) If a constant d.c. voltage is applied to a series connected $C - R$ circuit, a transient curve of capacitor voltage $v_C$ is as shown in Fig. 18.2(a).

(b) With reference to Fig. 18.3, let the constant voltage supply be replaced by a variable voltage supply at time $t_1$ seconds. Let the voltage be varied so that the **current** flowing in the circuit is **constant**.

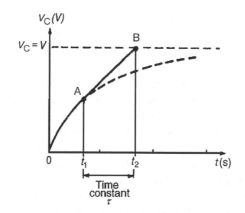

Figure 18.3

(c)   Since the current flowing is a constant, the curve will follow a tangent, AB, drawn to the curve at point A.

(d)   Let the capacitor voltage $v_C$ reach its final value of $V$ at time $t_2$ seconds.

(e)   The time corresponding to $(t_2 - t_1)$ seconds is called the **time constant** of the circuit, denoted by the Greek letter 'tau', $\tau$. The value of the time constant is $CR$ seconds, i.e. for a series connected $C - R$ circuit,

$$\text{time constant } \tau = CR \text{ seconds}$$

Since the variable voltage mentioned in paragraph (b) above can be applied at any instant during the transient change, it may be applied at $t = 0$, i.e. at the instant of connecting the circuit to the supply. If this is done, then the time constant of the circuit may be defined as: *'the time taken for a transient to reach its final state if the initial rate of change is maintained'*.

## 18.4   Transient curves for a $C - R$ circuit

There are two main methods of drawing transient curves graphically, these being:

(a)   the **tangent method** – this method is shown in Problem 1

(b)   the **initial slope and three point method**, which is shown in Problem 2, and is based on the following properties of a transient exponential curve:

   (i)   for a growth curve, the value of a transient at a time equal to one time constant is 0.632 of its steady-state value (usually taken as 63 per cent of the steady-state value), at a time equal to two and a half time constants is 0.918 of its steady-state value (usually taken as 92 per cent of its steady-state value) and at a time equal to five time constants is equal to its steady-state value,

   (ii)   for a decay curve, the value of a transient at a time equal to one time constant is 0.368 of its initial value (usually taken as 37 per cent of its initial value), at a time equal to two and a half time constants is 0.082 of its initial value (usually taken as 8 per cent of its initial value) and at a time equal to five time constants is equal to zero.

The transient curves shown in Fig. 18.2 have mathematical equations, obtained by solving the differential equations representing the circuit. The equations of the curves are:

growth of capacitor voltage,

$$v_C = V(1 - e^{-t/CR}) = V(1 - e^{-t/\tau})$$

decay of resistor voltage,

$$v_R = Ve^{-t/CR} = Ve^{-t/\tau} \quad \text{and}$$

decay of resistor voltage,

$$i = Ie^{-t/CR} = Ie^{-t/\tau}$$

**Problem 1.**   A 15 $\mu$F uncharged capacitor is connected in series with a 47 k$\Omega$ resistor across a 120 V, d.c. supply. Use the tangential graphical method to draw the capacitor voltage/time characteristic of the circuit. From the characteristic, determine the capacitor voltage at a time equal to one time constant after being connected to the supply, and also two seconds after being connected to the supply. Also, find the time for the capacitor voltage to reach one half of its steady-state value.

To construct an exponential curve, the time constant of the circuit and steady-state value need to be determined.

$$\text{Time constant} = CR = 15\,\mu\text{F} \times 47\,\text{k}\Omega$$
$$= 15 \times 10^{-6} \times 47 \times 10^3$$
$$= 0.705\,\text{s}$$

Steady-state value of $v_C = V$, i.e. $v_C = 120$ V.

   With reference to Fig. 18.4, the scale of the horizontal axis is drawn so that it spans at least five time constants, i.e. $5 \times 0.705$ or about 3.5 seconds. The scale of the vertical axis spans the change in the capacitor voltage, that is, from 0 to 120 V. A broken line AB is drawn corresponding to the final value of $v_C$.

   Point C is measured along AB so that AC is equal to $1\tau$, i.e. AC $= 0.705$ s. Straight line OC is drawn. Assuming that about five intermediate points are needed to draw the curve accurately, a point D is selected on OC corresponding to a $v_C$ value of about 20 V. DE is drawn vertically. *EF* is made to correspond to $1\tau$, i.e. EF $= 0.705$ s. A straight line is drawn joining DF. This procedure of

(a)   drawing a vertical line through point selected,

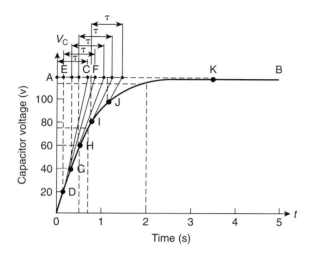

**Figure 18.4**

(b)   at the steady-state value, drawing a horizontal line corresponding to $1\tau$, and

(c)   joining the first and last points,

is repeated for $v_C$ values of 40, 60, 80 and 100 V, giving points G, H, I and J.

The capacitor voltage effectively reaches its steady-state value of 120 V after a time equal to five time constants, shown as point K. Drawing a smooth curve through points 0, D, G, H, I, J and K gives the exponential growth curve of capacitor voltage.

From the graph, the value of capacitor voltage at a time equal to the time constant is about **75 V**. It is a characteristic of all exponential growth curves, that after a time equal to one time constant, the value of the transient is 0.632 of its steady-state value. In this problem, $0.632 \times 120 = 75.84$ V. Also from the graph, when $t$ is two seconds, $v_C$ is about **115** Volts. [This value may be checked using the equation $v_C = V(1 - e^{-t/\tau})$, where $V = 120$ V, $\tau = 0.705$ s and $t = 2$ s. This calculation gives $v_C = 112.97$ V.]

The time for $v_C$ to rise to one half of its final value, i.e. 60 V, can be determined from the graph and is about **0.5 s**. [This value may be checked using $v_C = V(1 - e^{-t/\tau})$ where $V = 120$ V, $v_C = 60$ V and $\tau = 0.705$ s, giving $t = 0.489$ s.]

---

**Problem 2.**   A $4\,\mu$F capacitor is charged to 24 V and then discharged through a $220\,$k$\Omega$ resistor. Use the 'initial slope and three point' method to draw: (a) the capacitor voltage/time characteristic, (b) the resistor voltage/time characteristic, and (c) the current/time characteristic, for the transients which occur. From the characteristics determine the value

---

of capacitor voltage, resistor voltage and current 1.5 s after discharge has started.

To draw the transient curves, the time constant of the circuit and steady-state values are needed.

$$\text{Time constant, } \tau = CR$$
$$= 4 \times 10^{-6} \times 220 \times 10^3$$
$$= 0.88\,\text{s}$$

Initially, capacitor voltage $v_C = v_R = 24$ V,

$$i = \frac{V}{R} = \frac{24}{220 \times 10^3}$$
$$= 0.109\,\text{mA}$$

Finally, $v_C = v_R = i = 0$.

(a)   The exponential decay of capacitor voltage is from 24 V to 0 V in a time equal to five time constants, i.e. $5 \times 0.88 = 4.4$ s. With reference to Fig. 18.5, to construct the decay curve:

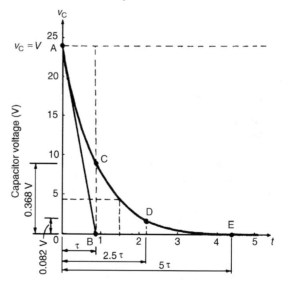

**Figure 18.5**

(i)   the horizontal scale is made so that it spans at least five time constants, i.e. 4.4 s,

(ii)   the vertical scale is made to span the change in capacitor voltage, i.e. 0 to 24 V,

(iii)   point A corresponds to the initial capacitor voltage, i.e. 24 V,

(iv)   OB is made equal to one time constant and line AB is drawn; this gives the initial slope of the transient,

(v)   the value of the transient after a time equal to one time constant is 0.368 of the initial value, i.e. $0.368 \times 24 = 8.83\,\text{V}$; a vertical line is drawn through B and distance BC is made equal to 8.83 V,

(vi)   the value of the transient after a time equal to two and a half time constants is 0.082 of the initial value, i.e. $0.082 \times 24 = 1.97\,\text{V}$, shown as point D in Fig. 18.5,

(vii)   the transient effectively dies away to zero after a time equal to five time constants, i.e. 4.4 s, giving point E.

The smooth curve drawn through points A, C, D and E represents the decay transient. At 1.5 s after decay has started, $v_C \approx \mathbf{4.4\,V}$.
[This may be checked using $v_C = V e^{-t/\tau}$, where $V = 24$, $t = 1.5$ and $\tau = 0.88$, giving $v_C = 4.36\,\text{V}$]

(b)   The voltage drop across the resistor is equal to the capacitor voltage when a capacitor is discharging through a resistor, thus the resistor voltage/time characteristic is identical to that shown in Fig. 18.5 Since $v_R = v_C$, then at 1.5 seconds after decay has started, $v_R \approx \mathbf{4.4\,V}$ (see (vii) above).

(c)   The current/time characteristic is constructed in the same way as the capacitor voltage/time characteristic, shown in part (a), and is as shown in Fig. 18.6. The values are:

point A: initial value of current $= 0.109\,\text{mA}$

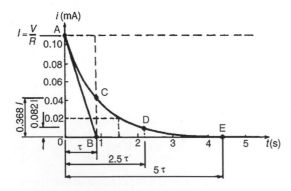

**Figure 18.6**

point C: at $1\,\tau$, $i = 0.368 \times 0.109 = 0.040\,\text{mA}$
point D: at $2.5\,\tau$, $i = 0.082 \times 0.109 = 0.009\,\text{mA}$
point E: at $5\,\tau$, $i = 0$

Hence the current transient is as shown. At a time of 1.5 s, the value of current, from the characteristic is **0.02 mA**

[This may be checked using $i = I e^{(-t/\tau)}$ where $I = 0.109$, $t = 1.5$ and $\tau = 0.88$, giving $i = 0.0198\,\text{mA}$ or $19.8\,\mu\text{A}$]

---

**Problem 3.**   A $20\,\mu\text{F}$ capacitor is connected in series with a $50\,\text{k}\Omega$ resistor and the circuit is connected to a 20 V, d.c. supply. Determine: (a) the initial value of the current flowing, (b) the time constant of the circuit, (c) the value of the current one second after connection, (d) the value of the capacitor voltage two seconds after connection, and (e) the time after connection when the resistor voltage is 15 V.

---

Parts (c), (d) and (e) may be determined graphically, as shown in Problems 1 and 2 or by calculation as shown below.

$V = 20\,\text{V}$, $C = 20\,\mu\text{F} = 20 \times 10^{-6}\,\text{F}$,
$R = 50\,\text{k}\Omega = 50 \times 10^3\,\text{V}$

(a)   The initial value of the current flowing is

$$I = \frac{V}{R} = \frac{20}{50 \times 10^3} = \mathbf{0.4\,mA}$$

(b)   From Section 18.3 the time constant,

$$\tau = CR = (20 \times 10^{-6})(50 \times 10^3) = \mathbf{1\,s}$$

(c)   Current, $i = I e^{-t/\tau}$ and working in mA units,

$$i = 0.4 e^{-1/1} = 0.4 \times 0.368 = \mathbf{0.147\,mA}$$

(d)   Capacitor voltage,

$$v_C = V(1 - e^{-t/\tau}) = 20(1 - e^{-2/1})$$
$$= 20(1 - 0.135) = 20 \times 0.865$$
$$= \mathbf{18.3\,V}$$

(e)   Resistor voltage, $v_R = V e^{-t/\tau}$
Thus $15 = 20 e^{-t/1}$, $15/20 = e^{-t}$ from which $e^t = 20/15 = 4/3$

Taking natural logarithms of each side of the equation gives

$$t = \ln \frac{4}{3} = \ln 1.3333 \text{ i.e. } \mathbf{time,\, t = 0.288\,s}$$

---

**Problem 4.**   A circuit consists of a resistor connected in series with a $0.5\,\mu\text{F}$ capacitor and has a time constant of 12 ms. Determine: (a) the value of the resistor, and (b) the capacitor voltage, 7 ms after connecting the circuit to a 10 V supply.

(a)   The time constant $\tau = CR$, hence

$$R = \frac{\tau}{C}$$

$$= \frac{12 \times 10^{-3}}{0.5 \times 10^{-6}}$$

$$= 24 \times 10^3 = \mathbf{24\,k\Omega}$$

(b)   The equation for the growth of capacitor voltage is: $v_C = V(1 - e^{-t/\tau})$
Since $\tau = 12\,\text{ms} = 12 \times 10^{-3}\,\text{s}$, $V = 10\,\text{V}$ and $t = 7\,\text{ms} = 7 \times 10^{-3}\,\text{s}$, then

$$v_C = 10(1 - e^{-7 \times 10^{-3}/12 \times 10^{-3}})$$

$$= 10(1 - e^{-0.583})$$

$$= 10(1 - 0.558) = \mathbf{4.42\,V}$$

Alternatively, the value of $v_C$ when $t$ is 7 ms may be determined using the growth characteristic as shown in Problem 1.

---

**Problem 5.**   A circuit consists of a $10\,\mu\text{F}$ capacitor connected in series with a $25\,\text{k}\Omega$ resistor with a switchable 100 V d.c. supply. When the supply is connected, calculate (a) the time constant, (b) the maximum current, (c) the voltage across the capacitor after 0.5 s, (d) the current flowing after one time constant, (e) the voltage across the resistor after 0.1 s, (f) the time for the capacitor voltage to reach 45 V, and (g) the initial rate of voltage rise.

---

(a)   **Time constant,**
$$\boldsymbol{\tau = C \times R = 10 \times 10^{-6} \times 25 \times 10^3 = 0.25\,s}$$

(b)   Current is a maximum when the circuit is first connected and is only limited by the value of resistance in the circuit, i.e.

$$\boldsymbol{I_m} = \frac{V}{R} = \frac{100}{25 \times 10^3} = \mathbf{4\,mA}$$

(c)   Capacitor voltage, $v_C = V_m(1 - e^{-t/\tau})$
When time, $t = 0.5\,\text{s}$, then
$$v_C = 100(1 - e^{-0.5/0.25}) = 100(0.8647) = \mathbf{86.47\,V}$$

(d)   Current, $i = I_m e^{-t/\tau}$
and when $t = \tau$,
**current, $i = 4\,e^{-\tau/\tau} = 4\,e^{-1} = \mathbf{1.472\,mA}$**

Alternatively, after one time constant the capacitor voltage will have risen to 63.2% of the supply voltage and the current will have fallen to 63.2% of

its final value, i.e. 36.8% of $I_m$. Hence, $i = 36.8\%$ of $4 = 0.368 \times 4 = \mathbf{1.472\,mA}$

(e)   The voltage across the resistor, $v_R = V\,e^{-t/\tau}$
When $t = 0.1\,\text{s}$,
**resistor voltage, $v_R = 100\,e^{-0.1/0.25} = \mathbf{67.03\,V}$**

(f)   Capacitor voltage, $v_C = V_m(1 - e^{-t/\tau})$
When the capacitor voltage reaches 45 V, then:

$$45 = 100(1 - e^{-t/0.25})$$

from which,   $$\frac{45}{100} = 1 - e^{-t/0.25}$$

and   $$e^{-t/0.25} = 1 - \frac{45}{100} = 0.55$$

Hence,   $$-\frac{t}{0.25} = \ln 0.55$$

and   **time, $t = -0.25 \ln 0.55 = \mathbf{0.149\,s}$**

(g)   **Initial rate of voltage rise** $= \dfrac{V}{\tau} = \dfrac{100}{0.25} = \mathbf{400\,V/s}$
(i.e. gradient of the tangent at $t = 0$)

## 18.5   Discharging a capacitor

When a capacitor is charged (i.e. with the switch in position A in Fig. 18.7), and the switch is then moved to position B, the electrons stored in the capacitor keep the current flowing for a short time. Initially, at the instant of moving from A to B, the current flow is such that the capacitor voltage $v_C$ is balanced by an equal and opposite voltage $v_R = iR$. Since initially $v_C = v_R = V$, then $i = I = V/R$. During the transient decay, by applying Kirchhoff's voltage law to Fig. 18.7, $v_C = v_R$.

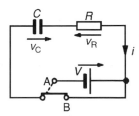

Figure 18.7

Finally the transients decay exponentially to zero, i.e. $v_C = v_R = 0$. The transient curves representing the voltages and current are as shown in Fig. 18.8.

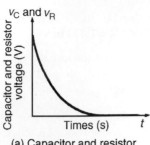

(a) Capacitor and resistor voltage transient

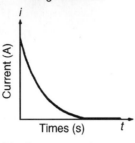

(b) Current transient

**Figure 18.8**

The equations representing the transient curves during the discharge period of a series connected $C - R$ circuit are:

decay of voltage,
$$v_C = v_R = Ve^{(-t/CR)} = Ve^{(-t/\tau)}$$

decay of current, $i = Ie^{(-t/CR)} = Ie^{(-t/\tau)}$

When a capacitor has been disconnected from the supply it may still be charged and it may retain this charge for some considerable time. Thus precautions must be taken to ensure that the capacitor is automatically discharged after the supply is switched off. This is done by connecting a high value resistor across the capacitor terminals.

> **Problem 6.** A capacitor is charged to 100 V and then discharged through a 50 kΩ resistor. If the time constant of the circuit is 0.8 s. Determine: (a) the value of the capacitor, (b) the time for the capacitor voltage to fall to 20 V, (c) the current flowing when the capacitor has been discharging for 0.5 s, and (d) the voltage drop across the resistor when the capacitor has been discharging for one second.

Parts (b), (c) and (d) of this problem may be solved graphically as shown in Problems 1 and 2 or by calculation as shown below.

$V = 100\,\text{V}, \tau = 0.8\,\text{s}, R = 50\,\text{k}\Omega = 50 \times 10^3\,\Omega$

(a) Since time constant, $\tau = CR$, capacitance,

$$C = \frac{\tau}{R} = \frac{0.8}{50 \times 10^3} = \mathbf{16\,\mu F}$$

(b) Since $v_C = Ve^{-t/\tau}$ then $20 = 100e^{-t/0.8}$ from which $1/5 = e^{-t/0.8}$
Thus $e^{t/0.8} = 5$ and taking natural logarithms of each side, gives $t/0.8 = \ln 5$ and time, $t = 0.8 \ln 5 = \mathbf{1.29\,s}$.

(c) $i = Ie^{-t/\tau}$ where the initial current flowing,

$$I = \frac{V}{R} = \frac{100}{50 \times 10^3} = 2\,\text{mA}$$

Working in mA units,

$$i = Ie^{-t/\tau} = 2e^{(-0.5/0.8)}$$
$$= 2e^{-0.625} = 2 \times 0.535 = \mathbf{1.07\,mA}$$

(d) $v_R = v_C = Ve^{-t/\tau} = 100e^{-1/0.8}$
$$= 100e^{-1.25} = 100 \times 0.287 = \mathbf{28.7\,V}$$

> **Problem 7.** A 0.1 μF capacitor is charged to 200 V before being connected across a 4 kΩ resistor. Determine (a) the initial discharge current, (b) the time constant of the circuit, and (c) the minimum time required for the voltage across the capacitor to fall to less than 2 V.

(a) Initial discharge current,

$$i = \frac{V}{R} = \frac{200}{4 \times 10^3} = \mathbf{0.05\,A} \quad \text{or} \quad \mathbf{50\,mA}$$

(b) Time constant $\tau = CR = 0.1 \times 10^{-6} \times 4 \times 10^3$
$$= \mathbf{0.0004\,s} \quad \text{or} \quad \mathbf{0.4\,ms}$$

(c) The minimum time for the capacitor voltage to fall to less than 2 V, i.e. less than 2/200 or 1 per cent of the initial value is given by $5\tau$. $5\tau = 5 \times 0.4 = \mathbf{2\,ms}$

In a d.c. circuit, a capacitor blocks the current except during the times that there are changes in the supply voltage.

For a practical laboratory experiment on the charging and discharging of a capacitor, see Chapter 24, page 419.

**Now try the following exercise**

**Exercise 104    Further problems on transients in series connected $C - R$ circuits**

1. An uncharged capacitor of $0.2\,\mu\text{F}$ is connected to a $100\,\text{V}$, d.c. supply through a resistor of $100\,\text{k}\Omega$. Determine, either graphically or by calculation the capacitor voltage $10\,\text{ms}$ after the voltage has been applied.
   [39.35 V]

2. A circuit consists of an uncharged capacitor connected in series with a $50\,\text{k}\Omega$ resistor and has a time constant of $15\,\text{ms}$. Determine either graphically or by calculation (a) the capacitance of the capacitor and (b) the voltage drop across the resistor $5\,\text{ms}$ after connecting the circuit to a $20\,\text{V}$, d.c. supply.
   [(a) $0.3\,\mu\text{F}$ (b) 14.33 V]

3. A $10\,\mu\text{F}$ capacitor is charged to $120\,\text{V}$ and then discharged through a $1.5\,\text{M}\Omega$ resistor. Determine either graphically or by calculation the capacitor voltage $2\,\text{s}$ after discharging has commenced. Also find how long it takes for the voltage to fall to $25\,\text{V}$.
   [105.0 V, 23.53 s]

4. A capacitor is connected in series with a voltmeter of resistance $750\,\text{k}\Omega$ and a battery. When the voltmeter reading is steady the battery is replaced with a shorting link. If it takes $17\,\text{s}$ for the voltmeter reading to fall to two-thirds of its original value, determine the capacitance of the capacitor.
   [$55.9\,\mu\text{F}$]

5. When a $3\,\mu\text{F}$ charged capacitor is connected to a resistor, the voltage falls by $70$ per cent in $3.9\,\text{s}$. Determine the value of the resistor.
   [$1.08\,\text{M}\Omega$]

6. A $50\,\mu\text{F}$ uncharged capacitor is connected in series with a $1\,\text{k}\Omega$ resistor and the circuit is switched to a $100\,\text{V}$, d.c. supply. Determine:
   (a) the initial current flowing in the circuit,
   (b) the time constant,
   (c) the value of current when $t$ is $50\,\text{ms}$ and
   (d) the voltage across the resistor $60\,\text{ms}$ after closing the switch.
   [(a) 0.1 A (b) 50 ms
   (c) 36.8 mA (d) 30.1 V]

7. An uncharged $5\,\mu\text{F}$ capacitor is connected in series with a $30\,\text{k}\Omega$ resistor across a $110\,\text{V}$, d.c. supply. Determine the time constant of the circuit, the initial charging current, the current flowing $120\,\text{ms}$ after connecting to the supply.
   [150 ms, 3.67 mA, 1.65 mA]

8. An uncharged $80\,\mu\text{F}$ capacitor is connected in series with a $1\,\text{k}\Omega$ resistor and is switched across a $110\,\text{V}$ supply. Determine the time constant of the circuit and the initial value of current flowing. Determine also the value of current flowing after (a) $40\,\text{ms}$ and (b) $80\,\text{ms}$.
   [80 ms, 0.11 A (a) 66.7 mA (b) 40.5 mA]

9. A resistor of $0.5\,\text{M}\Omega$ is connected in series with a $20\,\mu\text{F}$ capacitor and the capacitor is charged to $200\,\text{V}$. The battery is replaced instantaneously by a conducting link. Draw a graph showing the variation of capacitor voltage with time over a period of at least $6$ time constants. Determine from the graph the approximate time for the capacitor voltage to fall to $75\,\text{V}$.
   [9.8 s]

10. A $60\,\mu\text{F}$ capacitor is connected in series with a $10\,\text{k}\Omega$ resistor and connected to a $120\,\text{V}$ d.c. supply. Calculate (a) the time constant, (b) the initial rate of voltage rise, (c) the initial charging current, and (d) the time for the capacitor voltage to reach $50\,\text{V}$.
   [(a) 0.60 s (b) 200 V/s
   (c) 12 mA (d) 0.323 s]

11. If a $200\,\text{V}$ d.c. supply is connected to a $2.5\,\text{M}\Omega$ resistor and a $2\,\mu\text{F}$ capacitor in series. Calculate
   (a) the current flowing $4\,\text{s}$ after connecting,
   (b) the voltage across the resistor after $4\,\text{s}$, and
   (c) the energy stored in the capacitor after $4\,\text{s}$.
   [(a) $35.95\,\mu\text{A}$ (b) 89.87 V
   (c) 12.13 mJ]

12. (a) In the circuit shown in Fig. 18.9, with the switch in position 1, the capacitor is uncharged. If the switch is moved to position 2 at time $t = 0\,\text{s}$, calculate the (i) initial current through the $0.5\,\text{M}\Omega$ resistor, (ii) the voltage across the capacitor when $t = 1.5\,\text{s}$, and (iii) the time taken for the voltage across the capacitor to reach $12\,\text{V}$.

(b) If at the time $t = 1.5$ s, the switch is moved to position 3, calculate (i) the initial current through the 1 MΩ resistor, (ii) the energy stored in the capacitor 3.5 s later (i.e. when $t = 5$ s).

(c) Sketch a graph of the voltage across the capacitor against time from $t = 0$ to $t = 5$ s, showing the main points.

[(a)(i) 80 μA (ii) 18.05 V (iii) 0.892 s

(b)(i) 40 μA (ii) 48.30 μJ]

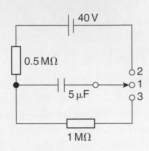

**Figure 18.9**

## 18.6 Camera flash

The internal workings of a camera flash are an example of the application of $C - R$ circuits. When a camera is first switched on, a battery slowly charges a capacitor to its full potential via a $C - R$ circuit. When the capacitor is fully charged, an indicator (red light) typically lets the photographer know that the flash is ready for use. Pressing the shutter button quickly discharges the capacitor through the flash (i.e. a resistor). The current from the capacitor is responsible for the bright light that is emitted. The flash rapidly draws current in order to emit the bright light. The capacitor must then be discharged before the flash can be used again.

## 18.7 Current growth in an $L - R$ circuit

(a) The circuit diagram for a series connected $L - R$ circuit is shown in Fig. 18.10. When switch S is closed, then by Kirchhoff's voltage law:

$$V = v_L + v_R \qquad (3)$$

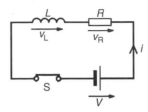

**Figure 18.10**

(b) The battery voltage $V$ is constant. The voltage across the inductance is the induced voltage, i.e.

$$v_L = L \times \frac{\text{change of current}}{\text{change of time}} = L\frac{di}{dt}$$

The voltage drop across $R$, $v_R$ is given by $iR$. Hence, at all times:

$$V = L\frac{di}{dt} + iR \qquad (4)$$

(c) At the instant of closing the switch, the rate of change of current is such that it induces an e.m.f. in the inductance which is equal and opposite to $V$, hence $V = v_L + 0$, i.e. $v_L = V$. From equation (3), because $v_L = V$, then $v_R = 0$ and $i = 0$.

(d) A short time later at time $t_1$ seconds after closing S, current $i_1$ is flowing, since there is a rate of change of current initially, resulting in a voltage drop of $i_1 R$ across the resistor. Since $V$ (which is constant) $= v_L + v_R$ the induced e.m.f. is reduced, and equation (4) becomes:

$$V = L\frac{di_1}{dt_1} + i_1 R$$

(e) A short time later still, say at time $t_2$ seconds after closing the switch, the current flowing is $i_2$, and the voltage drop across the resistor increases to $i_2 R$. Since $v_R$ increases, $v_L$ decreases.

(f) Ultimately, a few seconds after closing S, the current flow is entirely limited by $R$, the rate of change of current is zero and hence $v_L$ is zero. Thus $V = iR$. Under these conditions, steady-state current flows, usually signified by $I$. Thus, $I = V/R$, $v_R = IR$ and $v_L = 0$ at steady-state conditions.

(g) Curves showing the changes in $v_L$, $v_R$ and $i$ with time are shown in Fig. 18.11 and indicate that $v_L$ is a maximum value initially (i.e. equal to V), decaying exponentially to zero, whereas $v_R$ and $i$ grow exponentially from zero to their steady-state values of $V$ and $I = V/R$ respectively.

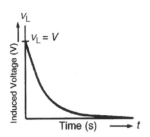
(a) Induced voltage transient

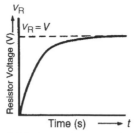
(b) Resistor voltage transient

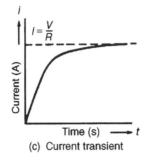
(c) Current transient

**Figure 18.11**

The application of these equations is shown in Problem 10.

Problem 8. A relay has an inductance of 100 mH and a resistance of 20 Ω. It is connected to a 60 V, d.c. supply. Use the 'initial slope and three point' method to draw the current/time characteristic and hence determine the value of current flowing at a time equal to two time constants and the time for the current to grow to 1.5 A.

Before the current/time characteristic can be drawn, the time constant and steady-state value of the current have to be calculated.
Time constant,

$$\tau = \frac{L}{R} = \frac{10 \times 10^{-3}}{20} = 5\,\text{ms}$$

Final value of current,

$$I = \frac{V}{R} = \frac{60}{20} = 3\,\text{A}$$

The method used to construct the characteristic is the same as that used in Problem 2

(a) The scales should span at least five time constants (horizontally), i.e. 25 ms, and 3 A (vertically)

(b) With reference to Fig. 18.12, the initial slope is obtained by making AB equal to 1 time constant, (i.e. 5 ms), and joining OB.

## 18.8 Time constant for an L − R circuit

With reference to Section 18.3, the time constant of a series connected L − R circuit is defined in the same way as the time constant for a series connected C − R circuit. Its value is given by:

$$\text{time constant, } \tau = \frac{L}{R}\ \text{seconds}$$

## 18.9 Transient curves for an L − R circuit

Transient curves representing the induced voltage/time, resistor voltage/time and current/time characteristics may be drawn graphically, as outlined in Section 18.4. A method of construction is shown in Problem 8.
Each of the transient curves shown in Fig. 18.11 have mathematical equations, and these are:

**decay of induced voltage,**

$$v_L = Ve^{(-Rt/L)} = Ve^{(-t/\tau)}$$

**growth of resistor voltage,**

$$v_R = V(1 - e^{-Rt/L}) = V(1 - e^{-t/\tau})$$

**growth of current flow,**

$$i = I(1 - e^{-Rt/L}) = I(1 - e^{-t/\tau})$$

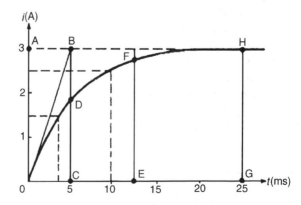

**Figure 18.12**

(c) At a time of 1 time constant,
CD is 0.632 × I = 0.632 × 3 = 1.896 A.
At a time of 2.5 time constants,
EF is 0.918 × I = 0.918 × 3 = 2.754 A.
At a time of 5 time constants, GH is I = 3 A.

(d)   A smooth curve is drawn through points $0, D, F$ and $H$ and this curve is the current/time characteristic.

From the characteristic, when $t = 2\tau, i \approx \textbf{2.6 A}$. [This may be checked by calculation using $i = I(1 - e^{-t/\tau})$, where $I = 3$ and $t = 2\tau$, giving $i = 2.59$ A.] Also, when the current is 1.5 A, the corresponding time is about **3.6 ms**. [Again, this may be checked by calculation, using $i = I(1 - e^{-t/\tau})$ where $i = 1.5$, $I = 3$ and $\tau = 5$ ms, giving $t = 3.466$ ms.]

> **Problem 9.**   A coil of inductance 0.04 H and resistance $10 \Omega$ is connected to a 120 V, d.c. supply. Determine (a) the final value of current, (b) the time constant of the circuit, (c) the value of current after a time equal to the time constant from the instant the supply voltage is connected, (d) the expected time for the current to rise to within 1 per cent of its final value.

(a)   Final steady current, $I = \dfrac{V}{R} = \dfrac{120}{10} = \textbf{12 A}$

(b)   Time constant of the circuit,

$$\tau = \frac{L}{R} = \frac{0.004}{10} = \textbf{0.004 s or 4 ms}$$

(c)   In the time $\tau$ s the current rises to 63.2 per cent of its final value of 12 A, i.e. in 4 ms the current rises to $0.632 \times 12 = \textbf{7.58 A}$.

(d)   The expected time for the current to rise to within 1 per cent of its final value is given by $5\tau$ s, i.e. $5 \times 4 = \textbf{20 ms}$.

> **Problem 10.**   The winding of an electromagnet has an inductance of 3 H and a resistance of $15 \Omega$. When it is connected to a 120 V, d.c. supply, calculate: (a) the steady-state value of current flowing in the winding, (b) the time constant of the circuit, (c) the value of the induced e.m.f. after 0.1 s, (d) the time for the current to rise to 85 per cent of its final value, and (e) the value of the current after 0.3 s.

(a)   The steady-state value of current,

$$I = \frac{V}{R} = \frac{120}{15} = \textbf{8 A}$$

(b)   The time constant of the circuit,

$$\tau = \frac{L}{R} = \frac{3}{15} = \textbf{0.2 s}$$

Parts (c), (d) and (e) of this problem may be determined by drawing the transients graphically, as shown in Problem 8 or by calculation as shown below.

(c)   The induced e.m.f., $v_L$ is given by $v_L = Ve^{-t/\tau}$. The d.c. voltage $V$ is 120 V, $t$ is 0.1 s and $\tau$ is 0.2 s, hence

$$v_L = 120e^{-0.1/0.2} = 120e^{-0.5}$$
$$= 120 \times 0.6065 = \textbf{72.78 V}$$

(d)   When the current is 85 per cent of its final value, $i = 0.85 I$. Also, $i = I(1 - e^{-t/\tau})$, thus

$$0.85 I = I(1 - e^{-t/\tau})$$
$$0.85 = 1 - e^{-t/\tau}$$

$\tau = 0.2$, hence

$$0.85 = 1 - e^{-t/0.2}$$
$$e^{-t/0.2} = 1 - 0.85 = 0.15$$
$$e^{t/0.2} = \frac{1}{0.15} = 6.\dot{6}$$

Taking natural logarithms of each side of this equation gives:

$$\ln e^{t/0.2} = \ln 6.\dot{6}$$

and by the laws of logarithms

$$\frac{t}{0.2} \ln e = \ln 6.\dot{6}$$

$\ln e = 1$, hence **time** $t = 0.2 \ln 6.\dot{6} = \textbf{0.379 s}$

(e)   The current at any instant is given by $i = I(1 - e^{-t/\tau})$. When $I = 8$, $t = 0.3$ and $\tau = 0.2$, then

$$\mathbf{i} = 8(1 - e^{-0.3/0.2}) = 8(1 - e^{-1.5})$$
$$= 8(1 - 0.2231) = 8 \times 0.7769 = \textbf{6.215 A}$$

## 18.10   Current decay in an $L - R$ circuit

When a series connected $L - R$ circuit is connected to a d.c. supply as shown with $S$ in position $A$ of Fig. 18.13, a current $I = V/R$ flows after a short time, creating a magnetic field ($\Phi \propto I$) associated with the inductor. When $S$ is moved to position B, the current value decreases, causing a decrease in the strength of the magnetic field. Flux linkages occur, generating a voltage $v_L$, equal to $L(di/dt)$. By Lenz's law, this voltage keeps current $i$ flowing in the circuit, its value being limited by $R$. Since $V = v_L + v_R$, $0 = v_L + v_R$ and $v_L = -v_R$, i.e. $v_L$ and

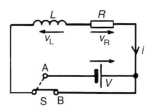

**Figure 18.13**

$v_R$ are equal in magnitude but opposite in direction. The current decays exponentially to zero and since $v_R$ is proportional to the current flowing, $v_R$ decays exponentially to zero. Since $v_L = v_R$, $v_L$ also decays exponentially to zero. The curves representing these transients are similar to those shown in Fig. 18.9.

The equations representing the decay transient curves are:

decay of voltages,

$$v_L = v_R = Ve^{(-Rt/L)} = Ve^{(-t/\tau)}$$

decay of current, $i = Ie^{(-Rt/L)} = Ie^{(-t/\tau)}$

---

**Problem 11.** The field winding of a 110 V, d.c. motor has a resistance of 15 $\Omega$ and a time constant of 2 s. Determine the inductance and use the tangential method to draw the current/time characteristic when the supply is removed and replaced by a shorting link. From the characteristic determine (a) the current flowing in the winding 3 s after being shorted-out and (b) the time for the current to decay to 5 A.

---

Since the time constant, $\tau = (L/R)$, $L = R\tau$ i.e. inductance $L = 15 \times 2 = \mathbf{30\,H}$

The current/time characteristic is constructed in a similar way to that used in Problem 1

(i) The scales should span at least five time constants horizontally, i.e. 10 s, and $I = V/R = 110/15 = 7.\dot{3}$ A vertically

(ii) With reference to Fig. 18.14, the initial slope is obtained by making OB equal to 1 time constant, (i.e. 2 s), and joining AB

(iii) At, say, $i = 6$ A, let C be the point on AB corresponding to a current of 6 A. Make DE equal to 1 time constant, (i.e. 2 s), and join CE

(iv) Repeat the procedure given in (iii) for current values of, say, 4 A, 2 A and 1 A, giving points F, G and H

(v) Point J is at five time constants, when the value of current is zero.

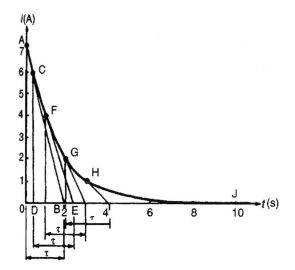

**Figure 18.14**

(vi) Join points A, C, F, G, H and J with a smooth curve. This curve is the current/time characteristic.

(a) From the current/time characteristic, when $t = 3$ s, $i = \mathbf{1.3\,A}$ [This may be checked by calculation using $i = Ie^{-t/\tau}$, where $I = 7.3$, $t = 3$ and $\tau = 2$, giving $i = 1.64$ A]. The discrepancy between the two results is due to relatively few values, such as C, F, G and H, being taken.

(b) From the characteristic, when $i = 5$ A, $t = \mathbf{0.70\,s}$ [This may be checked by calculation using $i = Ie^{-t/\tau}$, where $i = 5$, $I = 7.3$, $\tau = 2$, giving $t = 0.766$ s]. Again, the discrepancy between the graphical and calculated values is due to relatively few values such as C, F, G and H being taken.

---

**Problem 12.** A coil having an inductance of 6 H and a resistance of $R\,\Omega$ is connected in series with a resistor of 10 $\Omega$ to a 120 V, d.c. supply. The time constant of the circuit is 300 ms. When steady-state conditions have been reached, the supply is replaced instantaneously by a short-circuit. Determine: (a) the resistance of the coil, (b) the current flowing in the circuit one second after the shorting link has been placed in the circuit, and (c) the time taken for the current to fall to 10 per cent of its initial value.

---

(a) The time constant,

$$\tau = \frac{\text{circuit inductance}}{\text{total circuit resistance}} = \frac{L}{R + 10}$$

Thus $R = \dfrac{L}{\tau} - 10 = \dfrac{6}{0.3} - 10 = \mathbf{10\,\Omega}$

Parts (b) and (c) may be determined graphically as shown in Problems 8 and 11 or by calculation as shown below.

(b)    The steady-state current,

$$I = \frac{V}{R} = \frac{120}{10 + 10} = 6\,\text{A}$$

The transient current after 1 second,

$$i = I\mathrm{e}^{-t/\tau} = 6\mathrm{e}^{-1/0.3}$$

Thus    $i = 6\mathrm{e}^{-3.\dot{3}} = 6 \times 0.03567$

$$= \mathbf{0.214\,A}$$

(c)    10 per cent of the initial value of the current is $(10/100) \times 6$, i.e. $0.6\,\text{A}$. Using the equation

$$i = I\mathrm{e}^{-t/\tau} \text{ gives}$$

$$0.6 = 6\mathrm{e}^{-t/0.3}$$

i.e.    $\dfrac{0.6}{6} = \mathrm{e}^{-t/0.3}$

or    $\mathrm{e}^{t/0.3} = \dfrac{6}{0.6} = 10$

Taking natural logarithms of each side of this equation gives:

$$\frac{t}{0.3} = \ln 10$$

from which, time, $t = 0.3\ln 10 = \mathbf{0.691\,s}$

---

**Problem 13.**   An inductor has a negligible resistance and an inductance of 200 mH and is connected in series with a 1 kΩ resistor to a 24 V, d.c. supply. Determine the time constant of the circuit and the steady-state value of the current flowing in the circuit. Find (a) the current flowing in the circuit at a time equal to one time constant, (b) the voltage drop across the inductor at a time equal to two time constants, and (c) the voltage drop across the resistor after a time equal to three time constants.

---

The time constant,

$$\tau = \frac{L}{R} = \frac{0.2}{1000} = \mathbf{0.2\,ms}$$

The steady-state current

$$I = \frac{V}{R} = \frac{24}{1000} = \mathbf{24\,mA}$$

(a)    The transient current,

$$i = I(1 - \mathrm{e}^{-t/\tau}) \quad \text{and} \quad t = 1\tau.$$

Working in mA units gives,

$$i = 24(1 - \mathrm{e}^{-(1\tau/\tau)}) = 24(1 - \mathrm{e}^{-1})$$

$$= 24(1 - 0.368) = \mathbf{15.17\,mA}$$

(b)    The voltage drop across the inductor, $v_L = V\mathrm{e}^{-t/\tau}$

When $t = 2\tau$, $v_L = 24\mathrm{e}^{-2\tau/\tau} = 24\mathrm{e}^{-2}$

$$-\mathbf{3.248\,V}$$

(c)    The voltage drop across the resistor,
$v_R = V(1 - \mathrm{e}^{-t/\tau})$

When $t = 3\tau$, $v_R = 24(1 - \mathrm{e}^{-3\tau/\tau})$

$$= 24(1 - \mathrm{e}^{-3})$$

$$= \mathbf{22.81\,V}$$

---

**Now try the following exercise**

---

**Exercise 105    Further problems on transients in series $L - R$ circuits**

1.    A coil has an inductance of 1.2 H and a resistance of 40 Ω and is connected to a 200 V, d.c. supply. Either by drawing the current/time characteristic or by calculation determine the value of the current flowing 60 ms after connecting the coil to the supply.    [4.32 A]

2.    A 25 V d.c. supply is connected to a coil of inductance 1 H and resistance 5 Ω. Either by using a graphical method to draw the exponential growth curve of current or by calculation determine the value of the current flowing 100 ms after being connected to the supply.    [1.97 A]

3.    An inductor has a resistance of 20 Ω and an inductance of 4 H. It is connected to a 50 V d.c. supply. Calculate (a) the value of current flowing after 0.1 s and (b) the time for the current to grow to 1.5 A.    [(a) 0.984 A (b) 0.183 s]

4.    The field winding of a 200 V d.c. machine has a resistance of 20 Ω and an inductance of 500 mH. Calculate:

(a) the time constant of the field winding,
(b) the value of current flow one time constant after being connected to the supply, and
(c) the current flowing 50 ms after the supply has been switched on

[(a) 25 ms (b) 6.32 A (c) 8.65 A]

5. A circuit comprises an inductor of 9 H of negligible resistance connected in series with a 60 Ω resistor and a 240 V d.c. source. Calculate (a) the time constant, (b) the current after 1 time constant, (c) the time to develop maximum current, (d) the time for the current to reach 2.5 A, and (e) the initial rate of change of current.

[(a) 0.15 s (b) 2.528 A (c) 0.75 s
(d) 0.147 s (e) 26.67 A/s]

6. In the inductive circuit shown in Fig. 18.15, the switch is moved from position A to position B until maximum current is flowing. Calculate (a) the time taken for the voltage across the resistance to reach 8 volts, (b) the time taken for maximum current to flow in the circuit, (c) the energy stored in the inductor when maximum current is flowing, and (d) the time for current to drop to 750 mA after switching to position C.

[(a) 64.38 ms (b) 0.20 s
(c) 0.20 J (d) 7.67 ms]

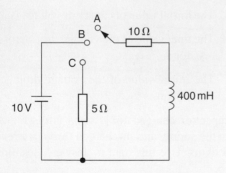

Figure 18.15

## 18.11  Switching inductive circuits

Energy stored in the magnetic field of an inductor exists because a current provides the magnetic field. When the d.c. supply is switched off the current falls rapidly, the magnetic field collapses causing a large induced e.m.f. which will either cause an arc across the switch contacts or will break down the insulation between adjacent turns of the coil. The high induced e.m.f. acts in a direction which tends to keep the current flowing, i.e. in the same direction as the applied voltage. The energy from the magnetic field will thus be aided by the supply voltage in maintaining an arc, which could cause severe damage to the switch. To reduce the induced e.m.f. when the supply switch is opened, a discharge resistor $R_D$ is connected in parallel with the inductor as shown in Fig. 18.16. The magnetic field energy is dissipated as heat in $R_D$ and $R$ and arcing at the switch contacts is avoided.

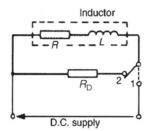

Figure 18.16

## 18.12  The effects of time constant on a rectangular waveform

### Integrator circuit

By varying the value of either $C$ or $R$ in a series-connected $C - R$ circuit, the time constant ($\tau = CR$), of a circuit can be varied. If a rectangular waveform varying from $+E$ to $-E$ is applied to a $C - R$ circuit as shown in Fig. 18.17, output waveforms of the capacitor voltage have various shapes, depending on the value of $R$. When $R$ is small, $\tau = CR$ is small and an output waveform such as that shown in Fig. 18.18(a) is obtained. As the value of $R$ is increased, the waveform changes to that shown in Fig. 18.18(b). When $R$ is large,

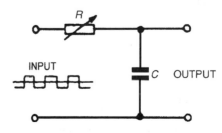

Figure 18.17

the waveform is as shown in Fig. 18.18(c), the circuit then being described as an **integrator circuit**.

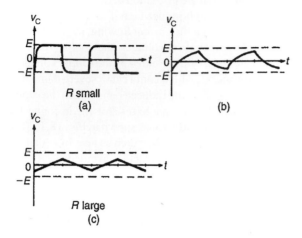

Figure 18.18

## Differentiator circuit

If a rectangular waveform varying from $+E$ to $-E$ is applied to a series connected $C - R$ circuit and the waveform of the voltage drop across the resistor is observed, as shown in Fig. 18.19, the output waveform alters as $R$ is varied due to the time constant, $(\tau = CR)$, altering.

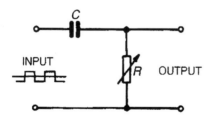

Figure 18.19

When $R$ is small, the waveform is as shown in Fig. 18.20(a), the voltage being generated across $R$ by the capacitor discharging fairly quickly. Since the change in capacitor voltage is from $+E$ to $-E$, the change in discharge current is $2E/R$, resulting in

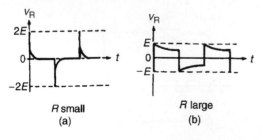

Figure 18.20

a change in voltage across the resistor of $2E$. This circuit is called a **differentiator circuit**. When $R$ is large, the waveform is as shown in Fig. 18.20(b).

---

### Now try the following exercises

**Exercise 106    Short answer questions on d.c. transients**

A capacitor of capacitance $C$ farads is connected in series with a resistor of $R$ ohms and is switched across a constant voltage d.c. supply of $V$ volts. After a time of $t$ seconds, the current flowing is $i$ amperes. Use this data to answer questions 1 to 10.

1.  The voltage drop across the resistor at time $t$ seconds is $v_R = \ldots\ldots$

2.  The capacitor voltage at time $t$ seconds is $v_C = \ldots\ldots$

3.  The voltage equation for the circuit is $V = \ldots\ldots$

4.  The time constant for the circuit is $\tau = \ldots\ldots$

5.  The final value of the current flowing is $\ldots\ldots$

6.  The initial value of the current flowing is $I = \ldots\ldots$

7.  The final value of capacitor voltage is $\ldots\ldots$

8.  The initial value of capacitor voltage is $\ldots\ldots$

9.  The final value of the voltage drop across the resistor is $\ldots\ldots$

10. The initial value of the voltage drop across the resistor is $\ldots\ldots$

A capacitor charged to $V$ volts is disconnected from the supply and discharged through a resistor of $R$ ohms. Use this data to answer questions 11 to 15.

11. The initial value of current flowing is $I = \ldots\ldots$

12. The approximate time for the current to fall to zero in terms of $C$ and $R$ is $\ldots\ldots$ seconds

13. If the value of resistance $R$ is doubled, the time for the current to fall to zero is $\ldots\ldots$ when compared with the time in question 12 above

14. The approximate fall in the value of the capacitor voltage in a time equal to one time constant is ...... per cent

15. The time constant of the circuit is given by ...... seconds

An inductor of inductance $L$ henrys and negligible resistance is connected in series with a resistor of resistance $R$ ohms and is switched across a constant voltage d.c. supply of $V$ volts. After a time interval of $t$ seconds, the transient current flowing is $i$ amperes. Use this data to answer questions 16 to 25.

16. The induced e.m.f., $v_L$, opposing the current flow when $t = 0$ is ......

17. The voltage drop across the resistor when $t = 0$ is $v_R =$ ......

18. The current flowing when $t = 0$ is ......

19. $V$, $v_R$ and $v_L$ are related by the equation $V =$ ......

20. The time constant of the circuit in terms of $L$ and $R$ is ......

21. The steady-state value of the current is reached in practise in a time equal to ...... seconds

22. The steady-state voltage across the inductor is ...... volts

23. The final value of the current flowing is ...... amperes

24. The steady-state resistor voltage is ...... volts

25. The e.m.f. induced in the inductor during the transient in terms of current, time and inductance is ...... volts

A series-connected $L - R$ circuit carrying a current of $I$ amperes is suddenly short-circuited to allow the current to decay exponentially. Use this data to answer questions 26 to 30.

26. The current will fall to ...... per cent of its final value in a time equal to the time constant

27. The voltage equation of the circuit is ......

28. The time constant of the circuit in terms of $L$ and $R$ is ......

29. The current reaches zero in a time equal to ...... seconds

30. If the value of $R$ is halved, the time for the current to fall to zero is ...... when compared with the time in question 29

31. With the aid of a circuit diagram, explain briefly the effects on the waveform of the capacitor voltage of altering the value of resistance in a series connected $C - R$ circuit, when a rectangular wave is applied to the circuit

32. What do you understand by the term 'integrator circuit' ?

33. With reference to a rectangular wave applied to a series connected $C - R$ circuit, explain briefly the shape of the waveform when $R$ is small and hence what you understand by the term 'differentiator circuit'

**Exercise 107   Multi-choice questions on d.c. transients**
**(Answers on page 421)**

An uncharged $2\mu F$ capacitor is connected in series with a $5 M\Omega$ resistor to a $100\,V$, constant voltage, d.c. supply. In questions 1 to 7, use this data to select the correct answer from those given below:

(a)  10 ms        (b)  100 V        (c)  10 s
(d)  10 V         (e)  20 μA        (f)  1 s
(g)  0 V          (h)  50 V         (i)  1 ms
(j)  50 μA        (k)  20 mA        (l)  0 A

1. Determine the time constant of the circuit

2. Determine the final voltage across the capacitor

3. Determine the initial voltage across the resistor

4. Determine the final voltage across the resistor

5. Determine the initial voltage across the capacitor

Section 2

6. Determine the initial current flowing in the circuit

7. Determine the final current flowing in the circuit

   In questions 8 and 9, a series connected $C - R$ circuit is suddenly connected to a d.c. source of $V$ volts. Which of the statements is false ?

8. (a) The initial current flowing is given by $V/R$
   (b) The time constant of the circuit is given by $CR$
   (c) The current grows exponentially
   (d) The final value of the current is zero

9. (a) The capacitor voltage is equal to the voltage drop across the resistor
   (b) The voltage drop across the resistor decays exponentially
   (c) The initial capacitor voltage is zero
   (d) The initial voltage drop across the resistor is $IR$, where $I$ is the steady-state current

10. A capacitor which is charged to $V$ volts is discharged through a resistor of $R$ ohms. Which of the following statements is false?
    (a) The initial current flowing is $V/R$ amperes
    (b) The voltage drop across the resistor is equal to the capacitor voltage
    (c) The time constant of the circuit is $CR$ seconds
    (d) The current grows exponentially to a final value of $V/R$ amperes

    An inductor of inductance 0.1 H and negligible resistance is connected in series with a 50 Ω resistor to a 20 V d.c. supply. In questions 11 to 15, use this data to determine the value required, selecting your answer from those given below:

    (a) 5 ms     (b) 12.6 V     (c) 0.4 A
    (d) 500 ms    (e) 7.4 V      (f) 2.5 A
    (g) 2 ms     (h) 0 V        (i) 0 A
    (j) 20 V

11. The value of the time constant of the circuit

12. The approximate value of the voltage across the resistor after a time equal to the time constant

13. The final value of the current flowing in the circuit

14. The initial value of the voltage across the inductor

15. The final value of the steady-state voltage across the inductor

16. The time constant for a circuit containing a capacitance of 100 nF in series with a 5 Ω resistance is:
    (a) 0.5 μs   (b) 20 ns   (c) 5 μs   (d) 50 μs

17. The time constant for a circuit containing an inductance of 100 mH in series with a resistance of 4 Ω is:
    (a) 25 ms   (b) 400 s   (c) 0.4 s   (d) 40 s

18. The graph shown in Fig. 18.21 represents the growth of current in an $L - R$ series circuit connected to a d.c. voltage $V$ volts. The equation for the graph is:
    (a) $i = I(1 - e^{-Rt/L})$    (b) $i = Ie^{-Li/t}$
    (c) $i = Ie^{-Rt/L}$         (d) $i = I(1 - e^{RL/t})$

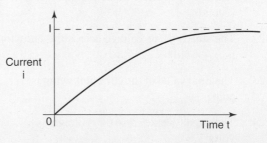

**Figure 18.21**

# Chapter 19

# Operational amplifiers

At the end of this chapter you should be able to:

- recognize the main properties of an operational amplifier
- understand op amp parameters input bias current and offset current and voltage
- define and calculate common-mode rejection ratio
- appreciate slew rate
- explain the principle of operation, draw the circuit diagram symbol and calculate gain for the following operational amplifiers:

  inverter

  non-inverter

  voltage follower (or buffer)

  summing

  voltage comparator

  integrator

  differentiator

- understand digital to analogue conversion
- understand analogue to digital conversion

## 19.1 Introduction to operational amplifiers

**Operational Amplifiers** (usually called **'op amps'**) were originally made from discrete components, being designed to solve mathematical equations electronically, by performing operations such as addition and division in analogue computers. Now produced in integrated-circuit (IC) form, op amps have many uses, with one of the most important being as a high-gain d.c. and a.c. voltage amplifier.

The **main properties** of an op amp include:

(i) a very high open-loop voltage gain $A_o$ of around $10^5$ for d.c. and low frequency a.c., which decreases with frequency increase

(ii) a very high input impedance, typically $10^6\,\Omega$ to $10^{12}\,\Omega$, such that current drawn from the device, or the circuit supplying it, is very small and the input voltage is passed on to the op amp with little loss

(iii) a very low output impedance, around $100\,\Omega$, such that its output voltage is transferred efficiently to any load greater than a few kilohms. The **circuit diagram symbol** for an op amp is shown in Fig. 19.1. It has one output, $V_o$, and two inputs; the **inverting input**, $V_1$ is marked $-$, and the non-**inverting input**, $V_2$, is marked $+$

The operation of an op amp is most convenient from a dual balanced d.c. power supply $\pm V_s$ (i.e. $+V_s$, $0$, $-V_s$); the centre point of the supply, i.e. $0\,V$, is

DOI: 10.1016/B978-0-08-089056-2.00019-X

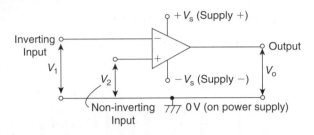

**Figure 19.1**

common to the input and output circuits and is taken as their voltage reference level. The power supply connections are not usually shown in a circuit diagram.

An op amp is basically a **differential** voltage amplifier, i.e. it amplifies the difference between input voltages $V_1$ and $V_2$. Three situations are possible:

(i)   if $V_2 > V_1$, $V_o$ is positive

(ii)   if $V_2 < V_1$, $V_o$ is negative

(iii)   if $V_2 = V_1$, $V_o$ is zero

In general,   $V_o = A_o(V_2 - V_1)$

or   $$A_o = \frac{V_o}{V_2 - V_1}$$   (1)

where $A_o$ is the open-loop voltage gain.

> **Problem 1.**   A differential amplifier has an open-loop voltage gain of 120. The input signals are 2.45 V and 2.35 V. Calculate the output voltage of the amplifier.

From equation (1), **output voltage**,

$$V_o = A_o(V_2 - V_1) = 120(2.45 - 2.35)$$

$$= (120)(0.1) = \mathbf{12\,V}$$

### Transfer characteristic

A typical **voltage characteristic** showing how the output $V_o$ varies with the input $(V_2 - V_1)$ is shown in Fig. 19.2.

It is seen from Fig. 19.2 that only within the very small input range P0Q is the output directly proportional to the input; it is in this range that the op amp behaves linearly and there is minimum distortion of the amplifier output. Inputs outside the linear range cause saturation and the output is then close to the maximum value, i.e. $+V_s$ or $-V_s$. The limited linear behaviour is due to the very high

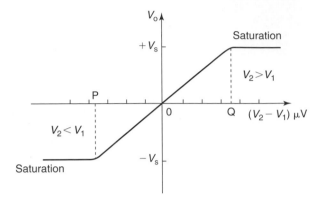

**Figure 19.2**

open-loop gain $A_o$, and the higher it is the greater is the limitation.

### Negative feedback

Operational amplifiers nearly always use **negative feedback**, obtained by feeding back some, or all, of the output to the inverting $(-)$ input (as shown in Fig. 19.5 in the next section). The feedback produces an output voltage that opposes the one from which it is taken. This reduces the new output of the amplifier and the resulting closed-loop gain $A$ is then less than the open-loop gain $A_o$. However, as a result, a wider range of voltages can be applied to the input for amplification. As long as $A_o \gg A$, negative feedback gives:

(i)   a constant and predictable voltage gain $A$,

(ii)   reduced distortion of the output, and

(iii)   better frequency response.

The advantages of using negative feedback outweigh the accompanying loss of gain which is easily increased by using two or more op amp stages.

### Bandwidth

The open-loop voltage gain of an op amp is not constant at all frequencies; because of capacitive effects it falls at high frequencies. Figure 19.3 shows the gain/bandwidth characteristic of a 741 op amp. At frequencies below 10 Hz the gain is constant, but at higher frequencies the gain falls at a constant rate of 6 dB/octave (equivalent to a rate of 20 dB per decade) to 0 dB. The gain-bandwidth product for any amplifier is the linear voltage gain multiplied by the bandwidth at that gain. The value of frequency at which the

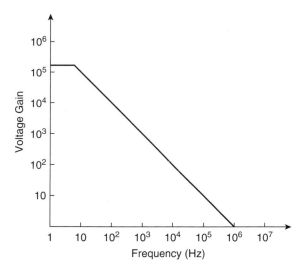

**Figure 19.3**

open-loop gain has fallen to unity is called the transition frequency $f_T$.

$$f_T = \text{closed-loop voltage gain} \times \text{bandwidth} \quad (2)$$

In Fig. 19.3, $f_T = 10^6$ Hz or 1 MHz; a gain of 20 dB (i.e. $20 \log_{10} 10$) gives a 100 kHz bandwidth, whilst a gain of 80 dB (i.e. $20 \log_{10} 10^4$) restricts the bandwidth to 100 Hz.

## 19.2 Some op amp parameters

### Input bias current

The input bias current, $I_B$, is the average of the currents into the two input terminals with the output at zero volts, which is typically around 80 nA (i.e. $80 \times 10^{-9}$ A) for a 741 op amp. The input bias current causes a volt drop across the equivalent source impedance seen by the op amp input.

### Input offset current

The input offset current, $I_{os}$, of an op amp is the difference between the two input currents with the output at zero volts. In a 741 op amp, $I_{os}$ is typically 20 nA.

### Input offset voltage

In the ideal op amp, with both inputs at zero there should be zero output. Due to imbalances within the amplifier this is not always the case and a small output voltage results. The effect can be nullified by applying a small offset voltage, $V_{os}$, to the amplifier. In a 741 op amp, $V_{os}$ is typically 1 mV.

## Common-mode rejection ratio

The output voltage of an op amp is proportional to the difference between the voltages applied to its two input terminals. Ideally, when the two voltages are equal, the output voltages should be zero. A signal applied to both input terminals is called a common-mode signal and it is usually an unwanted noise voltage. The ability of an op amp to suppress common-mode signals is expressed in terms of its common-mode rejection ratio (CMRR), which is defined by:

$$\text{CMRR} = 20 \log_{10}\left(\frac{\text{differential voltage gain}}{\text{common-mode gain}}\right) \text{dB} \quad (3)$$

In a 741 op amp, the CMRR is typically 90 dB. The common-mode gain, $A_{\text{com}}$, is defined as:

$$A_{\text{com}} = \frac{V_o}{V_{\text{com}}} \quad (4)$$

where $V_{\text{com}}$ is the common input signal.

**Problem 2.** Determine the common-mode gain of an op amp that has a differential voltage gain of $150 \times 10^3$ and a CMRR of 90 dB.

From equation (3),

$$\text{CMRR} = 20 \log_{10}\left(\frac{\text{differential voltage gain}}{\text{common-mode gain}}\right) \text{dB}$$

Hence $90 = 20 \log_{10}\left(\frac{150 \times 10^3}{\text{common-mode gain}}\right)$

from which

$$4.5 = \log_{10}\left(\frac{150 \times 10^3}{\text{common-mode gain}}\right)$$

and $10^{4.5} = \dfrac{150 \times 10^3}{\text{common-mode gain}}$

Hence, **common-mode gain** $= \dfrac{150 \times 10^3}{10^{4.5}} = \mathbf{4.74}$

**Problem 3.** A differential amplifier has an open-loop voltage gain of 120 and a common input signal of 3.0 V to both terminals. An output signal of 24 mV results. Calculate the common-mode gain and the CMRR.

From equation (4), the common-mode gain,

$$\mathbf{A}_{com} = \frac{V_o}{V_{com}} = \frac{24 \times 10^{-3}}{3.0} = 8 \times 10^{-3} = \mathbf{0.008}$$

From equation (3), the

$$\mathbf{CMRR} = 20 \log_{10} \left( \frac{\text{differential voltage gain}}{\text{common-mode gain}} \right) dB$$

$$= 20 \log_{10} \left( \frac{120}{0.008} \right)$$

$$= 20 \log_{10} 15\,000 = \mathbf{83.52\,dB}$$

## Slew rate

The slew rate of an op amp is the maximum rate of change of output voltage following a step input voltage. Figure 19.4 shows the effects of slewing; it causes the output voltage to change at a slower rate than the input, such that the output waveform is a distortion of the input waveform. 0.5 V/μs is a typical value for the slew rate.

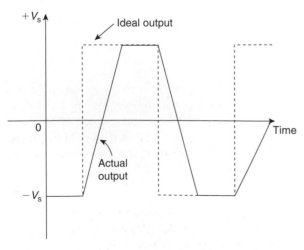

**Figure 19.4**

## 19.3 Op amp inverting amplifier

The basic circuit for an inverting amplifier is shown in Fig. 19.5 where the input voltage $V_i$ (a.c. or d.c.) to be amplified is applied via resistor $R_i$ to the inverting (−) terminal; the output voltage $V_o$ is therefore in anti-phase with the input. The non-inverting (+) terminal is held at 0 V. Negative feedback is provided by the feedback resistor, $R_f$, feeding back a certain fraction of the output voltage to the inverting terminal.

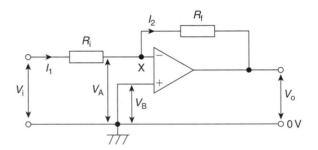

**Figure 19.5**

## Amplifier gain

In an **ideal op amp** two assumptions are made, these being that:

(i) each input draws zero current from the signal source, i.e. their input impedances are infinite, and

(ii) the inputs are both at the same potential if the op amp is not saturated, i.e. $V_A = V_B$ in Fig. 19.5

In Fig. 19.5, $V_B = 0$, hence $V_A = 0$ and point $X$ is called a **virtual earth**. Thus,

$$I_1 = \frac{V_i - 0}{R_i}$$

and

$$I_2 = \frac{0 - V_o}{R_f}$$

However, $I_1 = I_2$ from assumption (i) above. Hence

$$\frac{V_i}{R_i} = \frac{-V_o}{R_f}$$

the negative sign showing that $V_o$ is negative when $V_i$ is positive, and vice versa.
The **closed-loop gain $A$** is given by:

$$A = \frac{V_o}{V_i} = \frac{-R_f}{R_i} \qquad (5)$$

This shows that the gain of the amplifier depends only on the two resistors, which can be made with precise values, and not on the characteristics of the op amp, which may vary from sample to sample.

For example, if $R_i = 10\,k\Omega$ and $R_f = 100\,k\Omega$, then the closed-loop gain,

$$A = \frac{-R_f}{R_i} = \frac{-100 \times 10^3}{10 \times 10^3} = -10$$

Thus an input of 100 mV will cause an output change of 1 V.

## Input impedance

Since point X is a virtual earth (i.e. at 0 V), $R_i$ may be considered to be connected between the inverting $(-)$ input terminal and 0 V. The input impedance of the circuit is therefore $R_i$ in parallel with the much greater input impedance of the op amp, i.e. effectively $R_i$. The circuit input impedance can thus be controlled by simply changing the value of $R_i$.

---

**Problem 4.** In the inverting amplifier of Fig. 19.5, $R_i = 1\,k\Omega$ and $R_f = 2\,k\Omega$. Determine the output voltage when the input voltage is: (a) $+0.4\,V$ (b) $-1.2\,V$.

---

From equation (5),

$$V_o = \left(\frac{-R_f}{R_i}\right) V_i$$

(a) When $V_i = +0.4\,V$,

$$V_o = \left(\frac{-2000}{1000}\right)(+0.4) = -0.8\,V$$

(b) When $V_i = -1.2\,V$,

$$V_o = \left(\frac{-2000}{1000}\right)(-1.2) = +2.4\,V$$

---

**Problem 5.** The op amp shown in Fig. 19.6 has an input bias current of 100 nA at 20 °C. Calculate (a) the voltage gain, and (b) the output offset

---

voltage due to the input bias current. (c) How can the effect of input bias current be minimised?

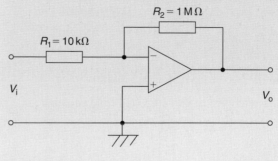

**Figure 19.6**

Comparing Fig. 19.6 with Fig. 19.5, gives $R_i = 10\,k\Omega$ and $R_f = 1\,M\Omega$

(a) From equation (5), **voltage gain,**

$$A = \frac{-R_f}{R_i} = \frac{-1 \times 10^6}{10 \times 10^3} = -100$$

(b) The input bias current, $I_B$, causes a volt drop across the equivalent source impedance seen by the op amp input, in this case, $R_i$ and $R_f$ in parallel. Hence, the offset voltage, $V_{os}$, at the input due to the 100 nA input bias current, $I_B$, is given by:

$$V_{os} = I_B\left(\frac{R_i R_f}{R_i + R_f}\right)$$

$$= (100 \times 10^{-9})\left(\frac{10 \times 10^3 \times 1 \times 10^6}{(10 \times 10^3) + (1 \times 10^6)}\right)$$

$$= (10^{-7})(9.9 \times 10^3) = 9.9 \times 10^{-4}$$

$$= 0.99\,mV$$

(c) The effect of input bias current can be minimised by ensuring that both inputs 'see' the same driving resistance. This means that **a resistance of value of 9.9 k$\Omega$** (from part (b)) **should be placed between the non-inverting (+) terminal and earth** in Fig. 19.6

---

**Problem 6.** Design an inverting amplifier to have a voltage gain of 40 dB, a closed-loop bandwidth of 5 kHz and an input resistance of 10 k$\Omega$.

---

The voltage gain of an op amp, in decibels, is given by:

$$\text{gain in decibels} = 20 \log_{10} (\text{voltage gain})$$

from Chapter 10.

Hence $\qquad 40 = 20 \log_{10} A$

from which, $\qquad 2 = \log_{10} A$

and $\qquad \boldsymbol{A = 10^2 = 100}$

With reference to Fig. 19.5, and from equation (5),

$$A = \left| \frac{R_f}{R_i} \right|$$

i.e. $\qquad 100 = \dfrac{R_f}{10 \times 10^3}$

Hence $\qquad \boldsymbol{R_f = 100 \times 10 \times 10^3 = 1\,M\Omega}$

From equation (2), Section 19.1,

$$\textbf{frequency} = \text{gain} \times \text{bandwidth}$$

$$= 100 \times 5 \times 10^3$$

$$= \textbf{0.5\,MHz or 500\,kHz}$$

**Now try the following exercise**

**Exercise 108   Further problems on operational amplifiers**

1. A differential amplifier has an open-loop voltage gain of 150 when the input signals are 3.55 V and 3.40 V. Determine the output voltage of the amplifier. [22.5 V]

2. Calculate the differential voltage gain of an op amp that has a common-mode gain of 6.0 and a CMRR of 80 dB. [$6 \times 10^4$]

3. A differential amplifier has an open-loop voltage gain of 150 and a common input signal of 4.0 V to both terminals. An output signal of 15 mV results. Determine the common-mode gain and the CMRR. [$3.75 \times 10^{-3}$, 92.04 dB]

4. In the inverting amplifier of Fig. 19.5 (on page 298), $R_i = 1.5\,k\Omega$ and $R_f = 2.5\,k\Omega$.

Determine the output voltage when the input voltage is: (a) $+0.6$ V (b) $-0.9$ V [(a) $-1.0$ V (b) $+1.5$ V]

5. The op amp shown in Fig. 19.7 has an input bias current of 90 nA at 20°C. Calculate (a) the voltage gain, and (b) the output offset voltage due to the input bias current. [(a) $-80$ (b) 1.33 mV]

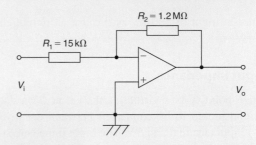

**Figure 19.7**

6. Determine (a) the value of the feedback resistor, and (b) the frequency for an inverting amplifier to have a voltage gain of 45 dB, a closed-loop bandwidth of 10 kHz and an input resistance of 20 kΩ. [(a) 3.56 MΩ (b) 1.78 MHz]

## 19.4   Op amp non-inverting amplifier

The basic circuit for a non-inverting amplifier is shown in Fig. 19.8 where the input voltage $V_i$ (a.c. or d.c.) is applied to the non-inverting ($+$) terminal of the op amp. This produces an output $V_o$ that is in phase with the input. Negative feedback is obtained by feeding back to the inverting ($-$) terminal, the fraction of $V_o$ developed across $R_i$ in the voltage divider formed by $R_f$ and $R_i$ across $V_o$.

**Amplifier gain**

In Fig. 19.8, let the feedback factor,

$$\beta = \frac{R_i}{R_i + R_f}$$

It may be shown that for an amplifier with open-loop gain $A_o$, the closed-loop voltage gain $A$ is given by:

$$A = \frac{A_o}{1 + \beta A_o}$$

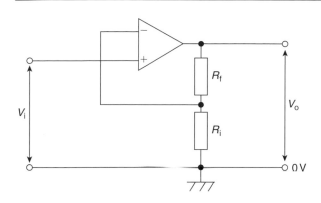

**Figure 19.8**

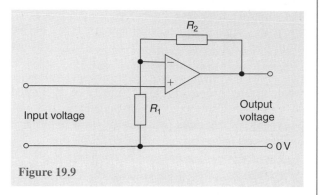

**Figure 19.9**

For a typical op amp, $A_o = 10^5$, thus $\beta A_o$ is large compared with 1, and the above expression approximates to:

$$A = \frac{A_o}{\beta A_o} = \frac{1}{\beta} \qquad (6)$$

Hence
$$A = \frac{V_o}{V_i} = \frac{R_i + R_f}{R_i} = 1 + \frac{R_f}{R_i} \qquad (7)$$

For example, if $R_i = 10\,\mathrm{k\Omega}$ and $R_f = 100\,\mathrm{k\Omega}$, then

$$A = 1 + \frac{100 \times 10^3}{10 \times 10^3} = 1 + 10 = \mathbf{11}$$

Again, the gain depends only on the values of $R_i$ and $R_f$ and is independent of the open-loop gain $A_o$.

**Input impedance**

Since there is no virtual earth at the non-inverting (+) terminal, the input impedance is much higher (– typically 50 MΩ) than that of the inverting amplifier. Also, it is unaffected if the gain is altered by changing $R_f$ and/or $R_i$. This non-inverting amplifier circuit gives good matching when the input is supplied by a high impedance source.

**Problem 7.** For the op amp shown in Fig. 19.9, $R_1 = 4.7\,\mathrm{k\Omega}$ and $R_2 = 10\,\mathrm{k\Omega}$. If the input voltage is $-0.4\,\mathrm{V}$, determine (a) the voltage gain (b) the output voltage.

The op amp shown in Fig. 19.9 is a non-inverting amplifier, similar to Fig. 19.8

(a) From equation (7), **voltage gain,**

$$A = 1 + \frac{R_f}{R_i} = 1 + \frac{R_2}{R_1} = 1 + \frac{10 \times 10^3}{4.7 \times 10^3}$$
$$= 1 + 2.13 = \mathbf{3.13}$$

(b) Also from equation (7), **output voltage,**

$$V_o = \left(1 + \frac{R_2}{R_1}\right) V_i = (3.13)(-0.4) = \mathbf{-1.25\,V}$$

## 19.5 Op amp voltage-follower

The **voltage-follower** is a special case of the non-inverting amplifier in which 100% negative feedback is obtained by connecting the output directly to the inverting (−) terminal, as shown in Fig. 19.10. Thus $R_f$ in Fig. 19.8 is zero and $R_i$ is infinite.

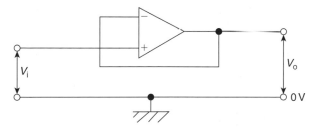

**Figure 19.10**

From equation (6), $A = 1/\beta$ (when $A_o$ is very large). Since all of the output is fed back, $\beta = 1$ and $A \approx 1$. Thus the voltage gain is nearly 1 and $V_o = V_i$ to within a few millivolts.

The circuit of Fig. 19.10 is called a voltage-follower since, as with its transistor emitter-follower equivalent,

$V_0$ follows $V_i$. It has an extremely high input impedance and a low output impedance. Its main use is as a **buffer amplifier,** giving current amplification, to match a high impedance source to a low impedance load. For example, it is used as the input stage of an analogue voltmeter where the highest possible input impedance is required so as not to disturb the circuit under test; the output voltage is measured by a relatively low impedance moving-coil meter.

## 19.6 Op amp summing amplifier

Because of the existence of the virtual earth point, an op amp can be used to add a number of voltages (d.c. or a.c.) when connected as a multi-input inverting amplifier. This, in turn, is a consequence of the high value of the open-loop voltage gain $A_0$. Such circuits may be used as 'mixers' in audio systems to combine the outputs of microphones, electric guitars, pick-ups, etc. They are also used to perform the mathematical process of addition in analogue computing.

The circuit of an op amp summing amplifier having three input voltages $V_1$, $V_2$ and $V_3$ applied via input resistors $R_1$, $R_2$ and $R_3$ is shown in Fig. 19.11. If it is assumed that the inverting ($-$) terminal of the op amp draws no input current, all of it passing through $R_f$, then:

$$I = I_1 + I_2 + I_3$$

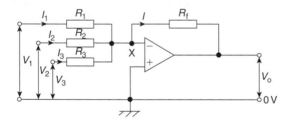

**Figure 19.11**

Since X is a virtual earth (i.e. at 0 V), it follows that:

$$\frac{-V_0}{R_f} = \frac{V_1}{R_1} + \frac{V_2}{R_2} + \frac{V_3}{R_3}$$

Hence

$$V_0 = -\left(\frac{R_f}{R_1}V_1 + \frac{R_f}{R_2}V_2 + \frac{R_f}{R_3}V_3\right)$$

$$= -R_f\left(\frac{V_1}{R_1} + \frac{V_2}{R_2} + \frac{V_3}{R_3}\right) \qquad (8)$$

The three input voltages are thus added and amplified if $R_f$ is greater than each of the input resistors; 'weighted'

summation is said to have occurred. Alternatively, the input voltages are added and attenuated if $R_f$ is less than each input resistor.

For example, if

$$\frac{R_f}{R_1} = 4 \quad \frac{R_f}{R_2} = 3$$

and

$$\frac{R_f}{R_3} = 1$$

and $V_1 = V_2 = V_3 = +1$ V, then

$$V_0 = -\left(\frac{R_f}{R_1}V_1 + \frac{R_f}{R_2}V_2 + \frac{R_f}{R_3}V_3\right)$$

$$= -(4 + 3 + 1) = -8\,\text{V}$$

If $R_1 = R_2 = R_3 = R_i$, the input voltages are amplified or attenuated equally, and

$$V_0 = -\frac{R_f}{R_i}(V_1 + V_2 + V_3)$$

If, also, $R_i = R_f$ then $V_0 = -(V_1 + V_2 + V_3)$.

The virtual earth is also called the **summing point** of the amplifier. It isolates the inputs from one another so that each behaves as if none of the others existed and none feeds any of the other inputs even though all the resistors are connected at the inverting ($-$) input.

---

**Problem 8.** For the summing op amp shown in Fig. 19.12, determine the output voltage, $V_0$.

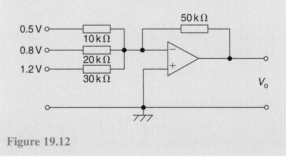

**Figure 19.12**

From equation (8),

$$V_0 = -R_f\left(\frac{V_1}{R_1} + \frac{V_2}{R_2} + \frac{V_3}{R_3}\right)$$

$$= -(50 \times 10^3)\left(\frac{0.5}{10 \times 10^3} + \frac{0.8}{20 \times 10^3} + \frac{1.2}{30 \times 10^3}\right)$$

$$= -(50 \times 10^3)\,(5 \times 10^{-5} + 4 \times 10^{-5} + 4 \times 10^{-5})$$

$$= -(50 \times 10^3)\,(13 \times 10^{-5})$$

$$= -6.5\,\text{V}$$

## 19.7 Op amp voltage comparator

If both inputs of the op amp shown in Fig. 19.13 are used simultaneously, then from equation (1), page 296, the output voltage is given by:

$$V_o = A_o(V_2 - V_1)$$

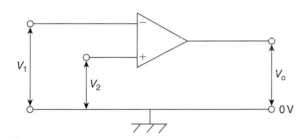

Figure 19.13

When $V_2 > V_1$ then $V_o$ is positive, its maximum value being the positive supply voltage $+V_s$, which it has when $(V_2 - V_1) \geq V_s/A_o$. The op amp is then saturated. For example, if $V_s = +9\,V$ and $A_o = 10^5$, then saturation occurs when

$$(V_2 - V_1) \geq \frac{9}{10^5}$$

i.e. when $V_2$ exceeds $V_1$ by $90\,\mu V$ and $V_o \approx 9\,V$.

When $V_1 > V_2$, then $V_o$ is negative and saturation occurs if $V_1$ exceeds $V_2$ by $V_s/A_o$ i.e. around $90\,\mu V$ in the above example; in this case, $V_o \approx -V_s = -9\,V$.

A small change in $(V_2 - V_1)$ therefore causes $V_o$ to switch between near $+V_s$ and near to $-V_s$ and enables the op amp to indicate when $V_2$ is greater or less than $V_1$, i.e. to act as a **differential amplifier** and compare two voltages. It does this in an electronic digital voltmeter.

> **Problem 9.** Devise a light-operated alarm circuit using an op amp, a LDR, a LED and a $\pm 15\,V$ supply.

A typical light-operated alarm circuit is shown in Fig. 19.14.

Resistor $R$ and the light dependent resistor (LDR) form a voltage divider across the $+15/0/-15\,V$ supply. The op amp compares the voltage $V_1$ at the voltage divider junction, i.e. at the inverting $(-)$ input, with that at the non-inverting $(+)$ input, i.e. with $V_2$, which is $0\,V$. In the dark the resistance of the LDR is much greater than that of $R$, so more of the $30\,V$ across the voltage divider is dropped across the LDR, causing $V_1$ to fall below $0\,V$. Now $V_2 > V_1$ and the output voltage $V_o$ switches from near $-15\,V$ to near $+15\,V$ and the light emitting diode (LED) lights.

## 19.8 Op amp integrator

The circuit for the op amp integrator shown in Fig. 19.15 is the same as for the op amp inverting amplifier shown in Fig. 19.5, but feedback occurs via a capacitor C, rather than via a resistor.

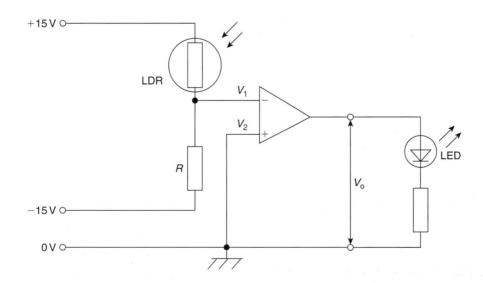

Figure 19.14

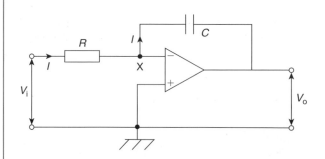

**Figure 19.15**

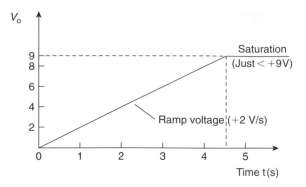

**Figure 19.16**

The output voltage is given by:

$$V_o = -\frac{1}{CR} \int V_i \, dt \qquad (9)$$

Since the inverting $(-)$ input is used in Fig. 19.15, $V_o$ is negative if $V_i$ is positive, and vice versa, hence the negative sign in equation (9).

Since $X$ is a virtual earth in Fig. 19.15, i.e. at 0 V, the voltage across $R$ is $V_i$ and that across C is $V_o$. Assuming again that none of the input current $I$ enters the op amp inverting $(-)$ input, then all of current $I$ flows through $C$ and charges it up. If $V_i$ is constant, $I$ will be a constant value given by $I = V_i/R$. Capacitor $C$ therefore charges at a constant rate and the potential of the output side of $C$ $(= V_o$, since its input side is zero) charges so that the feedback path absorbs $I$. If $Q$ is the charge on $C$ at time $t$ and the p.d. across it (i.e. the output voltage) changes from 0 to $V_o$ in that time then:

$$Q = -V_o C = It$$

(from Chapter 6)

i.e. $$-V_o C = \frac{V_i}{R} t$$

i.e. $$V = -\frac{1}{CR} V_i t$$

This result is the same as would be obtained from

$$V_o = -\frac{1}{CR} \int V_i \, dt$$

if $V_i$ is a constant value.

For example, if the input voltage $V_i = -2$ V and, say, $CR = 1$ s, then

$$V_o = -(-2)t = 2t$$

A graph of $V_o/t$ will be ramp function as shown in Fig. 19.16 ($V_o = 2t$ is of the straight line form $y = mx + c$; in this case $y = V_o$ and $x = t$, gradient, $m = 2$ and vertical axis intercept $c = 0$). $V_o$ rises steadily by $+2V/s$ in Fig. 19.16, and if the power supply is, say, $\pm 9$ V, then $V_o$ reaches $+9$ V after 4.5 s when the op amp saturates.

> **Problem 10.** A steady voltage of $-0.75$ V is applied to an op amp integrator having component values of $R = 200\,\text{k}\Omega$ and $C = 2.5\,\mu\text{F}$. Assuming that the initial capacitor charge is zero, determine the value of the output voltage 100 ms after application of the input.

From equation (9), output voltage,

$$V_o = -\frac{1}{CR} \int V_i \, dt$$

$$= -\frac{1}{(2.5 \times 10^{-6})(200 \times 10^3)} \int (-0.75) \, dt$$

$$= -\frac{1}{0.5} \int (-0.75) \, dt = -2[-0.75t]$$

$$= +1.5t$$

When time $t = 100$ ms,
**output voltage, $V_o = (1.5)(100 \times 10^{-3}) = \mathbf{0.15\,V}$.**

## 19.9 Op amp differential amplifier

The circuit for an op amp differential amplifier is shown in Fig. 19.17 where voltages $V_1$ and $V_2$ are applied to its two input terminals and the difference between these voltages is amplified.

(i) Let $V_1$ volts be applied to terminal 1 and 0 V be applied to terminal 2. The difference in the potentials at the inverting $(-)$ and non-inverting $(+)$ op amp inputs is practically zero and hence the inverting terminal must be at zero potential. Then

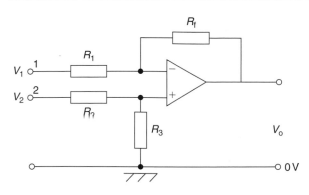

**Figure 19.17**

$I_1 = V_1/R_1$. Since the op amp input resistance is high, this current flows through the feedback resistor $R_f$. The volt drop across $R_f$, which is the output voltage

$$V_o = \frac{V_1}{R_1} R_f$$

hence, the closed loop voltage gain $A$ is given by:

$$A = \frac{V_o}{V_1} = -\frac{R_f}{R_1} \qquad (10)$$

(ii) By similar reasoning, if $V_2$ is applied to terminal 2 and 0 V to terminal 1, then the voltage appearing at the non-inverting terminal will be

$$\left(\frac{R_3}{R_2 + R_3}\right) V_2 \text{ volts.}$$

This voltage will also appear at the inverting $(-)$ terminal and thus the voltage across $R_1$ is equal to

$$-\left(\frac{R_3}{R_2 + R_3}\right) V_2 \text{ volts.}$$

Now the output voltage,

$$V_o = \left(\frac{R_3}{R_2 + R_3}\right) V_2$$

$$+ \left[-\left(\frac{R_3}{R_2 + R_3}\right) V_2\right]\left(\frac{-R_f}{R_1}\right)$$

and the voltage gain,

$$A = \frac{V_o}{V_2}$$

$$= \left(\frac{R_3}{R_2 + R_3}\right) + \left[-\left(\frac{R_3}{R_2 + R_3}\right)\right]\left(-\frac{R_f}{R_1}\right)$$

i.e. $\qquad A = \frac{V_o}{V_2} = \left(\frac{R_3}{R_2 + R_3}\right)\left(1 + \frac{R_f}{R_1}\right) \qquad (11)$

(iii) Finally, if the voltages applied to terminals 1 and 2 are $V_1$ and $V_2$ respectively, then the difference between the two voltages will be amplified.

If $V_1 > V_2$, then:

$$V_o = (V_1 - V_2)\left(-\frac{R_f}{R_1}\right) \qquad (12)$$

If $V_2 > V_1$, then:

$$V_o = (V_2 - V_1)\left(\frac{R_3}{R_2 + R_3}\right)\left(1 + \frac{R_f}{R_1}\right) \qquad (13)$$

**Problem 11.** In the differential amplifier shown in Fig. 19.17, $R_1 = 10\,k\Omega$, $R_2 = 10\,k\Omega$, $R_3 = 100\,k\Omega$ and $R_f = 100\,k\Omega$. Determine the output voltage $V_o$ if:
(a) $V_1 = 5\,mV$ and $V_2 = 0$
(b) $V_1 = 0$ and $V_2 = 5\,mV$
(c) $V_1 = 50\,mV$ and $V_2 = 25\,mV$
(d) $V_1 = 25\,mV$ and $V_2 = 50\,mV$.

(a) From equation (10),

$$V_o = -\frac{R_f}{R_1} V_1 = -\left(\frac{100 \times 10^3}{10 \times 10^3}\right)(5)\,mV$$

$$= -50\,mV$$

(b) From equation (11),

$$V_o = \left(\frac{R_3}{R_2 + R_3}\right)\left(1 + \frac{R_f}{R_1}\right) V_2$$

$$= \left(\frac{100}{110}\right)\left(1 + \frac{100}{10}\right)(5)\,mV = +50\,mV$$

(c) $V_1 > V_2$ hence from equation (12),

$$V_o = (V_1 - V_2)\left(-\frac{R_f}{R_1}\right)$$

$$= (50 - 25)\left(-\frac{100}{10}\right)mV = -250\,mV$$

(d) $V_2 > V_1$ hence from equation (13),

$$V_o = (V_2 - V_1)\left(\frac{R_3}{R_2 + R_3}\right)\left(1 + \frac{R_f}{R_1}\right)$$

$$= (50 - 25)\left(\frac{100}{100 + 10}\right)\left(1 + \frac{100}{10}\right)mV$$

$$= (25)\left(\frac{100}{110}\right)(11) = +250\,mV$$

Section 2

**Now try the following exercise**

### Exercise 109 Further problems on operational amplifiers

1. If the input voltage for the op amp shown in Fig. 19.18, is $-0.5$ V, determine (a) the voltage gain (b) the output voltage
   [(a) 3.21 (b) $-1.60$ V]

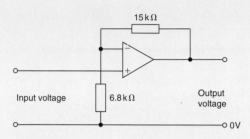

Figure 19.18

2. In the circuit of Fig. 19.19, determine the value of the output voltage, $V_o$, when (a) $V_1 = +1$ V and $V_2 = +3$ V (b) $V_1 = +1$ V and $V_2 = -3$ V
   [(a) $-10$ V (b) $+5$ V]

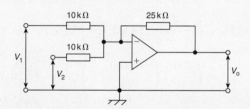

Figure 19.19

3. For the summing op amp shown in Fig. 19.20, determine the output voltage, $V_o$
   [$-3.9$ V]

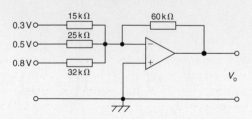

Figure 19.20

4. A steady voltage of $-1.25$ V is applied to an op amp integrator having component values of $R = 125$ k$\Omega$ and $C = 4.0 \mu$F. Calculate the value of the output voltage 120 ms after applying the input, assuming that the initial capacitor charge is zero. [0.3 V]

5. In the differential amplifier shown in Fig. 19.21, determine the output voltage, $V_o$, if: (a) $V_1 = 4$ mV and $V_2 = 0$ (b) $V_1 = 0$ and $V_2 = 6$ mV (c) $V_1 = 40$ mV and $V_2 = 30$ mV (d) $V_1 = 25$ mV and $V_2 = 40$ mV.
   [(a) $-60$ mV (b) $+90$ mV (c) $-150$ mV (d) $+225$ mV]

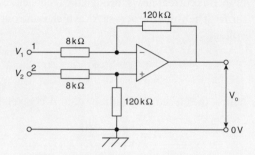

Figure 19.21

## 19.10 Digital to analogue (D/A) conversion

There are a number of situations when digital signals have to be converted to analogue ones. For example, a digital computer often needs to produce a graphical display on the screen; this involves using a D/A converter to change the two-level digital output voltage from the computer, into a continuously varying analogue voltage for the input to the cathode ray tube, so that it can deflect the electron beam to produce screen graphics.

A binary weighted resistor D/A converter is shown in Fig. 19.22 for a four-bit input. The values of the resistors, $R$, $2R$, $4R$, $8R$ increase according to the binary scale – hence the name of the converter. The circuit uses an op amp as a **summing amplifier** (see Section 19.6) with a feedback resistor $R_f$. Digitally controlled electronic switches are shown as $S_1$ to $S_4$. Each switch connects the resistor in series with it to a fixed reference voltage $V_{REF}$ when the input bit controlling it is a 1 and to ground (0 V) when it is a 0. The input voltages $V_1$ to $V_4$ applied to the op amp by the four-bit

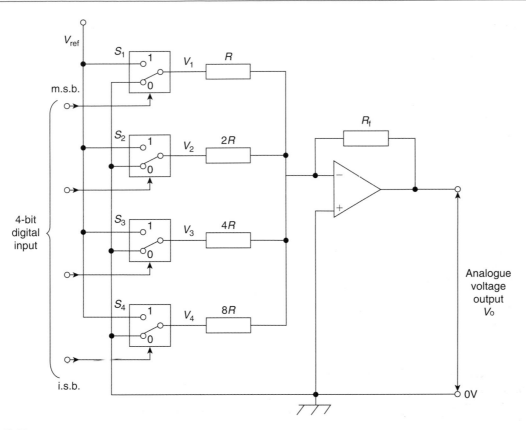

**Figure 19.22**

input via the resistors therefore have one of two values, i.e. either $V_{REF}$ or 0 V.

From equation (8), page 302, the analogue output voltage $V_o$ is given by:

$$V_o = -\left(\frac{R_f}{R}V_1 + \frac{R_f}{2R}V_2 + \frac{R_f}{4R}V_3 + \frac{R_f}{8R}V_4\right)$$

Let $R_f = R = 1\,k\Omega$, then:

$$V_o = -\left(V_1 + \frac{1}{2}V_2 + \frac{1}{4}V_3 + \frac{1}{8}V_4\right)$$

**With a four-bit input of 0001 (i.e. decimal 1)**, $S_4$ connects $8R$ to $V_{REF}$, i.e. $V_4 = V_{REF}$, and $S_1$, $S_2$ and $S_3$ connect $R$, $2R$ and $4R$ to 0 V, making $V_1 = V_2 = V_3 = 0$. Let $V_{REF} = -8\,V$, then output voltage,

$$V_o = -\left(0 + 0 + 0 + \frac{1}{8}(-8)\right) = +1\,V$$

**With a four-bit input of 0101 (i.e. decimal 5)**, $S_2$ and $S_4$ connects $2R$ and $8R$ to $V_{REF}$, i.e. $V_2 = V_4 = V_{REF}$, and $S_1$ and $S_3$ connect $R$ and $4R$ to 0 V, making $V_1 = V_3 = 0$.

Again, if $V_{REF} = -8\,V$, then output voltage,

$$V_o = -\left(0 + \frac{1}{2}(-8) + 0 + \frac{1}{8}(-8)\right) = +5\,V$$

If the input is 0111 (i.e. decimal 7), the output voltage will be 7 V, and so on. From these examples, it is seen that the analogue output voltage, $V_o$, is directly proportional to the digital input. $V_o$ has a 'stepped' waveform, the waveform shape depending on the binary input. A typical waveform is shown in Fig. 19.23.

## 19.11 Analogue to digital (A/D) conversion

In a digital voltmeter, its input is in analogue form and the reading is displayed digitally. This is an example where an analogue to digital converter is needed.

A block diagram for a four-bit counter type A/D conversion circuit is shown in Fig. 19.24. An op amp is again used, in this case as a **voltage comparator** (see Section 19.7). The analogue input voltage $V_2$, shown in

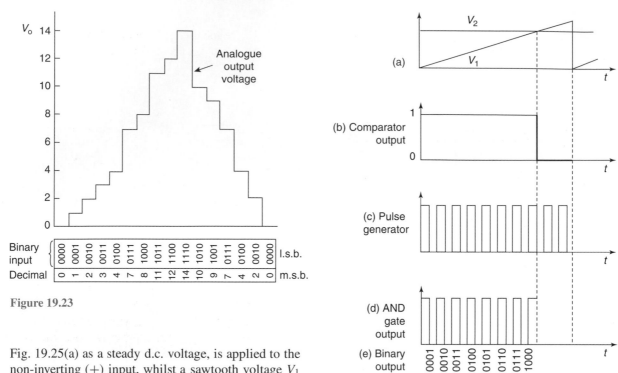

Figure 19.23

Figure 19.25

Fig. 19.25(a) as a steady d.c. voltage, is applied to the non-inverting (+) input, whilst a sawtooth voltage $V_1$ supplies the inverting (−) input.

The output from the comparator is applied to one input of an AND gate and is a 1 (i.e. 'high') until $V_1$ equals or exceeds $V_2$, when it then goes to 0 (i.e. 'low') as shown in Fig. 19.25(b). The other input of the AND gate is fed by a steady train of pulses from a pulse generator, as shown in Fig. 19.25(c). When both inputs to the AND gate are 'high', the gate 'opens' and gives a 'high' output, i.e. a pulse, as shown in Fig. 19.25(d). The time taken by $V_1$ to reach $V_2$ is proportional to the analogue voltage if the ramp is linear. The output pulses from the AND gate are recorded by a binary counter

and, as shown in Fig. 19.25(e), are the digital equivalent of the analogue input voltage $V_2$. In practise, the ramp generator is a D/A converter which takes its digital input from the binary counter, shown by the broken lines in Fig. 19.24. As the counter advances through its normal binary sequence, a staircase waveform with equal steps (i.e. a ramp) is built up at the output of the D/A converter (as shown by the first few steps in Fig. 19.23.

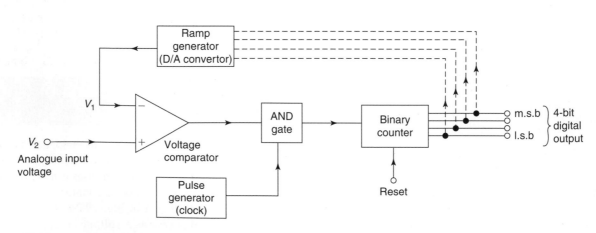

Figure 19.24

**Now try the following exercises**

### Exercise 110 Short answer questions on operational amplifiers

1. List three main properties of an op amp

2. Sketch a typical voltage characteristic showing how the output voltage varies with the input voltage for an op amp

3. What effect does negative feedback have when applied to an op amp

4. Sketch a typical gain/bandwidth characteristic for an op amp

5. With reference to an op amp explain the parameters input bias current, input offset current and input offset voltage

6. Define common-mode rejection ratio

7. Explain the principle of operation of an op amp inverting amplifier

8. In an inverting amplifier, the closed-loop gain $A$ is given by: $A = \ldots\ldots$

9. Explain the principle of operation of an op amp non-inverting amplifier

10. In a non-inverting amplifier, the closed-loop gain $A$ is given by: $A = \ldots\ldots$

11. Explain the principle of operation of an op amp voltage-follower (or buffer)

12. Explain the principle of operation of an op amp summing amplifier

13. In a summing amplifier having three inputs, the output voltage $V_o$ is given by: $V_o = \ldots\ldots$

14. Explain the principle of operation of an op amp voltage comparator

15. Explain the principle of operation of an op amp integrator

16. In an op amp integrator, the output voltage $V_o$ is given by: $V_o = \ldots\ldots$

17. Explain the principle of operation of an op amp differential amplifier

18. Explain the principle of operation of a binary weighted resistor digital to analogue converter using a four-bit input

19. Explain the principle of operation of a four-bit counter type analogue to digital converter

### Exercise 111 Multi-choice questions on operational amplifiers (Answers on page 421)

1. A differential amplifier has an open-loop voltage gain of 100. The input signals are 2.5 V and 2.4 V. The output voltage of the amplifier is:
   (a) $-10\,$V (b) $1\,$mV
   (c) $10\,$V (d) $1\,$kV

2. Which of the following statements relating to operational amplifiers is true?
   (a) It has a high open-loop voltage gain at low frequency, a low input impedance and low output impedance
   (b) It has a high open-loop voltage gain at low frequency, a high input impedance and low output impedance
   (c) It has a low open-loop voltage gain at low frequency, a high input impedance and low output impedance
   (d) It has a high open-loop voltage gain at low frequency, a low input impedance and high output impedance

3. A differential amplifier has a voltage gain of $120 \times 10^3$ and a common-mode rejection ratio of 100 dB. The common-mode gain of the operational amplifier is:
   (a) $1.2 \times 10^3$ (b) $1.2$
   (c) $1.2 \times 10^{10}$ (d) $1.2 \times 10^{-5}$

4. The output voltage, $V_o$, in the amplifier shown in Fig. 19.26 is:
   (a) $-0.2\,$V (b) $+1.8\,$V
   (c) $+0.2\,$V (d) $-1.8\,$V

5. The $3\,$k$\Omega$ resistor in Fig. 19.26 is replaced by one of value $0.1\,$M$\Omega$. If the op amp has an input bias current of 80 nA, the output offset voltage is:
   (a) $79.2\,\mu$V (b) $8\,\mu$V
   (c) $8\,$mV (d) $80.2\,$nV

Section 2

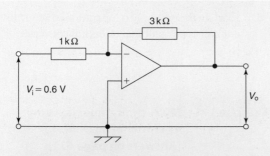

**Figure 19.26**

6. In the op amp shown in Fig. 19.27, the voltage gain is:
   (a) −3          (b) +4
   (c) +3          (d) −4

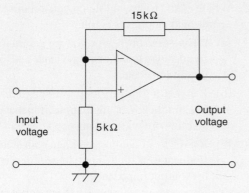

**Figure 19.27**

7. For the op amp shown in Fig. 19.28, the output voltage, $V_o$, is:
   (a) −1.2 V       (b) +5 V
   (c) +2 V         (d) −5 V

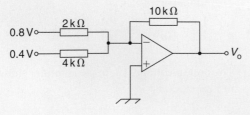

**Figure 19.28**

8. A steady voltage of −1.0 V is applied to an op amp integrator having component values of $R = 100\,k\Omega$ and $C = 10\,\mu F$. The value of the output voltage 10 ms after applying the input voltage is:
   (a) +10 mV       (b) −1 mV
   (c) −10 mV       (d) +1 mV

9. In the differential amplifier shown in Fig. 19.29, the output voltage, $V_o$, is:
   (a) +1.28 mV     (b) 1.92 mV
   (c) −1.28 mV     (d) +5 μV

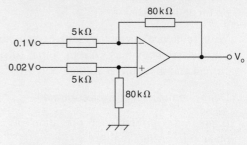

**Figure 19.29**

10. Which of the following statements is false?
    (a) A digital computer requires a D/A converter
    (b) When negative feedback is used in an op amp, a constant and predictable voltage gain results
    (c) A digital voltmeter requires a D/A converter
    (d) The value of frequency at which the open-loop gain has fallen to unity is called the transition frequency

This revision test covers the material contained in Chapters 15 to 19. *The marks for each question are shown in brackets at the end of each question.*

1. The power taken by a series inductive circuit when connected to a 100 V, 100 Hz supply is 250 W and the current is 5 A. Calculate (a) the resistance, (b) the impedance, (c) the reactance, (d) the power factor, and (e) the phase angle between voltage and current. (9)

2. A coil of resistance 20 Ω and inductance 200 mH is connected in parallel with a 4 µF capacitor across a 50 V, variable frequency supply. Calculate (a) the resonant frequency, (b) the dynamic resistance, (c) the current at resonance, and (d) the Q-factor at resonance. (10)

3. A series circuit comprises a coil of resistance 30 Ω and inductance 50 mH, and a 2500 pF capacitor. Determine the Q-factor of the circuit at resonance. (4)

4. The winding of an electromagnet has an inductance of 110 mH and a resistance of 5.5 Ω. When it is connected to a 110 V, d.c. supply, calculate (a) the steady-state value of current flowing in the winding, (b) the time constant of the circuit, (c) the value of the induced e.m.f. after 0.01 s, (d) the time for the current to rise to 75 per cent of it's final value, and (e) the value of the current after 0.02 s. (11)

5. A single-phase motor takes 30 A at a power factor of 0.65 lagging from a 300 V, 50 Hz supply. Calculate (a) the current taken by a capacitor connected in parallel with the motor to correct the power factor to unity, and (b) the value of the supply current after power factor correction. (7)

6. For the summing operational amplifier shown in Fig. RT5.1, determine the value of the output voltage, $V_0$ (3)

7. In the differential amplifier shown in Fig. RT5.2, determine the output voltage, $V_0$ when:
   (a) $V_1 = 4$ mV and $V_2 = 0$
   (b) $V_1 = 0$ and $V_2 = 5$ mV
   (c) $V_1 = 20$ mV and $V_2 = 10$ mV (6)

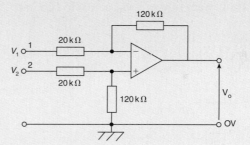

**Figure RT5.2**

8. A filter section is to have a characteristic impedance at zero frequency of 600 Ω and a cut-off frequency of 2.5 MHz. Design (a) a low-pass T-section filter, and (b) a low-pass $\pi$-section filter to meet these requirements. (6)

9. Determine the cut-off frequency and the nominal impedance for a high-pass $\pi$-connected section having a 5 nF capacitor in its series arm and inductances of 1 mH in each of its shunt arms. (4)

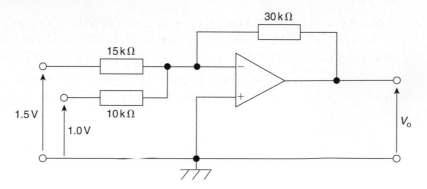

**Figure RT5.1**

## Formulae for further electrical and electronic principles

**Section 2**

### A.C. Theory:

$$T = \frac{1}{f} \quad \text{or} \quad f = \frac{1}{T}$$

$$I = \sqrt{\frac{i_1^2 + i_2^2 + i_2^2 + \cdots + i_n^2}{n}}$$

For a sine wave: $I_{AV} = \frac{2}{\pi} I_m$ or $0.637\, I_m$

$$I = \frac{1}{\sqrt{2}} I_m \quad \text{or} \quad 0.707\, I_m$$

Form factor $= \dfrac{\text{r.m.s.}}{\text{average}}$ Peak factor $= \dfrac{\text{maximum}}{\text{r.m.s.}}$

General sinusoidal voltage: $v = V_m \sin(\omega t \pm \phi)$

### Single-phase circuits:

$$X_L = 2\pi f L \qquad X_C = \frac{1}{2\pi f C}$$

$$Z = \frac{V}{I} = \sqrt{(R^2 + X^2)}$$

Series resonance: $f_r = \dfrac{1}{2\pi\sqrt{LC}}$

$$Q = \frac{V_L}{V} \quad \text{or} \quad \frac{V_C}{V} = \frac{2\pi f_r L}{R} = \frac{1}{2\pi f_r CR} = \frac{1}{R}\sqrt{\frac{L}{C}}$$

$$Q = \frac{f_r}{f_2 - f_1} \quad \text{or} \quad (f_2 - f_1) = \frac{f_r}{Q}$$

### Parallel resonance (LR-C circuit):

$$f_r = \frac{1}{2\pi}\sqrt{\frac{1}{LC} - \frac{R^2}{L^2}}$$

$$I_r = \frac{VRC}{L} \qquad R_D = \frac{L}{CR}$$

$$Q = \frac{2\pi f_r L}{R} = \frac{I_C}{I_r}$$

$$P = VI\cos\phi \quad \text{or} \quad I^2 R \qquad S = VI \qquad Q = VI\sin\phi$$

power factor $= \cos\phi = \dfrac{R}{Z}$

### Filter networks:

#### Low-pass T or π:

$$f_C = \frac{1}{\pi\sqrt{LC}} \qquad R_0 = \sqrt{\frac{L}{C}}$$

$$C = \frac{1}{\pi R_0 f_C} \qquad L = \frac{R_0}{\pi f_C}$$

See Fig. F1

#### High-pass T or π:

$$f_C = \frac{1}{4\pi\sqrt{LC}} \qquad R_0 = \sqrt{\frac{L}{C}}$$

$$C = \frac{1}{4\pi R_0 f_C} \qquad L = \frac{R_0}{4\pi f_C}$$

See Fig. F2

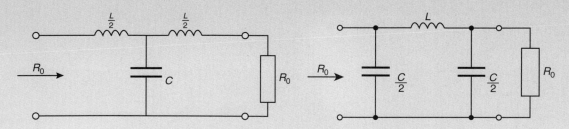

**Figure F1**

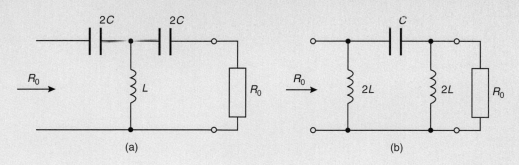

(a)                                              (b)

**Figure F2**

## D.C. Transients:

C–R circuit $\tau = CR$

Charging: $v_C = V(1 - e^{-t/CR})$

$v_r = Ve^{-t/CR}$

$i = Ie^{-t/CR}$

Discharging: $v_C = v_R = Ve^{-t/CR}$

$i = Ie^{-t/CR}$

L–R circuit $\tau = \dfrac{L}{R}$

Current growth: $v_L = Ve^{-Rt/L}$

$v_R = V(1 - e^{-Rt/L})$

$i = I(1 - e^{-Rt/L})$

Current decay: $v_L = v_R = Ve^{-Rt/L}$

$i = Ie^{-Rt/L}$

## Operational amplifiers:

$\text{CMRR} = 20\log_{10}\left(\dfrac{\text{differential voltage gain}}{\text{common-mode gain}}\right)\text{dB}$

Inverter: $A = \dfrac{V_o}{V_i} = \dfrac{-R_f}{R_i}$

Non-inverter: $A = \dfrac{V_o}{V_i} = 1 + \dfrac{R_f}{R_i}$

Summing: $V_o = -R_f\left(\dfrac{V_1}{R_1} + \dfrac{V_2}{R_2} + \dfrac{V_3}{R_3}\right)$

Integrator: $V_o = -\dfrac{1}{CR}\displaystyle\int V_i\,dt$

Differential:

If $V_1 > V_2 : V_o = (V_1 - V_2)\left(-\dfrac{R_f}{R_1}\right)$

If $V_2 > V_1 : V_o = (V_2 - V_1)\left(\dfrac{R_3}{R_2 + R_3}\right)\left(1 + \dfrac{R_f}{R_1}\right)$

Section 2

# Electrical Power Technology

# Chapter 20

# Three-phase systems

At the end of this chapter you should be able to:

- describe a single-phase supply
- describe a three-phase supply
- understand a star connection, and recognise that $I_L = I_p$ and $V_L = \sqrt{3}V_p$
- draw a complete phasor diagram for a balanced, star connected load
- understand a delta connection, and recognise that $V_L = V_p$ and $I_L = \sqrt{3}I_p$
- draw a phasor diagram for a balanced, delta connected load
- calculate power in three-phase systems using $P = \sqrt{3}\,V_L I_L \cos\phi$
- appreciate how power is measured in a three-phase system, by the one, two and three-wattmeter methods
- compare star and delta connections
- appreciate the advantages of three-phase systems

## 20.1 Introduction

Generation, transmission and distribution of electricity via the National Grid system is accomplished by three-phase alternating currents.

The voltage induced by a single coil when rotated in a uniform magnetic field is shown in Fig. 20.1 and is known as a **single-phase voltage**. Most consumers are fed by means of a single-phase a.c. supply. Two wires are used, one called the live conductor (usually coloured red) and the other is called the neutral conductor (usually coloured black). The neutral is usually connected via protective gear to earth, the earth wire being coloured green. The standard voltage for a single-phase a.c. supply is 240 V. The majority of single-phase supplies are obtained by connection to a three-phase supply (see Fig. 20.5, page 319).

## 20.2 Three-phase supply

**A three-phase supply** is generated when three coils are placed 120° apart and the whole rotated in a uniform magnetic field as shown in Fig. 20.2(a). The result is three independent supplies of equal voltages which are each displaced by 120° from each other as shown in Fig. 20.2(b).

(i) The convention adopted to identify each of the phase voltages is: R-red, Y-yellow, and B-blue, as shown in Fig. 20.2

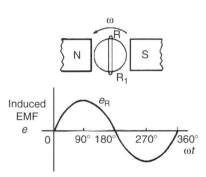

Figure 20.1

DOI: 10.1016/B978-0-08-089056-2.00020-6

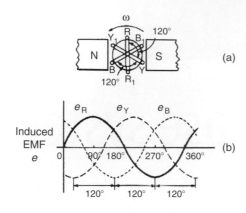

Figure 20.2

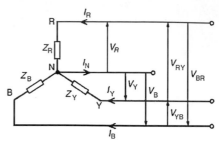

Figure 20.3

(ii) The **phase-sequence** is given by the sequence in which the conductors pass the point initially taken by the red conductor. The national standard phase sequence is R, Y, B.

A three-phase a.c. supply is carried by three conductors, called **'lines'** which are coloured red, yellow and blue. The currents in these conductors are known as line currents ($I_L$) and the p.d.'s between them are known as line voltages ($V_L$). A fourth conductor, called the **neutral** (coloured black, and connected through protective devices to earth) is often used with a three-phase supply. If the three-phase windings shown in Fig. 20.2 are kept independent then six wires are needed to connect a supply source (such as a generator) to a load (such as motor). To reduce the number of wires it is usual to interconnect the three phases. There are two ways in which this can be done, these being:

(a) a **star connection**, and (b) a **delta**, or **mesh, connection**. Sources of three-phase supplies, i.e. alternators, are usually connected in star, whereas three-phase transformer windings, motors and other loads may be connected either in star or delta.

## 20.3  Star connection

(i) A **star-connected load** is shown in Fig. 20.3 where the three line conductors are each connected to a load and the outlets from the loads are joined together at $N$ to form what is termed the **neutral point** or the **star point**.

(ii) The voltages, $V_R$, $V_Y$ and $V_B$ are called **phase voltages** or line to neutral voltages. Phase voltages are generally denoted by $V_p$.

(iii) The voltages, $V_{RY}$, $V_{YB}$ and $V_{BR}$ are called **line voltages**.

(iv) From Fig. 20.3 it can be seen that the phase currents (generally denoted by $I_p$) are equal to their respective line currents $I_R$, $I_Y$ and $I_B$, i.e. for a star connection:

$$I_L = I_p$$

(v) For a balanced system:

$$I_R = I_Y = I_B, \quad V_R = V_Y = V_B$$
$$V_{RY} = V_{YB} = V_{BR}, \quad Z_R = Z_Y = Z_B$$

and the current in the neutral conductor, $I_N = 0$. When a star-connected system is balanced, then the neutral conductor is unnecessary and is often omitted.

(vi) The line voltage, $V_{RY}$, shown in Fig. 20.4(a) is given by $V_{RY} = V_R - V_Y$ ($V_Y$ is negative since it is in the opposite direction to $V_{RY}$). In the phasor diagram of Fig. 20.4(b), phasor $V_Y$ is reversed (shown by the broken line) and then added phasorially to $V_R$ (i.e. $V_{RY} = V_R + (-V_Y)$). By trigonometry, or by measurement, $V_{RY} = \sqrt{3}\, V_R$, i.e. for a balanced star connection:

$$V_L = \sqrt{3}\, V_p$$

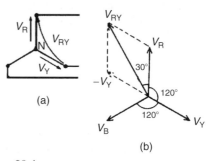

Figure 20.4

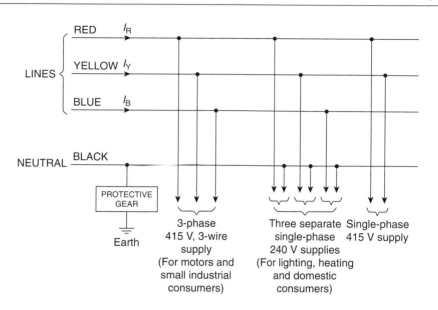

**Figure 20.5**

(See Problem 3 following for a complete phasor diagram of a star-connected system.)

(vii)   The star connection of the three phases of a supply, together with a neutral conductor, allows the use of two voltages – the phase voltage and the line voltage. A 4-wire system is also used when the load is not balanced. The standard electricity supply to consumers in Great Britain is 415/240 V, 50 Hz, 3-phase, 4-wire alternating current, and a diagram of connections is shown in Fig. 20.5.

For most of the 20th century, the **supply voltage in the UK in domestic premises has been 240 V a.c.** (r.m.s.) at 50 Hz. In 1988, a European-wide agreement was reached to change the various national voltages, which ranged at the time from 220 V to 240 V, to a common European standard of **230 V**.

As a result, the standard nominal supply voltage in domestic single-phase 50 Hz installations in the UK has been 230 V since 1995. However, as an interim measure, electricity suppliers can work with an asymmetric voltage tolerance of 230 V +10%/−6% (i.e. 216.2 V to 253 V). The old standard was 240 V ±6% (i.e. 225.6 V to 254.4 V), which is mostly contained within the new range, and so in practice suppliers have had no reason to actually change voltages.

Similarly, the **three-phase voltage** in the UK had been for many years **415 V** ±6% (i.e. 390 V to 440 V). European harmonisation required this to be changed to **400 V** +10%/−6% (i.e. 376 V to 440 V). Again, since the present supply voltage of 415 V lies within this range, supply companies are unlikely to reduce their voltages in the near future.

Many of the calculations following are based on the 240 V/415 V supply voltages which have applied for many years and are likely to continue to do so.

> **Problem 1.**   Three loads, each of resistance 30 Ω, are connected in star to a 415 V, 3-phase supply. Determine (a) the system phase voltage, (b) the phase current and (c) the line current.

A '415 V, 3-phase supply' means that 415 V is the line voltage, $V_L$

(a)   For a star connection, $V_L = \sqrt{3}\,V_p$. Hence phase voltage, $V_p = V_L/\sqrt{3} = 415/\sqrt{3} = \mathbf{239.6\,V}$ or **240 V**, correct to 3 significant figures.

(b)   Phase current, $I_p = V_p/R_p = 240/30 = \mathbf{8\,A}$

(c)   For a star connection, $I_p = I_L$ hence the line current, $I_L = \mathbf{8\,A}$

> **Problem 2.**   A star-connected load consists of three identical coils each of resistance 30 Ω and inductance 127.3 mH. If the line current is 5.08 A, calculate the line voltage if the supply frequency is 50 Hz.

Inductive reactance

$$X_L = 2\pi f L = 2\pi (50)(127.3 \times 10^{-3}) = 40\,\Omega$$

Impedance of each phase

$$Z_p = \sqrt{R^2 + X_L^2} = \sqrt{30^2 + 40^2} = 50\,\Omega$$

For a star connection

$$I_L = I_p = \frac{V_p}{Z_p}$$

Hence phase voltage,

$$V_p = I_p Z_p = (5.08)(50) = 254\,\text{V}$$

Line voltage

$$V_L = \sqrt{3}\,V_p = \sqrt{3}(254) = \mathbf{440\,V}$$

> **Problem 3.** A balanced, three-wire, star-connected, 3-phase load has a phase voltage of 240 V, a line current of 5 A and a lagging power factor of 0.966. Draw the complete phasor diagram.

The phasor diagram is shown in Fig. 20.6.

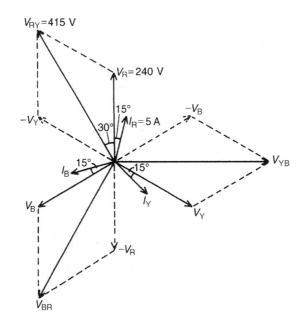

**Figure 20.6**

Procedure to construct the phasor diagram:

(i) Draw $V_R = V_Y = V_B = 240\,\text{V}$ and spaced 120° apart. (Note that $V_R$ is shown vertically upwards – this however is immaterial for it may be drawn in any direction.)

(ii) Power factor $= \cos\phi = 0.966$ lagging. Hence the load phase angle is given by $\cos^{-1}0.966$, i.e. 15°

lagging. Hence $I_R = I_Y = I_B = 5\,\text{A}$, lagging $V_R$, $V_Y$ and $V_B$ respectively by 15°.

(iii) $V_{RY} = V_R - V_Y$ (phasorially). Hence $V_Y$ is reversed and added phasorially to $V_R$. By measurement, $V_{RY} = 415\,\text{V}$ (i.e. $\sqrt{3} \times 240$) and leads $V_R$ by 30°. Similarly, $V_{YB} = V_Y - V_B$ and $V_{BR} = V_B - V_R$

> **Problem 4.** A 415 V, 3-phase, 4 wire, star-connected system supplies three resistive loads as shown in Fig. 20.7. Determine (a) the current in each line and (b) the current in the neutral conductor.

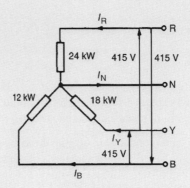

**Figure 20.7**

(a) For a star-connected system $V_L = \sqrt{3}\,V_p$, hence

$$V_p = \frac{V_L}{\sqrt{3}} = \frac{415}{\sqrt{3}} = 240\,\text{V}$$

Since current $I = $ power P/voltage V for a resistive load then

$$I_R = \frac{P_R}{V_R} = \frac{24\,000}{240} = \mathbf{100\,A}$$

$$I_Y = \frac{P_Y}{V_Y} = \frac{18\,000}{240} = \mathbf{75\,A}$$

and $$I_B = \frac{P_B}{V_B} = \frac{12\,000}{240} = \mathbf{50\,A}$$

(b) The three line currents are shown in the phasor diagram of Fig. 20.8. Since each load is resistive the currents are in phase with the phase voltages and are hence mutually displaced by 120°. The current in the neutral conductor is given by $I_N = I_R + I_Y + I_B$ phasorially.

Figure 20.9 shows the three line currents added phasorially. oa represents $I_R$ in magnitude and direction. From the nose of oa, ab is drawn representing $I_Y$ in

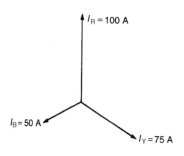

Figure 20.8

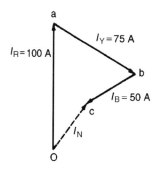

Figure 20.9

magnitude and direction. From the nose of ab, bc is drawn representing $I_B$ in magnitude and direction. oc represents the resultant, $I_N$ By measurement, $I_N = 43\,\text{A}$.

Alternatively, by calculation, considering $I_R$ at $90°$, $I_B$ at $210°$ and $I_Y$ at $330°$:

Total horizontal component

$= 100\cos 90° + 75\cos 330° + 50\cos 210° = 21.65$

Total vertical component

$= 100\sin 90° + 75\sin 330° + 50\sin 210° = 37.50$

Hence magnitude of $I_N = \sqrt{21.65^2 + 37.50^2} = \textbf{43.3 A}$

**Now try the following exercise**

**Exercise 112    Further problems on star connections**

1. Three loads, each of resistance $50\,\Omega$ are connected in star to a $400\,\text{V}$, 3-phase supply. Determine (a) the phase voltage, (b) the phase current, and (c) the line current.

    [(a) 231 V (b) 4.62 A (c) 4.62 A]

2. A star-connected load consists of three identical coils, each of inductance $159.2\,\text{mH}$ and resistance $50\,\Omega$. If the supply frequency is $50\,\text{Hz}$ and the line current is $3\,\text{A}$ determine (a) the phase voltage and (b) the line voltage.

    [(a) 212 V (b) 367 V]

3. Three identical capacitors are connected in star to a $400\,\text{V}$, $50\,\text{Hz}$ 3-phase supply. If the line current is $12\,\text{A}$ determine the capacitance of each of the capacitors.    [$165.4\,\mu\text{F}$]

4. Three coils each having resistance $6\,\Omega$ and inductance $L$ H are connected in star to a $415\,\text{V}$, $50\,\text{Hz}$, 3-phase supply. If the line current is $30\,\text{A}$, find the value of L.    [$16.78\,\text{mH}$]

5. A $400\,\text{V}$, 3-phase, 4 wire, star-connected system supplies three resistive loads of $15\,\text{kW}$, $20\,\text{kW}$ and $25\,\text{kW}$ in the red, yellow and blue phases respectively. Determine the current flowing in each of the four conductors.

    [$I_R = 64.95\,\text{A}$, $I_Y = 86.60\,\text{A}$,
    $I_B = 108.25\,\text{A}$, $I_N = 37.50\,\text{A}$]

## 20.4    Delta connection

(i)  A **delta (or mesh) connected load** is shown in Fig. 20.10 where the end of one load is connected to the start of the next load.

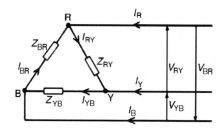

Figure 20.10

(ii)  From Fig. 20.10, it can be seen that the line voltages $V_{RY}$, $V_{YB}$ and $V_{BR}$ are the respective phase voltages, i.e. for a delta connection:

$$V_L = V_p$$

(iii)  Using Kirchhoff's current law in Fig. 20.10, $I_R = I_{RY} - I_{BR} = I_{RY} + (-I_{BR})$. From the phasor diagram shown in Fig. 20.11, by trigonometry or by measurement, $I_R = \sqrt{3}\,I_{RY}$, i.e. for a delta connection:

$$I_L = \sqrt{3}I_p$$

Problem 5.   Three identical coils each of resistance $30\,\Omega$ and inductance $127.3\,\text{mH}$ are connected in delta to a 440 V, 50 Hz, 3-phase supply. Determine (a) the phase current, and (b) the line current.

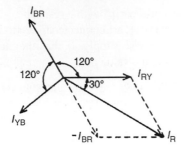

**Figure 20.11**

Phase impedance, $Z_p = 50\,\Omega$ (from Problem 2) and for a delta connection, $V_p = V_L$

(a)   Phase current,

$$I_p = \frac{V_p}{Z_p} = \frac{V_L}{Z_p} = \frac{440}{50} = \textbf{8.8 A}$$

(b)   For a delta connection,

$$I_L = \sqrt{3}\,I_p = \sqrt{3}(8.8) = \textbf{15.24 A}$$

*Thus when the load is connected in delta, three times the line current is taken from the supply than is taken if connected in star.*

**Problem 6.**   Three identical capacitors are connected in delta to a 415 V, 50 Hz, 3-phase supply. If the line current is 15 A, determine the capacitance of each of the capacitors.

For a delta connection $I_L = \sqrt{3}\,I_p$. Hence phase current,

$$I_p = \frac{I_L}{\sqrt{3}} = \frac{15}{\sqrt{3}} = 8.66\,\text{A}$$

Capacitive reactance per phase,

$$X_C = \frac{V_p}{I_p} = \frac{V_L}{I_p}$$

(since for a delta connection $V_L = V_p$). Hence

$$X_C = \frac{415}{8.66} = 47.92\,\Omega$$

$X_C = 1/2\pi f C$, from which capacitance,

$$C = \frac{1}{2\pi f X_C} = \frac{2}{2\pi (50)(47.92)}\,\text{F} = \textbf{66.43}\,\mathbf{\mu F}$$

**Problem 7.**   Three coils each having resistance $3\,\Omega$ and inductive reactance $4\,\Omega$ are connected (i) in star and (ii) in delta to a 415 V, 3-phase supply. Calculate for each connection (a) the line and phase voltages and (b) the phase and line currents.

(i)   **For a star connection:** $I_L = I_p$ and $V_L = \sqrt{3}\,V_p$.

   (a)   A 415 V, 3-phase supply means that the line voltage, $V_L = \textbf{415 V}$

   Phase voltage,

$$V_p = \frac{V_L}{\sqrt{3}} = \frac{415}{\sqrt{3}} = \textbf{240 V}$$

   (b)   Impedance per phase,

$$Z_p = \sqrt{R^2 + X_L^2} = \sqrt{3^2 + 4^2} = 5\,\Omega$$

   Phase current,

$$I_p = V_p/Z_p = 240/5 = \textbf{48 A}$$

   Line current,

$$I_L = I_p = \textbf{48 A}$$

(ii)   **For a delta connection:** $V_L = V_p$ and $I_L = \sqrt{3}\,I_p$.

   (a)   Line voltage, $V_L = 415\,\text{V}$

   Phase voltage, $V_p = V_L = \textbf{415 V}$

   (b)   Phase current,

$$I_p = \frac{V_p}{Z_p} = \frac{415}{5} = \textbf{83 A}$$

   Line current,

$$I_L = \sqrt{3}\,I_p = \sqrt{3}(83) = \textbf{144 A}$$

**Now try the following exercise**

**Exercise 113   Further problems on delta connections**

1.   Three loads, each of resistance $50\,\Omega$ are connected in delta to a 400 V, 3-phase supply. Determine (a) the phase voltage, (b) the phase current, and (c) the line current.
[(a) 400 V (b) 8 A (c) 13.86 A]

2. Three inductive loads each of resistance $75\,\Omega$ and inductance $318.4\,\text{mH}$ are connected in delta to a $415\,\text{V}$, $50\,\text{Hz}$, 3-phase supply. Determine (a) the phase voltage, (b) the phase current, and (c) the line current.

[(a) 415 V (b) 3.32 A (c) 5.75 A]

3. Three identical capacitors are connected in delta to a $400\,\text{V}$, $50\,\text{Hz}$, 3-phase supply. If the line current is $12\,\text{A}$ determine the capacitance of each of the capacitors. [55.13 $\mu$F]

4. Three coils each having resistance $6\,\Omega$ and inductance $L\,H$ are connected in delta, to a $415\,\text{V}$, $50\,\text{Hz}$, 3-phase supply. If the line current is $30\,\text{A}$, find the value of $L$.

[73.84 mH]

5. A 3-phase, star-connected alternator delivers a line current of $65\,\text{A}$ to a balanced delta-connected load at a line voltage of $380\,\text{V}$. Calculate (a) the phase voltage of the alternator, (b) the alternator phase current, and (c) the load phase current.

[(a) 219.4 V (b) 65 A (c) 37.53 A]

6. Three $24\,\mu\text{F}$ capacitors are connected in star across a $400\,\text{V}$, $50\,\text{Hz}$, 3-phase supply. What value of capacitance must be connected in delta in order to take the same line current?

[8 $\mu$F]

## 20.5 Power in three-phase systems

The power dissipated in a three-phase load is given by the sum of the power dissipated in each phase. If a load is balanced then the total power $P$ is given by: $P = 3 \times$ power consumed by one phase.

The power consumed in one phase $= I_p^2 R_p$ or $V_p I_p \cos\phi$ (where $\phi$ is the phase angle between $V_p$ and $I_p$).

For a star connection,

$$V_p = \frac{V_L}{\sqrt{3}} \quad \text{and} \quad I_p = I_L$$

hence

$$P = 3\frac{V_L}{\sqrt{3}} I_L \cos\phi = \sqrt{3}\, V_L I_L \cos\phi$$

For a delta connection,

$$V_p = V_L \quad \text{and} \quad I_p = \frac{I_L}{\sqrt{3}}$$

hence

$$P = 3 V_L \frac{I_L}{\sqrt{3}} \cos\phi = \sqrt{3}\, V_L I_L \cos\phi$$

Hence for either a star or a delta balanced connection the total power $P$ is given by:

$$P = \sqrt{3} V_L I_L \cos\phi \text{ watts}$$

$$\text{or } P = 3 I_p^2 R_p \text{ watts}$$

Total volt-amperes

$$S = \sqrt{3} V_L I_L \text{ volt-amperes}$$

**Problem 8.** Three $12\,\Omega$ resistors are connected in star to a $415\,\text{V}$, 3-phase supply. Determine the total power dissipated by the resistors.

Power dissipated, $P = \sqrt{3} V_L I_L \cos\phi$ or $P = 3 I_p^2 R_p$
Line voltage, $V_L = 415\,\text{V}$ and phase voltage

$$V_p = \frac{415}{\sqrt{3}} = 240\,\text{V}$$

(since the resistors are star-connected). Phase current,

$$I_p = \frac{V_p}{Z_p} = \frac{V_p}{R_p} = \frac{240}{12} = 20\,\text{A}$$

For a star connection

$$I_L = I_p = 20\,\text{A}$$

For a purely resistive load, the power

$$\text{factor} = \cos\phi = 1$$

Hence power

$$P = \sqrt{3} V_L I_L \cos\phi = \sqrt{3}(415)(20)(1)$$

$$= 14.4\,\text{kW}$$

or power

$$P = 3 I_p^2 R_p = 3(20)^2(12) = 14.4\,\text{kW}$$

**Problem 9.** The input power to a 3-phase a.c. motor is measured as 5 kW. If the voltage and current to the motor are 400 V and 8.6 A respectively, determine the power factor of the system.

Power $P = 5000$ W,

line voltage $V_L = 400$ V,

line current, $I_L = 8.6$ A and

power, $P = \sqrt{3} V_L I_L \cos\phi$

Hence

$$\textbf{power factor} = \cos\phi = \frac{P}{\sqrt{3} V_L I_L}$$

$$= \frac{5000}{\sqrt{3}(400)(8.6)} = \textbf{0.839}$$

**Problem 10.** Three identical coils, each of resistance 10 Ω and inductance 42 mH are connected (a) in star and (b) in delta to a 415 V, 50 Hz, 3-phase supply. Determine the total power dissipated in each case.

(a) **Star connection**

Inductive reactance,

$$X_L = 2\pi f L = 2\pi(50)(42 \times 10^{-3}) = 13.19\,\Omega.$$

Phase impedance,

$$Z_p = \sqrt{R^2 + X_L^2} = \sqrt{10^2 + 13.19^2} = 16.55\,\Omega.$$

Line voltage,

$$V_L = 415\ V$$

and phase voltage,

$$V_P = V_L/\sqrt{3} = 415/\sqrt{3} = 240\ V.$$

Phase current,

$$I_p = V_p/Z_p = 240/16.55 = 14.50\ A.$$

Line current,

$$I_L = I_p = 14.50\ A.$$

Power factor $= \cos\phi = R_p/Z_p = 10/16.55$
$= 0.6042$ lagging.

**Power dissipated,**

$$P = \sqrt{3} V_L I_L \cos\phi = \sqrt{3}(415)(14.50)(0.6042)$$

$$= \textbf{6.3 kW}$$

(Alternatively,

$$P = 3I_p^2 R_p = 3(14.50)^2(10) = \textbf{6.3 kW})$$

(b) **Delta connection**

$$V_L = V_p = 415\,\text{V},$$

$$Z_p = 16.55\,\Omega,\ \cos\phi = 0.6042$$

lagging (from above).

Phase current,

$$I_p = V_p/Z_p = 415/16.55 = 25.08\ A$$

Line current,

$$I_L = \sqrt{3} I_p = \sqrt{3}(25.08) = 43.44\ A$$

**Power dissipated,**

$$P = \sqrt{3} V_L I_L \cos\phi$$

$$= \sqrt{3}(415)(43.44)(0.6042) = \textbf{18.87 kW}$$

(Alternatively,

$$P = 3I_p^2 R_p = 3(25.08)^2(10) = \textbf{18.87 kW})$$

*Hence loads connected in delta dissipate three times the power than when connected in star, and also take a line current three times greater.*

**Problem 11.** A 415 V, 3-phase a.c. motor has a power output of 12.75 kW and operates at a power factor of 0.77 lagging and with an efficiency of 85 per cent. If the motor is delta-connected, determine (a) the power input, (b) the line current, and (c) the phase current.

(a) Efficiency = power output/power input. Hence 85/100 = 12 750/power input from which,

$$\textbf{power input} = \frac{12\,750 \times 100}{85}$$

$$= \textbf{15 000 W or 15 kW}$$

(b)   Power, $P = \sqrt{3}\,V_L I_L \cos\phi$, hence **line current**,

$$I_L = \frac{P}{\sqrt{3}(415)(0.77)}$$

$$= \frac{15\,000}{\sqrt{3}(415)(0.77)} = \mathbf{27.10\,A}$$

(c)   For a delta connection, $I_L = \sqrt{3}\,I_p$, hence **phase current**,

$$I_p = \frac{I_L}{\sqrt{3}} = \frac{27.10}{\sqrt{3}} = \mathbf{15.65\,A}$$

---

**Now try the following exercise**

**Exercise 114   Further problems on power in three-phase systems**

1.   Determine the total power dissipated by three $20\,\Omega$ resistors when connected (a) in star and (b) in delta to a 440 V, 3-phase supply.
[(a) 9.68 kW (b) 29.04 kW]

2.   Determine the power dissipated in the circuit of Problem 2, Exercise 112, page 321.
[1.35 kW]

3.   A balanced delta-connected load has a line voltage of 400 V, a line current of 8 A and a lagging power factor of 0.94. Draw a complete phasor diagram of the load. What is the total power dissipated by the load?    [5.21 kW]

4.   Three inductive loads, each of resistance $4\,\Omega$ and reactance $9\,\Omega$ are connected in delta. When connected to a 3-phase supply the loads consume 1.2 kW. Calculate (a) the power factor of the load, (b) the phase current, (c) the line current, and (d) the supply voltage.
[(a) 0.406 (b) 10 A (c) 17.32 A (d) 98.53 V]

5.   The input voltage, current and power to a motor is measured as 415 V, 16.4 A and 6 kW respectively. Determine the power factor of the system.    [0.509]

6.   A 440 V, 3-phase a.c. motor has a power output of 11.25 kW and operates at a power factor of 0.8 lagging and with an efficiency of 84 per cent. If the motor is delta connected determine (a) the power input, (b) the line current, and (c) the phase current.
[(a) 13.39 kW (b) 21.97 A (c) 12.68 A]

## 20.6   Measurement of power in three-phase systems

Power in three-phase loads may be measured by the following methods:

(i)   **One-wattmeter method for a balanced load**

Wattmeter connections for both star and delta are shown in Fig. 20.12.

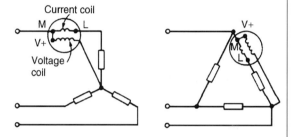

**Figure 20.12**

$$\text{Total power} = 3 \times \text{wattmeter reading}$$

(ii)   **Two-wattmeter method for balanced or unbalanced loads**

A connection diagram for this method is shown in Fig. 20.13 for a star-connected load. Similar connections are made for a delta-connected load.

$$\text{Total power} = \text{sum of wattmeter readings}$$
$$= P_1 + P_2$$

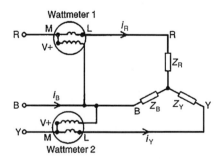

**Figure 20.13**

The power factor may be determined from:

$$\tan\phi = \sqrt{3}\left(\frac{P_1 - P_2}{P_1 + P_2}\right)$$

(see Problems 12 and 15 to 18).

It is possible, depending on the load power factor, for one wattmeter to have to be 'reversed' to obtain a reading. In this case it is taken as a negative reading (see Problem 17).

(iii) **Three-wattmeter method for a three-phase, 4-wire system for balanced and unbalanced loads** (see Fig. 20.14).

$$\text{Total power} = P_1 + P_2 + P_3$$

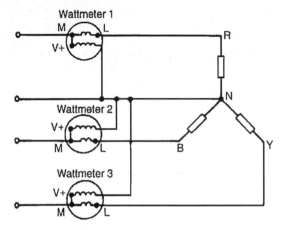

**Figure 20.14**

---

**Problem 12.** (a) Show that the total power in a 3-phase, 3-wire system using the two-wattmeter method of measurement is given by the sum of the wattmeter readings. Draw a connection diagram. (b) Draw a phasor diagram for the two-wattmeter method for a balanced load. (c) Use the phasor diagram of part (b) to derive a formula from which the power factor of a 3-phase system may be determined using only the wattmeter readings.

---

(a) A connection diagram for the two-wattmeter method of a power measurement is shown in Fig. 20.15 for a star-connected load.

Total instantaneous power, $p = e_R i_R + e_Y i_Y + e_B i_B$ and in any 3-phase system $i_R + i_Y + i_B = 0$; hence $i_B = -i_R - i_Y$ Thus,

$$p = e_R i_R + e_Y i_Y + e_B(-i_R - i_Y)$$
$$= (e_R - e_B)i_R + (e_Y - e_B)i_Y$$

However, $(e_R - e_B)$ is the p.d. across wattmeter 1 in Fig. 20.15 and $(e_Y - e_B)$ is the p.d. across

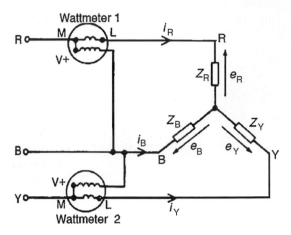

**Figure 20.15**

wattmeter 2. Hence total instantaneous power,

$$p = (\text{wattmeter 1 reading})$$
$$+ (\text{wattmeter 2 reading})$$
$$= p_1 + p_2$$

The moving systems of the wattmeters are unable to follow the variations which take place at normal frequencies and they indicate the mean power taken over a cycle. Hence the total power, $P = P_1 + P_2$ for balanced or unbalanced loads.

(b) The phasor diagram for the two-wattmeter method for a balanced load having a lagging current is shown in Fig. 20.16, where $V_{RB} = V_R - V_B$ and $V_{YB} = V_Y - V_B$ (phasorially).

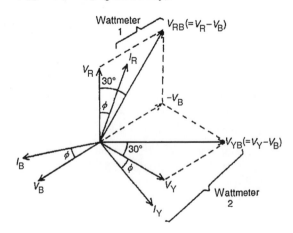

**Figure 20.16**

(c) Wattmeter 1 reads $V_{RB}I_R\cos(30° - \phi) = P_1$

Wattmeter 2 reads $V_{YB}I_Y\cos(30° + \phi) = P_2$

$$\frac{P_1}{P_2} = \frac{V_{RB}I_R\cos(30° - \phi)}{V_{YB}I_Y\cos(30° + \phi)} = \frac{\cos(30° - \phi)}{\cos(30° + \phi)}$$

since $I_R = I_Y$ and $V_{RB} = V_{YB}$ for a balanced load. Hence

$$\frac{P_1}{P_2} = \frac{\cos 30° \cos\phi + \sin 30° \sin\phi}{\cos 30° \cos\phi - \sin 30° \sin\phi}$$

(from compound angle formulae, see 'Engineering Mathematics').

Dividing throughout by $\cos 30° \cos\phi$ gives:

$$\frac{P_1}{P_2} = \frac{1 + \tan 30° \tan\phi}{1 - \tan 30° \tan\phi}$$

$$= \frac{1 + \frac{1}{\sqrt{3}} \tan\phi}{1 - \frac{1}{\sqrt{3}} \tan\phi}$$

$$\left( \text{since } \frac{\sin\phi}{\cos\phi} = \tan\phi \right)$$

Cross-multiplying gives:

$$P_1 - \frac{P_1}{\sqrt{3}} \tan\phi = P_2 + \frac{P_2}{\sqrt{3}} \tan\phi$$

Hence

$$P_1 - P_2 = (P_1 + P_2)\frac{\tan\phi}{\sqrt{3}}$$

from which

$$\tan\phi = \sqrt{3}\left(\frac{P_1 - P_2}{P_1 + P_2}\right)$$

$\phi$, $\cos\phi$ and thus power factor can be determined from this formula.

Problem 13.    A 400 V, 3-phase star connected alternator supplies a delta-connected load, each phase of which has a resistance of 30 Ω and inductive reactance 40 Ω. Calculate (a) the current supplied by the alternator and (b) the output power and the kVA of the alternator, neglecting losses in the line between the alternator and load.

A circuit diagram of the alternator and load is shown in Fig. 20.17.
(a)  Considering the load:

Phase current, $I_p = V_p/Z_p$

$V_p = V_L$ for a delta connection,

hence $V_p = 400$ V.

Phase impedance,

$$Z_p = \sqrt{R_p^2 + X_L^2} = \sqrt{30^2 + 40^2} = 50\,\Omega.$$

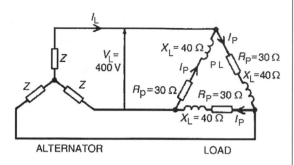

Figure 20.17

Hence $I_p = V_p/Z_p = 400/50 = 8$ A.

For a delta-connection, line current,

$$I_L = \sqrt{3}\,I_p = \sqrt{3}(8) = 13.86\ A.$$

Hence **13.86 A is the current supplied by the alternator**.

(b)  Alternator output power is equal to the power dissipated by the load i.e.

$$P = \sqrt{3}\,V_L I_L \cos\phi$$

where $\cos\phi = R_p/Z_p = 30/50 = 0.6$

Hence    $P = \sqrt{3}\,(400)(13.86)(0.6)$

$$= 5.76\,\text{kW}.$$

Alternator output kVA,

$$S = \sqrt{3}\,V_L I_L = \sqrt{3}(400)(13.86)$$

$$= 9.60\,\text{kVA}.$$

Problem 14.    Each phase of a delta-connected load comprises a resistance of 30 Ω and an 80 μF capacitor in series. The load is connected to a 400 V, 50 Hz, 3-phase supply. Calculate (a) the phase current, (b) the line current, (c) the total power dissipated, and (d) the kVA rating of the load. Draw the complete phasor diagram for the load.

(a)  Capacitive reactance,

$$X_C = \frac{1}{2\pi f C} = \frac{1}{2\pi (50)(80 \times 10^{-6})} = 39.79\,\Omega$$

Phase impedance,

$$Z_p = \sqrt{R_p^2 + X_C^2} = \sqrt{30^2 + 39.79^2} = 49.83\,\Omega.$$

Power factor $= \cos\phi = R_p/Z_p$

$$= 30/49.83 = 0.602$$

Hence $\phi = \cos^{-1} 0.602 = 52.99°$ leading.

Phase current,

$$I_p = V_p/Z_p \quad \text{and} \quad V_p = V_L$$

for a delta connection. Hence

$$I_p = 400/49.83 = \mathbf{8.027\,A}$$

(b) Line current, $I_L = \sqrt{3}\,I_p$ for a delta-connection. Hence $I_L = \sqrt{3}(8.027) = \mathbf{13.90\,A}$

(c) Total power dissipated,

$$P = \sqrt{3}\,V_L I_L \cos\phi$$

$$= \sqrt{3}(400)(13.90)(0.602) = \mathbf{5.797\,kW}$$

(d) Total kVA,

$$S = \sqrt{3}\,V_L I_L = \sqrt{3}(400)(13.90) = \mathbf{9.630\,kVA}$$

The phasor diagram for the load is shown in Fig. 20.18.

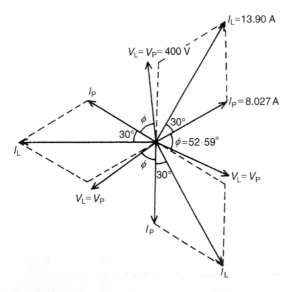

**Figure 20.18**

---

**Problem 15.** Two wattmeters are connected to measure the input power to a balanced 3-phase load by the two-wattmeter method. If the instrument readings are 8 kW and 4 kW, determine (a) the total power input and (b) the load power factor.

(a) Total input power,

$$P = P_1 + P_2 = 8 + 4 = \mathbf{12\,kW}$$

(b) $\tan\phi = \sqrt{3}\left(\dfrac{P_1 - P_2}{P_1 + P_2}\right) = \sqrt{3}\left(\dfrac{8 - 4}{8 + 4}\right)$

$$= \sqrt{3}\left(\frac{4}{12}\right) = \sqrt{3}\left(\frac{1}{3}\right) = \frac{1}{\sqrt{3}}$$

Hence $\phi = \tan^{-1}\dfrac{1}{\sqrt{3}} = 30°$

Power factor $= \cos\phi = \cos 30° = \mathbf{0.866}$

---

**Problem 16.** Two wattmeters connected to a 3-phase motor indicate the total power input to be 12 kW. The power factor is 0.6. Determine the readings of each wattmeter.

If the two wattmeters indicate $P_1$ and $P_2$ respectively then

$$P_1 + P_2 = 12\,\text{kW} \tag{1}$$

$$\tan\phi = \sqrt{3}\left(\frac{P_1 - P_2}{P_1 + P_2}\right)$$

and power factor $= 0.6 = \cos\phi$.
Angle $\phi = \cos^{-1} 0.6 = 53.13°$ and $\tan 53.13° = 1.3333$.
Hence

$$1.3333 = \sqrt{3}\left(\frac{P_1 - P_2}{12}\right)$$

from which,

$$P_1 - P_2 = \frac{12(1.3333)}{\sqrt{3}}$$

i.e. $\quad P_1 - P_2 = 9.237\,\text{kW} \tag{2}$

Adding equations (1) and (2) gives:

$$2P_1 = 21.237$$

i.e. $\qquad P_1 = \dfrac{21.237}{2}$

$$= 10.62\,\text{kW}$$

Hence **wattmeter 1 reads 10.62 kW**
From equation (1), **wattmeter 2 reads**
**(12 − 10.62) = 1.38 kW**

---

**Problem 17.** Two wattmeters indicate 10 kW and 3 kW respectively when connected to measure the input power to a 3-phase balanced load, the reverse switch being operated on the meter indicating the 3 kW reading. Determine (a) the input power and (b) the load power factor.

Since the reversing switch on the wattmeter had to be operated the 3 kW reading is taken as $-3$ kW

(a)    Total input power,

$$P = P_1 + P_2 = 10 + (-3) = \mathbf{7\,kW}$$

(b)    $\tan\phi = \sqrt{3}\left(\dfrac{P_1 - P_2}{P_1 + P_2}\right) = \sqrt{3}\left(\dfrac{10 - (-3)}{10 + (-3)}\right)$

$$= \sqrt{3}\left(\frac{13}{7}\right) = 3.2167$$

Angle $\phi = \tan^{-1} 3.2167 = 72.73°$

Power factor $= \cos\phi = \cos 72.73° = \mathbf{0.297}$

---

**Problem 18.** Three similar coils, each having a resistance of $8\,\Omega$ and an inductive reactance of $8\,\Omega$ are connected (a) in star and (b) in delta, across a 415 V, 3-phase supply. Calculate for each connection the readings on each of two wattmeters connected to measure the power by the two-wattmeter method.

---

(a)    **Star connection:** $V_L = \sqrt{3}\,V_p$ and $I_L = I_p$

Phase voltage, $V_p = \dfrac{V_L}{\sqrt{3}} = \dfrac{415}{\sqrt{3}}$

and phase impedance,

$$Z_p = \sqrt{R_p^2 + X_L^2} = \sqrt{8^2 + 8^2} = 11.31\,\Omega$$

Hence phase current,

$$I_p = \frac{V_p}{Z_p} = \frac{\dfrac{415}{\sqrt{3}}}{11.31} = 21.18\,\text{A}$$

Total power,

$$P = 3I_p^2 R_p = 3(21.18)^2(8) = 10\,766\,\text{W}$$

If wattmeter readings are $P_1$ and $P_2$ then:

$$P_1 + P_2 = 10\,766 \qquad (1)$$

Since $R_p = 8\,\Omega$ and $X_L = 8\,\Omega$, then phase angle $\phi = 45°$ (from impedance triangle).

$$\tan\phi = \sqrt{3}\left(\frac{P_1 - P_2}{P_1 + P_2}\right)$$

hence $\tan 45° = \dfrac{\sqrt{3}(P_1 - P_2)}{10\,766}$

from which

$$P_1 - P_2 = \frac{(10\,766)(1)}{\sqrt{3}} = 6216\,\text{W} \qquad (2)$$

Adding equations (1) and (2) gives:

$$2P_1 = 10\,766 + 6216 = 16\,982\,\text{W}$$

Hence $P_1 = 8491\,\text{W}$

From equation (1), $P_2 = 10\,766 - 8491 = 2275\,\text{W}$.

**When the coils are star-connected the wattmeter readings are thus 8.491 kW and 2.275 kW**

(b)    **Delta connection:** $V_L = V_p$   and   $I_L = \sqrt{3}\,I_p$

Phase current, $I_p = \dfrac{V_p}{Z_P} = \dfrac{415}{11.31} = 36.69\,\text{A}$.

Total power,

$$P = 3I_p^2 R_p = 3(36.69)^2(8) = 32\,310\,\text{W}$$

Hence $P_1 + P_2 = 32\,310\,\text{W}$ \qquad (3)

$\tan\phi = \sqrt{3}\left(\dfrac{P_1 - P_2}{P_1 + P_2}\right)$ thus $1 = \dfrac{\sqrt{3}(P_1 - P_2)}{32\,310}$

from which,

$$P_1 - P_2 = \frac{32\,310}{\sqrt{3}} = 18\,650\,\text{W} \qquad (4)$$

Adding equations (3) and (4) gives:

$$2P_1 = 50\,960 \text{ from which } P_1 = 25\,480\,\text{W}.$$

From equation (3), $P_2 = 32\,310 - 25\,480$
$$= 6830\,\text{W}$$

**When the coils are delta-connected the wattmeter readings are thus 25.48 kW and 6.83 kW**

---

**Now try the following exercise**

**Exercise 115    Further problems on the measurement of power in 3-phase circuits**

1.    Two wattmeters are connected to measure the input power to a balanced three-phase load. If the wattmeter readings are 9.3 kW and 5.4 kW determine (a) the total output power, and (b) the load power factor.
[(a) 14.7 kW (b) 0.909]

2. 8 kW is found by the two-wattmeter method to be the power input to a 3-phase motor. Determine the reading of each wattmeter if the power factor of the system is 0.85.

[5.431 kW, 2.569 kW]

3. When the two-wattmeter method is used to measure the input power of a balanced load, the readings on the wattmeters are 7.5 kW and 2.5 kW, the connections to one of the coils on the meter reading 2.5 kW having to be reversed. Determine (a) the total input power, and (b) the load power factor.

[(a) 5 kW (b) 0.277]

4. Three similar coils, each having a resistance of 4.0 Ω and an inductive reactance of 3.46 Ω are connected (a) in star and (b) in delta across a 400 V, 3-phase supply. Calculate for each connection the readings on each of two wattmeters connected to measure the power by the two-wattmeter method.

[(a) 17.15 kW, 5.73 kW
(b) 51.46 kW, 17.18 kW]

5. A 3-phase, star-connected alternator supplies a delta-connected load, each phase of which has a resistance of 15 Ω and inductive reactance 20 Ω. If the line voltage is 400 V, calculate (a) the current supplied by the alternator and (b) the output power and kVA rating of the alternator, neglecting any losses in the line between the alternator and the load.

[(a) 27.71 A (b) 11.52 kW, 19.20 kVA]

6. Each phase of a delta-connected load comprises a resistance of 40 Ω and a 40 μF capacitor in series. Determine, when connected to a 415 V, 50 Hz, 3-phase supply (a) the phase current, (b) the line current, (c) the total power dissipated, and (d) the kVA rating of the load.

[(a) 4.66 A (b) 8.07 A
(c) 2.605 kW (d) 5.80 kVA]

## 20.7 Comparison of star and delta connections

(i) Loads connected in delta dissipate three times more power than when connected in star to the same supply.

(ii) For the same power, the phase currents must be the same for both delta and star connections (since power $= 3I_p^2 R_p$), hence the line current in the delta-connected system is greater than the line current in the corresponding star-connected system. To achieve the same phase current in a star-connected system as in a delta-connected system, the line voltage in the star system is $\sqrt{3}$ times the line voltage in the delta system. Thus for a given power transfer, a delta system is associated with larger line currents (and thus larger conductor cross-sectional area) and a star system is associated with a larger line voltage (and thus greater insulation).

## 20.8 Advantages of three-phase systems

**Advantages of three-phase systems** over single-phase supplies include:

(i) For a given amount of power transmitted through a system, the three-phase system requires conductors with a smaller cross-sectional area. This means a saving of copper (or aluminium) and thus the original installation costs are less.

(ii) Two voltages are available (see Section 20.3 (vii))

(iii) Three-phase motors are very robust, relatively cheap, generally smaller, have self-starting properties, provide a steadier output and require little maintenance compared with single-phase motors.

---

**Now try the following exercises**

**Exercise 116  Short answer questions on three-phase systems**

1. Explain briefly how a three-phase supply is generated

2. State the national standard phase sequence for a three-phase supply

3. State the two ways in which phases of a three-phase supply can be interconnected to reduce the number of conductors used compared with three single-phase systems

4. State the relationships between line and phase currents and line and phase voltages for a star-connected system

5. When may the neutral conductor of a star-connected system be omitted?

6. State the relationships between line and phase currents and line and phase voltages for a delta-connected system

7. What is the standard electricity supply to domestic consumers in Great Britain?

8. State two formulae for determining the power dissipated in the load of a three-phase balanced system

9. By what methods may power be measured in a three-phase system?

10. State a formula from which power factor may be determined for a balanced system when using the two-wattmeter method of power measurement

11. Loads connected in star dissipate ...... the power dissipated when connected in delta and fed from the same supply

12. Name three advantages of three-phase systems over single-phase systems

---

**Exercise 117    Multi-choice questions on three-phase systems**
**(Answers on page 421)**

Three loads, each of $10\,\Omega$ resistance, are connected in star to a $400\,V$, 3-phase supply. Determine the quantities stated in questions 1 to 5, selecting answers from the following list:

(a) $\dfrac{40}{\sqrt{3}}\,A$    (b) $\sqrt{3}(16)\,kW$    (c) $\dfrac{400}{\sqrt{3}}\,V$

(d) $\sqrt{3}(40)\,A$    (e) $\sqrt{3}(400)\,V$    (f) $16\,kW$

(g) $400\,V$    (h) $48\,kW$    (i) $40\,A$

1. Line voltage

2. Phase voltage

3. Phase current

4. Line current

5. Total power dissipated in the load

6. Which of the following statements is false?
   (a) For the same power, loads connected in delta have a higher line voltage and a smaller line current than loads connected in star
   (b) When using the two-wattmeter method of power measurement the power factor is unity when the wattmeter readings are the same
   (c) A.c. may be distributed using a single-phase system with two wires, a three-phase system with three wires or a three-phase system with four wires
   (d) The national standard phase sequence for a three-phase supply is R, Y, B

Three loads, each of resistance $16\,\Omega$ and inductive reactance $12\,\Omega$ are connected in delta to a $400\,V$, 3-phase supply. Determine the quantities stated in questions 7 to 12, selecting the correct answer from the following list:

(a) $4\,\Omega$    (b) $\sqrt{3}(400)\,V$    (c) $\sqrt{3}(6.4)\,kW$

(d) $20\,A$    (e) $6.4\,kW$    (f) $\sqrt{3}(20)\,A$

(g) $20\,\Omega$    (h) $\dfrac{20}{\sqrt{3}}\,V$    (i) $\dfrac{400}{\sqrt{3}}\,V$

(j) $19.2\,kW$    (k) $100\,A$    (l) $400\,V$

(m) $28\,\Omega$

7. Phase impedance

8. Line voltage

9. Phase voltage

10. Phase current

11. Line current

12. Total power dissipated in the load

13. The phase voltage of a delta-connected three-phase system with balanced loads is $240\,V$. The line voltage is:
   (a) $720\,V$    (b) $440\,V$
   (c) $340\,V$    (d) $240\,V$

14. A 4-wire three-phase star-connected system has a line current of $10\,A$. The phase current is:
   (a) $40\,A$    (b) $10\,A$
   (c) $20\,A$    (d) $30\,A$

15. The line voltage of a 4-wire three-phase star-connected system is $11\,kV$. The phase voltage is:
   (a) $19.05\,kV$    (b) $11\,kV$
   (c) $6.35\,kV$    (d) $7.78\,kV$

16. In the two-wattmeter method of measurement power in a balanced three-phase system readings of $P_1$ and $P_2$ watts are obtained. The power factor may be determined from:

(a) $\sqrt{3}\left(\dfrac{P_1+P_2}{P_1-P_2}\right)$   (b) $\sqrt{3}\left(\dfrac{P_1-P_2}{P_1+P_2}\right)$

(c) $\dfrac{(P_1-P_2)}{\sqrt{3}(P_1+P_2)}$   (d) $\dfrac{(P_1+P_2)}{\sqrt{3}(P_1-P_2)}$

17. The phase voltage of a 4-wire three-phase star-connected system is 110 V. The line voltage is:

(a) 440 V   (b) 330 V
(c) 191 V   (d) 110 V

# Chapter 21

# Transformers

At the end of this chapter you should be able to:

- understand the principle of operation of a transformer

- understand the term 'rating' of a transformer

- use $V_1/V_2 = N_1/N_2 = I_2/I_1$ in calculations on transformers

- construct a transformer no-load phasor diagram and calculate magnetising and core loss components of the no-load current

- state the e.m.f. equation for a transformer $E = 4.44\,f\,\Phi_m\,N$ and use it in calculations

- construct a transformer on-load phasor diagram for an inductive circuit assuming the volt drop in the windings is negligible

- describe transformer construction

- derive the equivalent resistance, reactance and impedance referred to the primary of a transformer

- understand voltage regulation

- describe losses in transformers and calculate efficiency

- appreciate the concept of resistance matching and how it may be achieved

- perform calculations using $R_1 = (N_1/N_2)^2 R_L$

- describe an auto transformer, its advantages/disadvantages and uses

- describe an isolating transformer, stating uses

- describe a three-phase transformer

- describe current and voltage transformers

## 21.1 Introduction

A transformer is a device which uses the phenomenon of mutual induction (see Chapter 9) to change the values of alternating voltages and currents. In fact, one of the main advantages of a.c. transmission and distribution is the ease with which an alternating voltage can be increased or decreased by transformers.

Losses in transformers are generally low and thus efficiency is high. Being static they have a long life and are very stable.

Transformers range in size from the miniature units used in electronic applications to the large power transformers used in power stations; the principle of operation is the same for each.

A transformer is represented in Fig. 21.1(a) as consisting of two electrical circuits linked by a common

DOI: 10.1016/B978-0-08-089056-2.00021-8

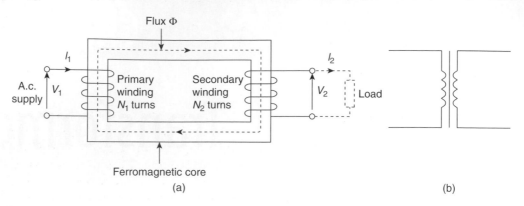

Figure 21.1

ferromagnetic core. One coil is termed the **primary winding** which is connected to the supply of electricity, and the other the **secondary winding**, which may be connected to a load. A circuit diagram symbol for a transformer is shown in Fig. 21.1(b).

## 21.2 Transformer principle of operation

When the secondary is an open-circuit and an alternating voltage $V_1$ is applied to the primary winding, a small current – called the no-load current $I_0$ – flows, which sets up a magnetic flux in the core. This alternating flux links with both primary and secondary coils and induces in them e.m.f.'s of $E_1$ and $E_2$ respectively by mutual induction.

The induced e.m.f. $E$ in a coil of $N$ turns is given by $E = -N(d\Phi/dt)$ volts, where $\frac{d\Phi}{dt}$ is the rate of change of flux. In an ideal transformer, the rate of change of flux is the same for both primary and secondary and thus $E_1/N_1 = E_2/N_2$ i.e. **the induced e.m.f. per turn is constant**.

Assuming no losses, $E_1 = V_1$ and $E_2 = V_2$

Hence 
$$\frac{V_1}{N_1} = \frac{V_2}{N_2} \quad \text{or} \quad \frac{V_1}{V_2} = \frac{N_1}{N_2} \tag{1}$$

$(V_1/V_2)$ is called the voltage ratio and $(N_1/N_2)$ the turns ratio, or the '**transformation ratio**' of the transformer. If $N_2$ is less than $N_1$ then $V_2$ is less than $V_1$ and the device is termed a **step-down transformer**. If $N_2$ is greater then $N_1$ then $V_2$ is greater than $V_1$ and the device is termed a **step-up transformer**.

When a load is connected across the secondary winding, a current $I_2$ flows. In an ideal transformer losses are neglected and a transformer is considered to be 100 per cent efficient. Hence input power = output power,

or $V_1 I_1 = V_2 I_2$ i.e. in an ideal transformer, the **primary and secondary ampere-turns are equal**

Thus 
$$\frac{V_1}{V_2} = \frac{I_2}{I_1} \tag{2}$$

Combining equations (1) and (2) gives:

$$\frac{V_1}{V_2} = \frac{N_1}{N_2} = \frac{I_2}{I_1} \tag{3}$$

The **rating** of a transformer is stated in terms of the volt-amperes that it can transform without overheating. With reference to Fig. 21.1(a), the transformer rating is either $V_1 I_1$ or $V_2 I_2$, where $I_2$ is the full-load secondary current.

Problem 1. A transformer has 500 primary turns and 3000 secondary turns. If the primary voltage is 240 V, determine the secondary voltage, assuming an ideal transformer.

For an ideal transformer, voltage ratio = turns ratio i.e.

$$\frac{V_1}{V_2} = \frac{N_1}{N_2} \quad \text{hence} \quad \frac{240}{V_2} = \frac{500}{3000}$$

Thus secondary voltage

$$V_2 = \frac{(240)(3000)}{500} = \mathbf{1440\,V} \text{ or } \mathbf{1.44\,kV}$$

Problem 2. An ideal transformer with a turns ratio of 2:7 is fed from a 240 V supply. Determine its output voltage.

A turns ratio of 2:7 means that the transformer has 2 turns on the primary for every 7 turns on the secondary (i.e. a step-up transformer); thus $(N_1/N_2) = (2/7)$.

For an ideal transformer, $(N_1/N_2)=(V_1/V_2)$ hence $(2/7)=(240/V_2)$. Thus the secondary voltage

$$V_2 = \frac{(240)(7)}{2} = \textbf{840 V}$$

**Problem 3.** An ideal transformer has a turns ratio of 8:1 and the primary current is 3 A when it is supplied at 240 V. Calculate the secondary voltage and current.

A turns ratio of 8:1 means $(N_1/N_2)=(1/8)$ i.e. a step-down transformer.

$$\left(\frac{N_1}{N_2}\right) = \left(\frac{V_1}{V_2}\right) \text{ or secondary voltage}$$

$$V_2 = V_1\left(\frac{N_1}{N_2}\right) = 240\left(\frac{1}{8}\right) = \textbf{30 volts}$$

Also, $\left(\frac{N_1}{N_2}\right) = \left(\frac{I_2}{I_1}\right)$ hence secondary current

$$I_2 = I_1\left(\frac{N_1}{N_2}\right) = 3\left(\frac{8}{1}\right) = \textbf{24 A}$$

**Problem 4.** An ideal transformer, connected to a 240 V mains, supplies a 12 V, 150 W lamp. Calculate the transformer turns ratio and the current taken from the supply.

$V_1 = 240$ V, $V_2 = 12$ V,
$I_2 = (P/V_2) = (150/12) = 12.5$ A.

$$\textbf{Turns ratio} = \frac{N_1}{N_2} = \frac{V_1}{V_2} = \frac{240}{12} = \textbf{20}$$

$$\left(\frac{V_1}{V_2}\right) = \left(\frac{I_2}{I_1}\right), \text{ from which,}$$

$$I_1 = I_2\left(\frac{V_2}{V_1}\right) = 12.5\left(\frac{12}{240}\right)$$

Hence current taken from the supply,

$$I_1 = \frac{12.5}{20} = \textbf{0.625 A}$$

**Problem 5.** A 12 Ω resistor is connected across the secondary winding of an ideal transformer whose secondary voltage is 120 V. Determine the primary voltage if the supply current is 4 A.

Secondary current $I_2=(V_2/R_2)=(120/12)=10$ A. $(V_1/V_2)=(I_2/I_1)$, from which the primary voltage

$$V_1 = V_2\left(\frac{I_2}{I_1}\right) = 120\left(\frac{10}{4}\right) = \textbf{300 volts}$$

**Problem 6.** A 5 kVA single-phase transformer has a turns ratio of 10:1 and is fed from a 2.5 kV supply. Neglecting losses, determine (a) the full-load secondary current, (b) the minimum load resistance which can be connected across the secondary winding to give full load kVA, (c) the primary current at full load kVA.

(a) $N_1/N_2 = 10/1$ and $V_1 = 2.5$ kV $= 2500$ V.
Since $\left(\frac{N_1}{N_2}\right) = \left(\frac{V_1}{V_2}\right)$, secondary voltage

$$V_2 = V_1\left(\frac{N_2}{N_1}\right) = 2500\left(\frac{1}{10}\right) = 250 \text{ V}$$

The transformer rating in volt-amperes $= V_2I_2$ (at full load) i.e. $5000 = 250I_2$
Hence full-load secondary current,
$I_2 = (5000/250) = \textbf{20 A}$.

(b) Minimum value of load resistance,

$$R_L = \left(\frac{V_2}{I_2}\right) = \left(\frac{250}{20}\right) = \textbf{12.5 Ω}.$$

(c) $\left(\frac{N_1}{N_2}\right) = \left(\frac{I_2}{I_1}\right)$ from which primary current

$$I_1 = I_2\left(\frac{N_1}{N_2}\right) = 20\left(\frac{1}{10}\right) = \textbf{2 A}$$

Now try the following exercise

**Exercise 118 Further problems on the transformer principle of operation**

1. A transformer has 600 primary turns connected to a 1.5 kV supply. Determine the number of secondary turns for a 240 V output voltage, assuming no losses. [96]

2. An ideal transformer with a turns ratio of 2:9 is fed from a 220 V supply. Determine its output voltage. [990 V]

3. A transformer has 800 primary turns and 2000 secondary turns. If the primary voltage is 160 V, determine the secondary voltage assuming an ideal transformer. [400 V]

4. An ideal transformer with a turns ratio of 3:8 has an output voltage of 640 V. Determine its input voltage. [240 V]

5. An ideal transformer has a turns ratio of 12:1 and is supplied at 192 V. Calculate the secondary voltage. [16 V]

6. A transformer primary winding connected across a 415 V supply has 750 turns. Determine how many turns must be wound on the secondary side if an output of 1.66 kV is required. [3000 turns]

7. An ideal transformer has a turns ratio of 15:1 and is supplied at 180 V when the primary current is 4 A. Calculate the secondary voltage and current. [12 V, 60 A]

8. A step-down transformer having a turns ratio of 20:1 has a primary voltage of 4 kV and a load of 10 kW. Neglecting losses, calculate the value of the secondary current. [50 A]

9. A transformer has a primary to secondary turns ratio of 1:15. Calculate the primary voltage necessary to supply a 240 V load. If the load current is 3 A determine the primary current. Neglect any losses. [16 V, 45 A]

10. A 10 kVA, single-phase transformer has a turns ratio of 12:1 and is supplied from a 2.4 kV supply. Neglecting losses, determine (a) the full-load secondary current, (b) the minimum value of load resistance which can be connected across the secondary winding without the kVA rating being exceeded, and (c) the primary current. [(a) 50 A (b) 4 Ω (c) 4.17 A]

11. A 20 Ω resistance is connected across the secondary winding of a single-phase power transformer whose secondary voltage is 150 V. Calculate the primary voltage and the turns ratio if the supply current is 5 A, neglecting losses. [225 V, 3:2]

## 21.3 Transformer no-load phasor diagram

The core flux is common to both primary and secondary windings in a transformer and is thus taken as the reference phasor in a phasor diagram. On no-load the primary winding takes a small no-load current $I_0$ and since, with losses neglected, the primary winding is a pure inductor, this current lags the applied voltage $V_1$ by 90°. In the phasor diagram assuming no losses, shown in Fig. 21.2(a), current $I_0$ produces the flux and is drawn in phase with the flux. The primary induced e.m.f. $E_1$ is in phase opposition to $V_1$ (by Lenz's law) and is shown 180° out of phase with $V_1$ and equal in magnitude. The secondary induced e.m.f. is shown for a 2:1 turns ratio transformer.

A no-load phasor diagram for a practical transformer is shown in Fig. 21.2(b). If current flows then losses will occur. When losses are considered then the no-load current $I_0$ is the phasor sum of two components – (i) $I_M$, **the magnetising component**, in phase with the flux, and (ii) $I_C$, **the core loss component** (supplying the hysteresis and eddy current losses). From Fig. 21.2(b):

**No-load current, $I_0 = \sqrt{I_M^2 + I_C^2}$ where**

$I_M = I_0 \sin \phi_0$ and $I_C = I_0 \cos \phi_0$.

**Power factor on no-load $= \cos \phi_0 = (I_C/I_0)$.**
**The total core losses (i.e. iron losses) $= V_1 I_0 \cos \phi_0$**

**Problem 7.** A 2400 V/400 V single-phase transformer takes a no-load current of 0.5 A and the core loss is 400 W. Determine the values of the magnetising and core loss components of the no-load current. Draw to scale the no-load phasor diagram for the transformer.

$V_1 = 2400$ V, $V_2 = 400$ V and $I_0 = 0.5$ A Core loss (i.e. iron loss) $= 400 = V_1 I_0 \cos \phi_0$.

i.e. $400 = (2400)(0.5) \cos \phi_0$

Hence $\cos \phi_0 = \dfrac{400}{(2400)(0.5)} = 0.3333$

$\phi_0 = \cos^{-1} 0.3333 = 70.53°$

The no-load phasor diagram is shown in Fig. 21.3
Magnetising component,
$I_M = I_0 \sin \phi_0 = 0.5 \sin 70.53° = \mathbf{0.471\,A}$.
Core loss component, $I_C = I_0 \cos \phi_0 = 0.5 \cos 70.53°$
$= \mathbf{0.167\,A}$

**Problem 8.** A transformer takes a current of 0.8 A when its primary is connected to a 240 volt, 50 Hz supply, the secondary being on open circuit. If the power absorbed is 72 watts, determine (a) the iron loss current, (b) the power factor on no-load, and (c) the magnetising current.

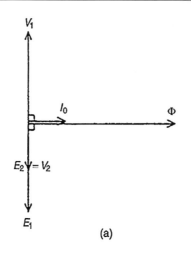

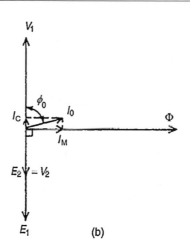

Figure 21.2

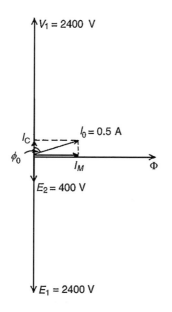

Figure 21.3

$I_0 = 0.8$ A and $V = 240$ V

(a)  Power absorbed = total core loss $= 72 = V_1 I_0 \cos\phi_0$. Hence $72 = 240 I_0 \cos\phi_0$ and iron loss current, $I_c = I_0 \cos\phi_0 = 72/240 = \mathbf{0.30\,A}$

(b)  Power factor at no load,

$$\cos\phi_0 = \frac{I_C}{I_0} = \frac{0.3}{0.8} = \mathbf{0.375}$$

(c)  From the right-angled triangle in Fig. 21.2(b) and using Pythagoras' theorem, $I_0^2 = I_C^2 + I_M^2$ from which, magnetising current,

$$I_M = \sqrt{I_0^2 - I_C^2} = \sqrt{0.8^2 - 0.3^2} = \mathbf{0.74\,A}$$

**Now try the following exercise**

**Exercise 119    Further problems on the no-load phasor diagram**

1.  A 500 V/100 V, single-phase transformer takes a full-load primary current of 4 A. Neglecting losses, determine (a) the full-load secondary current, and (b) the rating of the transformer.

    [(a) 20 A (b) 2 kVA]

2.  A 3300 V/440 V, single-phase transformer takes a no-load current of 0.8 A and the iron loss is 500 W. Draw the no-load phasor diagram and determine the values of the magnetising and core loss components of the no-load current.    [0.786 A, 0.152 A]

3.  A transformer takes a current of 1 A when its primary is connected to a 300 V, 50 Hz supply, the secondary being on open-circuit. If the power absorbed is 120 watts, calculate (a) the iron loss current, (b) the power factor on no-load, and (c) the magnetising current.

    [(a) 0.40 A (b) 0.40 (c) 0.917 A]

## 21.4    E.m.f. equation of a transformer

The magnetic flux $\Phi$ set up in the core of a transformer when an alternating voltage is applied to its primary winding is also alternating and is sinusoidal.

Let $\Phi_m$ be the maximum value of the flux and $f$ be the frequency of the supply. The time for 1 cycle of the alternating flux is the periodic time $T$, where $T = (1/f)$ seconds.

The flux rises sinusoidally from zero to its maximum value in $(1/4)$ cycle, and the time for $(1/4)$ cycle is $(1/4f)$ seconds. Hence the average rate of change of flux $= (\Phi_m/(1/4f)) = 4f\Phi_m$ Wb/s, and since 1 Wb/s = 1 volt, the average e.m.f. induced in each turn $= 4f\Phi_m$ volts. As the flux $\Phi$ varies sinusoidally, then a sinusoidal e.m.f. will be induced in each turn of both primary and secondary windings.

For a sine wave,

$$\text{form factor} = \frac{\text{r.m.s. value}}{\text{average value}}$$

$$= 1.11 \text{ (see Chapter 14)}$$

Hence r.m.s. value = form factor × average value = $1.11 \times$ average value. Thus r.m.s. e.m.f. induced in each turn

$$= 1.11 \times 4f\Phi_m \text{ volts}$$

$$= 4.44f\Phi_m \text{ volts}$$

Therefore, r.m.s. value of e.m.f. induced in primary,

$$E_1 = 4.44f\Phi_m N_1 \text{ volts} \qquad (4)$$

and r.m.s. value of e.m.f. induced in secondary,

$$E_2 = 4.44f\Phi_m N_2 \text{ volts} \qquad (5)$$

Dividing equation (4) by equation (5) gives:

$$\left(\frac{E_1}{E_2}\right) = \left(\frac{N_1}{N_2}\right)$$

as previously obtained in Section 21.2

**Problem 9.** A 100 kVA, 4000 V/200 V, 50 Hz single-phase transformer has 100 secondary turns. Determine (a) the primary and secondary current, (b) the number of primary turns, and (c) the maximum value of the flux.

$V_1 = 4000$ V, $V_2 = 200$ V, $f = 50$ Hz, $N_2 = 100$ turns

(a) Transformer rating $= V_1 I_1 = V_2 I_2 = 100000$ VA
Hence primary current,

$$I_1 = \frac{100000}{V_1} = \frac{100000}{4000} = 25 \text{ A}$$

and secondary current,

$$I_2 = \frac{100000}{V_2} = \frac{100000}{200} = 500 \text{ A}$$

(b) From equation (3), $\frac{V_1}{V_2} = \frac{N_1}{N_2}$ from which, primary turns,

$$N_1 = \left(\frac{V_1}{V_2}\right)(N_2) = \left(\frac{4000}{200}\right)(100) = 2000 \text{ turns}$$

(c) From equation (5), $E_2 = 4.44f\Phi_m N_2$ from which, maximum flux,

$$\Phi_m = \frac{E}{4.44fN_2}$$

$$= \frac{200}{(4.44)(50)(100)} \text{ (assuming } E_2 = V_2\text{)}$$

$$= 9.01 \times 10^{-3} \text{ Wb or } 9.01 \text{ mWb}$$

[Alternatively, equation (4) could have been used, where

$E_1 = 4.44f\Phi_m N_1$ from which,

$$\Phi_m = \frac{4000}{(4.44)(50)(2000)} \text{ (assuming } E_1 = V_1\text{)}$$

$$= 9.01 \text{ mWb as above]}$$

**Problem 10.** A single-phase, 50 Hz transformer has 25 primary turns and 300 secondary turns. The cross-sectional area of the core is 300 cm². When the primary winding is connected to a 250 V supply, determine (a) the maximum value of the flux density in the core, and (b) the voltage induced in the secondary winding.

(a) From equation (4),
e.m.f. $E_1 = 4.44f\Phi_m N_1$ volts
i.e. $250 = 4.44(50)\Phi_m(25)$ from which, maximum flux density,

$$\Phi_m = \frac{250}{(4.44)(50)(25)} \text{Wb} = 0.04505 \text{ Wb}$$

However, $\Phi_m = B_m \times A$, where $B_m =$ maximum flux density in the core and $A =$ cross-sectional area of the core (see Chapter 7). Hence
$B_m \times 300 \times 10^{-4} = 0.04505$ from which,

$$\textbf{maximum flux density, } B_m = \frac{0.04505}{300 \times 10^{-4}}$$

$$= 1.50 \text{ T}$$

(b) $\dfrac{V_1}{V_2}=\dfrac{N_1}{N_2}$ from which, $V_2=V_1\left(\dfrac{N_2}{N_1}\right)$ i.e. voltage induced in the secondary winding,

$$V_2=(250)\left(\frac{300}{25}\right)=\textbf{3000 V or 3 kV}$$

---

**Problem 11.** A single-phase 500 V/100 V, 50 Hz transformer has a maximum core flux density of 1.5 T and an effective core cross-sectional area of 50 cm$^2$. Determine the number of primary and secondary turns.

---

The e.m.f. equation for a transformer is $E=4.44\,f\,\Phi_m N$ and maximum flux, $\Phi_m=B\times A=(1.5)(50\times10^{-4})=75\times10^{-4}$ Wb

Since $E_1=4.44\,f\,\Phi_m N_1$ then primary turns,

$$N_1=\frac{E_1}{4.44\,f\,\Phi_m}=\frac{500}{(4.44)(50)(75\times10^{-4})}$$
$$=\textbf{300 turns}$$

Since $E_2=4.4\,f\,\Phi_m N_2$ then secondary turns,

$$N_2=\frac{E_2}{4.44\,f\,\Phi_m}=\frac{100}{(4.44)(50)(75\times10^{-4})}$$
$$=\textbf{60 turns}$$

---

**Problem 12.** A 4500 V/225 V, 50 Hz single-phase transformer is to have an approximate e.m.f. per turn of 15 V and operate with a maximum flux of 1.4 T. Calculate (a) the number of primary and secondary turns and (b) the cross-sectional area of the core.

---

(a) E.m.f. per turn $=\dfrac{E_1}{N_1}=\dfrac{E_2}{N_2}=15$

Hence primary turns, $N_1=\dfrac{E_1}{15}=\dfrac{4500}{15}=\textbf{300}$

and secondary turns, $N_2=\dfrac{E_2}{15}=\dfrac{255}{15}=\textbf{15}$

(b) E.m.f. $E_1=4.44\,f\,\Phi_m N_1$ from which,

$$\Phi_m=\frac{E_1}{4.44\,f\,N_1}=\frac{4500}{(4.44)(50)(300)}=0.0676\text{ Wb}$$

Now flux, $\Phi_m=B_m\times A$, where $A$ is the cross-sectional area of the core,

hence area, $A=\left(\dfrac{\Phi_m}{B_m}\right)=\left(\dfrac{0.0676}{1.4}\right)$

$$=\textbf{0.0483 m}^2 \text{ or } \textbf{483 cm}^2$$

---

**Now try the following exercise**

**Exercise 120  Further problems on the transformer e.m.f. equation**

1. A 60 kVA, 1600 V/100 V, 50 Hz, single-phase transformer has 50 secondary windings. Calculate (a) the primary and secondary current, (b) the number of primary turns, and (c) the maximum value of the flux.
   [(a) 37.5 A, 600 A (b) 800 (c) 9.0 mWb]

2. A single-phase, 50 Hz transformer has 40 primary turns and 520 secondary turns. The cross-sectional area of the core is 270 cm$^2$. When the primary winding is connected to a 300 volt supply, determine (a) the maximum value of flux density in the core, and (b) the voltage induced in the secondary winding.
   [(a) 1.25 T (b) 3.90 kV]

3. A single-phase 800 V/100 V, 50 Hz transformer has a maximum core flux density of 1.294 T and an effective cross-sectional area of 60 cm$^2$. Calculate the number of turns on the primary and secondary windings.
   [464, 58]

4. A 3.3 kV/110 V, 50 Hz, single-phase transformer is to have an approximate e.m.f. per turn of 22 V and operate with a maximum flux of 1.25 T. Calculate (a) the number of primary and secondary turns, and (b) the cross-sectional area of the core.
   [(a) 150, 5 (b) 792.8 cm$^2$]

## 21.5  Transformer on-load phasor diagram

If the voltage drop in the windings of a transformer are assumed negligible, then the terminal voltage $V_2$ is the same as the induced e.m.f. $E_2$ in the secondary. Similarly, $V_1=E_1$. Assuming an equal number of turns on primary and secondary windings, then $E_1=E_2$, and let the load have a lagging phase angle $\phi_2$.

In the phasor diagram of Fig. 21.4, current $I_2$ lags $V_2$ by angle $\phi_2$. When a load is connected across the secondary winding a current $I_2$ flows in the secondary winding. The resulting secondary e.m.f. acts so as to tend to reduce the core flux. However this does not happen since reduction of the core flux reduces $E_1$,

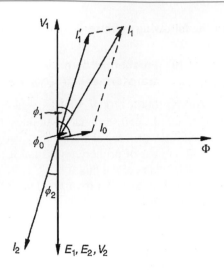

**Figure 21.4**

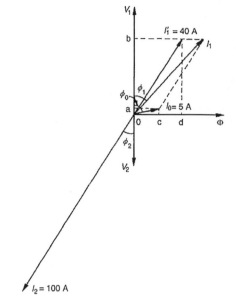

**Figure 21.5**

hence a reflected increase in primary current $I_1'$ occurs which provides a restoring m.m.f. Hence at all loads, primary and secondary m.m.f.'s are equal, but in opposition, and the core flux remains constant. $I_1'$ is sometimes called the 'balancing' current and is equal, but in the opposite direction, to current $I_2$ as shown in Fig. 21.4. $I_0$, shown at a phase angle $\phi_0$ to $V_1$, is the no-load current of the transformer (see Section 21.3).

The phasor sum of $I_1'$ and $I_0$ gives the supply current $I_1$ and the phase angle between $V_1$ and $I_1$ is shown as $\phi_1$.

> **Problem 13.** A single-phase transformer has 2000 turns on the primary and 800 turns on the secondary. Its no-load current is 5 A at a power factor of 0.20 lagging. Assuming the volt drop in the windings is negligible, determine the primary current and power factor when the secondary current is 100 A at a power factor of 0.85 lagging.

Let $I_1'$ be the component of the primary current which provides the restoring m.m.f. Then

$$I_1' N_1 = I_2 N_2$$

i.e. $\qquad I_1'(2000) = (100)(800)$

from which, $\qquad I_1' = \dfrac{(100)(800)}{2000}$

$$= 40 \text{ A}$$

If the power factor of the secondary is 0.85, then $\cos\phi_2 = 0.85$, from which, $\phi_2 = \cos^{-1} 0.85 = 31.8°$.

If the power factor on no-load is 0.20, then $\cos\phi_0 = 0.2$ and $\phi_0 = \cos^{-1} 0.2 = 78.5°$.

In the phasor diagram shown in Fig. 21.5, $I_2 = 100$ A is shown at an angle of $\phi = 31.8°$ to $V_2$ and $I_1' = 40$ A is shown in anti-phase to $I_2$.

The no-load current $I_0 = 5$ A is shown at an angle of $\phi_0 = 78.5°$ to $V_1$. Current $I_1$ is the phasor sum of $I_1'$ and $I_0$, and by drawing to scale, $I_1 = 44$ A and angle $\phi_1 = 37°$.

By calculation,

$$I_1 \cos\phi_1 = 0a + 0b$$

$$= I_0 \cos\phi_0 + I_1' \cos\phi_2$$

$$= (5)(0.2) + (40)(0.85)$$

$$= 35.0 \text{ A}$$

and $\qquad I_1 \sin\phi_1 = 0c + 0d$

$$= I_0 \sin\phi_0 + I_1' \sin\phi_2$$

$$= (5)\sin 78.5° + (40)\sin 31.8°$$

$$= 25.98 \text{ A}$$

Hence the magnitude of $\boldsymbol{I_1} = \sqrt{35.0^2 + 25.98^2} = \mathbf{43.59\,A}$ and $\tan\phi_1 = (25.98/35.0)$ from which, $\boldsymbol{\phi_1} = \tan^{-1}(25.98/35.0) = \mathbf{36.59°}$. Hence the power factor of the primary $= \cos\phi_1 = \cos 36.59° = \mathbf{0.80}$

**Now try the following exercise**

## 21.6 Transformer construction

(i)  There are broadly two types of single-phase double-wound transformer constructions – the **core type** and the **shell type**, as shown in Fig. 21.6. The low and high voltage windings are wound as shown to reduce leakage flux.

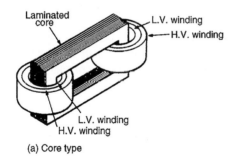

(a) Core type

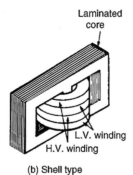

(b) Shell type

Figure 21.6

(ii) For **power transformers**, rated possibly at several MVA and operating at a frequency of 50 Hz in Great Britain, the core material used is usually laminated silicon steel or stalloy, the laminations reducing eddy currents and the silicon steel keeping hysteresis loss to a minimum.

Large power transformers are used in the main distribution system and in industrial supply circuits. Small power transformers have many applications, examples including welding and rectifier supplies, domestic bell circuits, imported washing machines, and so on.

(iii) For **audio frequency (a.f.) transformers**, rated from a few mVA to no more than 20 VA, and operating at frequencies up to about 15 kHz, the small core is also made of laminated silicon steel. A typical application of a.f. transformers is in an audio amplifier system.

(iv) **Radio frequency (r.f.) transformers**, operating in the MHz frequency region have either an air core, a ferrite core or a dust core. Ferrite is a ceramic material having magnetic properties similar to silicon steel, but having a high resistivity. Dust cores consist of fine particles of carbonyl iron or permalloy (i.e. nickel and iron), each particle of which is insulated from its neighbour. Applications of r.f. transformers are found in radio and television receivers.

(v) Transformer **windings** are usually of enamel-insulated copper or aluminium.

(vi) **Cooling** is achieved by air in small transformers and oil in large transformers.

## 21.7 Equivalent circuit of a transformer

Figure 21.7 shows an equivalent circuit of a transformer. $R_1$ and $R_2$ represent the resistances of the primary and secondary windings and $X_1$ and $X_2$ represent the reactances of the primary and secondary windings, due to leakage flux.

The core losses due to hysteresis and eddy currents are allowed for by resistance $R$ which takes a current $I_C$, the core loss component of the primary current. Reactance $X$ takes the magnetising component $I_m$. In a simplified equivalent circuit shown in Fig. 21.8, $R$ and $X$ are omitted since the no-load current $I_0$ is normally only about 3–5 per cent of the full-load primary current.

It is often convenient to assume that all of the resistance and reactance as being on one side of the transformer. Resistance $R_2$ in Fig. 21.8 can be replaced

Section 3

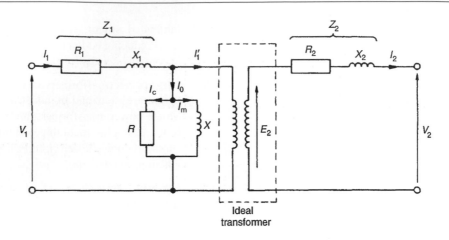

**Figure 21.7**

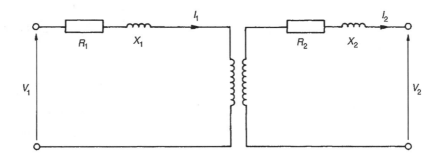

**Figure 21.8**

by inserting an additional resistance $R_2'$ in the primary circuit such that the power absorbed in $R_2'$ when carrying the primary current is equal to that in $R_2$ due to the secondary current, i.e.

$$I_1^2 R_2' = I_2^2 R_2$$

from which,    $R_2' = R_2 \left(\dfrac{I_2}{I_1}\right)^2 = R_2 \left(\dfrac{V_1}{V_2}\right)^2$

Then the total equivalent resistance in the primary circuit $R_e$ is equal to the primary and secondary resistances of the actual transformer.

Hence $R_e = R_1 + R_2'$

i.e.    $R_e = R_1 + R_2 \left(\dfrac{V_1}{V_2}\right)^2$    (6)

By similar reasoning, the equivalent reactance in the primary circuit is given by $X_e = X_1 + X_2'$

i.e.    $X_e = X_1 + X_2 \left(\dfrac{V_1}{V_2}\right)^2$    (7)

The equivalent impedance $Z_e$ of the primary and secondary windings referred to the primary is given by

$$Z_e = \sqrt{R_e^2 + X_e^2}$$    (8)

If $\phi_e$ is the phase angle between $I_1$ and the volt drop $I_1 Z_e$ then

$$\cos \phi_e = \frac{R_e}{Z_e}$$    (9)

The simplified equivalent circuit of a transformer is shown in Fig. 21.9.

**Problem 14.**   A transformer has 600 primary turns and 150 secondary turns. The primary and secondary resistances are 0.25 Ω and 0.01 Ω respectively and the corresponding leakage reactances are 1.0 Ω and 0.04 Ω respectively. Determine (a) the equivalent resistance referred to the primary winding, (b) the equivalent reactance referred to the primary winding, (c) the equivalent impedance referred to the primary winding, and (d) the phase angle of the impedance.

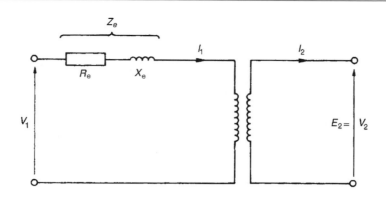

**Figure 21.9**

(a)  From equation (6), equivalent resistance

$$R_e = R_1 + R_2 \left(\frac{V_1}{V_2}\right)^2$$

i.e.  $R_e = 0.25 + 0.01 \left(\dfrac{600}{150}\right)^2$

$= \mathbf{0.41\,\Omega}$ since $\dfrac{N_1}{N_2} = \dfrac{V_1}{V_2}$

(b)  From equation (7), equivalent reactance,

$$X_e = X_1 + X_2 \left(\frac{V_1}{V_2}\right)^2$$

i.e.  $X_e = 1.0 + 0.04 \left(\dfrac{600}{150}\right)^2 = \mathbf{1.64\,\Omega}$

(c)  From equation (8), equivalent impedance,

$Z_e = \sqrt{R_e^2 + X_e^2} = \sqrt{0.41^2 + 1.64^2} = \mathbf{1.69\,\Omega}$

(d)  From equation (9),

$$\cos\phi_e = \frac{R_e}{Z_e} = \frac{0.41}{1.69}$$

Hence  $\phi_e = \cos^{-1}\dfrac{0.41}{1.69} = \mathbf{75.96°}$

---

**Now try the following exercise**

**Exercise 122    A further problem on the equivalent circuit of a transformer**

1.  A transformer has 1200 primary turns and 200 secondary turns. The primary and secondary resistances are $0.2\,\Omega$ and $0.02\,\Omega$ respectively and the corresponding leakage reactances are $1.2\,\Omega$ and $0.05\,\Omega$ respectively. Calculate

(a)  the equivalent resistance, reactance and impedance referred to the primary winding, and (b) the phase angle of the impedance.
    [(a) $0.92\,\Omega$, $3.0\,\Omega$, $3.14\,\Omega$ (b) $72.95°$]

## 21.8    Regulation of a transformer

When the secondary of a transformer is loaded, the secondary terminal voltage, $V_2$, falls. As the power factor decreases, this voltage drop increases. This is called the **regulation of the transformer** and it is usually expressed as a percentage of the secondary no-load voltage, $E_2$. For full-load conditions:

$$\mathbf{Regulation} = \left(\frac{E_2 - V_2}{E_2}\right) \times 100\% \qquad (10)$$

The fall in voltage, $(E_2 - V_2)$, is caused by the resistance and reactance of the windings. Typical values of voltage regulation are about 3% in small transformers and about 1% in large transformers.

**Problem 15.**    A 5 kVA, 200 V/400 V, single-phase transformer has a secondary terminal voltage of 387.6 volts when loaded. Determine the regulation of the transformer.

From equation (10):

$$\text{regulation} = \left(\frac{\begin{array}{c}\text{No-load secondary voltage} - \\ \text{terminal voltage on load}\end{array}}{\text{no-load secondary voltage}}\right) \times 100\%$$

$$= \left(\frac{400 - 387.6}{400}\right) \times 100\%$$

$$= \left(\frac{12.4}{400}\right) \times 100\%$$

$$= \mathbf{3.1\%}$$

**Problem 16.**   The open-circuit voltage of a transformer is 240 V. A tap-changing device is set to operate when the percentage regulation drops below 2.5%. Determine the load voltage at which the mechanism operates.

$$\text{Regulation} = \left(\frac{\begin{array}{c}\text{No-load secondary voltage} - \\ \text{terminal voltage on load}\end{array}}{\text{no-load secondary voltage}}\right) \times 100\%$$

Hence

$$2.5 = \left(\frac{240 - V_2}{240}\right) \times 100\%$$

$\therefore$

$$\frac{(2.5)(240)}{100} = 240 - V_2$$

i.e.

$$6 = 240 - V_2$$

from which, **load voltage, $V_2 = 240 - 6 = 234$ volts**

---

**Now try the following exercise**

**Exercise 123   Further problems on regulation**

1.  A 6 kVA, 100 V/500 V, single-phase transformer has a secondary terminal voltage of 487.5 volts when loaded. Determine the regulation of the transformer.   [2.5%]

2.  A transformer has an open-circuit voltage of 110 volts. A tap-changing device operates when the regulation falls below 3%. Calculate the load voltage at which the tap-changer operates.   [106.7 volts]

## 21.9   Transformer losses and efficiency

There are broadly two sources of **losses in transformers** on load, these being copper losses and iron losses.

(a)  **Copper losses** are variable and result in a heating of the conductors, due to the fact that they possess resistance. If $R_1$ and $R_2$ are the primary and secondary winding resistances then the total copper loss is $I_1^2 R_1 + I_2^2 R_2$

(b)  **Iron losses** are constant for a given value of frequency and flux density and are of two types – hysteresis loss and eddy current loss.

(i)  **Hysteresis loss** is the heating of the core as a result of the internal molecular structure reversals which occur as the magnetic flux alternates. The loss is proportional to the area of the hysteresis loop and thus low loss nickel iron alloys are used for the core since their hysteresis loops have small areas. (See Chapter 7)

(ii)  **Eddy current loss** is the heating of the core due to e.m.f.'s being induced not only in the transformer windings but also in the core. These induced e.m.f.'s set up circulating currents, called eddy currents. Owing to the low resistance of the core, eddy currents can be quite considerable and can cause a large power loss and excessive heating of the core. Eddy current losses can be reduced by increasing the resistivity of the core material or, more usually, by laminating the core (i.e. splitting it into layers or leaves) when very thin layers of insulating material can be inserted between each pair of laminations. This increases the resistance of the eddy current path, and reduces the value of the eddy current.

**Transformer efficiency,**

$$\eta = \frac{\text{output power}}{\text{input power}} = \frac{\text{input power} - \text{losses}}{\text{input power}}$$

i.e.

$$\eta = 1 - \frac{\text{losses}}{\text{input power}} \qquad (11)$$

and is usually expressed as a percentage. It is not uncommon for power transformers to have efficiencies of between 95% and 98%

**Output power** $= V_2 I_2 \cos\phi_2$
**Total losses** = copper loss + iron losses,
and input power = output power + losses

**Problem 17.**   A 200 kVA rated transformer has a full-load copper loss of 1.5 kW and an iron loss of 1 kW. Determine the transformer efficiency at full load and 0.85 power factor.

Efficiency, $\eta = \dfrac{\text{output power}}{\text{input power}}$

$= \dfrac{\text{input power} - \text{losses}}{\text{input power}}$

$= 1 - \dfrac{\text{losses}}{\text{input power}}$

Full-load output power $= VI\cos\phi = (200)(0.85)$
$= 170\,\text{kW}.$
Total losses $= 1.5 + 1.0 = 2.5\,\text{kW}$
Input power $=$ output power $+$ losses
$= 170 + 2.5 = 172.5\,\text{kW}.$

Hence efficiency $= \left(1 - \dfrac{2.5}{172.5}\right) = 1 - 0.01449$
$= 0.9855$ or **98.55%**

**Problem 18.** Determine the efficiency of the transformer in Problem 17 at half full load and 0.85 power factor.

Half full-load power output $= (1/2)(200)(0.85) = 85\,\text{kW}.$
Copper loss (or $I^2R$ loss) is proportional to current squared. Hence the copper loss at half full load is:
$\left(\frac{1}{2}\right)^2 (1500) = 375\,\text{W}$
Iron loss $= 1000\,\text{W}$ (constant)
Total losses $= 375 + 1000 = 1375\,\text{W}$ or $1.375\,\text{kW}.$
Input power at half full load
$=$ output power at half full load $+$ losses
$= 85 + 1.375 = 86.375\,\text{kW}.$
Hence

efficiency $= 1 - \dfrac{\text{losses}}{\text{input power}}$

$= \left(1 - \dfrac{1.375}{86.375}\right)$

$= 1 - 0.01592$

$= 0.9841$ or **98.41%**

**Problem 19.** A 400 kVA transformer has a primary winding resistance of 0.5 Ω and a secondary winding resistance of 0.001 Ω. The iron loss is 2.5 kW and the primary and secondary voltages are 5 kV and 320 V respectively. If the power factor of the load is 0.85, determine the efficiency of the transformer (a) on full load, and (b) on half load.

(a) Rating $= 400\,\text{kVA} = V_1 I_1 = V_2 I_2$. Hence primary current,

$$I_1 = \frac{400 \times 10^3}{V_1} = \frac{400 \times 10^3}{5000} = 80\,\text{A}$$

and secondary current,

$$I_2 = \frac{400 \times 10^3}{V_2} = \frac{400 \times 10^3}{320} = 1250\,\text{A}$$

Total copper loss $= I_1^2 R_1 + I_2^2 R_2,$
(where $R_1 = 0.5\,\Omega$ and $R_2 = 0.001\,\Omega$)

$= (80)^2 (0.5) + (1250)^2 (0.001)$

$= 3200 + 1562.5 = 4762.5$ watts

On full load, total loss $=$ copper loss $+$ iron loss

$= 4762.5 + 2500 = 7262.5\,\text{W} = 7.2625\,\text{kW}$

Total output power on full load

$= V_2 I_2 \cos\phi_2 = (400 \times 10^3)(0.85) = 340\,\text{kW}$

Input power $=$ output power $+$ losses
$= 340\,\text{kW} + 7.2625\,\text{kW}$
$= 347.2625\,\text{kW}$

Efficiency, $\eta = \left(1 - \dfrac{\text{losses}}{\text{input power}}\right) \times 100\%$

$= \left(1 - \dfrac{7.2625}{347.2625}\right) \times 100\%$

$= $ **97.91%**

(b) Since the copper loss varies as the square of the current, then total copper loss on half load $= 4762.5 \times \left(\frac{1}{2}\right)^2 = 1190.625\,\text{W}$. Hence total loss on half load $= 1190.625 + 2500 = 3690.625\,\text{W}$ or $3.691\,\text{kW}$.
Output power on half full load $= \left(\frac{1}{2}\right)(340)$
$= 170\,\text{kW}.$
Input power on half full load
$=$ output power $+$ losses
$= 170\,\text{kW} + 3.691\,\text{kW}$
$= 173.691\,\text{kW}$

Hence efficiency at half full load,

$$\eta = \left(1 - \frac{\text{losses}}{\text{input power}}\right) \times 100\%$$

$$= \left(1 - \frac{3.691}{173.691}\right) \times 100\% = \mathbf{97.87\%}$$

## Maximum efficiency

It may be shown that the efficiency of a transformer is a maximum when the variable copper loss (i.e. $I_1^2 R_1 + I_2^2 R_2$) is equal to the constant iron losses.

> **Problem 20.** A 500 kVA transformer has a full-load copper loss of 4 kW and an iron loss of 2.5 kW. Determine (a) the output kVA at which the efficiency of the transformer is a maximum, and (b) the maximum efficiency, assuming the power factor of the load is 0.75.

(a) Let $x$ be the fraction of full load kVA at which the efficiency is a maximum. The corresponding total copper loss $= (4\,\text{kW})(x^2)$. At maximum efficiency, copper loss = iron loss. Hence $4x^2 = 2.5$ from which $x^2 = 2.5/4$ and $x = \sqrt{2.5/4} = 0.791$. Hence **the output kVA at maximum efficiency**
$$= 0.791 \times 500 = \mathbf{395.5\,kVA}.$$

(b) Total loss at maximum efficiency
$$= 2 \times 2.5 = 5\,\text{kW}$$
Output power $= 395.5\,\text{kVA} \times \text{p.f.}$
$$= 395.5 \times 0.75 = 296.625\,\text{kW}$$
Input power = output power + losses
$$= 296.625 + 5 = 301.625\,\text{kW}$$

---

**Maximum efficiency,**

$$\eta = \left(1 - \frac{\text{losses}}{\text{input power}}\right) \times 100\%$$

$$= \left(1 - \frac{5}{301.625}\right) \times 100\% = \mathbf{98.34\%}$$

---

**Now try the following exercise**

> **Exercise 124 Further problems on losses and efficiency**
>
> 1. A single-phase transformer has a voltage ratio of 6:1 and the h.v. winding is supplied at 540 V.

The secondary winding provides a full load current of 30 A at a power factor of 0.8 lagging. Neglecting losses, find (a) the rating of the transformer, (b) the power supplied to the load, (c) the primary current.
$$\text{[(a) 2.7\,kVA (b) 2.16\,kW (c) 5\,A]}$$

2. A single-phase transformer is rated at 40 kVA. The transformer has full-load copper losses of 800 W and iron losses of 500 W. Determine the transformer efficiency at full load and 0.8 power factor. [96.10%]

3. Determine the efficiency of the transformer in problem 2 at half full load and 0.8 power factor. [95.81%]

4. A 100 kVA, 2000 V/400 V, 50 Hz, single-phase transformer has an iron loss of 600 W and a full-load copper loss of 1600 W. Calculate its efficiency for a load of 60 kW at 0.8 power factor. [97.56%]

5. Determine the efficiency of a 15 kVA transformer for the following conditions:
   (i) full-load, unity power factor
   (ii) 0.8 full-load, unity power factor
   (iii) half full-load, 0.8 power factor
   Assume that iron losses are 200 W and the full-load copper loss is 300 W.
   [(a) 96.77% (ii) 96.84% (iii) 95.62%]

6. A 300 kVA transformer has a primary winding resistance of 0.4 Ω and a secondary winding resistance of 0.0015 Ω. The iron loss is 2 kW and the primary and secondary voltages are 4 kV and 200 V respectively. If the power factor of the load is 0.78, determine the efficiency of the transformer (a) on full load, and (b) on half load. [(a) 96.84% (b) 97.17%]

7. A 250 kVA transformer has a full-load copper loss of 3 kW and an iron loss of 2 kW. Calculate (a) the output kVA at which the efficiency of the transformer is a maximum, and (b) the maximum efficiency, assuming the power factor of the load is 0.80.
$$\text{[(a) 204.1\,kVA (b) 97.61\%]}$$

## 21.10 Resistance matching

Varying a load resistance to be equal, or almost equal, to the source internal resistance is called **matching**. Examples where resistance matching is important include coupling an aerial to a transmitter or receiver, or in coupling a loudspeaker to an amplifier, where coupling transformers may be used to give maximum power transfer.

With d.c. generators or secondary cells, the internal resistance is usually very small. In such cases, if an attempt is made to make the load resistance as small as the source internal resistance, overloading of the source results.

A method of achieving maximum power transfer between a source and a load (see Section 13.9, page 204), is to adjust the value of the load resistance to 'match' the source internal resistance. A transformer may be used as a **resistance matching device** by connecting it between the load and the source.

The reason why a transformer can be used for this is shown below. With reference to Fig. 21.10:

$$R_{\mathrm{L}} = \frac{V_2}{I_2} \quad \text{and} \quad R_1 = \frac{V_1}{I_1}$$

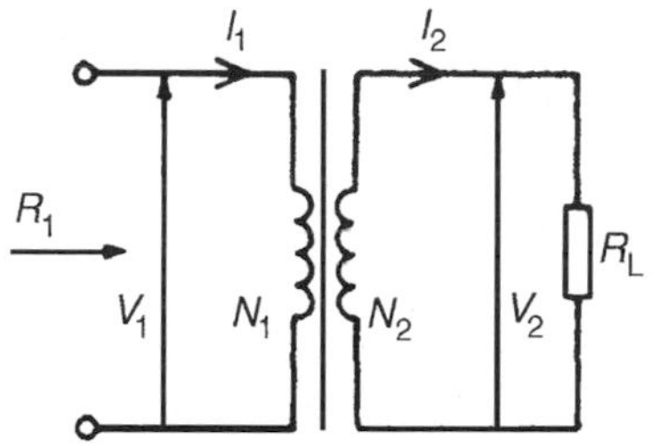

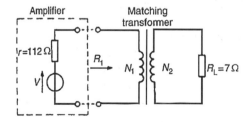

**Figure 21.10**

For an ideal transformer,

$$V_1 = \left(\frac{N_1}{N_2}\right) V_2$$

and

$$I_1 = \left(\frac{N_2}{N_1}\right) I_2$$

Thus the equivalent input resistance $R_1$ of the transformer is given by:

$$R_1 = \frac{V_1}{I_1} = \frac{\left(\dfrac{N_1}{N_2}\right) V_2}{\left(\dfrac{N_2}{N_1}\right) I_2}$$

$$= \left(\frac{N_1}{N_2}\right)^2 \left(\frac{V_2}{I_2}\right) = \left(\frac{N_1}{N_2}\right)^2 R_{\mathrm{L}}$$

i.e.

$$R_1 = \left(\frac{N_1}{N_2}\right)^2 R_{\mathrm{L}}$$

Hence by varying the value of the turns ratio, the equivalent input resistance of a transformer can be 'matched' to the internal resistance of a load to achieve maximum power transfer.

**Problem 21.** A transformer having a turns ratio of 4:1 supplies a load of resistance $100\,\Omega$. Determine the equivalent input resistance of the transformer.

From above, the equivalent input resistance,

$$R_1 = \left(\frac{N_1}{N_2}\right)^2 R_{\mathrm{L}}$$

$$= \left(\frac{4}{1}\right)^2 (100) = \mathbf{1600\,\Omega}$$

**Problem 22.** The output stage of an amplifier has an output resistance of $112\,\Omega$. Calculate the optimum turns ratio of a transformer which would match a load resistance of $7\,\Omega$ to the output resistance of the amplifier.

The circuit is shown in Fig. 21.11.

**Figure 21.11**

The equivalent input resistance, $R_1$ of the transformer needs to be $112\,\Omega$ for maximum power transfer.

$$R_1 = \left(\frac{N_1}{N_2}\right)^2 R_{\mathrm{L}}$$

Hence

$$\left(\frac{N_1}{N_2}\right)^2 = \frac{R_1}{R_{\mathrm{L}}} = \frac{112}{7} = 16$$

i.e.

$$\frac{N_1}{N_2} = \sqrt{16} = 4$$

**Hence the optimum turns ratio is 4:1**

Section 3

**Problem 23.**   Determine the optimum value of load resistance for maximum power transfer if the load is connected to an amplifier of output resistance $150\,\Omega$ through a transformer with a turns ratio of 5:1

The equivalent input resistance $R_1$ of the transformer needs to be $150\,\Omega$ for maximum power transfer.

$$R_1 = \left(\frac{N_1}{N_2}\right)^2 R_L$$

from which,   $R_L = R_1 \left(\frac{N_2}{N_1}\right)^2$

$$= 150 \left(\tfrac{1}{5}\right)^2 = 6\,\Omega$$

**Problem 24.**   A single-phase, $220\,V/1760\,V$ ideal transformer is supplied from a $220\,V$ source through a cable of resistance $2\,\Omega$. If the load across the secondary winding is $1.28\,k\Omega$ determine (a) the primary current flowing and (b) the power dissipated in the load resistor.

The circuit diagram is shown in Fig. 21.12

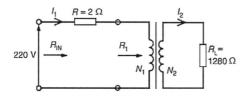

**Figure 21.12**

(a)   Turns ratio

$$\left(\frac{N_1}{N_2}\right) = \left(\frac{V_1}{V_2}\right) = \left(\frac{220}{1760}\right) = \left(\frac{1}{8}\right)$$

Equivalent input resistance of the transformer.

$$R_1 = \left(\frac{N_1}{N_2}\right)^2 R_L = \left(\frac{1}{8}\right)^2 (1.28 \times 10^3) = 20\,\Omega$$

Total input resistance,

$$R_{IN} = R + R_1 = 2 + 20 = 22\,\Omega$$

Primary current,

$$I_1 = \frac{V_1}{R_{IN}} = \frac{220}{22} = 10\,A$$

(b)   For an ideal transformer

$$\frac{V_1}{V_2} = \frac{I_2}{I_1}$$

from which,

$$I_2 = I_1 \left(\frac{V_1}{V_2}\right) = 10\left(\frac{220}{1760}\right) = 1.25\,A$$

Power dissipated in load resistor $R_L$,

$$\mathbf{P} = I_2^2 R_L = (1.25)^2 (1.28 \times 10^3)$$

$$= \mathbf{2000\ watts\ or\ 2\,kW}$$

**Problem 25.**   An a.c. source of $24\,V$ and internal resistance $15\,k\Omega$ is matched to a load by a 25:1 ideal transformer. Determine (a) the value of the load resistance and (b) the power dissipated in the load.

The circuit diagram is shown in Fig. 21.13

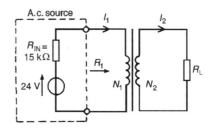

**Figure 21.13**

(a)   For maximum power transfer $R_1$ needs to be equal to $15\,k\Omega$.

$$R_1 = \left(\frac{N_1}{N_2}\right)^2 R_L$$

from which, load resistance,

$$R_L = R_1 \left(\frac{N_2}{N_1}\right)^2 = (15\,000)\left(\frac{1}{25}\right)^2 = \mathbf{24\,\Omega}$$

(b)   The total input resistance when the source is connected to the matching transformer is $R_{IN} + R_1$ i.e. $15\,k\Omega + 15\,k\Omega = 30\,k\Omega$.

Primary current,

$$I_1 = \frac{V}{30\,000} = \frac{24}{30\,000} = 0.8\,mA$$

$N_1/N_2 = I_2/I_1$   from   which,   $I_2 = I_1(N_1/N_2) = (0.8 \times 10^{-3})(25/1) = 20 \times 10^{-3}\,A$.

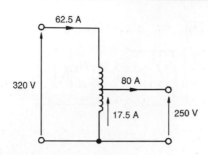

**Figure 21.16**

The volume, and hence weight, of copper required in a winding is proportional to the number of turns and to the cross-sectional area of the wire. In turn this is proportional to the current to be carried, i.e. volume of copper is proportional to $NI$.

Volume of copper in an auto transformer

$$\propto (N_1 - N_2)I_1 + N_2(I_2 - I_1)$$

$$\text{see Fig. 21.14(b)}$$

$$\propto N_1 I_1 - N_2 I_1 + N_2 I_2 - N_2 I_1$$

$$\propto N_1 I_1 + N_2 I_2 - 2N_2 I_1$$

$$\propto 2N_1 I_1 - 2N_2 I_1 \quad \text{(since } N_2 I_2 = N_1 I_1\text{)}$$

Volume of copper in a double-wound transformer

$$\propto N_1 I_1 + N_2 I_2 \propto 2N_1 I_1$$

(again, since $N_2 I_2 = N_1 I_1$). Hence

$$\frac{\begin{array}{c}\text{volume of copper in}\\ \text{an auto transformer}\end{array}}{\begin{array}{c}\text{volume of copper in a}\\ \text{double-wound transformer}\end{array}} = \frac{2N_1 I_1 - 2N_2 I_1}{2N_1 I_1}$$

$$= \frac{2N_1 I_1}{2N_1 I_1} - \frac{2N_2 I_1}{2N_1 I_1}$$

$$= 1 - \frac{N_2}{N_1}$$

If $(N_2/N_1) = x$ then

**(volume of copper in an auto transformer)**
**$= (1 - x)$ (volume of copper in a double-wound transformer)**    (12)

If, say, $x = (4/5)$ then (volume of copper in auto transformer)

$$= \left(1 - \tfrac{4}{5}\right) \begin{array}{l}\text{(volume of copper in a}\\ \text{double-wound transformer)}\end{array}$$

$$= \tfrac{1}{5} \text{ (volume in double-wound transformer)}$$

i.e. a saving of 80%.

Similarly, if $x = (1/4)$, the saving is 25 per cent, and so on. The closer $N_2$ is to $N_1$, the greater the saving in copper.

**Problem 27.** Determine the saving in the volume of copper used in an auto transformer compared with a double-wound transformer for (a) a 200 V:150 V transformer, and (b) a 500 V:100 V transformer.

(a) For a 200 V:150 V transformer,

$$x = \frac{V_2}{V_1} = \frac{150}{200} = 0.75$$

Hence from equation (12), (volume of copper in auto transformer)

$$= (1 - 0.75) \text{ (volume of copper in double-wound transformer)}$$

$$= (0.25) \text{ (volume of copper in double-wound transformer)}$$

$$= 25\% \text{ (of copper in a double-wound transformer)}$$

**Hence the saving is 75%**

(b) For a 500 V:100 V transformer,

$$x = \frac{V_2}{V_1} = \frac{100}{500} = 0.2$$

Hence, (volume of copper in auto transformer)

$$= (1 - 0.2) \text{ (volume of copper in double-wound transformer)}$$

$$= (0.8) \text{ (volume in double-wound transformer)}$$

$$= 80\% \text{ of copper in a double-wound transformer}$$

**Hence the saving is 20%.**

**Now try the following exercise**

**Exercise 126   Further problems on the auto transformer**

1. A single-phase auto transformer has a voltage ratio of 480 V:300 V and supplies a load of 30 kVA at 300 V. Assuming an ideal

transformer, calculate the current in each section of the winding.
$$[I_1 = 62.5\,\text{A},\ I_2 = 100\,\text{A},\ (I_2 - I_1) = 37.5\,\text{A}]$$

2.  Calculate the saving in the volume of copper used in an auto transformer compared with a double-wound transformer for (a) a 300 V: 240 V transformer, and (b) a 400 V:100 V transformer.                   [(a) 80% (b) 25%]

## Advantages of auto transformers

The advantages of auto transformers over double-wound transformers include:

1.  a saving in cost since less copper is needed (see above)

2.  less volume, hence less weight

3.  a higher efficiency, resulting from lower $I^2R$ losses

4.  a continuously variable output voltage is achievable if a sliding contact is used

5.  a smaller percentage voltage regulation.

## Disadvantages of auto transformers

The primary and secondary windings are not electrically separate, hence if an open-circuit occurs in the secondary winding the full primary voltage appears across the secondary.

## Uses of auto transformers

Auto transformers are used for reducing the voltage when starting induction motors (see Chapter 23) and for interconnecting systems that are operating at approximately the same voltage.

## 21.12    Isolating transformers

Transformers not only enable current or voltage to be transformed to some different magnitude but provide a means of isolating electrically one part of a circuit from another when there is no electrical connection between primary and secondary windings. An **isolating transformer** is a 1:1 ratio transformer with several important applications, including bathroom shaver-sockets, portable electric tools, model railways, and so on.

## 21.13    Three-phase transformers

Three-phase double-wound transformers are mainly used in power transmission and are usually of the core type. They basically consist of three pairs of single-phase windings mounted on one core, as shown in Fig. 21.17, which gives a considerable saving in the amount of iron used. The primary and secondary windings in Fig. 21.17 are wound on top of each other in the form of concentric cylinders, similar to that shown in Fig. 21.6(a). The windings may be with the primary delta-connected and the secondary star-connected, or star-delta, star-star or delta-delta, depending on its use.

Section 3

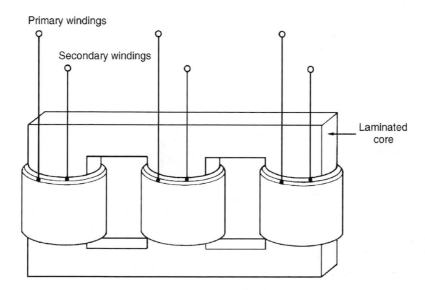

Primary windings

Secondary windings

Laminated core

**Figure 21.17**

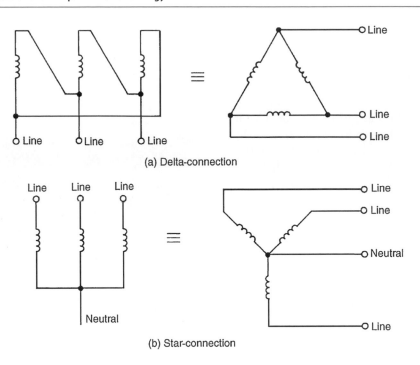

**Figure 21.18**

A delta-connection is shown in Fig. 21.18(a) and a star-connection in Fig. 21.18(b).

---

**Problem 28.** A three-phase transformer has 500 primary turns and 50 secondary turns. If the supply voltage is 2.4 kV find the secondary line voltage on no-load when the windings are connected (a) star-delta, (b) delta-star.

---

(a) For a star-connection, $V_L = \sqrt{3}\,V_p$ (see Chapter 20). Primary phase voltage,

$$V_p = \frac{V_{L1}}{\sqrt{3}} = \frac{2400}{\sqrt{3}} = 1385.64 \text{ volts}.$$

For a delta-connection, $V_L = V_p$
$N_1/N_2 = V_1/V_2$ from which, secondary phase voltage,

$$V_{p2} = V_{p1}\left(\frac{N_2}{N_1}\right) = (1385.64)\left(\frac{50}{500}\right)$$

$$= 138.6 \text{ volts}$$

(b) For a delta-connection, $V_L = V_p$ hence, primary phase voltage $V_{p1} = 2.4 \text{ kV} = 2400$ volts. Secondary phase voltage,

$$V_{p2} = V_{p1}\left(\frac{N_2}{N_1}\right) = (2400)\left(\frac{50}{500}\right) = 240 \text{ volts}$$

For a star-connection, $V_L = \sqrt{3}\,V_p$ hence, the secondary line voltage, $V_{L2} = \sqrt{3}(240) = \textbf{416 volts}$.

---

**Now try the following exercise**

---

**Exercise 127   A further problem on the three-phase transformer**

1.  A three-phase transformer has 600 primary turns and 150 secondary turns. If the supply voltage is 1.5 kV determine the secondary line voltage on no-load when the windings are connected (a) delta-star (b) star-delta.

[(a) 649.5 V (b) 216.5 V]

---

## 21.14   Current transformers

For measuring currents in excess of about 100 A a current transformer is normally used. With a d.c. moving-coil ammeter the current required to give full-scale deflection is very small – typically a few milliamperes. When larger currents are to be measured a shunt resistor is added to the circuit (see Chapter 10). However, even with shunt resistors added it is not possible to measure very large currents. When a.c. is being measured a shunt

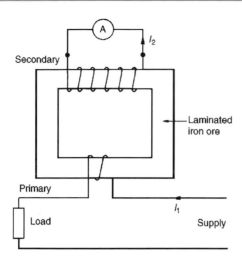

Figure 21.19

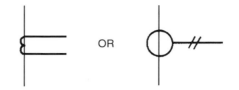

Figure 21.20

cannot be used since the proportion of the current which flows in the meter will depend on its impedance, which varies with frequency.

In a double-wound transformer:

$$\frac{I_1}{I_2} = \frac{N_2}{N_1}$$

from which,

**secondary current** $I_2 = I_1 \left( \dfrac{N_2}{N_1} \right)$

In current transformers the primary usually consists of one or two turns whilst the secondary can have several hundred turns. A typical arrangement is shown in Fig. 21.19.

If, for example, the primary has 2 turns and the secondary 200 turns, then if the primary current is 500 A,

$$\text{secondary current, } I_2 = I_1 \left( \frac{N_2}{N_1} \right) = (500) \left( \frac{2}{200} \right)$$

$$= 5\,\text{A}$$

Current transformers isolate the ammeter from the main circuit and allow the use of a standard range of ammeters giving full-scale deflections of 1 A, 2 A or 5 A.

For very large currents the transformer core can be mounted around the conductor or bus-bar. Thus the primary then has just one turn.

It is very important to short-circuit the secondary winding before removing the ammeter. This is because if current is flowing in the primary, dangerously high voltages could be induced in the secondary should it be open-circuited.

Current transformer circuit diagram symbols are shown in Fig. 21.20.

**Problem 29.** A current transformer has a single turn on the primary winding and a secondary winding of 60 turns. The secondary winding is connected to an ammeter with a resistance of $0.15\,\Omega$. The resistance of the secondary winding is $0.25\,\Omega$. If the current in the primary winding is 300 A, determine (a) the reading on the ammeter, (b) the potential difference across the ammeter and (c) the total load (in VA) on the secondary.

(a) Reading on the ammeter,

$$I_2 = I_1 \left( \frac{N_1}{N_2} \right) = 300 \left( \frac{1}{60} \right) = \textbf{5 A}.$$

(b) P.d. across the ammeter $= I_2 R_A$, (where $R_A$ is the ammeter resistance) $= (5)(0.15) = \textbf{0.75 volts}$.

(c) Total resistance of secondary circuit

$$= 0.15 + 0.25 = 0.40\,\Omega.$$

Induced e.m.f. in secondary $= (5)(0.40) = 2.0\,\text{V}$.
Total load on secondary $= (2.0)(5) = \textbf{10 VA}$.

**Now try the following exercise**

**Exercise 128  A further problem on the current transformer**

1. A current transformer has two turns on the primary winding and a secondary winding of 260 turns. The secondary winding is connected to an ammeter with a resistance of $0.2\,\Omega$. The resistance of the secondary winding is $0.3\,\Omega$. If the current in the primary winding is 650 A, determine (a) the reading on the ammeter, (b) the potential difference across the ammeter, and (c) the total load in VA on the secondary.

[(a) 5 A (b) 1 V (c) 7.5 VA]

Section 3

## 21.15   Voltage transformers

For measuring voltages in excess of about 500 V it is often safer to use a voltage transformer. These are normal double-wound transformers with a large number of turns on the primary, which is connected to a high voltage supply, and a small number of turns on the secondary. A typical arrangement is shown in Fig. 21.21.

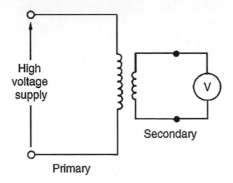

**Figure 21.21**

Since

$$\frac{V_1}{V_2} = \frac{N_1}{N_2}$$

the **secondary voltage**,

$$V_2 = \frac{V_1 N_2}{V_1}$$

Thus if the arrangement in Fig. 21.21 has 4000 primary turns and 20 secondary turns then for a voltage of 22 kV on the primary, the voltage on the secondary,

$$V_2 = V_1 \left(\frac{N_2}{N_1}\right) = (22\,000)\left(\frac{20}{4000}\right) = \textbf{110 volts}$$

**Now try the following exercises**

### Exercise 129   Short answer questions on transformers

1.  What is a transformer?

2.  Explain briefly how a voltage is induced in the secondary winding of a transformer

3.  Draw the circuit diagram symbol for a transformer

4.  State the relationship between turns and voltage ratios for a transformer

5.  How is a transformer rated?

6.  Briefly describe the principle of operation of a transformer

7.  Draw a phasor diagram for an ideal transformer on no-load

8.  State the e.m.f. equation for a transformer

9.  Draw an on-load phasor diagram for an ideal transformer with an inductive load

10.  Name two types of transformer construction

11.  What core material is normally used for power transformers

12.  Name three core materials used in r.f. transformers

13.  State a typical application for (a) a.f. transformers (b) r.f. transformers

14.  How is cooling achieved in transformers?

15.  State the expressions for equivalent resistance and reactance of a transformer, referred to the primary

16.  Define regulation of a transformer

17.  Name two sources of loss in a transformer

18.  What is hysteresis loss? How is it minimised in a transformer?

19.  What are eddy currents? How may they be reduced in transformers?

20.  How is efficiency of a transformer calculated?

21.  What is the condition for maximum efficiency of a transformer?

22.  What does 'resistance matching' mean?

23.  State a practical application where matching would be used

24.  Derive a formula for the equivalent resistance of a transformer having a turns ratio of $N_1{:}N_2$ and load resistance $R_L$

25.  What is an auto transformer?

26.  State three advantages and one disadvantage of an auto transformer compared with a double-wound transformer

27. In what applications are auto transformers used?

28. What is an isolating transformer? Give two applications

29. Describe briefly the construction of a three-phase transformer

30. For what reason are current transformers used?

31. Describe how a current transformer operates

32. For what reason are voltage transformers used?

33. Describe how a voltage transformer operates

**Exercise 130 Multi-choice questions on transformers**

**(Answers on page 421)**

1. The e.m.f. equation of a transformer of secondary turns $N_2$, magnetic flux density $B_{\mathrm{m}}$, magnetic area of core a, and operating at frequency $f$ is given by:

   (a) $E_2 = 4.44 N_2 B_{\mathrm{m}} a f$ volts

   (b) $E_2 = 4.44 \dfrac{N_2 B_{\mathrm{m}} f}{a}$ volts

   (c) $E_2 = \dfrac{N_2 B_{\mathrm{m}} f}{a}$ volts

   (d) $E_2 = 1.11 N_2 B_{\mathrm{m}} a f$ volts

2. In the auto-transformer shown in Fig. 21.22, the current in section PQ is:
   (a) 3.3 A (b) 1.7 A (c) 5 A (d) 1.6 A

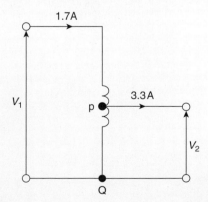

**Figure 21.22**

3. A step-up transformer has a turns ratio of 10. If the output current is 5 A, the input current is:
   (a) 50 A          (b) 5 A
   (c) 2.5 A        (d) 0.5 A

4. A 440 V/110 V transformer has 1000 turns on the primary winding. The number of turns on the secondary is:
   (a) 550          (b) 250
   (c) 4000       (d) 25

5. An advantage of an auto-transformer is that:
   (a) it gives a high step-up ratio
   (b) iron losses are reduced
   (c) copper loss is reduced
   (d) it reduces capacitance between turns

6. A 1 kV/250 V transformer has 500 turns on the secondary winding. The number of turns on the primary is:
   (a) 2000       (b) 125
   (c) 1000       (d) 250

7. The core of a transformer is laminated to:
   (a) limit hysteresis loss
   (b) reduce the inductance of the windings
   (c) reduce the effects of eddy current loss
   (d) prevent eddy currents from occurring

8. The power input to a mains transformer is 200 W. If the primary current is 2.5 A, the secondary voltage is 2 V and assuming no losses in the transformer, the turns ratio is:
   (a) 40:1 step down    (b) 40:1 step up
   (c) 80:1 step down    (d) 80:1 step up

9. A transformer has 800 primary turns and 100 secondary turns. To obtain 40 V from the secondary winding the voltage applied to the primary winding must be:
   (a) 5 V        (b) 320 V
   (c) 2.5 V     (d) 20 V

A 100 kVA, 250 V/10 kV, single-phase transformer has a full-load copper loss of 800 W and an iron loss of 500 W. The primary winding contains 120 turns. For the statements in questions 10 to 16, select the correct answer from the following list:

| | | |
|---|---|---|
| (a) 81.3 kW | (b) 800 W | (c) 97.32% |
| (d) 80 kW | (e) 3 | (f) 4800 |
| (g) 1.3 kW | (h) 98.40% | (i) 100 kW |
| (j) 98.28% | (k) 200 W | (l) 101.3 kW |
| (m) 96.38% | (n) 400 W | |

10. The total full-load losses

11. The full-load output power at 0.8 power factor

12. The full-load input power at 0.8 power factor

13. The full-load efficiency at 0.8 power factor

14. The half full-load copper loss

15. The transformer efficiency at half full load, 0.8 power factor

16. The number of secondary winding turns

17. Which of the following statements is false?
    (a) In an ideal transformer, the volts per turn are constant for a given value of primary voltage
    (b) In a single-phase transformer, the hysteresis loss is proportional to frequency
    (c) A transformer whose secondary current is greater than the primary current is a step-up transformer
    (d) In transformers, eddy current loss is reduced by laminating the core

18. An ideal transformer has a turns ratio of 1:5 and is supplied at 200 V when the primary current is 3 A. Which of the following statements is false?
    (a) The turns ratio indicates a step-up transformer
    (b) The secondary voltage is 40 V
    (c) The secondary current is 15 A
    (d) The transformer rating is 0.6 kVA
    (e) The secondary voltage is 1 kV
    (f) The secondary current is 0.6 A

19. Iron losses in a transformer are due to:
    (a) eddy currents only
    (b) flux leakage
    (c) both eddy current and hysteresis losses
    (d) the resistance of the primary and secondary windings

20. A load is to be matched to an amplifier having an effective internal resistance of 10 Ω via a coupling transformer having a turns ratio of 1:10. The value of the load resistance for maximum power transfer is:
    (a) 100 Ω     (b) 1 kΩ
    (c) 100 mΩ    (d) 1 mΩ

## Revision Test 6

This revision test covers the material contained in Chapters 20 to 21. *The marks for each question are shown in brackets at the end of each question.*

1.  Three identical coils each of resistance $40\,\Omega$ and inductive reactance $30\,\Omega$ are connected (i) in star, and (ii) in delta to a 400 V, three-phase supply. Calculate for each connection (a) the line and phase voltages, (b) the phase and line currents, and (c) the total power dissipated. (12)

2.  Two wattmeters are connected to measure the input power to a balanced three-phase load by the two-wattmeter method. If the instrument readings are 10 kW and 6 kW, determine (a) the total power input, and (b) the load power factor. (5)

3.  An ideal transformer connected to a 250 V mains, supplies a 25 V, 200 W lamp. Calculate the transformer turns ratio and the current taken from the supply. (5)

4.  A 200 kVA, 8000 V/320 V, 50 Hz single-phase transformer has 120 secondary turns. Determine (a) the primary and secondary currents, (b) the number of primary turns, and (c) the maximum value of flux. (9)

5.  Determine the percentage regulation of an 8 kVA, 100 V/200 V, single-phase transformer when its secondary terminal voltage is 194 V when loaded. (3)

6.  A 500 kVA rated transformer has a full-load copper loss of 4 kW and an iron loss of 3 kW. Determine the transformer efficiency (a) at full load and 0.80 power factor, and (b) at half full load and 0.80 power factor. (10)

7.  Determine the optimum value of load resistance for maximum power transfer if the load is connected to an amplifier of output resistance $288\,\Omega$ through a transformer with a turns ratio 6:1 (3)

8.  A single-phase auto transformer has a voltage ratio of 250 V:200 V and supplies a load of 15 kVA at 200 V. Assuming an ideal transformer, determine the current in each section of the winding. (3)

# Chapter 22

# D.C. machines

At the end of this chapter you should be able to:

- distinguish between the function of a motor and a generator
- describe the action of a commutator
- describe the construction of a d.c. machine
- distinguish between wave and lap windings
- understand shunt, series and compound windings of d.c. machines
- understand armature reaction
- calculate generated e.m.f. in an armature winding using $E = 2p\Phi nZ/c$
- describe types of d.c. generator and their characteristics
- calculate generated e.m.f. for a generator using $E = V + I_a R_a$
- state typical applications of d.c. generators
- list d.c. machine losses and calculate efficiency
- calculate back e.m.f. for a d.c. motor using $E = V - I_a R_a$
- calculate the torque of a d.c. motor using $T = EI_a/2\pi n$ and $T = p\Phi ZI_a/\pi c$
- describe types of d.c. motor and their characteristics
- state typical applications of d.c. motors
- describe a d.c. motor starter
- describe methods of speed control of d.c. motors
- list types of enclosure for d.c. motors

## 22.1 Introduction

When the input to an electrical machine is electrical energy, (seen as applying a voltage to the electrical terminals of the machine), and the output is mechanical energy, (seen as a rotating shaft), the machine is called an electric **motor**. Thus an electric motor converts electrical energy into mechanical energy.

The principle of operation of a motor is explained in Section 8.4, page 97. When the input to an electrical machine is mechanical energy, (seen as, say, a diesel motor, coupled to the machine by a shaft), and the output is electrical energy, (seen as a voltage appearing at the electrical terminals of the machine), the machine is called a **generator**. Thus, a generator converts mechanical energy to electrical energy.

The principle of operation of a generator is explained in Section 9.2, page 103.

## 22.2 The action of a commutator

In an electric motor, conductors rotate in a uniform magnetic field. A single-loop conductor mounted between

DOI: 10.1016/B978-0-08-089056-2.00022-X

permanent magnets is shown in Fig. 22.1. A voltage is applied at points A and B in Fig. 22.1(a).

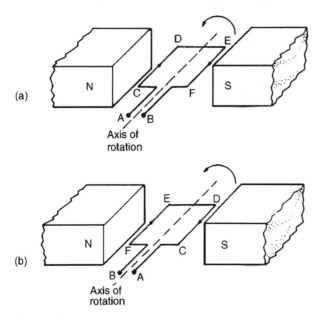

**Figure 22.1**

A force, $F$, acts on the loop due to the interaction of the magnetic field of the permanent magnets and the magnetic field created by the current flowing in the loop. This force is proportional to the flux density, B, the current flowing, $I$, and the effective length of the conductor, $l$, i.e. $F = BIl$. The force is made up of two parts, one acting vertically downwards due to the current flowing from C to D and the other acting vertically upwards due to the current flowing from E to F (from Fleming's left-hand rule). If the loop is free to rotate, then when it has rotated through 180°, the conductors are as shown in Fig. 22.1(b). For rotation to continue in the same direction, it is necessary for the current flow to be as shown in Fig. 22.1(b), i.e. from D to C and from F to E. This apparent reversal in the direction of current flow is achieved by a process called **commutation**.

With reference to Fig. 22.2(a), when a direct voltage is applied at A and B, then as the single-loop conductor rotates, current flow will always be away from the commutator for the part of the conductor adjacent to the N-pole and towards the commutator for the part of the conductor adjacent to the S-pole. Thus the forces act to give continuous rotation in an anticlockwise direction. The arrangement shown in Fig. 22.2(a) is called a 'two-segment' commutator and the voltage is applied to the rotating segments by stationary **brushes**, (usually carbon blocks), which slide on the commutator material, (usually copper), when rotation takes place.

In practice, there are many conductors on the rotating part of a d.c. machine and these are attached to many commutator segments. A schematic diagram of a multi-segment commutator is shown in Fig. 22.2(b).

Poor commutation results in sparking at the trailing edge of the brushes. This can be improved by using **interpoles** (situated between each pair of main poles), high resistance brushes, or using brushes spanning several commutator segments.

## 22.3 D.C. machine construction

The basic parts of any d.c. machine are shown in Fig. 22.3, and comprise:

(a) a stationary part called the **stator** having,

    (i) a steel ring called the **yoke**, to which are attached

    (ii) the magnetic **poles**, around which are the

    (iii) **field windings**, i.e. many turns of a conductor wound round the pole core; current passing through this conductor creates an electromagnet, (rather than the permanent magnets shown in Figs 22.1 and 22.2),

(b) a rotating part called the **armature** mounted in bearings housed in the stator and having,

    (iv) a laminated cylinder of iron or steel called the **core**, on which teeth are cut to house the

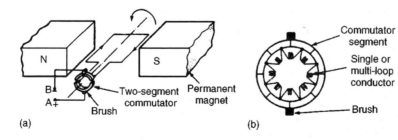

**Figure 22.2**

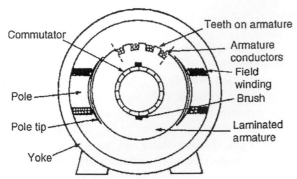

**Figure 22.3**

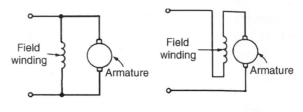

(a) Shunt-wound machine    (b) Series-wound machine

**Figure 22.4**

(v) **armature winding**, i.e. a single or multi-loop conductor system, and

(vi) the **commutator**, (see Section 22.2)

Armature windings can be divided into two groups, depending on how the wires are joined to the commutator. These are called **wave windings** and **lap windings**.

(a) In **wave windings** there are two paths in parallel irrespective of the number of poles, each path supplying half the total current output. Wave wound generators produce high voltage, low current outputs.

(b) In **lap windings** there are as many paths in parallel as the machine has poles. The total current output divides equally between them. Lap wound generators produce high current, low voltage output.

## 22.4  Shunt, series and compound windings

When the field winding of a d.c. machine is connected in parallel with the armature, as shown in Fig. 22.4(a), the machine is said to be **shunt** wound. If the field winding is connected in series with the armature, as shown in Fig. 22.4(b), then the machine is said to be **series** wound. A **compound** wound machine has a combination of series and shunt windings.

Depending on whether the electrical machine is series wound, shunt wound or compound wound, it behaves differently when a load is applied. The behaviour of a d.c. machine under various conditions is shown by means of graphs, called characteristic curves or just **characteristics**. The characteristics shown in the following sections are theoretical, since they neglect the effects of armature reaction.

**Armature reaction** is the effect that the magnetic field produced by the armature current has on the magnetic field produced by the field system. In a generator, armature reaction results in a reduced output voltage, and in a motor, armature reaction results in increased speed.

A way of overcoming the effect of armature reaction is to fit compensating windings, located in slots in the pole face.

## 22.5  E.m.f. generated in an armature winding

Let    $Z$ = number of armature conductors,
    $\Phi$ = useful flux per pole, in webers,
    $p$ = number of **pairs** of poles
and    $n$ = armature speed in rev/s

The e.m.f. generated by the armature is equal to the e.m.f. generated by one of the parallel paths. Each conductor passes $2p$ poles per revolution and thus cuts $2p\Phi$ webers of magnetic flux per revolution. Hence flux cut by one conductor per second $= 2p\Phi n$ Wb and so the average e.m.f. $E$ generated per conductor is given by:

$$E2p\Phi n \text{ volts}$$

(since 1 volt = 1 Weber per second)

Let    $c$ = number of parallel paths through the winding between positive and negative brushes

$$c = 2 \quad \textbf{for a wave winding}$$

$$c = 2p \quad \textbf{for a lap winding}$$

The number of conductors in series in each path $= Z/c$
The total e.m.f. between

brushes = (average e.m.f./conductor) (number of conductors in series per path)

$$= 2p\Phi nZ/c$$

i.e.    **generated e.m.f. $E = \dfrac{2p\Phi nZ}{c}$ volts** (1)

Since $Z$, $p$ and $c$ are constant for a given machine, then $E \propto \Phi n$. However $2\pi n$ is the angular velocity $\omega$ in radians per second, hence the generated e.m.f. is proportional to $\Phi$ and $\omega$,

i.e.    **generated e.m.f. $E \propto \Phi\omega$**    (2)

---

**Problem 1.** An 8-pole, wave-connected armature has 600 conductors and is driven at 625 rev/min. If the flux per pole is 20 mWb, determine the generated e.m.f.

$Z = 600$, $c = 2$ (for a wave winding), $p = 4$ pairs, $n = 625/60$ rev/s and $\Phi = 20 \times 10^{-3}$ Wb.

**Generated e.m.f.**

$$E = \frac{2p\Phi nZ}{c}$$

$$= \frac{2(4)(20 \times 10^{-3})\left(\frac{625}{60}\right)(600)}{2}$$

$$= \mathbf{500\,volts}$$

---

**Problem 2.** A 4-pole generator has a lap-wound armature with 50 slots with 16 conductors per slot. The useful flux per pole is 30 mWb. Determine the speed at which the machine must be driven to generate an e.m.f. of 240 V.

$E = 240$ V, $c = 2p$ (for a lap winding), $Z = 50 \times 16 = 800$ and $\Phi = 30 \times 10^{-3}$ Wb.

**Generated e.m.f.**

$$E = \frac{2p\Phi nZ}{c} = \frac{2p\Phi nZ}{2p} = \Phi nZ$$

Rearranging gives, speed,

$$n = \frac{E}{\Phi Z} = \frac{240}{(30 \times 10^{-3})(800)}$$

$$= \mathbf{10\,rev/s\ or\ 600\,rev/min}$$

---

**Problem 3.** An 8-pole, lap-wound armature has 1200 conductors and a flux per pole of 0.03 Wb. Determine the e.m.f. generated when running at 500 rev/min.

**Generated e.m.f.,**

$$E = \frac{2p\Phi nZ}{c}$$

$$= \frac{2p\Phi nZ}{2p} \text{ for a lap-wound machine,}$$

i.e.    $E = \Phi nZ$

$$= (0.03)\left(\frac{500}{60}\right)(1200)$$

$$= \mathbf{300\,volts}$$

---

**Problem 4.** Determine the generated e.m.f. in Problem 3 if the armature is wave-wound.

**Generated e.m.f.**

$$E = \frac{2p\Phi nZ}{c}$$

$$= \frac{2p\Phi nZ}{2} \quad \text{(since } c = 2 \text{ for wave-wound)}$$

$$= p\Phi nZ = (4)(\Phi nZ)$$

$$= (4)(300) \text{ from Problem 3}$$

$$= \mathbf{1200\,volts}$$

---

**Problem 5.** A d.c. shunt-wound generator running at constant speed generates a voltage of 150 V at a certain value of field current. Determine the change in the generated voltage when the field current is reduced by 20 per cent, assuming the flux is proportional to the field current.

The generated e.m.f. $E$ of a generator is proportional to $\Phi\omega$, i.e. is proportional to $\Phi n$, where $\Phi$ is the flux and $n$ is the speed of rotation. It follows that $E = k\Phi n$, where $k$ is a constant.

At speed $n_1$ and flux $\Phi_1$, $E_1 = k\Phi_1 n_1$
At speed $n_2$ and flux $\Phi_2$, $E_2 = k\Phi_2 n_2$

Thus, by division:

$$\frac{E_1}{E_2} = \frac{k\Phi_1 n_1}{k\Phi_2 n_2} = \frac{\Phi_1 n_1}{\Phi_2 n_2}$$

The initial conditions are $E_1 = 150$ V, $\Phi = \Phi_1$ and $n = n_1$. When the flux is reduced by 20 per cent, the new value of flux is $80/100$ or $0.8$ of the initial value, i.e. $\Phi_2 = 0.8\Phi_1$. Since the generator is running at constant speed, $n_2 = n_1$

Thus    $\dfrac{E_1}{E_2} = \dfrac{\Phi_1 n_1}{\Phi_2 n_2} = \dfrac{\Phi_1 n_1}{0.8\Phi_1 n_2} = \dfrac{1}{0.8}$

that is,    $E_2 = 150 \times 0.8 = 120$ V

Thus, a reduction of 20 per cent in the value of the flux **reduces the generated voltage to 120 V** at constant speed.

**Problem 6.** A d.c. generator running at 30 rev/s generates an e.m.f. of 200 V. Determine the percentage increase in the flux per pole required to generate 250 V at 20 rev/s.

From equation (2), generated e.m.f., $E \propto \Phi \omega$ and since $\omega = 2\pi n$, $E \propto \Phi n$

Let $E_1 = 200$ V, $n_1 = 30$ rev/s

and flux per pole at this speed be $\Phi_1$

Let $E_2 = 250$ V, $n_2 = 20$ rev/s

and flux per pole at this speed be $\Phi_2$

Since $E \propto \Phi n$ then $\dfrac{E_1}{E_2} = \dfrac{\Phi_1 n_1}{\Phi_2 n_2}$

Hence $\dfrac{200}{250} = \dfrac{\Phi_1(30)}{\Phi_2(20)}$

from which, $\Phi_2 = \dfrac{\Phi_1(30)(250)}{(20)(200)}$

$= 1.875\Phi_1$

Hence **the increase in flux per pole needs to be 87.5 per cent.**

---

**Now try the following exercise**

**Exercise 131    Further problems on generator e.m.f.**

1.  A 4-pole, wave-connected armature of a d.c. machine has 750 conductors and is driven at 720 rev/min. If the useful flux per pole is 15 mWb, determine the generated e.m.f.
    [270 volts]

2.  A 6-pole generator has a lap-wound armature with 40 slots with 20 conductors per slot. The flux per pole is 25 mWb. Calculate the speed at which the machine must be driven to generate an e.m.f. of 300 V.   [15 rev/s or 900 rev/min]

3.  A 4-pole armature of a d.c. machine has 1000 conductors and a flux per pole of 20 mWb. Determine the e.m.f. generated when running at 600 rev/min when the armature is (a) wave-wound (b) lap-wound.
    [(a) 400 volts (b) 200 volts]

4.  A d.c. generator running at 25 rev/s generates an e.m.f. of 150 V. Determine the percentage increase in the flux per pole required to generate 180 V at 20 rev/s.    [50%]

## 22.6   D.C. generators

D.C. generators are classified according to the method of their field excitation. These groupings are:

(i)   **Separately-excited generators**, where the field winding is connected to a source of supply other than the armature of its own machine.

(ii)  **Self-excited generators**, where the field winding receives its supply from the armature of its own machine, and which are sub-divided into (a) shunt, (b) series, and (c) compound wound generators.

## 22.7   Types of d.c. generator and their characteristics

### (a) Separately-excited generator

A typical separately-excited generator circuit is shown in Fig. 22.5.

When a load is connected across the armature terminals, a load current $I_a$ will flow. The terminal voltage $V$ will fall from its open-circuit e.m.f. $E$ due to a volt drop caused by current flowing through the armature resistance, shown as $R_a$.

i.e.     **terminal voltage, $V = E - I_a R_a$**

or       **generated e.m.f., $E = V + I_a R_a$**     (3)

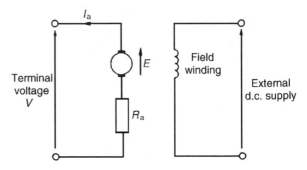

**Figure 22.5**

**Problem 7.** Determine the terminal voltage of a generator which develops an e.m.f. of 200 V and has an armature current of 30 A on load. Assume the armature resistance is 0.30 Ω.

With reference to Fig. 22.5, terminal voltage,

$$V = E - I_a R_a$$
$$= 200 - (30)(0.30)$$
$$= 200 - 9 = \mathbf{191\,volts}$$

**Problem 8.** A generator is connected to a 60 Ω load and a current of 8 A flows. If the armature resistance is 1 Ω determine (a) the terminal voltage, and (b) the generated e.m.f.

(a) Terminal voltage, $V = I_a R_L = (8)(60) = \mathbf{480\,volts}$

(b) Generated e.m.f.,

$$E = V + I_a R_a \quad \text{from equation (3)}$$
$$= 480 + (8)(1) = 480 + 8 = \mathbf{488\,volts}$$

**Problem 9.** A separately-excited generator develops a no-load e.m.f. of 150 V at an armature speed of 20 rev/s and a flux per pole of 0.10 Wb. Determine the generated e.m.f. when (a) the speed increases to 25 rev/s and the pole flux remains unchanged, (b) the speed remains at 20 rev/s and the pole flux is decreased to 0.08 Wb, and (c) the speed increases to 24 rev/s and the pole flux is decreased to 0.07 Wb.

(a) From Section 22.5, generated e.m.f. $E \propto \Phi n$

from which, $\dfrac{E_1}{E_2} = \dfrac{\Phi_1 N_1}{\Phi_2 N_2}$

Hence $\dfrac{150}{E_2} = \dfrac{(0.10)(20)}{(0.1)(25)}$

from which, $E_2 = \dfrac{(150)(0.10)(25)}{(0.10)(20)}$

$$= \mathbf{187.5\,volts}$$

(b) $\dfrac{150}{E_3} = \dfrac{(0.10)(20)}{(0.08)(20)}$

from which, e.m.f., $E_3 = \dfrac{(150)(0.08)(20)}{(0.10)(20)}$

$$= \mathbf{120\,volts}$$

(c) $\dfrac{150}{E_4} = \dfrac{(0.10)(20)}{(0.07)(24)}$

from which, e.m.f., $E_4 = \dfrac{(150)(0.07)(24)}{(0.10)(20)}$

$$= \mathbf{126\,volts}$$

### Characteristics

The two principal generator characteristics are the generated voltage/field current characteristics, called the **open-circuit characteristic** and the terminal voltage/load current characteristic, called the **load characteristic**. A typical separately-excited generator **open-circuit characteristic** is shown in Fig. 22.6(a) and a typical **load characteristic** is shown in Fig. 22.6(b).

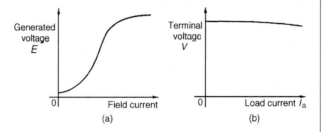

**Figure 22.6**

A separately-excited generator is used only in special cases, such as when a wide variation in terminal p.d. is required, or when exact control of the field current is necessary. Its disadvantage lies in requiring a separate source of direct current.

### (b) Shunt-wound generator

In a shunt-wound generator the field winding is connected in parallel with the armature as shown in Fig. 22.7. The field winding has a relatively high resistance and therefore the current carried is only a fraction of the armature current.

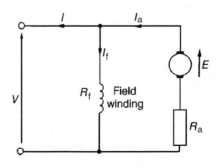

**Figure 22.7**

For the circuit shown in Fig. 22.7

$$\textbf{terminal voltage, } V = E - I_a R_a$$

or    $$\textbf{generated e.m.f., } E = V + I_a R_a$$

$I_a = I_f + I$ from Kirchhoff's current law, where $I_a$ = armature current, $I_f$ = field current ($= V/R_f$) and $I$ = load current.

> **Problem 10.**   A shunt generator supplies a 20 kW load at 200 V through cables of resistance, $R = 100\,\text{m}\Omega$. If the field winding resistance, $R_f = 50\,\Omega$ and the armature resistance, $R_a = 40\,\text{m}\Omega$, determine (a) the terminal voltage, and (b) the e.m.f. generated in the armature.

(a)   The circuit is as shown in Fig. 22.8

Load current, $I = \dfrac{20\,000\,\text{watts}}{200\,\text{volts}} = 100\,\text{A}$

Volt drop in the cables to the load
$= IR = (100)(100 \times 10^{-3}) = 10\,\text{V}$.
Hence **terminal voltage, $V = 200 + 10 = 210$ volts**.

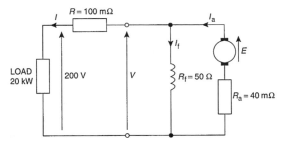

**Figure 22.8**

(b)   Armature current $I_a = I_f + I$

Field current, $I_f = \dfrac{V}{R_f} = \dfrac{210}{50} = 4.2\,\text{A}$

Hence $I_a = I_f + I = 4.2 + 100 = 104.2\,\text{A}$

**Generated e.m.f. $E = V + I_a R_a$**

$= 210 + (104.2)(40 \times 10^{-3})$

$= 210 + 4.168$

$= \textbf{214.17 volts}$

## Characteristics

The generated e.m.f., $E$, is proportional to $\Phi\omega$, (see Section 22.5), hence at constant speed, since $\omega = 2\pi n$,

$E \propto \Phi$. Also the flux $\Phi$ is proportional to field current $I_f$ until magnetic saturation of the iron circuit of the generator occurs. Hence the open circuit characteristic is as shown in Fig. 22.9(a).

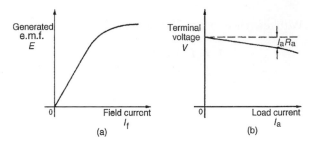

**Figure 22.9**

As the load current on a generator having constant field current and running at constant speed increases, the value of armature current increases, hence the armature volt drop, $I_a R_a$ increases. The generated voltage $E$ is larger than the terminal voltage $V$ and the voltage equation for the armature circuit is $V = E - I_a R_a$. Since $E$ is constant, $V$ decreases with increasing load. The load characteristic is as shown in Fig. 22.9(b). In practice, the fall in voltage is about 10 per cent between no-load and full-load for many d.c. shunt-wound generators.

The shunt-wound generator is the type most used in practice, but the load current must be limited to a value that is well below the maximum value. This then avoids excessive variation of the terminal voltage. Typical applications are with battery charging and motor car generators.

## (c) Series-wound generator

In the series-wound generator the field winding is connected in series with the armature as shown in Fig. 22.10.

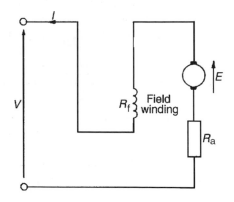

**Figure 22.10**

## Characteristics

The load characteristic is the terminal voltage/current characteristic. The generated e.m.f. $E$ is proportional to $\Phi\omega$ and at constant speed $\omega(=2\pi n)$ is a constant. Thus $E$ is proportional to $\Phi$. For values of current below magnetic saturation of the yoke, poles, air gaps and armature core, the flux $\Phi$ is proportional to the current, hence $E \propto I$. For values of current above those required for magnetic saturation, the generated e.m.f. is approximately constant. The values of field resistance and armature resistance in a series wound machine are small, hence the terminal voltage $V$ is very nearly equal to $E$. A typical load characteristic for a series generator is shown in Fig. 22.11.

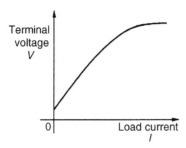

Figure 22.11

In a series-wound generator, the field winding is in series with the armature and it is not possible to have a value of field current when the terminals are open circuited, thus it is not possible to obtain an open-circuit characteristic.

Series-wound generators are rarely used in practise, but can be used as a 'booster' on d.c. transmission lines.

## (d) Compound-wound generator

In the compound-wound generator two methods of connection are used, both having a mixture of shunt and series windings, designed to combine the advantages of each. Figure 22.12(a) shows what is termed a **long-shunt** compound generator, and Fig. 22.12(b) shows a

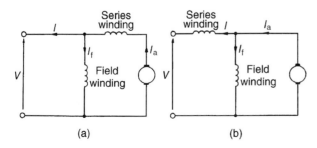

Figure 22.12

**short-shunt** compound generator. The latter is the most generally used form of d.c. generator.

> **Problem 11.** A short-shunt compound generator supplies 80 A at 200 V. If the field resistance, $R_f = 40\,\Omega$, the series resistance, $R_{Se} = 0.02\,\Omega$ and the armature resistance, $R_a = 0.04\,\Omega$, determine the e.m.f. generated.

The circuit is shown in Fig. 22.13.
Volt drop in series winding $= IR_{Se} = (80)(0.02)$
$$= 1.6\,V$$

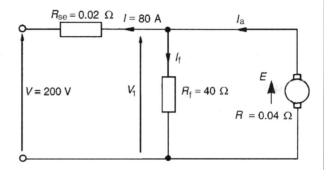

Figure 22.13

P.d. across the field winding $=$ p.d. across armature $= V_1 = 200 + 1.6 = 201.6\,V$

$$\text{Field current } I_f = \frac{V_1}{R_f} = \frac{201.6}{40} = 5.04\,A$$

Armature current, $I_a = I + I_f = 80 + 5.04 = 85.04\,A$

$$\textbf{Generated e.m.f., } E = V_1 + I_a R_a$$
$$= 201.6 + (85.04)(0.04)$$
$$= 201.6 + 3.4016$$
$$= \textbf{205 volts}$$

## Characteristics

In cumulative-compound machines the magnetic flux produced by the series and shunt fields are additive. Included in this group are **over-compounded**, **level-compounded** and **under-compounded machines** – the degree of compounding obtained depending on the number of turns of wire on the series winding.

A large number of series winding turns results in an over-compounded characteristic, as shown in Fig. 22.14, in which the full-load terminal voltage exceeds the no-load voltage. A level-compound machine gives a

Section 3

full-load terminal voltage which is equal to the no-load voltage, as shown in Fig. 22.14.

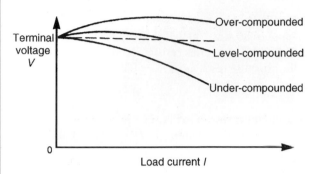

**Figure 22.14**

An under-compounded machine gives a full-load terminal voltage which is less than the no-load voltage, as shown in Fig. 22.14. However even this latter characteristic is a little better than that for a shunt generator alone. Compound-wound generators are used in electric arc welding, with lighting sets and with marine equipment.

**Now try the following exercise**

**Exercise 132 Further problems on the d.c. generator**

1. Determine the terminal voltage of a generator which develops an e.m.f. of 240 V and has an armature current of 50 A on load. Assume the armature resistance is 40 mΩ.    [238 volts]

2. A generator is connected to a 50 Ω load and a current of 10 A flows. If the armature resistance is 0.5 Ω, determine (a) the terminal voltage, and (b) the generated e.m.f.
   [(a) 500 volts (b) 505 volts]

3. A separately excited generator develops a no-load e.m.f. of 180 V at an armature speed of 15 rev/s and a flux per pole of 0.20 Wb. Calculate the generated e.m.f. when:
   (a) the speed increases to 20 rev/s and the flux per pole remains unchanged
   (b) the speed remains at 15 rev/s and the pole flux is decreased to 0.125 Wb
   (c) the speed increases to 25 rev/s and the pole flux is decreased to 0.18 Wb.
      [(a) 240 volts (b) 112.5 volts (c) 270 volts]

4. A shunt generator supplies a 50 kW load at 400 V through cables of resistance 0.2 Ω. If the field winding resistance is 50 Ω and the armature resistance is 0.05 Ω, determine (a) the terminal voltage, (b) the e.m.f. generated in the armature.    [(a) 425 volts (b) 431.68 volts]

5. A short-shunt compound generator supplies 50 A at 300 V. If the field resistance is 30 Ω, the series resistance 0.03 Ω and the armature resistance 0.05 Ω, determine the e.m.f. generated.    [304.5 volts]

6. A d.c. generator has a generated e.m.f. of 210 V when running at 700 rev/min and the flux per pole is 120 mWb. Determine the generated e.m.f.
   (a) at 1050 rev/min, assuming the flux remains constant,
   (b) if the flux is reduced by one-sixth at constant speed, and
   (c) at a speed of 1155 rev/min and a flux of 132 mWb.
      [(a) 315 V (b) 175 V (c) 381.2 V]

7. A 250 V d.c. shunt-wound generator has an armature resistance of 0.1 Ω. Determine the generated e.m.f. when the generator is supplying 50 kW, neglecting the field current of the generator.    [270 V]

## 22.8 D.C. machine losses

As stated in Section 22.1, a generator is a machine for converting mechanical energy into electrical energy and a motor is a machine for converting electrical energy into mechanical energy. When such conversions take place, certain losses occur which are dissipated in the form of heat.

The principal **losses of machines** are:

(i) **Copper loss**, due to $I^2R$ heat losses in the armature and field windings.

(ii) **Iron (or core) loss**, due to hysteresis and eddy-current losses in the armature. This loss can be reduced by constructing the armature of silicon steel laminations having a high resistivity and low hysteresis loss. At constant speed, the iron loss is assumed constant.

(iii) **Friction and windage losses**, due to bearing and brush contact friction and losses due to air resistance against moving parts (called windage).

At constant speed, these losses are assumed to be constant.

(iv) **Brush contact loss** between the brushes and commutator. This loss is approximately proportional to the load current.

The total losses of a machine can be quite significant and operating efficiencies of between 80 per cent and 90 per cent are common.

## 22.9 Efficiency of a d.c. generator

The efficiency of an electrical machine is the ratio of the output power to the input power and is usually expressed as a percentage. The Greek letter, '$\eta$' (eta) is used to signify efficiency and since the units are, power/power, then efficiency has no units. Thus

$$\text{efficiency, } \eta = \left(\frac{\text{output power}}{\text{input power}}\right) \times 100\%$$

If the total resistance of the armature circuit (including brush contact resistance) is $R_a$, then **the total loss in the armature circuit is $I_a^2 R_a$**

If the terminal voltage is $V$ and the current in the shunt circuit is $I_f$, then **the loss in the shunt circuit is $I_f V$**

If the sum of the iron, friction and windage losses is $C$ then **the total losses is given by: $I_a^2 R_a + I_f V + C$** ($I_a^2 R_a + I_f V$ is, in fact, the 'copper loss').

If the output current is $I$, then **the output power is $VI$**. Total input power $= VI + I_a^2 R_a + I_f V + C$. Hence

$$\text{efficiency, } \eta = \frac{\text{output}}{\text{input}}, \text{i.e.}$$

$$\eta = \left(\frac{VI}{VI + I_a^2 R_a + I_f V + C}\right) \times 100\% \qquad (4)$$

The **efficiency of a generator is a maximum** when the load is such that:

$$I_a^2 R_a = V I_f + C$$

i.e. when the variable loss = the constant loss

**Problem 12.** A 10 kW shunt generator having an armature circuit resistance of 0.75 $\Omega$ and a field

resistance of 125 $\Omega$, generates a terminal voltage of 250 V at full load. Determine the efficiency of the generator at full load, assuming the iron, friction and windage losses amount to 600 W.

The circuit is shown in Fig. 22.15

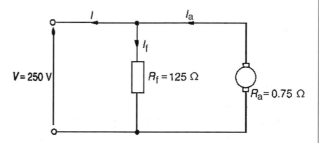

**Figure 22.15**

Output power $= 10\,000\,\text{W} = VI$ from which, load current $I = 10\,000/V = 10\,000/250 = 40\,\text{A}$.
Field current, $I_f = V/R_f = 250/125 = 2\,\text{A}$.
Armature current, $I_a = I_f + I = 2 + 40 = 42\,\text{A}$

$$\text{Efficiency, } \eta = \left(\frac{VI}{\begin{array}{c}VI + I_a^2 R \\ + I_f V + C\end{array}}\right) \times 100\%$$

$$= \left(\frac{10\,000}{\begin{array}{c}10\,000 + (42)^2(0.75) \\ + (2)(250) + 600\end{array}}\right) \times 100\%$$

$$= \left(\frac{10\,000}{12\,423}\right) \times 100\%$$

$$= \mathbf{80.50\%}$$

**Now try the following exercise**

**Exercise 133    A further problem on the efficiency of a d.c. generator**

1.  A 15 kW shunt generator having an armature circuit resistance of 0.4 $\Omega$ and a field resistance of 100 $\Omega$, generates a terminal voltage of 240 V at full load. Determine the efficiency of the generator at full load, assuming the iron, friction and windage losses amount to 1 kW.

[82.14%]

## 22.10 D.C. motors

The construction of a d.c. motor is the same as a d.c. generator. The only difference is that in a generator the generated e.m.f. is greater than the terminal voltage, whereas in a motor the generated e.m.f. is less than the terminal voltage.

D.C. motors are often used in power stations to drive emergency stand-by pump systems which come into operation to protect essential equipment and plant should the normal a.c. supplies or pumps fail.

### Back e.m.f.

When a d.c. motor rotates, an e.m.f. is induced in the armature conductors. By Lenz's law this induced e.m.f. $E$ opposes the supply voltage $V$ and is called a **back e.m.f.**, and the supply voltage, $V$ is given by:

$$V = E + I_a R_a \quad \text{or} \quad E = V - I_a R_a \qquad (5)$$

**Problem 13.** A d.c. motor operates from a 240 V supply. The armature resistance is $0.2\,\Omega$. Determine the back e.m.f. when the armature current is 50 A.

For a motor, $V = E + I_a R_a$ hence back e.m.f.,

$$E = V - I_a R_a$$
$$= 240 - (50)(0.2)$$
$$= 240 - 10 = \mathbf{230\,volts}$$

**Problem 14.** The armature of a d.c. machine has a resistance of $0.25\,\Omega$ and is connected to a 300 V supply. Calculate the e.m.f. generated when it is running: (a) as a generator giving 100 A, and (b) as a motor taking 80 A.

(a) As a generator, generated e.m.f.,

$$E = V + I_a R_a, \text{ from equation (3)},$$
$$= 300 + (100)(0.25)$$
$$= 300 + 25$$
$$= \mathbf{325\,volts}$$

(b) As a motor, generated e.m.f. (or back e.m.f.),

$$E = V - I_a R_a, \text{ from equation (5)},$$
$$= 300 - (80)(0.25)$$
$$= \mathbf{280\,volts}$$

Now try the following exercise

**Exercise 134 Further problems on back e.m.f.**

1. A d.c. motor operates from a 350 V supply. If the armature resistance is $0.4\,\Omega$ determine the back e.m.f. when the armature current is 60 A.
   [326 volts]

2. The armature of a d.c. machine has a resistance of $0.5\,\Omega$ and is connected to a 200 V supply. Calculate the e.m.f. generated when it is running (a) as a motor taking 50 A, and (b) as a generator giving 70 A.
   [(a) 175 volts (b) 235 volts]

3. Determine the generated e.m.f. of a d.c. machine if the armature resistance is $0.1\,\Omega$ and it (a) is running as a motor connected to a 230 V supply, the armature current being 60 A, and (b) is running as a generator with a terminal voltage of 230 V, the armature current being 80 A.
   [(a) 224 V (b) 238 V]

## 22.11 Torque of a d.c. motor

From equation (5), for a d.c. motor, the supply voltage $V$ is given by

$$V = E + I_a R_a$$

Multiplying each term by current $I_a$ gives:

$$V I_a = E I_a + I_a^2 R_a$$

The term $\mathbf{V I_a}$ is the **total electrical power supplied to the armature**, the term $\mathbf{I_a^2 R_a}$ is the **loss due to armature resistance**, and the term $\mathbf{E I_a}$ is the **mechanical power developed by the armature**. If $T$ is the torque, in newton metres, then the mechanical power developed is given by $T\omega$ watts (see '*Science for Engineering*')

Hence $\qquad T\omega = 2\pi n T = E I_a$

from which,

$$\text{torque } T = \frac{E I_a}{2\pi n} \text{ newton metres} \qquad (6)$$

From Section 22.5, equation (1), the e.m.f. $E$ generated is given by

$$E = \frac{2p\Phi nZ}{c}$$

Hence

$$2\pi nT = EI_a = \left(\frac{2p\Phi nZ}{c}\right)I_a$$

Hence torque

$$T = \frac{\left(\frac{2p\Phi nZ}{c}\right)}{2\pi n}I_a$$

i.e.

$$T = \frac{p\Phi ZI_a}{\pi c} \text{ newton metres} \tag{7}$$

For a given machine, $Z$, $c$ and $p$ are fixed values

Hence

$$\text{torque, } T \propto \Phi I_a \tag{8}$$

**Problem 15.** An 8-pole d.c. motor has a wave-wound armature with 900 conductors. The useful flux per pole is 25 mWb. Determine the torque exerted when a current of 30 A flows in each armature conductor.

$p = 4$, $c = 2$ for a wave winding,
$\Phi = 25 \times 10^{-3}$ Wb, $Z = 900$ and $I_a = 30$ A.
From equation (7),

$$\text{torque, } T = \frac{p\Phi ZI_a}{\pi c}$$

$$= \frac{(4)(25 \times 10^{-3})(900)(30)}{\pi(2)}$$

$$= \textbf{429.7 Nm}$$

**Problem 16.** Determine the torque developed by a 350 V d.c. motor having an armature resistance of $0.5\,\Omega$ and running at 15 rev/s. The armature current is 60 A.

$V = 350$ V, $R_a = 0.5\,\Omega$, $n = 15$ rev/s and $I_a = 60$ A.
Back e.m.f. $E = V - I_a R_a = 350 - (60)(0.5) = 320$ V.
From equation (6),

$$\text{torque, } T = \frac{EI_a}{2\pi n} = \frac{(320)(60)}{2\pi(15)} = \textbf{203.7 Nm}$$

**Problem 17.** A six-pole lap-wound motor is connected to a 250 V d.c. supply. The armature has 500 conductors and a resistance of $1\,\Omega$. The flux per pole is 20 mWb. Calculate (a) the speed and (b) the torque developed when the armature current is 40 A.

$V = 250$ V, $Z = 500$, $R_a = 1\,\Omega$, $\Phi = 20 \times 10^{-3}$ Wb, $I_a = 40$ A and $c = 2p$ for a lap winding

(a) Back e.m.f. $E = V - I_a R_a = 250 - (40)(1)$
$= 210$ V

$$\text{E.m.f. } E = \frac{2p\Phi nZ}{c}$$

i.e. $210 = \dfrac{2p(20 \times 10^{-3})n(500)}{2p} = 10n$

Hence **speed** $n = \dfrac{210}{10} = \textbf{21 rev/s}$ or $(21 \times 60)$

$$= \textbf{1260 rev/min}$$

(b) **Torque** $T = \dfrac{EI_a}{2\pi n} = \dfrac{(210)(40)}{2\pi(21)} = \textbf{63.66 Nm}$

**Problem 18.** The shaft torque of a diesel motor driving a 100 V d.c. shunt-wound generator is 25 Nm. The armature current of the generator is 16 A at this value of torque. If the shunt field regulator is adjusted so that the flux is reduced by 15 per cent, the torque increases to 35 Nm. Determine the armature current at this new value of torque.

From equation (8), the shaft torque $T$ of a generator is proportional to $\Phi I_a$, where $\Phi$ is the flux and $I_a$ is the armature current, or, $T = k\Phi I_a$, where $k$ is a constant.
The torque at flux $\Phi_1$ and armature current $I_{a1}$ is $T_1 = k\Phi_1 I_{a1}$ Similarly, $T_2 = k\Phi_2 I_{a2}$

By division $\dfrac{T_1}{T_2} = \dfrac{k\Phi_1 I_{a1}}{k\Phi_2 I_{a2}} = \dfrac{\Phi_1 I_{a1}}{\Phi_2 I_{a2}}$

Hence $\dfrac{25}{35} = \dfrac{\Phi_1 \times 16}{0.85\Phi_1 \times I_{a2}}$

i.e. $I_{a2} = \dfrac{16 \times 35}{0.85 \times 25} = 26.35$ A

That is, **the armature current at the new value of torque is 26.35 A**

**Problem 19.** A 100 V d.c. generator supplies a current of 15 A when running at 1500 rev/min. If the torque on the shaft driving the generator is 12 Nm, determine (a) the efficiency of the generator and (b) the power loss in the generator.

(a)   From Section 22.9, the efficiency of a generator = output power/input power × 100 per cent. The output power is the electrical output, i.e. VI watts. The input power to a generator is the mechanical power in the shaft driving the generator, i.e. $T\omega$ or $T(2\pi n)$ watts, where $T$ is the torque in Nm and $n$ is speed of rotation in rev/s. Hence, for a generator,

$$\text{efficiency, } \eta = \frac{VI}{T(2\pi n)} \times 100\%$$

$$= \frac{(100)(15)(100)}{(12)(2\pi)\left(\dfrac{1500}{60}\right)}$$

i.e. **efficiency = 79.6%**

(b)   The input power = output power + losses

Hence, $T(2\pi n) = VI + $ losses

i.e. losses $= T(2\pi n) - VI$

$$= \left[(12)(2\pi)\left(\frac{1500}{60}\right)\right]$$

$$- [(100)(15)]$$

i.e. **power loss** $= 1885 - 1500 = $ **385 W**

---

**Now try the following exercise**

**Exercise 135   Further problems on losses, efficiency, and torque**

1.   The shaft torque required to drive a d.c. generator is 18.7 Nm when it is running at 1250 rev/min. If its efficiency is 87 per cent under these conditions and the armature current is 17.3 A, determine the voltage at the terminals of the generator.   [123.1 V]

2.   A 220 V, d.c. generator supplies a load of 37.5 A and runs at 1550 rev/min. Determine the shaft torque of the diesel motor driving the generator, if the generator efficiency is 78 per cent.   [65.2 Nm]

3.   A 4-pole d.c. motor has a wave-wound armature with 800 conductors. The useful flux per pole is 20 mWb. Calculate the torque exerted when a current of 40 A flows in each armature conductor.   [203.7 Nm]

4.   Calculate the torque developed by a 240 V d.c. motor whose armature current is 50 A, armature resistance is 0.6 Ω and is running at 10 rev/s.   [167.1 Nm]

5.   An 8-pole lap-wound d.c. motor has a 200 V supply. The armature has 800 conductors and a resistance of 0.8 Ω. If the useful flux per pole is 40 mWb and the armature current is 30 A, calculate (a) the speed and (b) the torque developed.
   [(a) 5.5 rev/s or 330 rev/min (b) 152.8 Nm]

6.   A 150 V d.c. generator supplies a current of 25 A when running at 1200 rev/min. If the torque on the shaft driving the generator is 35.8 Nm, determine (a) the efficiency of the generator, and (b) the power loss in the generator.
   [(a) 83.4 per cent (b) 748.8 W]

## 22.12   Types of d.c. motor and their characteristics

### (a) Shunt-wound motor

In the shunt-wound motor the field winding is in parallel with the armature across the supply as shown in Fig. 22.16.

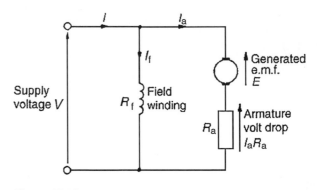

**Figure 22.16**

For the circuit shown in Fig. 22.16,

$$\text{Supply voltage, } V = E + I_a R_a$$

$$\text{or generated e.m.f., } E = V - I_a R_a$$

$$\text{Supply current, } I = I_a + I_f$$

from Kirchhoff's current law

---

**Problem 20.** A 240 V shunt motor takes a total current of 30 A. If the field winding resistance $R_f = 150\,\Omega$ and the armature resistance $R_a = 0.4\,\Omega$ determine (a) the current in the armature, and (b) the back e.m.f.

---

(a) Field current $I_f = \dfrac{V}{R_f} = \dfrac{240}{150} = 1.6\,\text{A}$

Supply current $I = I_a + I_f$
Hence armature current, $I_a = I - I_f = 30 - 1.6$
$$= \mathbf{28.4\,A}$$

(b) Back e.m.f.

$$E = V - I_a R_a = 240 - (28.4)(0.4) = \mathbf{228.64\,volts}$$

---

### Characteristics

The two principal characteristics are the torque/armature current and speed/armature current relationships. From these, the torque/speed relationship can be derived.

(i)   The theoretical torque/armature current characteristic can be derived from the expression $T \propto \Phi I_a$, (see Section 22.11). For a shunt-wound motor, the field winding is connected in parallel with the armature circuit and thus the applied voltage gives a constant field current, i.e. a shunt-wound motor is a constant flux machine. Since $\Phi$ is constant, it follows that $T \propto I_a$, and the characteristic is as shown in Fig. 22.17.

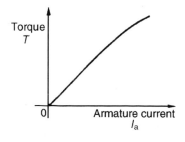

Figure 22.17

(ii)  The armature circuit of a d.c. motor has resistance due to the armature winding and brushes, $R_a$ ohms, and when armature current $I_a$ is flowing through it, there is a voltage drop of $I_a R_a$ volts. In Fig. 22.16 the armature resistance is shown as a separate resistor in the armature circuit to help understanding. Also, even though the machine is a motor, because conductors are rotating in a magnetic field, a voltage, $E \propto \Phi\omega$, is generated by the armature conductors. From equation (5), $V = E + I_a R_a$ or $E = V - I_a R_a$ However, from Section 22.5, $E \propto \Phi n$, hence $n \propto E/\Phi$ i.e.

$$\text{speed of rotation, } n \propto \frac{E}{\Phi} \propto \frac{V - I_a R_a}{\Phi} \qquad (9)$$

For a shunt motor, $V$, $\Phi$ and $R_a$ are constants, hence as armature current $I_a$ increases, $I_a R_a$ increases and $V - I_a R_a$ decreases, and the speed is proportional to a quantity which is decreasing and is as shown in Fig. 22.18. As the load on the shaft of the motor increases, $I_a$ increases and the speed drops slightly. In practice, the speed falls by about 10 per cent between no-load and full-load on many d.c. shunt-wound motors. Due to this relatively small drop in speed, the d.c. shunt-wound motor is taken as basically being a constant-speed machine and may be used for driving lathes, lines of shafts, fans, conveyor belts, pumps, compressors, drilling machines and so on.

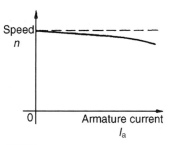

Figure 22.18

(iii) Since torque is proportional to armature current, (see (i) above), the theoretical speed/torque characteristic is as shown in Fig. 22.19.

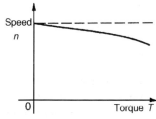

Figure 22.19

Power dissipated in the load $R_L$,

$$P = I_2^2 R_L = (20 \times 10^{-3})^2 (24)$$

$$= 9600 \times 10^{-6} \, \text{W} = \mathbf{9.6\,mW}$$

**Now try the following exercise**

**Exercise 125  Further problems on resistance matching**

1. A transformer having a turns ratio of 8:1 supplies a load of resistance $50\,\Omega$. Determine the equivalent input resistance of the transformer.
   [$3.2\,k\Omega$]

2. What ratio of transformer turns is required to make a load of resistance $30\,\Omega$ appear to have a resistance of $270\,\Omega$?
   [3:1]

3. Determine the optimum value of load resistance for maximum power transfer if the load is connected to an amplifier of output resistance $147\,\Omega$ through a transformer with a turns ratio of 7:2.
   [$12\,\Omega$]

4. A single-phase, 240 V/2880 V ideal transformer is supplied from a 240 V source through a cable of resistance $3\,\Omega$. If the load across the secondary winding is $720\,\Omega$ determine (a) the primary current flowing and (b) the power dissipated in the load resistance.
   [(a) 30 A (b) 4.5 kW]

5. A load of resistance $768\,\Omega$ is to be matched to an amplifier which has an effective output resistance of $12\,\Omega$. Determine the turns ratio of the coupling transformer.
   [1:8]

6. An a.c. source of 20 V and internal resistance $20\,k\Omega$ is matched to a load by a 16:1 single-phase transformer. Determine (a) the value of the load resistance and (b) the power dissipated in the load.
   [(a) $78.13\,\Omega$ (b) 5 mW]

## 21.11  Auto transformers

An auto transformer is a transformer which has part of its winding common to the primary and secondary circuits. Fig. 21.14(a) shows the circuit for a double-wound transformer and Fig. 21.14(b) that for an auto

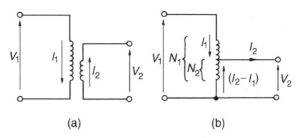

(a)                    (b)

Figure 21.14

transformer. The latter shows that the secondary is actually part of the primary, the current in the secondary being $(I_2 - I_1)$. Since the current is less in this section, the cross-sectional area of the winding can be reduced, which reduces the amount of material necessary. Figure 21.15 shows the circuit diagram symbol for an auto transformer.

Figure 21.15

**Problem 26.**  A single-phase auto transformer has a voltage ratio 320 V:250 V and supplies a load of 20 kVA at 250 V. Assuming an ideal transformer, determine the current in each section of the winding.

Rating $= 20\,\text{kVA} = V_1 I_1 = V_2 I_2$
Hence primary current,

$$I_1 = \frac{20 \times 10^3}{V_1} = \frac{20 \times 10^3}{320} = \mathbf{62.5\,A}$$

and secondary current,

$$I_2 = \frac{20 \times 10^3}{V_2} = \frac{20 \times 10^3}{250} = \mathbf{80\,A}$$

Hence current in common part of the winding
$= 80 - 62.5 = \mathbf{17.5\,A}$
The current flowing in each section of the transformer is shown in Fig. 21.16.

### Saving of copper in an auto transformer

For the same output and voltage ratio, the auto transformer requires less copper than an ordinary double-wound transformer. This is explained below.

**Problem 21.** A 200 V, d.c. shunt-wound motor has an armature resistance of $0.4\,\Omega$ and at a certain load has an armature current of 30 A and runs at 1350 rev/min. If the load on the shaft of the motor is increased so that the armature current increases to 45 A, determine the speed of the motor, assuming the flux remains constant.

The relationship $E \propto \Phi n$ applies to both generators and motors. For a motor, $E = V - I_a R_a$, (see equation (5))

Hence        $E_1 = 200 - 30 \times 0.4 = 188\,\text{V}$

and          $E_2 = 200 - 45 \times 0.4 = 182\,\text{V}$

The relationship

$$\frac{E_1}{E_2} = \frac{\Phi_1 n_1}{\Phi_2 n_2}$$

applies to both generators and motors. Since the flux is constant, $\Phi_1 = \Phi_2$. Hence

$$\frac{188}{182} = \frac{\Phi_1 \times \left(\dfrac{1350}{60}\right)}{\Phi_1 \times n_2}$$

i.e.        $n_2 = \dfrac{22.5 \times 182}{188} = 21.78\,\text{rev/s}$

Thus **the speed of the motor when the armature current is 45 A** is $21.78 \times 60$ rev/min i.e. **1307 rev/min**.

**Problem 22.** A 220 V, d.c. shunt-wound motor runs at 800 rev/min and the armature current is 30 A. The armature circuit resistance is $0.4\,\Omega$. Determine (a) the maximum value of armature current if the flux is suddenly reduced by 10 per cent and (b) the steady-state value of the armature current at the new value of flux, assuming the shaft torque of the motor remains constant.

(a) For a d.c. shunt-wound motor, $E = V - I_a R_a$. Hence initial generated e.m.f., $E_1 = 220 = 30 \times 0.4 = 208\,\text{V}$. The generated e.m.f. is also such that $E \propto \Phi n$, so at the instant the flux is reduced, the speed has not had time to change, and $E = 208 \times 90/100 - 187.2\,\text{V}$. Hence, the voltage drop due to the armature resistance is $220 - 187.2$ i.e. 32.8 V. The **instantaneous value of the current** $= 32.8/0.4 = $ **82 A**. This increase

in current is about three times the initial value and causes an increase in torque, $(T \propto \Phi I_a)$. The motor accelerates because of the larger torque value until steady-state conditions are reached.

(b) $T \propto \Phi I_a$ and, since the torque is constant, $\Phi_1 I_{a1} = \Phi_2 I_{a2}$. The flux $\Phi$ is reduced by 10 per cent, hence $\Phi_2 = 0.9 \Phi_1$. Thus, $\Phi_1 \times 30 = 0.9 \Phi_1 \times I_{a2}$ i.e. the steady-state value of armature current, $I_{a2} = 30/0.9 = $ **33.33 A**.

## (b) Series-wound motor

In the series-wound motor the field winding is in series with the armature across the supply as shown in Fig. 22.20.

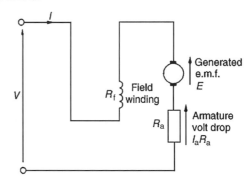

**Figure 22.20**

For the series motor shown in Fig. 22.20,

$$\text{Supply voltage } V = E + I(R_a + R_f)$$

$$\text{or generated e.m.f. } E = V - I(R_a + R_f)$$

### *Characteristics*

In a series motor, the armature current flows in the field winding and is equal to the supply current, $I$.

(i) **The torque/current characteristic**
It is shown in Section 22.11 that torque $T \propto \Phi I_a$. Since the armature and field currents are the same current, $I$, in a series machine, then $T \propto \Phi I$ over a limited range, before magnetic saturation of the magnetic circuit of the motor is reached, (i.e. the linear portion of the B–H curve for the yoke, poles, air gap, brushes and armature in series). Thus $\Phi \propto I$ and $T \propto I^2$. After magnetic saturation, $\Phi$ almost becomes a constant and $T \propto I$. Thus the theoretical torque/current characteristic is as shown in Fig. 22.21.

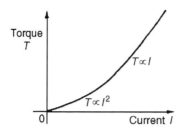

**Figure 22.21**

### (ii) The speed/current characteristic

It is shown in equation (9) that

$$n \propto \frac{V - I_a R_a}{\Phi}$$

In a series motor, $I_a = I$ and below the magnetic saturation level, $\Phi \propto I$. Thus $n \propto (V - IR)/I$ where $R$ is the combined resistance of the series field and armature circuit. Since $IR$ is small compared with $V$, then an approximate relationship for the speed is $n \propto V/I \propto 1/I$ since $V$ is constant. Hence the theoretical speed/current characteristic is as shown in Fig. 22.22. The high speed at small values of current indicate that this type of motor must not be run on very light loads and invariably, such motors are permanently coupled to their loads.

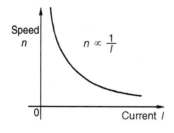

**Figure 22.22**

### (iii) The theoretical **speed/torque characteristic**

may be derived from (i) and (ii) above by obtaining the torque and speed for various values of current and plotting the co-ordinates on the speed/torque characteristics. A typical speed/torque characteristic is shown in Fig. 22.23.

A d.c. series motor takes a large current on starting and the characteristic shown in Fig. 22.21 shows that the series-wound motor has a large torque when the current is large. Hence these motors are used for traction (such as trains, milk delivery vehicles, etc.), driving fans and for cranes and hoists, where a large initial torque is required.

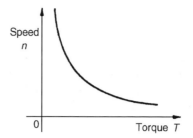

**Figure 22.23**

Problem 23.    A series motor has an armature resistance of $0.2\,\Omega$ and a series field resistance of $0.3\,\Omega$. It is connected to a 240 V supply and at a particular load runs at 24 rev/s when drawing 15 A from the supply. (a) Determine the generated e.m.f. at this load. (b) Calculate the speed of the motor when the load is changed such that the current is increased to 30 A. Assume that this causes a doubling of the flux.

(a)  With reference to Fig. 22.20, generated e.m.f., $E_1$ at initial load, is given by

$$E_1 = V - I_a(R_a + R_f)$$

$$= 240 - (15)(0.2 + 0.3)$$

$$= 240 - 7.5 = \mathbf{232.5\,volts}$$

(b)  When the current is increased to 30 A, the generated e.m.f. is given by:

$$E_2 = V - I_2(R_a + R_f)$$

$$= 240 - (30)(0.2 + 0.3)$$

$$= 240 - 15 = 225\,volts$$

Now e.m.f. $E \propto \Phi n$ thus

$$\frac{E_1}{E_2} = \frac{\Phi_1 n_1}{\Phi_2 n_2}$$

i.e. $\dfrac{232.5}{22.5} = \dfrac{\Phi_1(24)}{(2\Phi_1)n_2}$  since $\Phi_2 = 2\Phi_1$

Hence

**speed of motor,** $n_2 = \dfrac{(24)(225)}{(232.5)(2)} = \mathbf{11.6\,rev/s}$

As the current has been increased from 15 A to 30 A, the speed has decreased from 24 rev/s to 11.6 rev/s. Its speed/current characteristic is similar to Fig. 22.22.

### (c) Compound-wound motor

There are two types of compound-wound motor:

(i) **Cumulative compound**, in which the series winding is so connected that the field due to it assists that due to the shunt winding.

(ii) **Differential compound**, in which the series winding is so connected that the field due to it opposes that due to the shunt winding.

Figure 22.24(a) shows a **long-shunt** compound motor and Fig. 22.24(b) a **short-shunt** compound motor.

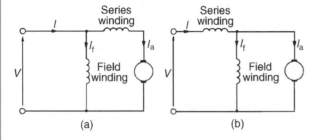

(a)                    (b)

**Figure 22.24**

### Characteristics

A compound-wound motor has both a series and a shunt field winding, (i.e. one winding in series and one in parallel with the armature), and is usually wound to have a characteristic similar in shape to a series-wound motor (see Figs 22.21–22.23). A limited amount of shunt winding is present to restrict the no-load speed to a safe value. However, by varying the number of turns on the series and shunt windings and the directions of the magnetic fields produced by these windings (assisting or opposing), families of characteristics may be obtained to suit almost all applications. Generally, compound-wound motors are used for heavy duties, particularly in applications where sudden heavy load may occur such as for driving plunger pumps, presses, geared lifts, conveyors, hoists and so on.

Typical compound motor torque and speed characteristics are shown in Fig. 22.25.

### 22.13   The efficiency of a d.c. motor

It was stated in Section 22.9, that the efficiency of a d.c. machine is given by:

$$\text{efficiency, } \eta = \frac{\text{output power}}{\text{input power}} \times 100\%$$

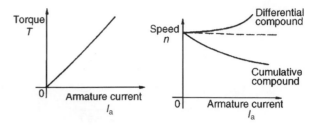

**Figure 22.25**

Also, the total losses $= I_a^2 R_a + I_f V + C$ (for a shunt motor) where $C$ is the sum of the iron, friction and windage losses.

For a motor,

$$\text{the input power} = VI$$

$$\text{and the output power} = VI - \text{losses}$$

$$= VI - I_a^2 R_a - I_f V - C$$

Hence **efficiency**,

$$\eta = \left( \frac{VI - I_a^2 R_a - I_f V - C}{VI} \right) \times 100\% \qquad (10)$$

The **efficiency of a motor is a maximum** when the load is such that:

$$I_a^2 R_a = I_f V + C$$

> **Problem 24.**   A 320 V shunt motor takes a total current of 80 A and runs at 1000 rev/min. If the iron, friction and windage losses amount to 1.5 kW, the shunt field resistance is 40 Ω and the armature resistance is 0.2 Ω, determine the overall efficiency of the motor.

The circuit is shown in Fig. 22.26.
Field current, $I_f = V/R_f = 320/40 = 8$ A.
Armature current $I_a = I - I_f = 80 - 8 = 72$ A.
$C =$ iron, friction and windage losses $= 1500$ W.

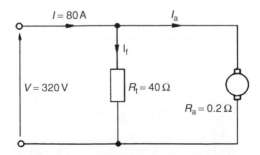

**Figure 22.26**

Efficiency,

$$\eta = \left(\frac{VI - I_a^2 R_a - I_f V - C}{VI}\right) \times 100\%$$

$$= \left(\frac{\begin{array}{c}(320)(80) - (72)^2\,(0.2) \\ -\,(8)\,(320) - 1500\end{array}}{(320)(80)}\right) \times 100\%$$

$$= \left(\frac{25\,600 - 1036.8 - 2560 - 1500}{25\,600}\right) \times 100\%$$

$$= \left(\frac{20\,503.2}{25\,600}\right) \times 100\%$$

$$= \mathbf{80.1\%}$$

**Problem 25.** A 250 V series motor draws a current of 40 A. The armature resistance is $0.15\,\Omega$ and the field resistance is $0.05\,\Omega$. Determine the maximum efficiency of the motor.

The circuit is as shown in Fig. 22.27.
From equation (10), efficiency,

$$\eta = \left(\frac{VI - I_a^2 R_a - I_f V - C}{VI}\right) \times 100\%$$

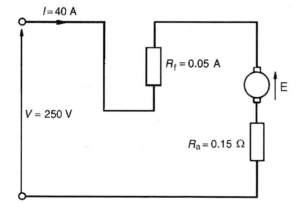

**Figure 22.27**

However for a series motor, $I_f = 0$ and the $I_a^2 R_a$ loss needs to be $I^2(R_a + R_f)$. Hence efficiency,

$$\eta = \left(\frac{VI - I^2(R_a + R_f) - C}{VI}\right) \times 100\%$$

For maximum efficiency $I^2(R_a + R_f) = C$. Hence efficiency,

$$\eta = \left(\frac{VI - 2I^2(R_a + R_f)}{VI}\right) \times 100\%$$

$$= \left(\frac{(250)(40) - 2(40)^2\,(0.15 + 0.05)}{(250)(40)}\right) \times 100\%$$

$$= \left(\frac{10\,000 - 640}{10\,000}\right) \times 100\%$$

$$= \left(\frac{9360}{10\,000}\right) \times 100\% = \mathbf{93.6\%}$$

**Problem 26.** A 200 V d.c. motor develops a shaft torque of 15 Nm at 1200 rev/min. If the efficiency is 80 per cent, determine the current supplied to the motor.

The efficiency of a motor $= \dfrac{\text{output power}}{\text{input power}} \times 100\%$.

The output power of a motor is the power available to do work at its shaft and is given by $T\omega$ or $T(2\pi n)$ watts, where $T$ is the torque in Nm and $n$ is the speed of rotation in rev/s. The input power is the electrical power in watts supplied to the motor, i.e. $VI$ watts.
Thus for a motor,

efficiency, $\qquad \eta = \dfrac{T(2\pi n)}{VI} \times 100\%$

i.e. $\qquad 80 = \left[\dfrac{(15)(2\pi n)\left(\dfrac{1200}{60}\right)}{(200)(I)}\right] \times 100$

Thus the current supplied,

$$I = \frac{(15)(2\pi)(20)(100)}{(200)(80)}$$

$$= \mathbf{11.8\,A}$$

**Problem 27.** A d.c. series motor drives a load at 30 rev/s and takes a current of 10 A when the supply voltage is 400 V. If the total resistance of the motor is $2\,\Omega$ and the iron, friction and windage losses amount to 300 W, determine the efficiency of the motor.

Efficiency,

$$\eta = \left( \frac{VI - I^2R - C}{VI} \right) \times 100\%$$

$$= \left( \frac{(400)(10) - (10)^2(2) - 300}{(400)(10)} \right) \times 100\%$$

$$= \left( \frac{4000 - 200 - 300}{4000} \right) \times 100\%$$

$$= \left( \frac{3500}{4000} \right) \times 100\% = \mathbf{87.5\%}$$

**Now try the following exercise**

**Exercise 136    Further problems on d.c. motors**

1.  A 240 V shunt motor takes a total current of 80 A. If the field winding resistance is 120 Ω and the armature resistance is 0.4 Ω, determine (a) the current in the armature, and (b) the back e.m.f.                    [(a) 78 A (b) 208.8 V]

2.  A d.c. motor has a speed of 900 rev/min when connected to a 460 V supply. Find the approximate value of the speed of the motor when connected to a 200 V supply, assuming the flux decreases by 30 per cent and neglecting the armature volt drop.            [559 rev/min]

3.  A series motor having a series field resistance of 0.25 Ω and an armature resistance of 0.15 Ω, is connected to a 220 V supply and at a particular load runs at 20 rev/s when drawing 20 A from the supply. Calculate the e.m.f. generated at this load. Determine also the speed of the motor when the load is changed such that the current increases to 25 A. Assume the flux increases by 25 per cent.
[212 V, 15.85 rev/s]

4.  A 500 V shunt motor takes a total current of 100 A and runs at 1200 rev/min. If the shunt field resistance is 50 Ω, the armature resistance is 0.25 Ω and the iron, friction and windage losses amount to 2 kW, determine the overall efficiency of the motor.        [81.95 per cent]

5.  A 250 V, series-wound motor is running at 500 rev/min and its shaft torque is 130 Nm.

If its efficiency at this load is 88 per cent, find the current taken from the supply.
[30.94 A]

6.  In a test on a d.c. motor, the following data was obtained. Supply voltage: 500 V, current taken from the supply: 42.4 A, speed: 850 rev/min, shaft torque: 187 Nm. Determine the efficiency of the motor correct to the nearest 0.5 per cent.            [78.5 per cent]

7.  A 300 V series motor draws a current of 50 A. The field resistance is 40 mΩ and the armature resistance is 0.2 Ω. Determine the maximum efficiency of the motor.        [92 per cent]

8.  A series motor drives a load at 1500 rev/min and takes a current of 20 A when the supply voltage is 250 V. If the total resistance of the motor is 1.5 Ω and the iron, friction and windage losses amount to 400 W, determine the efficiency of the motor.        [80 per cent]

9.  A series-wound motor is connected to a d.c. supply and develops full-load torque when the current is 30 A and speed is 1000 rev/min. If the flux per pole is proportional to the current flowing, find the current and speed at half full-load torque, when connected to the same supply.            [21.2 A, 1415 rev/min]

## 22.14    D.C. motor starter

If a d.c. motor whose armature is stationary is switched directly to its supply voltage, it is likely that the fuses protecting the motor will burn out. This is because the armature resistance is small, frequently being less than one ohm. Thus, additional resistance must be added to the armature circuit at the instant of closing the switch to start the motor.

As the speed of the motor increases, the armature conductors are cutting flux and a generated voltage, acting in opposition to the applied voltage, is produced, which limits the flow of armature current. Thus the value of the additional armature resistance can then be reduced.

When at normal running speed, the generated e.m.f. is such that no additional resistance is required in the armature circuit. To achieve this varying resistance in the armature circuit on starting, a d.c. motor starter is used, as shown in Fig. 22.28.

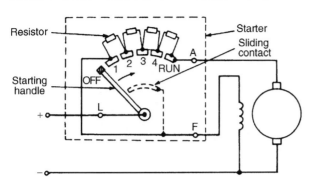

**Figure 22.28**

The starting handle is moved **slowly** in a clockwise direction to start the motor. For a shunt-wound motor, the field winding is connected to stud 1 or to $L$ via a sliding contact on the starting handle, to give maximum field current, hence maximum flux, hence maximum torque on starting, since $T \propto \Phi I_a$. A similar arrangement without the field connection is used for series motors.

## 22.15 Speed control of d.c. motors

### Shunt-wound motors

The speed of a shunt-wound d.c. motor, $n$, is proportional to

$$\frac{V - I_a R_a}{\Phi}$$

(see equation (9)). The speed is varied either by varying the value of flux, $\Phi$, or by varying the value of $R_a$. The former is achieved by using a variable resistor in series with the field winding, as shown in Fig. 22.29(a) and such a resistor is called the **shunt field regulator**.

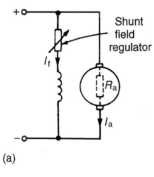
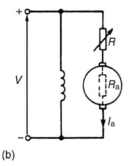

(a)                              (b)

**Figure 22.29**

As the value of resistance of the shunt field regulator is increased, the value of the field current, $I_f$, is decreased.

This results in a decrease in the value of flux, $\Phi$, and hence an increase in the speed, since $n \propto 1/\Phi$. Thus only speeds **above** that given without a shunt field regulator can be obtained by this method. Speeds **below** those given by

$$\frac{V - I_a R_a}{\Phi}$$

are obtained by increasing the resistance in the armature circuit, as shown in Fig. 22.29(b), where

$$n \propto \frac{V - I_a(R_a + R)}{\Phi}$$

Since resistor $R$ is in series with the armature, it carries the full armature current and results in a large power loss in large motors where a considerable speed reduction is required for long periods.

These methods of speed control are demonstrated in the following worked problem.

**Problem 28.** A 500 V shunt motor runs at its normal speed of 10 rev/s when the armature current is 120 A. The armature resistance is 0.2 Ω.
(a) Determine the speed when the current is 60 A and a resistance of 0.5 Ω is connected in series with the armature, the shunt field remaining constant.
(b) Determine the speed when the current is 60 A and the shunt field is reduced to 80 per cent of its normal value by increasing resistance in the field circuit.

(a)  With reference to Fig. 22.29(b), back e.m.f. at 120 A, $E_1 = V - I_a R_a = 500 - (120)(0.2)$

$$= 500 - 24 = 476 \text{ volts.}$$

When $I_a = 60$ A,

$$E_2 = 500 - (60)(0.2 + 0.5)$$

$$= 500 - (60)(0.7)$$

$$= 500 - 42 = 458 \text{ volts}$$

Now $\dfrac{E_1}{E_2} = \dfrac{\Phi_1 n_1}{\Phi_2 n_2}$

i.e. $\dfrac{476}{458} = \dfrac{\Phi_1 (10)}{\Phi_1 n_2}$  since $\Phi_2 = \Phi_1$

from which,

**speed $n_2$** $= \dfrac{(10)(458)}{476} = \mathbf{9.62\,rev/s}$

(b) Back e.m.f. when $I_a = 60\,\text{A}$,

$$E_3 = 500 - (60)(0.2)$$

$$= 500 - 12 = 488\,\text{volts}$$

Now $\dfrac{E_1}{E_3} = \dfrac{\Phi_1 n_1}{\Phi_3 n_3}$

i.e. $\dfrac{476}{488} = \dfrac{\Phi_1(10)}{0.8\Phi_1 n_3}$   since $\Phi_3 = 0.8\,\Phi_1$

from which,

**speed** $n_3 = \dfrac{(10)(488)}{(0.8)(476)} = \mathbf{12.82\,rev/s}$

## Series-wound motors

The speed control of series-wound motors is achieved using either (a) field resistance, or (b) armature resistance techniques.

(a) The speed of a d.c. series-wound motor is given by:

$$n = k\left(\frac{V - IR}{\Phi}\right)$$

where $k$ is a constant, $V$ is the terminal voltage, $R$ is the combined resistance of the armature and series field and $\Phi$ is the flux. Thus, a reduction in flux results in an increase in speed. This is achieved by putting a variable resistance in parallel with the field winding and reducing the field current, and hence flux, for a given value of supply current. A circuit diagram of this arrangement is shown in Fig. 22.30(a). A variable resistor connected in parallel with the series-wound field to control speed is called a **diverter**. Speeds above those given with no diverter are obtained by this method. Problem 29 below demonstrates this method.

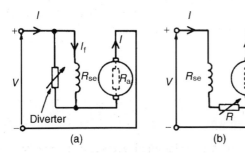

Diverter

(a)        (b)

**Figure 22.30**

(b) Speeds below normal are obtained by connecting a variable resistor in series with the field winding and armature circuit, as shown in Fig. 22.30(b). This effectively increases the value of $R$ in the equation

$$n = k\left(\frac{V - IR}{\Phi}\right)$$

and thus reduces the speed. Since the additional resistor carries the full supply current, a large power loss is associated with large motors in which a considerable speed reduction is required for long periods. This method is demonstrated in problem 30.

> **Problem 29.** On full-load a 300 V series motor takes 90 A and runs at 15 rev/s. The armature resistance is $0.1\,\Omega$ and the series winding resistance is $50\,\text{m}\Omega$. Determine the speed when developing full load torque but with a $0.2\,\Omega$ diverter in parallel with the field winding. (Assume that the flux is proportional to the field current.)

At 300 V, e.m.f.

$$E_1 = V - IR = V - I(R_a + R_{se})$$

$$= 300 - (90)(0.1 + 0.05)$$

$$= 300 - (90)(0.15)$$

$$= 300 - 13.5 = 286.5\,\text{volts}$$

With the $0.2\,\Omega$ diverter in parallel with $R_{se}$ (see Fig. 22.30(a)), the equivalent resistance,

$$R = \frac{(0.2)(0.05)}{0.2 + 0.05} = \frac{(0.2)(0.05)}{0.25} = 0.04\,\Omega$$

By current division, current

$$I_1\,(\text{in Fig. 22.30(a)}) = \left(\frac{0.2}{0.2 + 0.05}\right)I = 0.8\,I$$

Torque, $T \propto I_a \Phi$ and for full load torque, $I_{a1}\Phi_1 = I_{a2}\Phi_2$.

Since flux is proportional to field current $\Phi_1 \propto I_{a1}$ and $\Phi_2 \propto 0.8\,I_{a2}$ then $(90)(90) = (I_{a2})(0.8\,I_{a2})$

from which,   $I_{a2}^2 = \dfrac{90^2}{0.8}$

and   $I_{a2} = \dfrac{90}{\sqrt{0.8}} = 100.62\,\text{A}$

Hence e.m.f. 
$$E_2 = V - I_{a2}(R_a + R)$$
$$= 300 - (100.62)(0.1 + 0.04)$$
$$= 300 - (100.62)(0.14)$$
$$= 300 - 14.087 = 285.9 \text{ volts}$$

Now e.m.f., $E \propto \Phi n$, from which,

$$\frac{E_1}{E_2} = \frac{\Phi_1 n_1}{\Phi_2 n_2} = \frac{I_{a1} n_1}{0.8 I_{a2} n_2}$$

Hence 
$$\frac{286.5}{285.9} = \frac{(90)(15)}{(0.8)(100.62)n_2}$$

and **new speed, $n_2$** $= \dfrac{(285.9)(90)(15)}{(286.5)(0.8)(100.62)}$

$$= 16.74 \text{ rev/s}$$

Thus the speed of the motor has increased from 15 rev/s (i.e. 900 rev/min) to 16.74 rev/s (i.e. 1004 rev/min) by inserting a 0.2 Ω diverter resistance in parallel with the series winding.

**Problem 30.** A series motor runs at 800 rev/min when the voltage is 400 V and the current is 25 A. The armature resistance is 0.4 Ω and the series field resistance is 0.2 Ω. Determine the resistance to be connected in series to reduce the speed to 600 rev/min with the same current.

With reference to Fig. 22.30(b), at 800 rev/min,

e.m.f., 
$$E_1 = V - I(R_a + R_{se})$$
$$= 400 - (25)(0.4 + 0.2)$$
$$= 400 - (25)(0.6)$$
$$= 400 - 15 = 385 \text{ volts}$$

At 600 rev/min, since the current is unchanged, the flux is unchanged.
Thus $E \propto \Phi n$ or $E \propto n$ and

$$\frac{E_1}{E_2} = \frac{n_1}{n_2}$$

Hence 
$$\frac{385}{E_2} = \frac{800}{600}$$

from which, 
$$E_2 = \frac{(385)(600)}{800} = 288.75 \text{ volts}$$

and 
$$E_2 = V - I(R_a + R_{se} + R)$$
Hence 
$$288.75 = 400 - 25(0.4 + 0.2 + R)$$
Rearranging gives:

$$0.6 + R = \frac{400 - 288.75}{25} = 4.45$$

from which, extra series resistance, $R = 4.45 - 0.6$ i.e.
**R = 3.85 Ω.**
Thus the addition of a series resistance of 3.85 Ω has reduced the speed from 800 rev/min to 600 rev/min.

**Now try the following exercise**

**Exercise 137  Further problems on the speed control of d.c. motors**

1. A 350 V shunt motor runs at its normal speed of 12 rev/s when the armature current is 90 A. The resistance of the armature is 0.3 Ω.
   (a) Find the speed when the current is 45 A and a resistance of 0.4 Ω is connected in series with the armature, the shunt field remaining constant.
   (b) Find the speed when the current is 45 A and the shunt field is reduced to 75 per cent of its normal value by increasing resistance in the field circuit.
   [(a) 11.83 rev/s (b) 16.67 rev/s]

2. A series motor runs at 900 rev/min when the voltage is 420 V and the current is 40 A. The armature resistance is 0.3 Ω and the series field resistance is 0.2 Ω. Calculate the resistance to be connected in series to reduce the speed to 720 rev/min with the same current. [2 Ω]

3. A 320 V series motor takes 80 A and runs at 1080 rev/min at full load. The armature resistance is 0.2 Ω and the series winding resistance is 0.05 Ω. Assuming the flux is proportional to the field current, calculate the speed when developing full-load torque, but with a 0.15 Ω diverter in parallel with the field winding.
   [1239 rev/min]

## 22.16  Motor cooling

Motors are often classified according to the type of enclosure used, the type depending on the conditions under which the motor is used and the degree of ventilation required.

The most common type of protection is the **screen-protected type**, where ventilation is achieved by fitting a fan internally, with the openings at the end of the motor fitted with wire mesh.

A **drip-proof type** is similar to the screen-protected type but has a cover over the screen to prevent drips of water entering the machine.

A **flame-proof type** is usually cooled by the conduction of heat through the motor casing.

With a **pipe-ventilated type**, air is piped into the motor from a dust-free area, and an internally fitted fan ensures the circulation of this cool air.

**Now try the following exercises**

**Exercise 138    Short answer questions on d.c. machines**

1. A ...... converts mechanical energy into electrical energy

2. A ...... converts electrical energy into mechanical energy

3. What does 'commutation' achieve?

4. Poor commutation may cause sparking. How can this be improved?

5. State any five basic parts of a d.c. machine

6. State the two groups armature windings can be divided into

7. What is armature reaction? How can it be overcome?

8. The e.m.f. generated in an armature winding is given by $E = 2p\Phi nZ/c$ volts. State what $p$, $\Phi$, $n$, $Z$ and $c$ represent

9. In a series-wound d.c. machine, the field winding is in ...... with the armature circuit

10. In a d.c. generator, the relationship between the generated voltage, terminal voltage, current and armature resistance is given by $E = $ ......

11. A d.c. machine has its field winding in parallel with the armatures circuit. It is called a ...... wound machine

12. Sketch a typical open-circuit characteristic for (a) a separately excited generator (b) a shunt generator (c) a series generator

13. Sketch a typical load characteristic for (a) a separately excited generator (b) a shunt generator

14. State one application for (a) a shunt generator (b) a series generator (c) a compound generator

15. State the principle losses in d.c. machines

16. The efficiency of a d.c. machine is given by the ratio (......) per cent

17. The equation relating the generated e.m.f., $E$, terminal voltage, armature current and armature resistance for a d.c. motor is $E = $ ......

18. The torque $T$ of a d.c. motor is given by $T = p\Phi ZI_a/\pi c$ newton metres. State what $p$, $\Phi$, $Z$, $I$ and $c$ represent

19. Complete the following. In a d.c. machine
    (a) generated e.m.f. $\propto$ ...... $\times$ ......
    (b) torque $\propto$ ...... $\times$ ......

20. Sketch typical characteristics of torque/armature current for
    (a) a shunt motor
    (b) a series motor
    (c) a compound motor

21. Sketch typical speed/torque characteristics for a shunt and series motor

22. State two applications for each of the following motors:
    (a) shunt    (b) series    (c) compound

In questions 23 to 26, an electrical machine runs at $n$ rev/s, has a shaft torque of $T$, and takes a current of $I$ from a supply voltage $V$

23. The power input to a generator is ...... watts

24. The power input to a motor is ...... watts

25. The power output from a generator is ...... watts

26. The power output from a motor is ...... watts

27. The generated e.m.f. of a d.c machine is proportional to ...... volts

28. The torque produced by a d.c. motor is proportional to ...... Nm

29. A starter is necessary for a d.c. motor because the generated e.m.f. is ...... at low speeds

30. The speed of a d.c. shunt-wound motor will ...... if the value of resistance of the shunt field regulator is increased

31. The speed of a d.c. motor will ...... if the value of resistance in the armature circuit is increased

32. The value of the speed of a d.c. shunt-wound motor ...... as the value of the armature current increases

33. At a large value of torque, the speed of a d.c. series-wound motor is ......

34. At a large value of field current, the generated e.m.f. of a d.c. shunt-wound generator is approximately ......

35. In a series-wound generator, the terminal voltage increases as the load current ......

36. One type of d.c. motor uses resistance in series with the field winding to obtain speed variations and another type uses resistance in parallel with the field winding for the same purpose. Explain briefly why these two distinct methods are used and why the field current plays a significant part in controlling the speed of a d.c. motor

37. Name three types of motor enclosure

**Exercise 139  Multi-choice questions on d.c. machines**
**(Answers on page 421)**

1. Which of the following statements is false?
   (a) A d.c. motor converts electrical energy to mechanical energy
   (b) The efficiency of a d.c. motor is the ratio input power to output power
   (c) A d.c. generator converts mechanical power to electrical power
   (d) The efficiency of a d.c. generator is the ratio output power to input power
   A shunt-wound d.c. machine is running at n rev/s and has a shaft torque of $T$ Nm. The

supply current is $I$A when connected to d.c. bus-bars of voltage $V$ volts. The armature resistance of the machine is $R_a$ ohms, the armature current is $I_aA$ and the generated voltage is $E$ volts. Use this data to find the formulae of the quantities stated in questions 2 to 9, selecting the correct answer from the following list:
(a) $V - I_aR_a$     (b) $E + I_aR_a$
(c) $VI$     (d) $E - I_aR_a$
(e) $T(2\pi n)$     (f) $V + I_aR_a$

2. The input power when running as a generator

3. The output power when running as a motor

4. The input power when running as a motor

5. The output power when running as a generator

6. The generated voltage when running as a motor

7. The terminal voltage when running as a generator

8. The generated voltage when running as a generator

9. The terminal voltage when running as a motor

10. Which of the following statements is false?
    (a) A commutator is necessary as part of a d.c. motor to keep the armature rotating in the same direction
    (b) A commutator is necessary as part of a d.c. generator to produce unidirectional voltage at the terminals of the generator
    (c) The field winding of a d.c. machine is housed in slots on the armature
    (d) The brushes of a d.c. machine are usually made of carbon and do not rotate with the armature

11. If the speed of a d.c. machine is doubled and the flux remains constant, the generated e.m.f. (a) remains the same (b) is doubled (c) is halved

12. If the flux per pole of a shunt-wound d.c. generator is increased, and all other variables are kept the same, the speed (a) decreases (b) stays the same (c) increases

13. If the flux per pole of a shunt-wound d.c. generator is halved, the generated e.m.f. at

Section 3

constant speed (a) is doubled (b) is halved (c) remains the same

14. In a series-wound generator running at constant speed, as the load current increases, the terminal voltage
    (a) increases       (b) decreases
    (c) stays the same

15. Which of the following statements is false for a series-wound d.c. motor?
    (a) The speed decreases with increase of resistance in the armature circuit
    (b) The speed increases as the flux decreases
    (c) The speed can be controlled by a diverter
    (d) The speed can be controlled by a shunt field regulator

16. Which of the following statements is false?
    (a) A series-wound motor has a large starting torque
    (b) A shunt-wound motor must be permanently connected to its load
    (c) The speed of a series-wound motor drops considerably when load is applied
    (d) A shunt-wound motor is essentially a constant-speed machine

17. The speed of a d.c. motor may be increased by
    (a) increasing the armature current
    (b) decreasing the field current
    (c) decreasing the applied voltage
    (d) increasing the field current

18. The armature resistance of a d.c. motor is $0.5\,\Omega$, the supply voltage is 200 V and the back e.m.f. is 196 V at full speed. The armature current is:
    (a) 4 A       (b) 8 A
    (c) 400 A       (d) 392 A

19. In d.c. generators iron losses are made up of:
    (a) hysteresis and friction losses
    (b) hysteresis, eddy current and brush contact losses
    (c) hysteresis and eddy current losses
    (d) hysteresis, eddy current and copper losses

20. The effect of inserting a resistance in series with the field winding of a shunt motor is to:
    (a) increase the magnetic field
    (b) increase the speed of the motor

(c) decrease the armature current
(d) reduce the speed of the motor

21. The supply voltage to a d.c. motor is 240 V. If the back e.m.f. is 230 V and the armature resistance is $0.25\,\Omega$, the armature current is:
    (a) 10 A       (b) 40 A
    (c) 960 A       (d) 920 A

22. With a d.c. motor, the starter resistor:
    (a) limits the armature current to a safe starting value
    (b) controls the speed of the machine
    (c) prevents the field current flowing through and damaging the armature
    (d) limits the field current to a safe starting value

23. From Fig. 22.31, the expected characteristic for a shunt-wound d.c. generator is:
    (a) P       (b) Q
    (c) R       (d) S

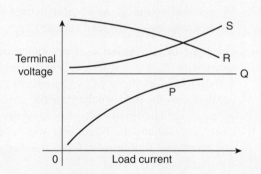

**Figure 22.31**

24. A commutator is a device fitted to a generator. Its function is:
    (a) to prevent sparking when the load changes
    (b) to convert the a.c. generated into a d.c. output
    (c) to convey the current to and from the windings
    (d) to generate a direct current

# Chapter 23

# Three-phase induction motors

At the end of this chapter you should be able to:

- appreciate the merits of three-phase induction motors
- understand how a rotating magnetic field is produced
- state the synchronous speed, $n_s = (f/p)$ and use in calculations
- describe the principle of operation of a three-phase induction motor
- distinguish between squirrel-cage and wound-rotor types of motor
- understand how a torque is produced causing rotor movement
- understand and calculate slip
- derive expressions for rotor e.m.f., frequency, resistance, reactance, impedance, current and copper loss, and use them in calculations
- state the losses in an induction motor and calculate efficiency
- derive the torque equation for an induction motor, state the condition for maximum torque, and use in calculations
- describe torque-speed and torque-slip characteristics for an induction motor
- state and describe methods of starting induction motors
- state advantages of cage rotor and wound rotor types of induction motor
- describe the double cage induction motor
- state typical applications of three-phase induction motors

## 23.1 Introduction

In d.c. motors, introduced in Chapter 22, conductors on a rotating armature pass through a stationary magnetic field. In a **three-phase induction motor**, the magnetic field rotates and this has the advantage that no external electrical connections to the rotor need be made. Its name is derived from the fact that the current in the rotor is **induced** by the magnetic field instead of being supplied through electrical connections to the supply.

The result is a motor which: (i) is cheap and robust, (ii) is explosion proof, due to the absence of a commutator or slip-rings and brushes with their associated sparking, (iii) requires little or no skilled maintenance, and (iv) has self-starting properties when switched to a supply with no additional expenditure on auxiliary equipment. The principal disadvantage of a three-phase induction motor is that its speed cannot be readily adjusted.

DOI: 10.1016/B978-0-08-089056-2.00023-1

## 23.2 Production of a rotating magnetic field

When a three-phase supply is connected to symmetrical three-phase windings, the currents flowing in the windings produce a magnetic field. This magnetic field is constant in magnitude and rotates at constant speed as shown below, and is called the **synchronous speed**.

With reference to Fig. 23.1, the windings are represented by three single-loop conductors, one for each phase, marked $R_S R_F$, $Y_S Y_F$ and $B_S B_F$, the S and F signifying start and finish. In practice, each phase winding comprises many turns and is distributed around the stator; the single-loop approach is for clarity only.

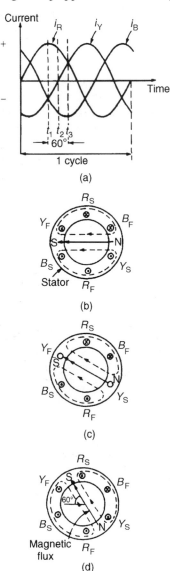

Figure 23.1

When the stator windings are connected to a three-phase supply, the current flowing in each winding varies with time and is as shown in Fig. 23.1(a). If the value of current in a winding is positive, the assumption is made that it flows from start to finish of the winding, i.e. if it is the red phase, current flows from $R_S$ to $R_F$, i.e. away from the viewer in $R_S$ and towards the viewer in $R_F$. When the value of current is negative, the assumption is made that it flows from finish to start, i.e. towards the viewer in an 'S' winding and away from the viewer in an 'F' winding. At time, say $t_1$, shown in Fig. 23.1(a), the current flowing in the red phase is a maximum positive value. At the same time $t_1$, the currents flowing in the yellow and blue phases are both 0.5 times the maximum value and are negative.

The current distribution in the stator windings is therefore as shown in Fig. 23.1(b), in which current flows away from the viewer, (shown as $\otimes$) in $R_S$ since it is positive, but towards the viewer (shown as $\odot$) in $Y_S$ and $B_S$, since these are negative. The resulting magnetic field is as shown, due to the 'solenoid' action and application of the corkscrew rule.

A short time later at time $t_2$, the current flowing in the red phase has fallen to about 0.87 times its maximum value and is positive, the current in the yellow phase is zero and the current in the blue phase is about 0.87 times its maximum value and is negative. Hence the currents and resultant magnetic field are as shown in Fig. 23.1(c). At time $t_3$, the currents in the red and yellow phases are 0.5 of their maximum values and the current in the blue phase is a maximum negative value. The currents and resultant magnetic field are as shown in Fig. 23.1(d).

Similar diagrams to Fig. 23.1(b), (c) and (d) can be produced for all time values and these would show that the magnetic field travels through one revolution for each cycle of the supply voltage applied to the stator windings.

By considering the flux values rather than the current values, it is shown below that the rotating magnetic field has a constant value of flux. The three coils shown in Fig. 23.2(a), are connected in star to a three-phase supply. Let the positive directions of the fluxes produced by currents flowing in the coils, be $\phi_A$, $\phi_B$ and $\phi_C$ respectively. The directions of $\phi_A$, $\phi_B$ and $\phi_C$ do not alter, but their magnitudes are proportional to the currents flowing in the coils at any particular time. At time $t_1$, shown in Fig. 23.2(b), the currents flowing in the coils are:

$i_B$, a maximum positive value, i.e. the flux is towards point P; $i_A$ and $i_C$, half the maximum value and negative, i.e. the flux is away from point P.

These currents give rise to the magnetic fluxes $\phi_A$, $\phi_B$ and $\phi_C$, whose magnitudes and directions are as shown

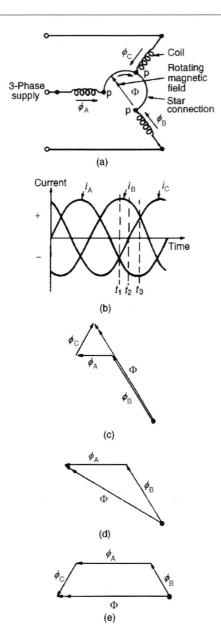

**Figure 23.2**

in Fig. 23.2(c). The resultant flux is the phasor sum of $\phi_A$, $\phi_B$ and $\phi_C$, shown as $\Phi$ in Fig. 23.2(c). At time $t_2$, the currents flowing are:

$i_B$, $0.866 \times$ maximum positive value, $i_C$, zero, and $i_A$, $0.866 \times$ maximum negative value.

The magnetic fluxes and the resultant magnetic flux are as shown in Fig. 23.2(d).

At time $t_3$,

$i_B$ is $0.5 \times$ maximum value and is positive

$i_A$ is a maximum negative value, and

$i_C$ is $0.5 \times$ maximum value and is positive.

The magnetic fluxes and the resultant magnetic flux are as shown in Fig. 23.2(e).

Inspection of Fig. 23.2(c), (d) and (e) shows that the magnitude of the resultant magnetic flux, $\Phi$, in each case is constant and is $1\frac{1}{2} \times$ the maximum value of $\phi_A$, $\phi_B$ or $\phi_C$, but that its direction is changing. The process of determining the resultant flux may be repeated for all values of time and shows that the magnitude of the resultant flux is constant for all values of time and also that it rotates at constant speed, making one revolution for each cycle of the supply voltage.

## 23.3    Synchronous speed

The rotating magnetic field produced by three-phase windings could have been produced by rotating a permanent magnet's north and south pole at synchronous speed, (shown as N and S at the ends of the flux phasors in Fig. 23.1(b), (c) and (d)). For this reason, it is called a 2-pole system and an induction motor using three-phase windings only is called a 2-pole induction motor. If six windings displaced from one another by $60°$ are used, as shown in Fig. 23.3(a), by drawing the current and resultant magnetic field diagrams at various time values, it may be shown that one cycle of the supply

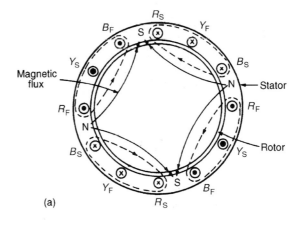

(a)

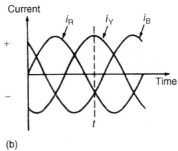

(b)

**Figure 23.3**

Section 3

current to the stator windings causes the magnetic field to move through half a revolution. The current distribution in the stator windings are shown in Fig. 23.3(a), for the time $t$ shown in Fig. 23.3(b).

It can be seen that for six windings on the stator, the magnetic flux produced is the same as that produced by rotating two permanent magnet north poles and two permanent magnet south poles at synchronous speed. This is called a 4-pole system and an induction motor using six phase windings is called a 4-pole induction motor. By increasing the number of phase windings the number of poles can be increased to any even number.

In general, if $f$ is the frequency of the currents in the stator windings and the stator is wound to be equivalent to $p$ **pairs** of poles, the speed of revolution of the rotating magnetic field, i.e. the synchronous speed, $n_s$ is given by:

$$n_s = \frac{f}{p}\text{rev/s}$$

**Problem 1.**  A three-phase 2-pole induction motor is connected to a 50 Hz supply. Determine the synchronous speed of the motor in rev/min.

From above, $n_s = (f/p)$ rev/s, where $n_s$ is the synchronous speed, $f$ is the frequency in hertz of the supply to the stator and $p$ is the number of **pairs** of poles. Since the motor is connected to a 50 hertz supply, $f = 50$.

The motor has a two-pole system, hence $p$, the number of pairs of poles, is 1. Thus, synchronous speed, $n_s = (50/1) = 50$ rev/s $= 50 \times 60$ rev/min
$$= 3000 \text{ rev/min}.$$

**Problem 2.**  A stator winding supplied from a three-phase 60 Hz system is required to produce a magnetic flux rotating at 900 rev/min. Determine the number of poles.

Synchronous speed,

$$n_s = 900 \text{ rev/min} = \frac{900}{60} \text{ rev/s} = 15 \text{ rev/s}$$

Since

$$n_s = \left(\frac{f}{p}\right) \text{ then } p = \left(\frac{f}{n_s}\right) = \left(\frac{60}{15}\right) = 4$$

Hence **the number of pole pairs is 4** and thus **the number of poles is 8**

**Problem 3.**  A three-phase 2-pole motor is to have a synchronous speed of 6000 rev/min. Calculate the frequency of the supply voltage.

Since $n_s = \left(\frac{f}{p}\right)$ then

$$\textbf{frequency, } f = (n_s)(p)$$
$$= \left(\frac{6000}{60}\right)\left(\frac{2}{2}\right) = \textbf{100 Hz}$$

Now try the following exercise

**Exercise 140  Further problems on synchronous speed**

1. The synchronous speed of a 3-phase, 4-pole induction motor is 60 rev/s. Determine the frequency of the supply to the stator windings. [120 Hz]

2. The synchronous speed of a 3-phase induction motor is 25 rev/s and the frequency of the supply to the stator is 50 Hz. Calculate the equivalent number of pairs of poles of the motor. [2]

3. A 6-pole, 3-phase induction motor is connected to a 300 Hz supply. Determine the speed of rotation of the magnetic field produced by the stator. [100 rev/s]

## 23.4 Construction of a three-phase induction motor

The stator of a three-phase induction motor is the stationary part corresponding to the yoke of a d.c. machine. It is wound to give a 2-pole, 4-pole, 6-pole, ...... rotating magnetic field, depending on the rotor speed required. The rotor, corresponding to the armature of a d.c. machine, is built up of laminated iron, to reduce eddy currents.

In the type most widely used, known as a **squirrel-cage rotor**, copper or aluminium bars are placed in slots cut in the laminated iron, the ends of the bars being welded or brazed into a heavy conducting ring, (see Fig. 23.4(a)). A cross-sectional view of a three-phase induction motor is shown in Fig. 23.4(b).

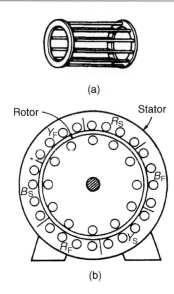

(a)

(b)

**Figure 23.4**

The conductors are placed in slots in the laminated iron rotor core. If the slots are skewed, better starting and quieter running is achieved. This type of rotor has no external connections which means that slip-rings and brushes are not needed. The squirrel-cage motor is cheap, reliable and efficient. Another type of rotor is the **wound rotor**. With this type there are phase windings in slots, similar to those in the stator. The windings may be connected in star or delta and the connections made to three slip rings. The slip-rings are used to add external resistance to the rotor circuit, particularly for starting (see Section 23.13), but for normal running the slip-rings are short-circuited.

The principle of operation is the same for both the squirrel-cage and the wound rotor machines.

## 23.5 Principle of operation of a three-phase induction motor

When a three-phase supply is connected to the stator windings, a rotating magnetic field is produced. As the magnetic flux cuts a bar on the rotor, an e.m.f. is induced in it and since it is joined, via the end conducting rings, to another bar one pole pitch away, a current flows in the bars. The magnetic field associated with this current flowing in the bars interacts with the rotating magnetic field and a force is produced, tending to turn the rotor in the same direction as the rotating magnetic field, (see Fig. 23.5). Similar forces are applied to all the conductors on the rotor, so that a torque is produced causing the rotor to rotate.

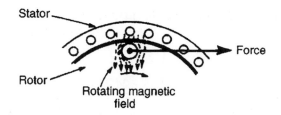

**Figure 23.5**

## 23.6 Slip

The force exerted by the rotor bars causes the rotor to turn in the direction of the rotating magnetic field. As the rotor speed increases, the rate at which the rotating magnetic field cuts the rotor bars is less and the frequency of the induced e.m.f.'s in the rotor bars is less. If the rotor runs at the same speed as the rotating magnetic field, no e.m.f.'s are induced in the rotor, hence there is no force on them and no torque on the rotor. Thus the rotor slows down. For this reason the rotor can never run at synchronous speed.

When there is no load on the rotor, the resistive forces due to windage and bearing friction are small and the rotor runs very nearly at synchronous speed. As the rotor is loaded, the speed falls and this causes an increase in the frequency of the induced e.m.f.'s in the rotor bars and hence the rotor current, force and torque increase. The difference between the rotor speed, $n_r$, and the synchronous speed, $n_s$, is called the **slip speed**, i.e.

$$\text{slip speed} = n_s - n_r \text{ rev/s}$$

The ratio $(n_s - n_r)/n_s$ is called the **fractional slip** or just the **slip**, $s$, and is usually expressed as a percentage. Thus

$$\text{slip, s} = \left(\frac{n_s - n_r}{n_s}\right) \times 100\%$$

Typical values of slip between no load and full load are about 4 to 5 per cent for small motors and 1.5 to 2 per cent for large motors.

Problem 4. The stator of a three-phase, 4-pole induction motor is connected to a 50 Hz supply. The rotor runs at 1455 rev/min at full load. Determine (a) the synchronous speed and (b) the slip at full load.

(a) The number of pairs of poles, $p = (4/2) = 2$. The supply frequency $f = 50$ Hz. The **synchronous speed**, $n_s = (f/p) = (50/2) = 25$ **rev/s**.

(b)   The rotor speed, $n_r = (1455/60) = 24.25$ rev/s.

$$\textbf{Slip, s} = \left(\frac{n_s - n_r}{n_s}\right) \times 100\%$$

$$= \left(\frac{25 - 24.25}{25}\right) \times 100\%$$

$$= \textbf{3\%}$$

> **Problem 5.**   A three-phase, 60 Hz induction motor has 2 poles. If the slip is 2 per cent at a certain load, determine (a) the synchronous speed, (b) the speed of the rotor, and (c) the frequency of the induced e.m.f.'s in the rotor.

(a)  $f = 60$ Hz and $p = (2/2) = 1$. Hence **synchronous speed**, $n_s = (f/p) = (60/1) = \textbf{60 rev/s}$ or $60 \times 60 = \textbf{3600 rev/min}$.

(b)  Since slip,

$$s = \left(\frac{n_s - n_r}{n_s}\right) \times 100\%$$

$$2 = \left(\frac{60 - n_r}{60}\right) \times 100$$

Hence

$$\frac{2 \times 60}{100} = 60 - n_r$$

i.e.

$$n_r = 60 - \frac{2 \times 60}{100} = 58.8 \text{ rev/s}$$

i.e. the rotor runs at $58.8 \times 60 = \textbf{3528 rev/min}$

(c)  Since the synchronous speed is 60 rev/s and that of the rotor is 58.8 rev/s, the rotating magnetic field cuts the rotor bars at $(60 - 58.8) = 1.2$ rev/s.

**Thus the frequency of the e.m.f.'s induced in the rotor bars, is** $f = n_s p = (1.2)(\frac{2}{2}) = \textbf{1.2 Hz}$.

> **Problem 6.**   A three-phase induction motor is supplied from a 50 Hz supply and runs at 1200 rev/min when the slip is 4 per cent. Determine the synchronous speed.

$$\text{Slip, } s = \left(\frac{n_s - n_r}{n_s}\right) \times 100\%$$

Rotor speed, $n_r = (1200/60) = 20$ rev/s and $s = 4$.

Hence

$$4 = \left(\frac{n_s - 20}{n_s}\right) \times 100\% \text{ or } 0.04 = \frac{n_s - 20}{n_s}$$

from which, $n_s(0.04) = n_s - 20$ and
$$20 = n_s - 0.04 n_s = n_s(1 - 0.04).$$

Hence **synchronous speed,**

$$n_s = \frac{20}{1 - 0.04} = 20.8\dot{3} \text{ rev/s}$$

$$= (20.8\dot{3} \times 60) \text{ rev/min}$$

$$= \textbf{1250 rev/min}$$

---

**Now try the following exercise**

**Exercise 141   Further problems on slip**

1.  A 6-pole, 3-phase induction motor runs at 970 rev/min at a certain load. If the stator is connected to a 50 Hz supply, find the percentage slip at this load.                     [3%]

2.  A 3-phase, 50 Hz induction motor has 8 poles. If the full load slip is 2.5 per cent, determine
    (a)  the synchronous speed,
    (b)  the rotor speed, and
    (c)  the frequency of the rotor e.m.f.'s.
        [(a) 750 rev/min (b) 731 rev/min (c) 1.25 Hz]

3.  A three-phase induction motor is supplied from a 60 Hz supply and runs at 1710 rev/min when the slip is 5 per cent. Determine the synchronous speed.                     [1800 rev/min]

4.  A 4-pole, 3-phase, 50 Hz induction motor runs at 1440 rev/min at full load. Calculate
    (a)  the synchronous speed,
    (b)  the slip, and
    (c)  the frequency of the rotor induced e.m.f.'s.
        [(a) 1500 rev/min (b) 4% (c) 2 Hz]

## 23.7   Rotor e.m.f. and frequency

### Rotor e.m.f.

When an induction motor is stationary, the stator and rotor windings form the equivalent of a transformer as shown in Fig. 23.6.

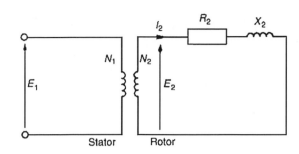

**Figure 23.6**

The rotor e.m.f. at standstill is given by

$$E_2 = \left(\frac{N_2}{N_1}\right) E_1 \qquad (1)$$

where $E_1$ is the supply voltage per phase to the stator.

When an induction motor is running, the induced e.m.f. in the rotor is less since the relative movement between conductors and the rotating field is less. The induced e.m.f. is proportional to this movement, hence it must be proportional to the slip, $s$. Hence **when running**, rotor e.m.f. per phase $= E_r = sE_2$

i.e. rotor e.m.f. per phase $= s \left(\dfrac{N_2}{N_1}\right) E_1 \qquad (2)$

## Rotor frequency

The rotor e.m.f. is induced by an alternating flux and the rate at which the flux passes the conductors is the slip speed. Thus the frequency of the rotor e.m.f. is given by:

$$f_r = (n_s - n_r)p = \left(\frac{n_s - n_r}{n_s}\right)(n_s p)$$

However $(n_s - n_r)/n_s$ is the slip $s$ and $(n_s p)$ is the supply frequency $f$, hence

$$f_r = sf \qquad (3)$$

> **Problem 7.** The frequency of the supply to the stator of an 8-pole induction motor is 50 Hz and the rotor frequency is 3 Hz. Determine (a) the slip, and (b) the rotor speed.

(a)  From equation (3), $f_r = sf$. Hence $3 = (s)(50)$ from which,

$$\textbf{slip, } \textbf{\textit{s}} = \frac{3}{50} = \textbf{0.06 or 6\%}$$

(b)  Synchronous speed, $n_s = f/p = 50/4 = 12.5$ rev/s or $(12.5 \times 60) = 750$ rev/min

$$\text{Slip, } s = \left(\frac{n_s - n_r}{n_s}\right)$$

hence $\qquad 0.06 = \left(\dfrac{12.5 - n_r}{12.5}\right)$

$$(0.06)(12.5) = 12.5 - n_r$$

and **rotor speed,**

$$\boldsymbol{n_r} = 12.5 - (0.06)(12.5)$$

$$= \textbf{11.75 rev/s or 705 rev/min}$$

**Now try the following exercise**

> **Exercise 142   Further problems on rotor frequency**
>
> 1.  A 12-pole, 3-phase, 50 Hz induction motor runs at 475 rev/min. Determine
>     (a)  the slip speed,
>     (b)  the percentage slip, and
>     (c)  the frequency of rotor currents.
>           [(a) 25 rev/min (b) 5% (c) 2.5 Hz]
>
> 2.  The frequency of the supply to the stator of a 6-pole induction motor is 50 Hz and the rotor frequency is 2 Hz. Determine
>     (a)  the slip, and
>     (b)  the rotor speed, in rev/min.
>           [(a) 0.04 or 4% (b) 960 rev/min]

## 23.8   Rotor impedance and current

### Rotor resistance

The rotor resistance $R_2$ is unaffected by frequency or slip, and hence remains constant.

### Rotor reactance

Rotor reactance varies with the frequency of the rotor current. At standstill, reactance per phase, $X_2 = 2\pi f L$. When running, reactance per phase,

$$X_r = 2\pi f_r L$$

$$= 2\pi (sf)L \quad \text{from equation (3)}$$

$$= s(2\pi f L)$$

i.e.  $\qquad \boldsymbol{X_r = sX_2} \qquad (4)$

Figure 23.7 represents the rotor circuit when running.

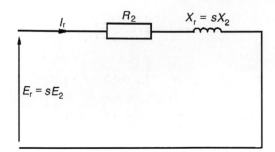

**Figure 23.7**

### Rotor impedance

Rotor impedance per phase,

$$Z_r = \sqrt{R_2^2 + (sX_2)^2} \qquad (5)$$

At standstill, slip $s = 1$, then

$$Z_2 = \sqrt{R_2^2 + X_2^2} \qquad (6)$$

### Rotor current

From Fig. 23.6 and 23.7, **at standstill, starting current,**

$$I_2 = \frac{E_2}{Z_2} = \frac{\left(\dfrac{N_2}{N_1}\right)E_1}{\sqrt{R_2^2 + X_2^2}} \qquad (7)$$

and when running, current,

$$I_r = \frac{E_r}{Z_r} = \frac{s\left(\dfrac{N_2}{N_1}\right)E_1}{\sqrt{R_2^2 + (sX_2)^2}} \qquad (8)$$

## 23.9   Rotor copper loss

Power $P = 2\pi nT$, where $T$ is the torque in newton metres, hence torque $T = (P/2\pi n)$. If $P_2$ is the power input to the rotor from the rotating field, and $P_m$ is the mechanical power output (including friction losses)

then

$$T = \frac{P_2}{2\pi n_s} = \frac{P_m}{2\pi n_r}$$

from which,

$$\frac{P_2}{n_s} = \frac{P_m}{n_r} \quad \text{or} \quad \frac{P_m}{P_2} = \frac{n_r}{n_s}$$

Hence

$$1 - \frac{P_m}{P_2} = 1 - \frac{n_r}{n_s}$$

$$\frac{P_2 - P_m}{P_2} = \frac{n_s - n_r}{n_s} = s$$

$P_2 - P_m$ is the electrical or copper loss in the rotor, i.e. $P_2 - P_m = I_r^2 R_2$. Hence

$$\text{slip}, s = \frac{\text{rotor copper loss}}{\text{rotor input}} = \frac{I_r^2 R_2}{P_2} \qquad (9)$$

or power input to the rotor,

$$P_2 = \frac{I_r^2 R_2}{s} \qquad (10)$$

## 23.10   Induction motor losses and efficiency

Figure 23.8 summarises losses in induction motors. Motor efficiency,

$$\eta = \frac{\text{output power}}{\text{input power}} = \frac{P_m}{P_1} \times 100\%$$

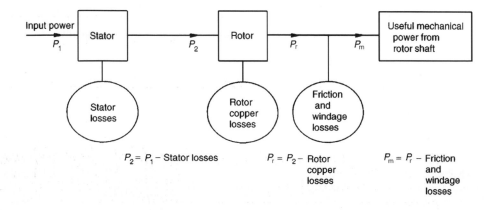

**Figure 23.8**

**Problem 8.** The power supplied to a three-phase induction motor is 32 kW and the stator losses are 1200 W. If the slip is 5 per cent, determine (a) the rotor copper loss, (b) the total mechanical power developed by the rotor, (c) the output power of the motor if friction and windage losses are 750 W, and (d) the efficiency of the motor, neglecting rotor iron loss.

(a) Input power to rotor = stator input power
$$- \text{ stator losses}$$
$$= 32 \text{ kW} - 1.2 \text{ kW}$$
$$= 30.8 \text{ kW}$$

From equation (9),

$$\text{slip} = \frac{\text{rotor copper loss}}{\text{rotor input}}$$

i.e. $$\frac{5}{100} = \frac{\text{rotor copper loss}}{30.8}$$

from which, **rotor copper loss** = (0.05)(30.8)
$$= \textbf{1.54 kW}$$

(b) Total mechanical power developed by the rotor

$$= \text{rotor input power} - \text{rotor losses}$$
$$- 30.8 - 1.54 = \textbf{29.26 kW}$$

(c) Output power of motor

$$= \text{power developed by the rotor}$$
$$- \text{friction and windage losses}$$
$$= 29.26 - 0.75 = \textbf{28.51 kW}$$

(d) Efficiency of induction motor,

$$\eta = \left(\frac{\text{output power}}{\text{input power}}\right) \times 100\%$$

$$= \left(\frac{28.51}{32}\right) \times 100\%$$

$$= \textbf{89.10\%}$$

**Problem 9.** The speed of the induction motor of Problem 8 is reduced to 35 per cent of its synchronous speed by using external rotor resistance. If the torque and stator losses are unchanged, determine (a) the rotor copper loss, and (b) the efficiency of the motor.

(a) Slip, $s = \left(\dfrac{n_s - n_r}{n_s}\right) \times 100\%$

$$\left(\frac{n_s - 0.35n_s}{n_s}\right) \times 100\%$$

$$= (0.65)(100) = 65\%$$

Input power to rotor = 30.8 kW (from Problem 8)

Since $$s = \frac{\text{rotor copper loss}}{\text{rotor input}}$$

then **rotor copper loss** = (s)(rotor input)

$$= \left(\frac{65}{100}\right)(30.8)$$

$$= \textbf{20.02 kW}$$

(b) Power developed by rotor

$$= \text{input power to rotor} - \text{rotor copper loss}$$
$$= 30.8 - 20.02 = 10.78 \text{ kW}$$

Output power of motor

$$= \text{power developed by rotor}$$
$$- \text{friction and windage losses}$$
$$= 10.78 - 0.75 = 10.03 \text{ kW}$$

Efficiency,

$$\eta = \left(\frac{\text{output power}}{\text{input power}}\right) \times 100\%$$

$$= \left(\frac{10.03}{32}\right) \times 100\%$$

$$= \textbf{31.34\%}$$

**Now try the following exercise**

**Exercise 143 Further problems on losses and efficiency**

1. The power supplied to a three-phase induction motor is 50 kW and the stator losses are 2 kW. If the slip is 4 per cent, determine
   (a) the rotor copper loss,
   (b) the total mechanical power developed by the rotor,

Section 3

(c) the output power of the motor if friction and windage losses are 1 kW, and

(d) the efficiency of the motor, neglecting rotor iron losses.

[(a) 1.92 kW (b) 46.08 kW (c) 45.08 kW (d) 90.16%]

2. By using external rotor resistance, the speed of the induction motor in Problem 1 is reduced to 40 per cent of its synchronous speed. If the torque and stator losses are unchanged, calculate

(a) the rotor copper loss, and

(b) the efficiency of the motor.

[(a) 28.80 kW (b) 36.40%]

## 23.11 Torque equation for an induction motor

Torque

$$T = \frac{P_2}{2\pi n_s} = \left(\frac{1}{2\pi n_s}\right)\left(\frac{I_r^2 R_2}{s}\right)$$

(from equation (10))

From equation (8),
$$I_r = \frac{s\left(\frac{N_2}{N_1}\right)E_1}{\sqrt{R_2^2 + (sX_2)^2}}$$

Hence torque per phase,

$$T = \left(\frac{1}{2\pi n_s}\right)\left(\frac{s^2\left(\frac{N_2}{N_1}\right)^2 E_1^2}{R_2^2 + (sX_2)^2}\right)\left(\frac{R_2}{s}\right)$$

i.e.

$$T = \left(\frac{1}{2\pi n_s}\right)\left(\frac{s\left(\frac{N_2}{N_2}\right)^2 E_1^2 R_2}{R_2^2 + (sX_2)^2}\right)$$

If there are $m$ phases then torque,

$$T = \left(\frac{m}{2\pi n_s}\right)\left(\frac{s\left(\frac{N_2}{N_1}\right)^2 E_1^2 R_2}{R_2^2 + (sX_2)^2}\right)$$

i.e.

$$T = \left(\frac{m\left(\frac{N_2}{N_1}\right)^2}{2\pi n_s}\right)\left(\frac{sE_1^2 R_2}{R_2^2 + (sX_2)^2}\right) \quad (11)$$

$$= k\left(\frac{sE_1^2 R_2}{R_2^2 + (sX_2)^2}\right)$$

where $k$ is a constant for a particular machine, i.e.

$$\text{torque, } T \propto \left(\frac{sE_1^2 R_2}{R_2^2 + (sX_2)^2}\right) \quad (12)$$

Under normal conditions, the supply voltage is usually constant, hence equation (12) becomes:

$$T \propto \frac{sR_2}{R_2^2 + (sX_2)^2}$$

$$\propto \frac{R_2}{\frac{R_2^2}{s} + sX_2^2}$$

The torque will be a maximum when the denominator is a minimum and this occurs when

$$\frac{R_2^2}{s} = sX_2^2$$

i.e. when

$$s = \frac{R_2}{X_2} \quad \text{or} \quad R_2 = sX_2 = X_r$$

from equation (4). Thus **maximum torque** occurs when rotor resistance and rotor reactance are equal, i.e. when $R_2 = X_r$

Problems 10 to 13 following illustrate some of the characteristics of three-phase induction motors.

**Problem 10.** A 415 V, three-phase, 50 Hz, 4-pole, star-connected induction motor runs at 24 rev/s on full load. The rotor resistance and reactance per phase are 0.35 Ω and 3.5 Ω respectively, and the effective rotor-stator turns ratio is 0.85:1. Calculate (a) the synchronous speed, (b) the slip, (c) the full load torque, (d) the power output if mechanical losses amount to 770 W, (e) the maximum torque, (f) the speed at which maximum torque occurs, and (g) the starting torque.

(a) Synchronous speed, $n_s = (f/p) = (50/2) = $ **25 rev/s** or $(25 \times 60) = $ **1500 rev/min**

(b) Slip, $s = \left(\dfrac{n_s - n_r}{n_s}\right) = \dfrac{25 - 24}{25} = $ **0.04** or **4%**

(c) Phase voltage,

$$E_1 = \frac{415}{\sqrt{3}} = 239.6 \text{ volts}$$

Full load torque,

$$T = \left(\frac{m\left(\dfrac{N_2}{N_1}\right)^2}{2\pi n_s}\right)\left(\frac{sE_1^2 R_2}{R_2^2 + (sX_2)^2}\right)$$

from equation (11)

$$= \left(\frac{3(0.85)^2}{2\pi(25)}\right)\left(\frac{(0.04)(239.6)^2(0.35)}{(0.35)^2 + (0.04 \times 3.5)^2}\right)$$

$$= (0.01380)\left(\frac{803.71}{0.1421}\right)$$

$$= \mathbf{78.05\,Nm}$$

(d) Output power, including friction losses,

$$P_m = 2\pi n_r T$$

$$= 2\pi(24)(78.05)$$

$$= 11\,770 \text{ watts}$$

Hence, **power output** $= P_m - $ mechanical losses

$$= 11\,770 - 770$$

$$= 11\,000\,W$$

$$= \mathbf{11\,kW}$$

(e) Maximum torque occurs when $R_2 = X_r = 0.35\,\Omega$

Slip, $\qquad s = \dfrac{R_2}{X_2} = \dfrac{0.35}{3.5} = 0.1$

Hence **maximum torque,**

$$\mathbf{T_m} = (0.01380)\left(\frac{sE_1^2 R_2}{R_2^2 + (sX_2)^2}\right) \text{ from part (c)}$$

$$= (0.01380)\left(\frac{0.1(239.6)^2 0.35}{0.35^2 + 0.35^2}\right)$$

$$= (0.01380)\left(\frac{2009.29}{0.245}\right) = \mathbf{113.18\,Nm}$$

(f) For maximum torque, slip $s = 0.1$

Slip, $\qquad s = \left(\dfrac{n_s - n_r}{n_s}\right)$

i.e.

$$0.1 = \left(\frac{25 - n_s}{25}\right)$$

Hence $(0.1)(25) = 25 - n_r$ and
$$n_r = 25 - (0.1)(25)$$

Thus speed at which maximum torque occurs, $n_r = 25 - 2.5 = $ **22.5 rev/s** or **1350 rev/min**

(g) At the start, i.e. at standstill, slip $s = 1$. Hence,

$$\text{starting torque} = \left(\frac{m\left(\dfrac{N_2}{N_1}\right)^2}{2\pi n_s}\right)\left(\frac{E_1^2 R_2}{R_2^2 + X_2^2}\right)$$

from equation (11) with $s = 1$

$$= (0.01380)\left(\frac{(239.6)^2 0.35}{0.35^2 + 3.5^2}\right)$$

$$= (0.01380)\left(\frac{20\,092.86}{12.3725}\right)$$

i.e. **starting torque = 22.41 Nm**

(Note that the full load torque (from part (c)) is 78.05 Nm but the starting torque is only 22.41 Nm)

**Problem 11.** Determine for the induction motor in Problem 10 at full load, (a) the rotor current, (b) the rotor copper loss, and (c) the starting current.

(a)  From equation (8), **rotor current**,

$$I_r = \frac{s\left(\frac{N_2}{N_1}\right)E_1}{\sqrt{R_2^2 + (sX_2)^2}}$$

$$= \frac{(0.04)(0.85)(239.6)}{\sqrt{0.35^2 + (0.04 \times 3.5)^2}}$$

$$= \frac{8.1464}{0.37696} = \mathbf{21.61\,A}$$

(b)  Rotor copper

$$\text{loss per phase} = I_r^2 R_2$$

$$= (21.61)^2 (0.35)$$

$$= 163.45\,W$$

**Total copper loss** (for 3 phases)
$$= 3 \times 163.45$$

$$= \mathbf{490.35\,W}$$

(c)  From equation (7), starting current,

$$I_2 = \frac{\left(\frac{N_2}{N_1}\right)E_1}{\sqrt{R_2^2 + X_2^2}} = \frac{(0.85)(239.5)}{\sqrt{0.35^2 + 3.5^2}} = \mathbf{57.90\,A}$$

(Note that the starting current of 57.90 A is considerably higher than the full load current of 21.61 A)

---

**Problem 12.**  For the induction motor in Problems 10 and 11, if the stator losses are 650 W, determine (a) the power input at full load, (b) the efficiency of the motor at full load, and (c) the current taken from the supply at full load, if the motor runs at a power factor of 0.87 lagging.

---

(a)  Output power $P_m = 11.770\,kW$ from part (d), Problem 10. Rotor copper loss $= 490.35\,W = 0.49035\,kW$ from part (b), Problem 11.
**Stator input power,**

$$P_1 = P_m + \text{rotor copper loss} + \text{rotor stator loss}$$

$$= 11.770 + 0.49035 + 0.650$$

$$= \mathbf{12.91\,kW}$$

(b)  Net power output $= 11\,kW$ from part (d), Problem 10. Hence efficiency,

$$\eta = \frac{\text{output}}{\text{input}} \times 100\% = \left(\frac{11}{12.91}\right) \times 100\%$$

$$= \mathbf{85.21\%}$$

(c)  Power input, $P_1 = \sqrt{3}\,V_L I_L \cos\phi$ (see Chapter 20) and $\cos\phi = \text{p.f.} = 0.87$ hence, **supply current,**

$$I_L = \frac{P_1}{\sqrt{3}\,V_L \cos\phi} = \frac{12.91 \times 1000}{\sqrt{3}(415)0.87} = \mathbf{20.64\,A}$$

---

**Problem 13.**  For the induction motor of Problems 10 to 12, determine the resistance of the rotor winding required for maximum starting torque.

---

From equation (4), rotor reactance $X_r = sX_2$. At the moment of starting, slip, $s = 1$. Maximum torque occurs when rotor reactance equals rotor resistance hence for **maximum torque,**
$R_2 = X_r = sX_2 = X_2 = \mathbf{3.5\,\Omega}$.

Thus if the induction motor was a wound rotor type with slip-rings then an external star-connected resistance of $(3.5 - 0.35)\,\Omega = 3.15\,\Omega$ per phase could be added to the rotor resistance to give maximum torque at starting (see Section 23.13).

---

**Now try the following exercise**

**Exercise 144  Further problems on the torque equation**

1.  A 400 V, three-phase, 50 Hz, 2-pole, star-connected induction motor runs at 48.5 rev/s on full load. The rotor resistance and reactance per phase are $0.4\,\Omega$ and $4.0\,\Omega$ respectively, and the effective rotor-stator turns ratio is 0.8:1. Calculate
    (a)  the synchronous speed,
    (b)  the slip,
    (c)  the full load torque,
    (d)  the power output if mechanical losses amount to 500 W,
    (e)  the maximum torque,
    (f)  the speed at which maximum torque occurs, and
    (g)  the starting torque.
    [(a) 50 rev/s or 3000 rev/min (b) 0.03 or 3%
    (c) 22.43 Nm (d) 6.34 kW (e) 40.74 Nm
    (f) 45 rev/s or 2700 rev/min (g) 8.07 Nm]

2. For the induction motor in Problem 1, calculate at full load
   (a) the rotor current,
   (b) the rotor copper loss, and
   (c) the starting current.
   [(a) 13.27 A (b) 211.3 W (c) 45.96 A]

3. If the stator losses for the induction motor in Problem 1 are 525 W, calculate at full load
   (a) the power input,
   (b) the efficiency of the motor, and
   (c) the current taken from the supply if the motor runs at a power factor of 0.84.
   [(a) 7.57 kW (b) 83.75% (c) 13.0 A]

4. For the induction motor in Problem 1, determine the resistance of the rotor winding required for maximum starting torque.   [4.0 Ω]

## 23.12 Induction motor torque-speed characteristics

From Problem 10, parts (c) and (g), it is seen that the normal starting torque may be less than the full load torque. Also, from Problem 10, parts (e) and (f), it is seen that the speed at which maximum torque occurs is determined by the value of the rotor resistance. At synchronous speed, slip $s = 0$ and torque is zero. From these observations, the torque-speed and torque-slip characteristics of an induction motor are as shown in Fig. 23.9.

The rotor resistance of an induction motor is usually small compared with its reactance (for example, $R_2 = 0.35\,\Omega$ and $X_2 = 3.5\,\Omega$ in the above Problems), so that maximum torque occurs at a high speed, typically about 80 per cent of synchronous speed.

Curve P in Fig. 23.9 is a typical characteristic for an induction motor. The curve P cuts the full load torque line at point X, showing that at full load the slip is about 4–5 per cent. The normal operating conditions are between 0 and X, thus it can be seen that for normal operation the speed variation with load is quite small – the induction motor is an almost constant-speed machine. Redrawing the speed-torque characteristic between 0 and X gives the characteristic shown in Fig. 23.10, which is similar to a d.c. shunt motor as shown in Chapter 22.

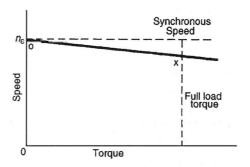

Figure 23.10

If maximum torque is required at starting then a high resistance rotor is necessary, which gives characteristic Q in Fig. 23.9. However, as can be seen, the motor has a full load slip of over 30 per cent, which results in a drop in efficiency. Also such a motor has a large

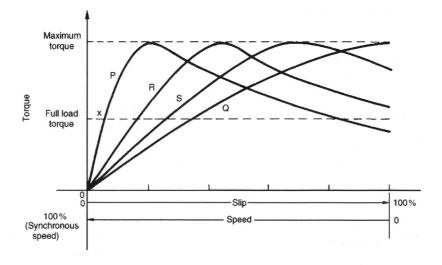

Figure 23.9

speed variation with variations of load. Curves R and S of Fig. 23.9 are characteristics for values of rotor resistances between those of P and Q. Better starting torque than for curve P is obtained, but with lower efficiency and with speed variations under operating conditions.

A **squirrel-cage induction motor** would normally follow characteristic P. This type of machine is highly efficient and about constant-speed under normal running conditions. However it has a poor starting torque and must be started off-load or very lightly loaded (see Section 23.13 below). Also, on starting, the current can be four or five times the normal full load current, due to the motor acting like a transformer with secondary short-circuited. In Problem 11, for example, the current at starting was nearly three times the full load current.

A **wound rotor induction motor** would follow characteristic P when the slip-rings are short-circuited, which is the normal running condition. However, the slip-rings allow for the addition of resistance to the rotor circuit externally and, as a result, for starting, the motor can have a characteristic similar to curve Q in Fig. 23.9 and the high starting current experienced by the cage induction motor can be overcome.

In general, for three-phase induction motors, the power factor is usually between about 0.8 and 0.9 lagging, and the full load efficiency is usually about 80–90 per cent.

From equation (12), it is seen that torque is proportional to the square of the supply voltage. Any voltage variations therefore would seriously affect the induction motor performance.

## 23.13 Starting methods for induction motors

### Squirrel-cage rotor

(i) **Direct-on-line starting**

With this method, starting current is high and may cause interference with supplies to other consumers.

(ii) **Auto transformer starting**

With this method, an auto transformer is used to reduce the stator voltage, $E_1$, and thus the starting current (see equation (7)). However, the starting torque is seriously reduced (see equation (12)), so the voltage is reduced only sufficiently to give the required reduction of the starting current. A typical arrangement is shown in Fig. 23.11. A double-throw switch connects the auto transformer in circuit for starting, and when the motor is

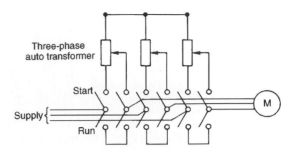

**Figure 23.11**

up to speed the switch is moved to the run position which connects the supply directly to the motor.

(iii) **Star-delta starting**

With this method, for starting, the connections to the stator phase winding are star-connected, so that the voltage across each phase winding is $(1/\sqrt{3})$ (i.e. 0.577) of the line voltage. For running, the windings are switched to delta-connection. A typical arrangement is shown in Fig. 23.12. This method of starting is less expensive than by auto transformer.

### Wound rotor

When starting on load is necessary, a wound rotor induction motor must be used. This is because maximum torque at starting can be obtained by adding external resistance to the rotor circuit via slip-rings, (see Problem 13). A face-plate type starter is used, and as the resistance is gradually reduced, the machine characteristics at each stage will be similar to Q, S, R and P of Fig. 23.13. At each resistance step, the motor operation will transfer from one characteristic to the next so that the overall starting characteristic will be as shown by the bold line in Fig. 23.13. For very large induction motors, very gradual and smooth starting is achieved by a liquid type resistance.

## 23.14 Advantages of squirrel-cage induction motors

The advantages of squirrel-cage motors compared with the wound rotor type are that they:

(i)  are cheaper and more robust

(ii)  have slightly higher efficiency and power factor

(iii)  are explosion-proof, since the risk of sparking is eliminated by the absence of slip-rings and brushes.

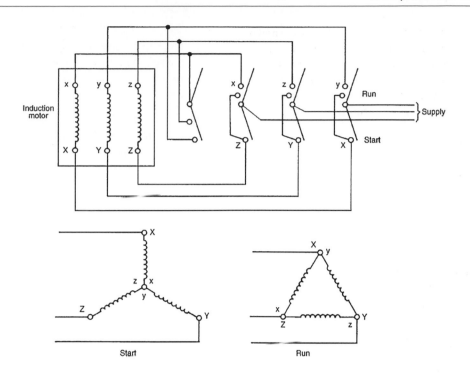

Figure 23.12

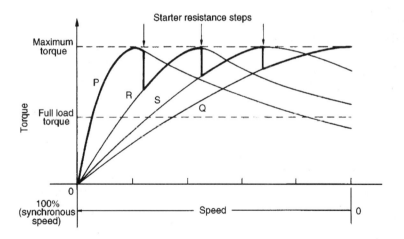

Figure 23.13

## 23.15 Advantages of wound rotor induction motors

The advantages of the wound rotor motor compared with the cage type are that they:

(i)   have a much higher starting torque

(ii)  have a much lower starting current

(iii) have a means of varying speed by use of external rotor resistance.

## 23.16 Double cage induction motor

The advantages of squirrel-cage and wound rotor induction motors are combined in the double cage induction motor. This type of induction motor is specially constructed with the rotor having two cages, one inside the other. The outer cage has high resistance conductors so that maximum torque is achieved at or near starting. The inner cage has normal low resistance copper conductors but high reactance since it is embedded deep in the iron core. The torque-speed characteristic of the inner cage is that of a normal induction motor, as shown in Fig. 23.14. At starting, the outer cage produces the torque, but when running the inner cage produces the torque. The combined characteristic of inner and outer cages is shown in Fig. 23.14. The double cage induction motor is highly efficient when running.

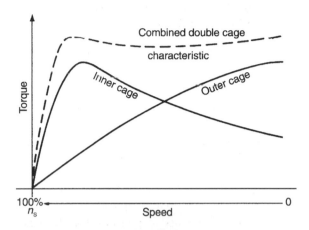

**Figure 23.14**

## 23.17 Uses of three-phase induction motors

Three-phase induction motors are widely used in industry and constitute almost all industrial drives where a nearly constant speed is required, from small workshops to the largest industrial enterprises.

Typical applications are with machine tools, pumps and mill motors. The squirrel-cage rotor type is the most widely used of all a.c. motors.

**Now try the following exercises**

**Exercise 145    Short answer questions on three-phase induction motors**

1. Name three advantages that a three-phase induction motor has when compared with a d.c. motor

2. Name the principal disadvantage of a three-phase induction motor when compared with a d.c. motor

3. Explain briefly, with the aid of sketches, the principle of operation of a 3-phase induction motor

4. Explain briefly how slip-frequency currents are set up in the rotor bars of a 3-phase induction motor and why this frequency varies with load

5. Explain briefly why a 3-phase induction motor develops no torque when running at synchronous speed. Define the slip of an induction motor and explain why its value depends on the load on the rotor

6. Write down the two properties of the magnetic field produced by the stator of a three-phase induction motor

7. The speed at which the magnetic field of a three-phase induction motor rotates is called the ...... speed

8. The synchronous speed of a three-phase induction motor is ...... proportional to supply frequency

9. The synchronous speed of a three-phase induction motor is ...... proportional to the number of pairs of poles

10. The type of rotor most widely used in a three-phase induction motor is called a ......

11. The slip of a three-phase induction motor is given by: $s = \dfrac{\cdots}{\cdots} \times 100\%$

12. A typical value for the slip of a small three-phase induction motor is ... %

13. As the load on the rotor of a three-phase induction motor increases, the slip ......

14. $\dfrac{\text{Rotor copper loss}}{\text{Rotor input power}} = \ldots\ldots$

15. State the losses in an induction motor

16. Maximum torque occurs when ...... = ......

17. Sketch a typical speed-torque characteristic for an induction motor

18. State two methods of starting squirrel-cage induction motors

19. Which type of induction motor is used when starting on-load is necessary?

20. Describe briefly a double cage induction motor

21. State two advantages of cage rotor machines compared with wound rotor machines

22. State two advantages of wound rotor machines compared with cage rotor machines

23. Name any three applications of three-phase induction motors

### Exercise 146    Multi-choice questions on three-phase induction motors (Answers on page 421)

1. Which of the following statements about a three-phase squirrel-cage induction motor is false?
   (a) It has no external electrical connections to its rotor
   (b) *A three-phase supply is connected to its stator*
   (c) *A magnetic flux which alternates is produced*
   (d) It is cheap, robust and requires little or no skilled maintenance

2. Which of the following statements about a three-phase induction motor is false?
   (a) The speed of rotation of the magnetic field is called the synchronous speed
   (b) A three-phase supply connected to the rotor produces a rotating magnetic field
   (c) The rotating magnetic field has a constant speed and constant magnitude
   (d) It is essentially a constant speed type machine

3. Which of the following statements is false when referring to a three-phase induction motor?
   (a) The synchronous speed is half the supply frequency when it has four poles
   (b) In a 2-pole machine, the synchronous speed is equal to the supply frequency
   (c) If the number of poles is increased, the synchronous speed is reduced
   (d) The synchronous speed is inversely proportional to the number of poles

4. A 4-pole three-phase induction motor has a synchronous speed of 25 rev/s. The frequency of the supply to the stator is:
   (a) 50 Hz      (b) 100 Hz
   (c) 25 Hz      (d) 12.5 Hz

Questions 5 and 6 refer to a three-phase induction motor. Which statements are false?

5. (a) The slip speed is the synchronous speed minus the rotor speed
   (b) As the rotor is loaded, the slip decreases
   (c) The frequency of induced rotor e.m.f.'s increases with load on the rotor
   (d) The torque on the rotor is due to the interaction of magnetic fields

6. (a) If the rotor is running at synchronous speed, there is no torque on the rotor
   (b) If the number of poles on the stator is doubled, the synchronous speed is halved
   (c) At no-load, the rotor speed is very nearly equal to the synchronous speed
   (d) The direction of rotation of the rotor is opposite to the direction of rotation of the magnetic field to give maximum current induced in the rotor bars

A three-phase, 4-pole, 50 Hz induction motor runs at 1440 rev/min. In questions 7 to 10, determine the correct answers for the quantities stated, selecting your answer from the list given below:
   (a) 12.5 rev/s   (b) 25 rev/s   (c) 1 rev/s
   (d) 50 rev/s     (e) 1%         (f) 4%
   (g) 50%          (h) 4 Hz       (i) 50 Hz
   (j) 2 Hz

7. The synchronous speed

8. The slip speed

9. The percentage slip

10. The frequency of induced e.m.f.'s in the rotor

11. The slip speed of an induction motor may be defined as the:
    (a) number of pairs of poles ÷ frequency
    (b) rotor speed − synchronous speed
    (c) rotor speed + synchronous speed
    (d) synchronous speed − rotor speed

12. The slip speed of an induction motor depends upon:
   (a) armature current   (b) supply voltage
   (c) mechanical load   (d) eddy currents

13. The starting torque of a simple squirrel-cage motor is:
   (a) low
   (b) increases as rotor current rises
   (c) decreases as rotor current rises
   (d) high

14. The slip speed of an induction motor:
   (a) is zero until the rotor moves and then rises slightly
   (b) is 100 per cent until the rotor moves and then decreases slightly
   (c) is 100 per cent until the rotor moves and then falls to a low value
   (d) is zero until the rotor moves and then rises to 100 per cent

15. A four-pole induction motor when supplied from a 50 Hz supply experiences a 5 per cent slip. The rotor speed will be:
   (a) 25 rev/s     (b) 23.75 rev/s
   (c) 26.25 rev/s  (d) 11.875 rev/s

16. A stator winding of an induction motor supplied from a three-phase, 60 Hz system is required to produce a magnetic flux rotating at 900 rev/min. The number of poles is:
   (a) 2          (b) 8
   (c) 6          (d) 4

17. The stator of a three-phase, 2-pole induction motor is connected to a 50 Hz supply. The rotor runs at 2880 rev/min at full load. The slip is:
   (a) 4.17%     (b) 92%
   (c) 4%       (d) 96%

18. An 8-pole induction motor, when fed from a 60 Hz supply, experiences a 5 per cent slip. The rotor speed is:
   (a) 427.5 rev/min  (b) 855 rev/min
   (c) 900 rev/min    (d) 945 rev/min

This revision test covers the material contained in Chapters 22 and 23. *The marks for each question are shown in brackets at the end of each question.*

1. A 6-pole armature has 1000 conductors and a flux per pole of 40 mWb. Determine the e.m.f. generated when running at 600 rev/min when (a) lap wound (b) wave wound. (6)

2. The armature of a d.c. machine has a resistance of 0.3 Ω and is connected to a 200 V supply. Calculate the e.m.f. generated when it is running (a) as a generator giving 80 A (b) as a motor taking 80 A. (4)

3. A 15 kW shunt generator having an armature circuit resistance of 1 Ω and a field resistance of 160 Ω generates a terminal voltage of 240 V at full load. Determine the efficiency of the generator at full load assuming the iron, friction and windage losses amount to 544 W. (6)

4. A 4-pole d.c. motor has a wave-wound armature with 1000 conductors. The useful flux per pole is 40 mWb. Calculate the torque exerted when a current of 25 A flows in each armature conductor. (4)

5. A 400 V shunt motor runs at its normal speed of 20 rev/s when the armature current is 100 A. The armature resistance is 0.25 Ω. Calculate the speed, in rev/min when the current is 50 A and a resistance of 0.40 Ω is connected in series with the armature, the shunt field remaining constant. (7)

6. The stator of a three-phase, 6-pole induction motor is connected to a 60 Hz supply. The rotor runs at 1155 rev/min at full load. Determine (a) the synchronous speed, and (b) the slip at full load. (6)

7. The power supplied to a three-phase induction motor is 40 kW and the stator losses are 2 kW. If the slip is 4 per cent determine (a) the rotor copper loss, (b) the total mechanical power developed by the rotor, (c) the output power of the motor if frictional and windage losses are 1.48 kW, and (d) the efficiency of the motor, neglecting rotor iron loss. (9)

8. A 400 V, three-phase, 100 Hz, 8-pole induction motor runs at 24.25 rev/s on full load. The rotor resistance and reactance per phase are 0.2 Ω and 2 Ω respectively and the effective rotor-stator turns ratio is 0.80:1. Calculate (a) the synchronous speed, (b) the slip, and (c) the full load torque. (8)

**Section 3**

## Three-phase systems:

Star $I_L = I_p$  $V_L = \sqrt{3}\,V_p$

Delta $V_L = V_p$  $I_L = \sqrt{3}\,I_p$

$P = \sqrt{3}\,V_L I_L \cos\phi$  or  $P = 3I_p^2 R_p$

Two-wattmeter method

$P = P_1 + P_2$  $\tan\phi = \sqrt{3}\dfrac{(P_1 - P_2)}{(P_1 + P_2)}$

## Transformers:

$\dfrac{V_1}{V_2} = \dfrac{N_1}{N_2} = \dfrac{I_2}{I_1}$  $I_0 = \sqrt{(I_M^2 + I_C^2)}$

$I_M = I_0 \sin\phi_0$  $I_c = I_0 \cos\phi_0$

$E = 4.44\, f\,\Phi_m N$

$\text{Regulation} = \left(\dfrac{E_2 - E_1}{E_2}\right) \times 100\%$

Equivalent circuit: $R_e = R_1 + R_2 \left(\dfrac{V_1}{V_2}\right)^2$

$X_e = X_1 + X_2 \left(\dfrac{V_1}{V_2}\right)^2$  $Z_e = \sqrt{(R_e^2 + X_e^2)}$

Efficiency, $\eta = 1 - \dfrac{\text{losses}}{\text{input power}}$

Output power $= V_2 I_2 \cos\phi_2$

Total loss $=$ copper loss $+$ iron loss

Input power $=$ output power $+$ losses

Resistance matching: $R_1 = \left(\dfrac{N_1}{N_2}\right)^2 R_L$

## D.C. Machines:

Generated e.m.f. $E = \dfrac{2p\Phi n Z}{c} \propto \Phi\omega$

($c = 2$ for wave winding, $c = 2p$ for lap winding)

Generator: $E = V + I_a R_a$

Efficiency, $\eta = \left(\dfrac{VI}{VI + I_a^2 R_a + I_f V + C}\right) \times 100\%$

Motor:  $E = V - I_a R_a$

Efficiency, $\eta = \left(\dfrac{VI - I_a^2 R_a - I_f V - C}{VI}\right) \times 100\%$

$\text{Torque} = \dfrac{E I_a}{2\pi n} = \dfrac{p\Phi Z I_a}{\pi c} \propto I_a \Phi$

## Three-phase induction motors:

$n_S = \dfrac{f}{p}$  $s = \left(\dfrac{n_s - n_r}{n_s}\right) \times 100$

$f_r = sf$  $X_r = s X_2$

$I_r = \dfrac{E_r}{Z_r} = \dfrac{s\left(\dfrac{N_2}{N_1}\right) E_1}{\sqrt{[R_2^2 + (s X_2)^2]}}$  $s = \dfrac{I_r^2 R_2}{P_2}$

Efficiency,

$\eta = \dfrac{P_m}{P_1} = \dfrac{\text{input} - \text{stator loss} - \text{rotor copper loss} - \text{friction \& windage loss}}{\text{input power}}$

Torque,

$T = \left(\dfrac{m\left(\dfrac{N_2}{N_1}\right)^2}{2\pi n_s}\right)\left(\dfrac{s E_1^2 R_2}{R_2^2 + (s X_2)^2}\right) \propto \dfrac{s E_1^2 R_2}{R_2^2 + (s X_2)^2}$

# Section 4

# Laboratory Experiments

# Chapter 24

# Some practical laboratory experiments

This chapter contains 10 straightforward practical laboratory experiments to help supplement and enhance academic studies. Copies of these exercises have been made available on line at http://www.booksite.elsevier. com/newnes/bird and may be edited by tutors to suit availability of equipment and components.

The list of experiments is not exhaustive, but covers some of the more important aspects of early electrical engineering studies.

Experiments covered are:

- **Ohm's law** (see Chapter 2)
- **Series-parallel d.c. circuit** (see Chapter 5)
- **Superposition theorem** (see Chapter 13)
- **Thévenin's theorem** (see Chapter 13)
- **Use of CRO to measure voltage, frequency and phase** (see Chapter 14)
- **Use of CRO with a bridge rectifier circuit** (see Chapter 14)
- **Measurement of the inductance of a coil** (see Chapter 15)
- **Series a.c. circuit and resonance** (see Chapter 15)
- **Parallel a.c. circuit and resonance** (see Chapter 16)
- **Charging and discharging a capacitor** (see Chapter 18)

DOI: 10.1016/B978-0-08-089056-2.00024-3

## 24.1   Ohm's law

**Objectives:**

1.   To determine the voltage-current relationship in a d.c. circuit and relate it to Ohm's law.

**Equipment required:**

1.   D.C. Power Supply Unit (PSU).

2.   Constructor board (for example, 'Feedback' EEC470).

3.   An ammeter and voltmeter or two Flukes (for example, 89).

4.   LCR Data bridge.

**Procedure:**

1.   Construct the circuit shown below with $R = 470\Omega$.

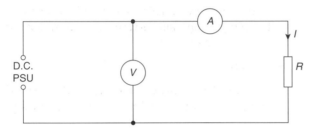

2.   Check the colour coding of the resistor and then measure its value accurately using an LCR data bridge or a Fluke.

3.   Initially set the d.c. power supply unit to 1 V.

4.   Measure the value of the current in the circuit and record the reading in the table below.

5.   Increase the value of voltage in 1 V increments, measuring the current for each value. Complete the table of values below.

**Resistance $R = 470\,\Omega$**
**[colour code is]:**

| Voltage $V$ (V) | 1 | 2 | 3 | 4 | 5 | 6 | 7 | 8 |
|---|---|---|---|---|---|---|---|---|
| Current $I$ (mA) | | | | | | | | |

6.   Repeat procedures 1 to 5 for a resistance value of $R = 2.2\,k\Omega$ and complete the table below.

**Resistance $R = 2.2\,k\Omega$**
**[colour code is]:**

| Voltage $V$ (V) | 1 | 2 | 3 | 4 | 5 | 6 | 7 | 8 |
|---|---|---|---|---|---|---|---|---|
| Current $I$ (mA) | | | | | | | | |

7.   Repeat procedures 1 to 5 for a resistance value of $R = 10\,k\Omega$ and complete the table below.

**Resistance $R = 10\,k\Omega$**
**[colour code is]:**

| Voltage $V$ (V) | 1 | 2 | 3 | 4 | 5 | 6 | 7 | 8 |
|---|---|---|---|---|---|---|---|---|
| Current $I$ (mA) | | | | | | | | |

8.   Plot graphs of $V$ (vertically) against $I$ (horizontally) for $R = 470\Omega$, $R = 2.2\,k\Omega$ and $R = 10\,k\Omega$ respectively.

**Conclusions:**

1.   What is the nature of the graphs plotted?

2.   If the graphs plotted are straight lines, determine their gradients. Can you draw any conclusions from the gradient values?

3.   State Ohm's law. Has this experiment proved Ohm's law to be true?

## 24.2  Series-parallel d.c. circuit

**Objectives:**

1. To compare calculated with measured values of voltages and currents in a series-parallel d.c. circuit.

**Equipment required:**

1. D.C. Power Supply Unit (PSU).

2. Constructor board (for example, 'Feedback' EEC470).

3. An ammeter and voltmeter or a Fluke (for example, 89).

4. LCR Data bridge.

**Procedure:**

1. Construct the circuit as shown below.

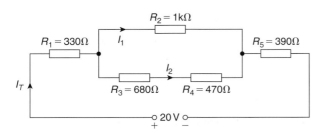

2. State the colour code for each of the five resistors in the above circuit and record them in the table below.

3. Using a Fluke or LCR bridge, measure accurately the value of each resistor and note their values in the table below.

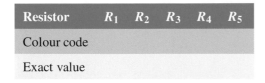

| Resistor | $R_1$ | $R_2$ | $R_3$ | $R_4$ | $R_5$ |
|---|---|---|---|---|---|
| Colour code | | | | | |
| Exact value | | | | | |

4. Calculate, using the exact values of resistors, the voltage drops and currents and record them in the table below.

| Quantity | Calculated value | Measured value |
|---|---|---|
| $V_{R_1}$ | | |
| $V_{R_2}$ | | |
| $V_{R_3}$ | | |
| $V_{R_4}$ | | |
| $V_{R_5}$ | | |
| $I_T$ | | |
| $I_1$ | | |
| $I_2$ | | |

5. With an ammeter, a voltmeter or a Fluke, measure the voltage drops and currents and record them in the above table.

**Conclusions:**

1. Compare the calculated and measured values of voltages and currents and comment on any discrepancies.

2. Calculate the total circuit power and the power dissipated in each resistor.

3. If the circuit was connected for 2 weeks, calculate the energy used.

## 24.3 Superposition theorem

**Objectives:**

1. To measure and calculate the current in each branch of a series-parallel circuit.

2. To verify the superposition theorem.

**Equipment required:**

1. Constructor board (for example, 'Feedback' EEC470).

2. D.C. Power Supply Units.

3. Digital Multimeter, such as a Fluke (for example, 89).

4. LCR Data bridge.

**Procedure:**

1. Construct the circuit as shown below, measuring and noting in the table below the exact values of the resistors using a Fluke or LCR bridge.

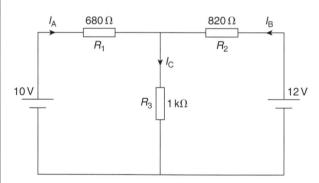

2. **Measure** the values of $I_A$, $I_B$ and $I_C$ and record the values in the table below.

| $R_1 (\Omega)$ | $R_2 (\Omega)$ | $R_3 (\Omega)$ |
|---|---|---|
|  |  |  |

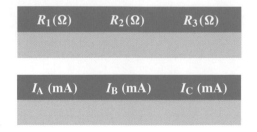

| $I_A$ (mA) | $I_B$ (mA) | $I_C$ (mA) |
|---|---|---|
|  |  |  |

3. Remove the 12 V source from the above circuit and replace with a link, giving the circuit shown on the next column.

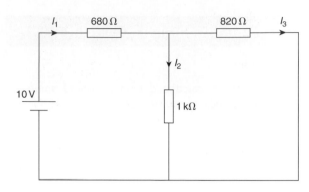

4. **Measure** the values of $I_1$, $I_2$ and $I_3$ and record the values in the table below.

| Measured $I_1$ (mA) | Measured $I_2$ (mA) | Measured $I_3$ (mA) |
|---|---|---|
|  |  |  |

| Calculated $I_1$ (mA) | Calculated $I_2$ (mA) | Calculated $I_3$ (mA) |
|---|---|---|
|  |  |  |

5. **Calculate** the values of $I_1$, $I_2$ and $I_3$ and record the values in the above table.

6. Replace the 12 V source in the original circuit and then replace the 10 V source with a link, giving the circuit shown below.

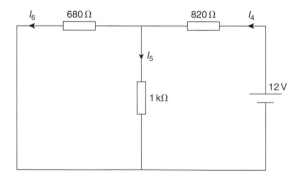

7. **Measure** the values of $I_4$, $I_5$ and $I_6$ and record the values in the table below.

| Measured $I_4$ (mA) | Measured $I_5$ (mA) | Measured $I_6$ (mA) |
|---|---|---|
|  |  |  |

| Calculated $I_4$ (mA) | Calculated $I_5$ (mA) | Calculated $I_6$ (mA) |
|---|---|---|
|  |  |  |

8. **Calculate** the values of $I_4$, $I_5$ and $I_6$ and record the values in the above table.

9. By superimposing the latter two diagrams on top of each other, calculate the algebraic sum of the currents in each branch and record them in the table below.

| Measured $I_A = I_1 - I_6$ | Measured $I_B = I_4 - I_3$ | Measured $I_C = I_2 + I_5$ |
|---|---|---|
|  |  |  |

| Calculated $I_A = I_1 - I_6$ | Calculated $I_B = I_4 - I_3$ | Calculated $I_B = I_2 + I_5$ |
|---|---|---|
|  |  |  |

**Conclusions:**

1. State in your own words the superposition theorem.

2. Compare the measured and calculated values of $I_A$, $I_B$ and $I_C$ in procedure 9 and comment on any discrepancies.

3. Compare these values of $I_A$, $I_B$ and $I_C$ with those measured in procedure 2 and comment on any discrepancies.

4. Can the principle of superposition be applied in a circuit having more than two sources?

Section 4

## 24.4 Thévenin's theorem

**Objectives:**

1. To calculate Thévenin's equivalent of a given circuit.

2. To verify Thévenin's theorem.

**Equipment required:**

1. Constructor board (for example, 'Feedback' EEC470).

2. D.C. Power Supply Units.

3. Digital Multimeter, such as a Fluke (for example, 89).

4. LCR Data bridge.

**Procedure:**

1. Construct the circuit as shown below, measuring and noting in the table below the exact values of the resistors using a Fluke or LCR bridge.

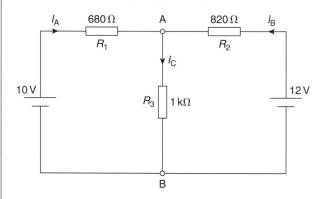

2. **Measure** the values of $I_A$, $I_B$ and $I_C$ and record the values in the table below.

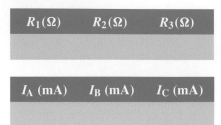

| $R_1 (\Omega)$ | $R_2 (\Omega)$ | $R_3 (\Omega)$ |
|---|---|---|
| | | |

| $I_A$ (mA) | $I_B$ (mA) | $I_C$ (mA) |
|---|---|---|
| | | |

3. Remove the 1 kΩ resistor from the above circuit and **measure** the open-circuit voltage $V_{OC}$ at the terminals AB. Record the value in the table in the next column.

4. With the 1 kΩ resistor still removed, remove the two voltage sources replacing each with a link.

Now **measure** the resistance $r_{OC}$ across the open circuited terminals AB and record the value in the table below.

| Measured $V_{OC}$ (V) | Measured $r_{OC}$ (Ω) | Calculated $V_{OC}$ (V) | Calculated $r_{OC}$ (Ω) |
|---|---|---|---|
| | | | |

5. **Calculate** values of $V_{OC}$ and $r_{OC}$ and record the values in the above table.

6. Compare the measured and calculated values of $V_{OC}$ and $r_{OC}$.

7. Using the calculated values of $V_{OC}$ and $r_{OC}$ calculate and record the current $I_C$ from the circuit below.

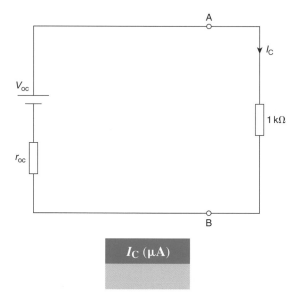

| $I_C$ (μA) |
|---|
| |

8. Compare this value of $I_C$ with that initially measured in the original circuit (i.e. procedure 2).

9. Calculate the voltage $V$ shown in the circuit below, using your calculated value of $I_C$, and record the value in the table below.

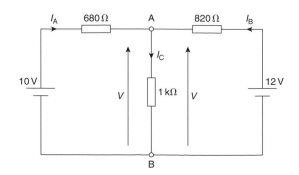

10. The terminal voltage of a source, $V = E - I \times r$. Using this, calculate and record the values of $I_A$ and $I_B$, i.e. transpose the equations:
$V = 10 - I_A \times 680$ and $V = 12 - I_B \times 820$.

| V (V) | $I_A$ (mA) | $I_B$(mA) |
|-------|------------|-----------|
|       |            |           |

11. Compare these values of $I_A$ and $I_B$ with those initially measured in the original circuit (i.e. procedure 2).

**Conclusions:**

1. State in your own words Thévenin's theorem.

2. Compare the measured and calculated values of $I_A$, $I_B$ and $I_C$ and comment on any discrepancies.

3. Can Thévenin's theorem be applied in a circuit having more than two sources?

4. If the $1\,k\Omega$ resistor is replaced with (a) $470\,\Omega$ (b) $2.2\,k\Omega$, calculate the current flowing between the terminals A and B.

## 24.5 Use of a CRO to measure voltage, frequency and phase

**Objectives:**

1. To measure a d.c. voltage using an oscilloscope.
2. To measure the peak-to-peak voltage of a waveform and then calculate its r.m.s. value.
3. To measure the periodic time of a waveform and then calculate its frequency.
4. To measure the phase angle between two waveforms.

**Equipment required:**

1. Cathode ray oscilloscope (for example, 'Phillips' digital Fluke PM3082).
2. Constructor board (for example, 'Feedback' EEC470).
3. Function Generator ('Escort' EFG 3210).
4. D.C. Power Supply Unit.
5. Fluke (for example, 89).

**Procedure:**

1. Switch on the oscilloscope and place the trace at the bottom of the screen.
2. Set the d.c. power supply unit to 20 V, making sure the output switch is in the off position.
3. Connect a test lead from channel 1 of the CRO to the d.c. PSU.
4. Switch on the output of the d.c. PSU.
5. Measure the d.c. voltage output on the CRO.

**d.c voltage** 

6. Connect up the circuit as shown below.

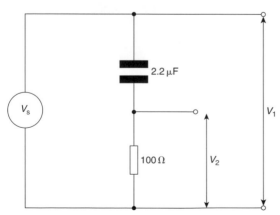

7. Set the function generator to output a voltage of 5 V at 500 Hz.

8. Measure the peak-to-peak voltages at $V_1$ and $V_2$ using the CRO and record in the table below.
9. Calculate the r.m.s. values corresponding to $V_1$ and $V_2$ and record in the table below.
10. Measure the voltages $V_1$ and $V_2$ using a Fluke.
11. Measure the periodic time of the waveforms obtained at $V_1$ and $V_2$ and record in the table below.
12. Calculate the frequency of the two waveforms and record in the table below.

| Voltage | Peak-to-peak voltage | r.m.s. value |
|---------|----------------------|--------------|
| $V_1$   |                      |              |
| $V_2$   |                      |              |

| Voltage | Periodic time | Frequency |
|---------|---------------|-----------|
| $V_1$   |               |           |
| $V_2$   |               |           |

13. Measure the phase angle $\phi$ between the two waveforms using:

$$\phi = \frac{\textbf{displacement between waveforms}}{\textbf{periodic time}} \times 360°$$

$$= \frac{\mathbf{t}}{\mathbf{T}} \times \mathbf{360°}$$

(For example, if $t = 0.6\,\text{ms}$ and $T = 4\,\text{ms}$, then $\phi = \frac{0.6}{4} \times 360° = \mathbf{54°}$)

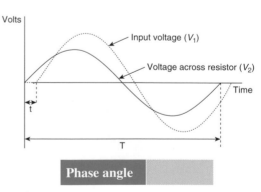

**Phase angle** 

**Conclusions:**

1. Is a measurement of voltage or current with a Fluke an r.m.s. value or a peak value?
2. Write expressions for the instantaneous values of voltages $V_1$ and $V_2$ (i.e. in the form $V = A \sin(\omega t \pm \phi)$ where $\phi$ is in radians).

## 24.6  Use of a CRO with a bridge rectifier

**Objectives:**

1. To measure and observe the input and output waveforms of a bridge rectifier circuit using a CRO.

2. To investigate smoothing of the output waveform

**Equipment required:**

1. Cathode Ray Oscilloscope (for example, 'Phillips' digital Fluke PM3082).

2. Constructor board (for example, 'Feedback' EEC470).

3. Transformer (for example, IET 464).

4. Bridge rectifier.

5. Fluke (for example, 89).

**Procedure:**

1. Construct the circuit shown below with a mains transformer stepping down to a voltage $V_1$ between 15 V and 20 V.

2. Measure the output voltage $V_1$ of the transformer using a Fluke and a CRO, noting the value in the table below. Sketch the waveform.

3. Measure the output voltage $V_2$ of the bridge rectifier using a Fluke and observe the waveform using a CRO, noting the value in the table below. Sketch the waveform.

4. Place a 100 μF capacitor across the terminals AB and observe the waveform across these terminals using a CRO. Measure the voltage across terminals AB, $V_3$, noting the value in the table below. Sketch the waveform.

5. Place a second 100 μF capacitor in parallel with the first across the terminals AB. What is the effect on the waveform? Measure the voltage across terminals AB, $V_4$, noting the value in the table below. Sketch the waveform.

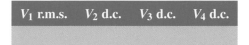

| $V_1$ r.m.s. | $V_2$ d.c. | $V_3$ d.c. | $V_4$ d.c. |
|---|---|---|---|
|  |  |  |  |

**Conclusions:**

1. What is the effect of placing a capacitor across the full-wave rectifier output?

2. What is the total capacitance of two 100 μF capacitors connected in parallel?

3. What is meant by ripple? Comment on the ripple when (a) one capacitor is connected, (b) both capacitors are connected.

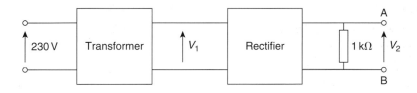

## 24.7 Measurement of the inductance of a coil

**Objectives:**

1. To measure the inductance of a coil.

**Equipment required:**

1. Constructor board (for example, 'Feedback' EEC470).

2. D.C. Power Supply Unit.

3. Function Generator (for example, 'Escort' EFG 3210).

4. Unknown inductor.

5. Digital Multimeter, such as a Fluke (for example, 89).

6. LCR Data bridge.

**Procedure:**

1. Construct the circuit, with the inductance of unknown value, as shown below.

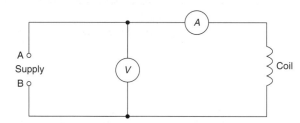

2. Connect a d.c. power supply unit set at 1 V to the terminals AB.

3. Measure the voltage $V$ and current $I$ in the above circuit.

4. Calculate the resistance $R$ of the coil, using $R = \dfrac{V}{I}$ recording the value in the table below.

5. Remove the d.c. PSU and connect an a.c. function generator set at 1 V, 50 Hz to the terminals AB.

6. Measure the voltage $V$ and current $I$ in the above circuit.

7. Calculate the impedance $Z$ of the coil, using $Z = \dfrac{V}{I}$, recording the value in the table below.

8. From the impedance triangle, $Z^2 = R^2 + X_L^2$, from which, $X_L = \sqrt{Z^2 - R^2}$. Calculate $X_L$ and record the value in the table below.

| $R\,(\Omega)$ | $Z\,(\Omega)$ | $X_L = \sqrt{Z^2 - R^2}\,(\Omega)$ | $L = \frac{X_L}{2\pi f}\,(\mathrm{H})$ |
|---|---|---|---|
| | | | |

9. Since $X_L = 2\pi f L$ then $L = \dfrac{X_L}{2\pi f}$; calculate inductance $L$ and record the value in the table above.

10. Hence, **for the coil, $L = \ldots$ H** and **resistance, $R = \ldots \Omega$**.

11. Measure the inductance of the coil using an LCR data bridge.

12. Using an ammeter, a voltmeter or a Fluke, measure the resistance of the coil.

**Conclusions:**

1. Compare the measured values of procedures 11 and 12 with those stated in procedure 10 and comment on any discrepancies.

## 24.8    Series a.c. circuit and resonance

**Objectives:**

1.  To measure and record current and voltages in an a.c. series circuit at varying frequencies.

2.  To investigate the relationship between voltage and current at resonance.

3.  To investigate the value of current and impedance at resonance.

4.  To compare measured values with theoretical calculations.

**Equipment required:**

1.  Cathode Ray Oscilloscope (for example, 'Phillips' digital Fluke PM3082).

2.  Constructor board (for example, 'Feedback' EEC470).

3.  Function Generator (for example, 'Escort' EFG 3210).

4.  Digital Multimeter, such as a Fluke (for example, 89).

5.  LCR Data bridge.

**Procedure:**

1.  Construct the series RCL circuit as shown below, measuring and noting the exact values of $R$, $C$ and $L$.

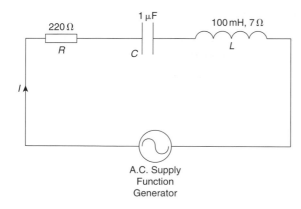

2.  Set the a.c. supply (function generator) to 2 V at 100 Hz.

3.  Measure the magnitude of the current in the circuit using an ammeter or Fluke and record it in the table on the next column.

4.  Measure the magnitudes of $V_R$, $V_C$ and $V_L$ and record them in the table in the next column.

5.  Calculate the values of $X_L$ and $X_C$ and record them in the table below.

6.  Using the values of circuit resistance (which is $R+$ resistance of coil), $X_L$ and $X_C$, calculate impedance $Z$.

7.  Calculate current $I$ using $I = \dfrac{V}{Z}$

8.  Repeat the procedures 2 to 7 using frequencies of 200 Hz up to 800 Hz and record the results in the table below. Ensure that the voltage is kept constant at 2 V for each frequency.

| Supply voltage $V$ | Measured $I$ (mA) | Measured $V_R$ (V) | Measured $V_C$ (V) | Measured $V_L$ (V) |
|---|---|---|---|---|
| 2V, 100 Hz | | | | |
| 2V, 200 Hz | | | | |
| 2V, 300 Hz | | | | |
| 2V, 400 Hz | | | | |
| 2V, 500 Hz | | | | |
| 2V, 600 Hz | | | | |
| 2V, 700 Hz | | | | |
| 2V, 800 Hz | | | | |

| Supply voltage $V$ | Calculate $X_L$ ($\Omega$) | Calculate $X_C$ ($\Omega$) | Calculate $Z$ ($\Omega$) | Calculate $I = \dfrac{V}{Z}$ (mA) |
|---|---|---|---|---|
| 2V, 100 Hz | | | | |
| 2V, 200 Hz | | | | |
| 2V, 300 Hz | | | | |
| 2V, 400 Hz | | | | |
| 2V, 500 Hz | | | | |
| 2V, 600 Hz | | | | |
| 2V, 700 Hz | | | | |
| 2V, 800 Hz | | | | |

9.  Plot a graph of measured current $I$ (vertically) against frequency (horizontally).

10. Plot on the same axes a graph of impedance $Z$ (vertically) against frequency (horizontally).

11. Determine from the graphs the resonant frequency $f_r$.

12. State the formula for the resonant frequency of a series LCR circuit. Use this formula to calculate the resonant frequency $f_r$.

13. Set the supply voltage to 2 V at the resonant frequency and measure the current $I$ and voltages $V_R$, $V_C$ and $V_L$.

14. Connect a cathode ray oscilloscope such that channel 1 is across the whole circuit and channel 2 is across the inductor.

15. Adjust the oscilloscope to obtain both waveforms.

16. Adjust the function generator from 2 V, 100 Hz up to 2 V, 800 Hz. Check at what frequency the voltage across L (i.e. channel 2) is a maximum. Note any change of phase either side of this frequency.

**Conclusions:**

1. Compare measured values of current with the theoretical calculated values and comment on any discrepancies.

2. Comment on the values of current $I$ and impedance $Z$ at resonance.

3. Comment on the values of $V_R$, $V_C$ and $V_L$ at resonance.

4. What is the phase angle between the supply current and voltage at resonance?

5. Sketch the phasor diagrams for frequencies of (a) 300 Hz (b) $f_r$ (c) 700 Hz.

6. Define resonance.

7. Calculate the values of Q-factor and bandwidth for the above circuit.

## 24.9   Parallel a.c. circuit and resonance

### Objectives:

1.  To measure and record currents in an a.c. parallel circuit at varying frequencies.

2.  To investigate the relationship between voltage and current at resonance.

3.  To calculate the circuit impedance over a range of frequencies.

4.  To investigate the value of current and impedance at resonance and plot their graphs over a range of frequencies.

5.  To compare measured values with theoretical calculations.

### Equipment required:

1.  Constructor board (for example, 'Feedback' EEC470).

2.  Function Generator (for example, 'Escort' EFG 3210).

3.  Digital Multimeter, such as a Fluke (for example, 89).

4.  LCR Data bridge.

### Procedure:

1.  Construct the parallel LR–C circuit as shown below, measuring and noting the exact values of $R$, $C$ and $L$.

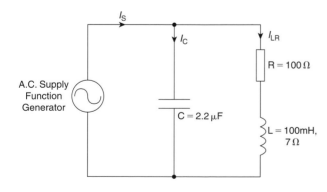

2.  Set the function generator to 3 V, 100 Hz using a Fluke.

3.  **Measure** the magnitude of the supply current, $I_S$, capacitor current, $I_C$, and inductor branch current, $I_{LR}$ and record the results in the table on the next column.

4.  Adjust the function generator to the other frequencies listed in the table ensuring that the voltage remains at 3 V. Record the values of the three currents for each value of frequency in the table below.

| Supply Voltage V | Measured $I_S$ (mA) | Measured $I_C$ (mA) | Measured $I_{LR}$ (mA) | Calculate $I_C = \frac{V}{-jX_C}$ |
|---|---|---|---|---|
| 3V, 100 Hz | | | | |
| 3V, 150 Hz | | | | |
| 3V, 200 Hz | | | | |
| 3V, 220 Hz | | | | |
| 3V, 240 Hz | | | | |
| 3V, 260 Hz | | | | |
| 3V, 280 Hz | | | | |
| 3V, 300 Hz | | | | |
| 3V, 320 Hz | | | | |
| 3V, 340 Hz | | | | |
| 3V, 360 Hz | | | | |
| 3V, 380 Hz | | | | |
| 3V, 400 Hz | | | | |
| 3V, 450 Hz | | | | |

| Supply Voltage V | Calculate $I_{LR} = \frac{V}{R+jX_{LR}}$ | Calculate $I_S = I_C + I_{LR}$ | Calculate $Z = \frac{V}{I_S}$ |
|---|---|---|---|
| 3V, 100 Hz | | | |
| 3V, 150 Hz | | | |
| 3V, 200 Hz | | | |
| 3V, 220 Hz | | | |
| 3V, 240 Hz | | | |
| 3V, 260 Hz | | | |
| 3V, 280 Hz | | | |
| 3V, 300 Hz | | | |
| 3V, 320 Hz | | | |
| 3V, 340 Hz | | | |
| 3V, 360 Hz | | | |
| 3V, 380 Hz | | | |
| 3V, 400 Hz | | | |
| 3V, 450 Hz | | | |

5. **Calculate** the magnitude and phase of $I_C$, $I_{LR}$ and $I_S (= I_C + I_{LR})$ for each frequency and record the values in the table on the previous page.

6. **Calculate** the magnitude and phase of the circuit impedance for each frequency and record the values in the table on the previous page.

7. Plot a graph of the magnitudes of $I_S$, $I_C$, $I_{LR}$ and $Z$ (vertically) against frequency (horizontally), all on the same axes.

8. Determine from the graphs the resonant frequency.

9. State the formula and calculate the resonant frequency for the LR–C parallel circuit.

**Conclusions:**

1. Compare measured values of the supply current $I_S$ with the theoretical calculated values and comment on any discrepancies.

2. Comment on the values of current $I$ and impedance $Z$ at resonance.

3. Compare the value of resonance obtained from the graphs to that calculated and comment on any discrepancy.

4. Compare the graphs of supply current and impedance against frequency with those for series resonance.

5. Calculate the value of dynamic resistance, $R_D$ and compare with the value obtained from the graph.

6. What is the phase angle between the supply current and voltage at resonance?

7. Sketch the phasor diagrams for frequencies of (a) 200 Hz (b) $f_r$ (c) 400 Hz.

8. Define resonance.

9. Calculate the values of Q-factor and bandwidth for the above circuit.

## 24.10 Charging and discharging a capacitor

**Objectives:**

1. To charge a capacitor and measure at intervals the current through and voltage across it.

2. To discharge a capacitor and measure at intervals the current through and voltage across it.

3. To plot graphs of voltage against time for both charging and discharging cycles.

4. To plot graphs of current against time for both charging and discharging cycles.

**Equipment required:**

1. Constructor board (for example, 'Feedback' EEC470).

2. D.c. Power Supply Unit.

3. Digital Multimeter, such as a Fluke (for example, 89).

4. LCR Data bridge.

5. Stop watch.

**Procedure:**

1. Construct the series CR circuit as shown below, measuring the exact values of $C$ and $R$.

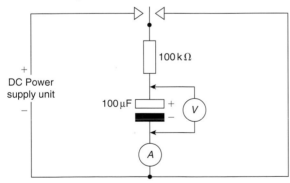

2. Set the D.C. Power Supply Unit to 10 V, making sure the output switch is in the off position.

3. Charge the capacitor, measuring the capacitor voltage (in volts) at 5 second intervals over a period of 60 seconds. Record results in the table on the next column.

4. Discharge the capacitor, measuring the capacitor voltage at 5 second intervals over a period of 60 seconds. Record results in the table on the next column.

| Times (s) | 0 | 5 | 10 | 15 | 20 | 25 | 30 |
|---|---|---|---|---|---|---|---|
| Charge $V_C$ (V) | | | | | | | |
| Discharge $V_C$ (V) | | | | | | | |

| Times (s) | 35 | 40 | 45 | 50 | 55 | 60 |
|---|---|---|---|---|---|---|
| Charge $V_C$ (V) | | | | | | |
| Discharge $V_C$ (V) | | | | | | |

5. Again, charge the capacitor, this time measuring the current (in $\mu$A) at 5 second intervals over a period of 60 seconds. Record results in the table below.

6. Discharge the capacitor, measuring the current at 5 second intervals over a period of 60 seconds. Record results in the table below.

| Times (s) | 0 | 5 | 10 | 15 | 20 | 25 | 30 |
|---|---|---|---|---|---|---|---|
| Current $I_C$ ($\mu$A) | | | | | | | |
| Discharge $I_C$ ($\mu$A) | | | | | | | |

| Times (s) | 35 | 40 | 45 | 50 | 55 | 60 |
|---|---|---|---|---|---|---|
| Current $I_C$ ($\mu$A) | | | | | | |
| Discharge $I_C$ ($\mu$A) | | | | | | |

7. Plot graphs of $V_C$ against time for both charge and discharge cycles.

8. Plot graphs of $I_C$ against time for both charge and discharge cycles.

9. Calculate the time constant of the circuit (using the measured values of $C$ and $R$).

10. Take a sample of the times and calculate values of $V_C$ and $I_C$ using the appropriate exponential formulae $V_C = V(1 - e^{-t/CR})$, $V_C = Ve^{-t/CR}$ and $I_C = Ie^{-t/CR}$.

**Conclusions:**

1. Compare theoretical and measured values of voltages and currents for the capacitor charging and discharging.

2. Discuss the charging and discharging characteristics of the capacitor.

3. Comment on reasons for any errors encountered.

4. What is the circuit time constant? What does this mean? Approximately, how long does the voltage and current take to reach their final values?

# Answers to multiple-choice questions

**Chapter 1.  Exercise 4  (page 7)**

| | | | | |
|---|---|---|---|---|
| 1 (c) | 4 (a) | 7 (b) | 9 (d) | 11 (b) |
| 2 (d) | 5 (c) | 8 (c) | 10 (a) | 12 (d) |
| 3 (c) | 6 (b) | | | |

**Chapter 2.  Exercise 10  (page 19)**

| | | | | |
|---|---|---|---|---|
| 1 (b) | 4 (b) | 7 (b) | 10 (c) | 12 (d) |
| 2 (b) | 5 (d) | 8 (c) | 11 (c) | 13 (a) |
| 3 (c) | 6 (d) | 9 (b) | | |

**Chapter 3.  Exercise 15  (page 28)**

| | | | | |
|---|---|---|---|---|
| 1 (c) | 3 (b) | 5 (d) | 7 (b) | 9 (d) |
| 2 (d) | 4 (d) | 6 (c) | 8 (c) | |

**Chapter 4.  Exercise 18  (page 40)**

| | | | | |
|---|---|---|---|---|
| 1 (d) | 4 (c) | 7 (d) | 10 (d) | 12 (a) |
| 2 (a) | 5 (b) | 8 (b) | 11 (c) | 13 (c) |
| 3 (b) | 6 (d) | 9 (c) | | |

**Chapter 5.  Exercise 25  (page 59)**

| | | | | |
|---|---|---|---|---|
| 1 (a) | 4 (c) | 7 (b) | 9 (b) | 11 (d) |
| 2 (a) | 5 (c) | 8 (d) | 10 (c) | 12 (d) |
| 3 (c) | 6 (a) | | | |

**Chapter 6.  Exercise 32  (page 76)**

| | | | | |
|---|---|---|---|---|
| 1 (b) | 4 (c) | 6 (b) | 8 (a) | 10 (c) |
| 2 (a) | 5 (a) | 7 (b) | 9 (c) | 11 (d) |
| 3 (b) | | | | |

**Chapter 7.  Exercise 38  (page 88)**

| | | | | |
|---|---|---|---|---|
| 1 (d) | 5 (c) | 8 (c) | 11 (a) and (d), | 12 (a) |
| 2 (b) | 6 (d) | 9 (c) | (b) and (f), | 13 (a) |
| 3 (b) | 7 (a) | 10 (c) | (c) and (e) | |
| 4 (c) | | | | |

**Chapter 8.  Exercise 42  (page 100)**

| | | | | |
|---|---|---|---|---|
| 1 (d) | 3 (d) | 5 (b) | 7 (d) | 9 (a) |
| 2 (c) | 4 (a) | 6 (c) | 8 (a) | 10 (b) |

**Chapter 9.  Exercise 50  (page 112)**

| | | | | |
|---|---|---|---|---|
| 1 (c) | 4 (b) | 7 (c) | 9 (c) | 11 (a) |
| 2 (b) | 5 (c) | 8 (d) | 10 (a) | 12 (b) |
| 3 (c) | 6 (a) | | | |

**Chapter 10.  Exercise 60  (page 141)**

| | | | |
|---|---|---|---|
| 1 (d) | 7 (c) | 13 (b) | 19 (d) |
| 2 (a) or (c) | 8 (a) | 14 (p) | 20 (a) |
| 3 (b) | 9 (i) | 15 (d) | 21 (d) |
| 4 (b) | 10 (j) | 16 (o) | 22 (c) |
| 5 (c) | 11 (g) | 17 (n) | 23 (a) |
| 6 (f) | 12 (c) | 18 (b) | |

**Chapter 11.  Exercise 64  (page 155)**

| | | | | |
|---|---|---|---|---|
| 1 (c) | 3 (d) | 5 (b) | 7 (c) | 9 (a) |
| 2 (a) | 4 (c) | 6 (b) | 8 (d) | 10 (b) |

**Chapter 12.  Exercise 68  (page 175)**

| | | | | |
|---|---|---|---|---|
| 1 (b) | 5 (a) | 9 (b) | 13 (b) | 17 (c) |
| 2 (b) | 6 (d) | 10 (c) | 14 (b) | 18 (b) |
| 3 (c) | 7 (b) | 11 (a) | 15 (b) | 19 (a) |
| 4 (a) | 8 (d) | 12 (b) | 16 (b) | 20 (b) |

**Chapter 13.  Exercise 76  (page 206)**

| | | | | |
|---|---|---|---|---|
| 1 (d) | 5 (a) | 8 (a) | 11 (b) | 14 (b) |
| 2 (c) | 6 (d) | 9 (c) | 12 (d) | 15 (c) |
| 3 (b) | 7 (c) | 10 (c) | 13 (d) | 16 (a) |
| 4 (c) | | | | |

**Chapter 14.  Exercise 82  (page 224)**

| | | | | |
|---|---|---|---|---|
| 1 (c) | 4 (a) | 7 (b) | 9 (b) | 11 (b) |
| 2 (d) | 5 (d) | 8 (c) | 10 (c) | 12 (d) |
| 3 (d) | 6 (c) | | | |

**Chapter 15.  Exercise 90  (page 245)**

| | | | | |
|---|---|---|---|---|
| 1 (c) | 5 (a) | 9 (d) | 13 (b) | 17 (c) |
| 2 (a) | 6 (b) | 10 (d) | 14 (c) | 18 (a) |
| 3 (b) | 7 (a) | 11 (b) | 15 (b) | 19 (d) |
| 4 (b) | 8 (d) | 12 (c) | 16 (b) | |

## Chapter 16. Exercise 98 (page 264)

| | | | |
|---|---|---|---|
| 1 (d) | 5 (h) | 9 (a) | 12 (d) |
| 2 (g) | 6 (b) | 10 (d), (g), (i) and (l) | 13 (c) |
| 3 (i) | 7 (k) | 11 (b) | 14 (b) |
| 4 (s) | 8 (l) | | |

## Chapter 17. Exercise 103 (page 276)

| | | | | |
|---|---|---|---|---|
| 1 (d) | 4 (c) | 7 (b) | 9 (d) | 11 (d) |
| 2 (b) | 5 (c) | 8 (a) | 10 (b) | 12 (c) |
| 3 (a) | 6 (a) | | | |

## Chapter 18. Exercise 107 (page 293)

| | | | | |
|---|---|---|---|---|
| 1 (c) | 5 (g) | 9 (a) | 13 (c) | 16 (c) |
| 2 (b) | 6 (e) | 10 (d) | 14 (j) | 17 (a) |
| 3 (b) | 7 (l) | 11 (g) | 15 (h) | 18 (a) |
| 4 (g) | 8 (c) | 12 (b) | | |

## Chapter 19. Exercise 111 (page 309)

| | | | | |
|---|---|---|---|---|
| 1 (c) | 3 (b) | 5 (a) | 7 (d) | 9 (c) |
| 2 (b) | 4 (d) | 6 (b) | 8 (a) | 10 (c) |

## Chapter 20. Exercise 117 (page 331)

| | | | | |
|---|---|---|---|---|
| 1 (g) | 5 (f) | 9 (l) | 12 (j) | 15 (c) |
| 2 (c) | 6 (a) | 10 (d) | 13 (d) | 16 (b) |
| 3 (a) | 7 (g) | 11 (f) | 14 (b) | 17 (c) |
| 4 (a) | 8 (l) | | | |

## Chapter 21. Exercise 130 (page 355)

| | | | | |
|---|---|---|---|---|
| 1 (a) | 5 (c) | 9 (b) | 13 (h) | 17 (c) |
| 2 (d) | 6 (a) | 10 (g) | 14 (k) | 18 (b) and (c) |
| 3 (a) | 7 (b) | 11 (d) | 15 (j) | 19 (c) |
| 4 (b) | 8 (a) | 12 (a) | 16 (f) | 20 (b) |

## Chapter 22. Exercise 139 (page 381)

| | | | | |
|---|---|---|---|---|
| 1 (b) | 6 (a) | 11 (b) | 16 (b) | 21 (b) |
| 2 (e) | 7 (d) | 12 (a) | 17 (b) | 22 (a) |
| 3 (e) | 8 (f) | 13 (b) | 18 (b) | 23 (c) |
| 4 (c) | 9 (b) | 14 (a) | 19 (c) | 24 (d) |
| 5 (c) | 10 (c) | 15 (d) | 20 (b) | |

## Chapter 23. Exercise 146 (page 399)

| | | | | |
|---|---|---|---|---|
| 1 (c) | 5 (b) | 9 (f) | 13 (a) | 16 (b) |
| 2 (b) | 6 (d) | 10 (j) | 14 (c) | 17 (c) |
| 3 (d) | 7 (b) | 11 (d) | 15 (b) | 18 (b) |
| 4 (a) | 8 (c) | 12 (c) | | |

# Index